The Silicon Web

The Silicon Web
Physics for the Internet Age

Michael G. Raymer
University of Oregon
Eugene, USA

CRC Press
Taylor & Francis Group
Boca Raton London New York

CRC Press is an imprint of the
Taylor & Francis Group, an **informa** business

A TAYLOR & FRANCIS BOOK

On the cover: Electron diffraction pattern, with 5-fold symmetry, observed when electrons pass through a AICoNi quasicrystal. Courtesy of Dr. Conradin Beeli, Solid State Physics Laboratory, Swiss Federal Institute of Technology (ETH), Zürich, Switzerland.

Taylor & Francis
6000 Broken Sound Parkway NW, Suite 300
Boca Raton, FL 33487-2742

First issued in paperback 2022

© 2009 by Taylor and Francis Group, LLC
CRC Press is an imprint of Taylor & Francis Group, an Informa business

No claim to original U.S. Government works

ISBN-13: 978-1-439-80311-0 (hbk)
ISBN-13: 978-1-03-234030-2 (pbk)
DOI: 10.1201/9781439803127

Library of Congress Cataloging-in-Publication Data

Raymer, Michael G.
The silicon web : physics for the Internet age / Michael G. Raymer.
p. cm.
Includes bibliographical references and index.
ISBN 978-1-4398-0311-0 (alk. paper)
1. Physics. 2. Solid state physics. 3. Electromagnetics. 4. Mathematical physics. 5. Telecommunication--Equipment and supplies.
6. Information technology--Equipment and supplies. 7. Internet. I. Title.

QC23.2.R36 2009
530.4'1--dc22   2008054580

Visit the Taylor & Francis Web site at
http://www.taylorandfrancis.com

and the CRC Press Web site at
http://www.crcpress.com

Dedicated to my wife, Kathie Lindlan

Contents

Foreword

The advent of the Internet launched a revolution of revolutions. In less than a generation, the web has fundamentally changed how we live and work, what we do with our time, and the ways in which we communicate. Online media are rapidly consuming all the media that preceded; who knows where the web will take us in the next decade or two? The only thing that's fairly certain is that there is no going back.

Nowadays, everything seems just keystrokes away. As more and more information becomes available online, it's tempting to know less and less. Why bother learning anything when you can always look it up? Alas, you won't know what you don't know and you won't be able to distinguish real information from nonsense without a solid background. The web is still a wild west, an unfiltered assortment of insight and idiocy, hype and hooey, sage advice and snake oil. With all these competing voices on the Internet, it takes an informed browser to figure out which ones are worthy of attention. In a paradoxical way, the very availability of knowledge has made the world more difficult rather than easier to understand.

Furthermore, the web is not neutral in its impact on humanity. As always, choices concerning technology have important ethical, social, and political implications, which are overlooked easily by people who don't understand technology or can't see beyond their immediate self-interest. We live in the best of times and the worst of times, not only with respect to the web but also because we face increasingly important societal choices regarding limited resources and a limited planet. It's critical to our future that all of us understand enough about science and technology to make informed decisions that will lead to a desirable and sustainable world years from now. The sky is still the limit, but opportunities for poor choices abound.

Unfortunately, students rarely study a topic simply because someone else says it's important. The biggest challenge in trying to teach physical science to a broad audience is to make the science immediately relevant. For several generations, physicists have taught physics mostly as an abstract academic subject and have hidden its relevance under a blizzard of massless pulleys and frictionless planes. Physics is truly beautiful to those who understand it well, but for those who are just beginning to learn its concepts, this sophisticated beauty often lies far beyond the horizon. It's no wonder that the phrase "I'm a physicist" is a notorious conversation killer at a party.

To make physics more immediately appealing and relevant to a broad audience, you have to teach it in a real context. Students need familiar hooks on which to hang each new scientific idea. That is exactly what Michael Raymer has accomplished in this book. *The Silicon Web* is an elegant and elaborate textbook, one that examines the science underlying the current revolution in communications technology. Each scientific concept arises on a need-to-know basis in the context of a particular networking issue or device. For example, instead of presenting the frequency spectrum of a wave as an abstract idea, Michael "discovers" its concepts in the process of explaining how radio waves can transmit sound. All the important physics is there, woven into an engaging story, so we're not simply checking off topics on a curriculum.

I'm a big fan of telling stories in order to explain science. Books and articles are much more approachable that way, and have a chance of being read rather than merely serving as a reference. In our frantic, overcommitted lives, we easily allow external constraints to guide how we teach and learn. If the only need-to-know aspect of a book is external to it—a homework assignment or a looming exam—then that book is going to be treated as a reference work. *The Silicon Web*, however, is a stunningly rich story about the Internet and its associated technologies, and that story motivates virtually all of its science. Here in a single text we have both a complete exploration of the most game-changing technology since World War II and a surprisingly thorough and well-motivated presentation of basic physics.

Lou Bloomfield
Professor of Physics
The University of Virginia

Preface

Have you ever wondered how words, music, and pictures are transmitted across the Internet? You probably know that the Internet operates on a huge collection of computers worldwide. But, what exactly is "The Internet," and how do computers work? How does a computer use "ones and zeros" to carry out its operations? What do "semiconductors" have to do with it, and what exactly is a semiconductor? How do cell phones really work? You know that phone conversations are carried on fiber optics. What exactly is an optical fiber? Why are lasers used as the light sources to "light the fiber?" How does a laser work? Why is the present-day information system based on *digital* communication, whereas the old system years ago was not digital, but *analog*? What is the difference, and why was the digital system invented? You have probably heard the term "bandwidth," and you know that it is good to have a lot of it. But, what is bandwidth? *The Silicon Web* aims to answer these and other questions.

From the point of view of physics, we are interested in knowing: What basic knowledge was required for inventing computers and the Internet, and how was it discovered? The technology behind computers, fiber optics, and networks did not originate in the minds of engineers attempting to build an Internet. The Internet is a culmination of intellectual work by tens of thousands of scientists and engineers spanning over 100 years. Before there was the Internet, there had to be network theory; before that, computers; before that, digital electronics; before that, semiconductor physics; before that, atomic physics; before that, electromagnetism; and so on. The important inventions underlying the Internet are rooted in discoveries in physics and mathematics. So, that is where we will begin to pursue the foundations of the network.

To appreciate the physical basis of computers and networks, one needs to understand the concepts of mechanics (force, acceleration, energy), electromagnetism (electric and magnetic forces), quantum physics (electrons and atoms), properties of materials (crystals, glass, metals, semiconductors), electronics (diodes, transistors), waves (light, radio), thermodynamics (heat), and optical physics (light sources, lasers, and fibers). Topics in information science and technology are needed to complete the picture: binary mathematics, mathematical logic, signal synthesis and analysis, structure of networks, and communication hardware. These topics are all intertwined, and their study illustrates a basic unity that ties together a wide range of topics.

The global information network (Internet) is made mostly of silicon. Silicon, both in its glassy form (as optical fibers) and its crystalline form (as semiconductors), is at the foundation of many of the components making up the network. Together these form a kind of delicate, tenuous, interconnected nervous system—a "Silicon Web"—transmitting information around the world.

The book also discusses some of the social impacts of science in general and information technology in particular. One can argue that scientific thought and methods have changed the course of human history as much as or more than anything else. To have a deeper understanding of where humanity is at present, it pays to study science and technology. For example, nowadays nearly everything seems to be growing exponentially. We are immersed in a sea of information. Computers and the Internet connect us to each other and to the physical world. This historic transition warrants some reflection on how we got here and where we are going. Therefore, many chapters contain essays on these social impacts, which are mere starting points for you to begin exploring these topics.

To the Student

You might be using this book for the only college-level physics course you will take. In this case, you are probably not planning a career as a scientist or technologist, but want to have some exposure to the most important scientific concepts related to information technology. What will you need to be successful in this course? Mostly, you should come with a curiosity and willingness to work and think carefully through new ideas. Some familiarity with high-school chemistry or physics is helpful, but not required.

This is not primarily a math-based course. Instead, it is a course that requires use of logical thinking to learn core physics concepts. Mathematical thinking comes into play only as a way to represent physical quantities, such as energy and power, by using numbers. A willingness to work with simple mathematics at the high-school level is necessary, but there will be no extensive use of algebra, trigonometry, or geometry.

I hope that you enjoy it.

To the Instructor

The Silicon Web offers a new approach to learning conceptual physics. The book is for a one- or two-semester college or university physics course for students majoring in a field outside of the physical sciences. These students may have less advanced preparation in high-school science and math than university science majors. The course is structured to satisfy a general science requirement. Such a course has been successfully offered at the University of Oregon since 2002 under the title *The Physics Behind the Internet*. Students taking this course include arts, humanities, journalism, and business majors. This course is offered with no prerequisites and is often the students' only college physics course. The material lends itself well to classroom lecture demonstrations because it covers mechanics, waves, electricity and magnetism, lasers, optics, electronics, and computers.

The course has a conceptual focus while still penetrating enough to gain a substantive appreciation of the physics involved. The immediate educational goal is to teach students about the physics underlying the revolutionary technologies of computers and communications. At the same time, the text develops the ideas behind the scientific method and the historical context of scientific discoveries and addresses their social impacts. *The Silicon Web* emphasizes physics over technology. It is written to teach science appreciation rather than train future scientists. On the other hand, quantitative thinking and problem solving are important components of the course.

Although it is not feasible to cover all of the material in the text in a one-semester course, the chapters are structured to enable you as the instructor to select various paths through the material. Three main tracks can be easily identified: the first emphasizes the basis of computers; the second emphasizes the basis of communications; and the third emphasizes important concepts of modern physics—electromagnetism, quantum physics, and semiconductor physics. The following are suggested chapter selections for each track:

> Track 1. The Physics behind Computers: Chapters 1, 2, 3, 5, 6, 9, 10, 11.
> Track 2. The Physics behind Communications: Chapters 1, 2, 3, 5, 7, 8, 12, 13, 14, 15, 16.
> Track 3. Modern Physics: Chapters 1, 2, 3, 4, 5, 7, 9, 10, 12, 13.

Note that in Track 2, Chapter 14 on lasers can be studied without having studied Chapters 9, 10, and 12, which cover quantum physics.

Of course, you may also design other paths, taking advantage of optional sections, which are identified by the labels "Real-World Example" and "In-Depth Look."

Technical Notes

The nonstandard notation "sec" has been used for seconds, rather than the SI unit "s." There is no detailed introduction to the concept of momentum. Although momentum is a central theme in physics, it is not essential for understanding the core concepts emphasized in this book.

I plan to maintain a companion website with chapter notes, including updates and additional resources. Please refer to the publisher's website for further information.

Finally, I invite readers to send me comments and corrections so that I might improve the book's quality. Please address correspondence to: Michael Raymer, Dept. of Physics, Univ. of Oregon, Eugene, OR 97403, or to raymer@uoregon.edu.

Acknowledgments

I thank my wife Kathie Lindlan for her support and encouragement at every stage, and for her expert help with editing and in finding source materials in computer science.

I thank my parents, Dorothy and Gordon Raymer, for critically reading and editing the entire manuscript and applying their writing and engineering expertise, respectively, to make it better.

I thank the many students who, over five years of classes, asked so many good questions, stimulating me to clarify explanations.

I also thank Andrew Funk and Brian Smith, then graduate students of the University of Oregon Physics Department. Andrew participated as a co-instructional assistant during the first year (2001) that this course was taught. Through many discussions with him, the course content and approach were developed. Brian later served as a co-instructional assistant and contributed greatly through his lectures and lecture demonstration preparation. Faculty colleague Stan Micklavzina also deserves thanks and credit for providing many of the lecture demonstrations used.

I am very grateful to those colleagues who volunteered their valuable time to critically read parts of the text. Some of the better ideas contained here are theirs, whereas any errors are mine. Paul Csonka read the core physics Chapters 3, 4, and 5 and provided terrific feedback. David Strom provided helpful comments on the semiconductor and electronics chapters. Robert Burger offered a critical reading of Chapter 11 on computers. Robert Hammerling at the Technische Universität in Vienna offered helpful comments on Chapter 5. Brian Jones at EDX Wireless provided helpful comments on Chapter 16 on networks. Dennis Hall and Daniel Steck made very helpful comments on several sections.

I also thank the many reviewers who bravely read early drafts of the book when it was more an idea than a useful product. Their encouragement and (sometimes fierce) criticism helped put me on the proper track.

Introduction: Physics and Its Relation to Computer and Internet Technologies

Physicists know that all scientific knowledge consists of models created by humans at particular times and places, and that those models evolve or are supplanted as new information becomes available.

Diane J. Grayson
(Professor at University of Pretoria)

Quantum technologies grow organically out of asking interesting physics questions.

David DiVincenzo
(Physics researcher at IBM)

David DiVincenzo, PhD, works at IBM's Watson Research Center. He is a leading researcher in the emerging field of quantum information technology.

Diane Grayson, PhD, is Professor Extraordinarius at the University of Pretoria in South Africa. She directs a program that promotes quality science education.

1.1 PHYSICS, SILICON, AND THE "MAGIC" BEHIND THE INTERNET AGE

Physics is not a body of incontrovertible facts. Physics is a way of looking at the world and trying to understand the world as it is. Physics is an activity of people, as exemplified by the two distinguished physicists in the photos on page 1. Their quotes above summarize the themes of this book: what physics is and how physics is related to the development of technology. This introductory chapter will set the stage for the remainder of the book.

The Internet is perhaps the biggest advance in human communications since the invention of the movable-type printing press in the sixth century in China and in the fifteenth century in Germany. For this reason, many people say we are now in the Information Age, following, in reverse order, the Space Age, the Nuclear Age, the Industrial Age, the Agrarian Age, the Iron Age, the Bronze Age, and the Stone Age.

The technology of the Internet might seem to some like technical "magic." But, as in the movie *The Wizard of Oz*, when Dorothy finally looks behind the curtain, she sees not a magician but a person pulling levers and controlling special effects. In this text we will attempt to pull back the curtain, behind which operates an amazing collection of inventions, devices, and applications that make the Internet operate and that allow you to send and receive text, images, audio, and video.

What role does physics play in all of this? The technology behind electronics, computers, lasers, and fiber optics did not arise straightforwardly from the minds of engineers attempting to build an Internet. The Internet is the culmination of intellectual work by tens of thousands of scientists and engineers spanning hundreds of years. Before there was the Internet, there had to be network theory, and before that computers, and before that digital electronics, and before that semiconductor physics, and before that atomic physics, and before that electromagnetism, and before that the understanding of motion and of heat. The important inventions that underlie the Internet are rooted in earlier discoveries in science, especially physics.

This book is entitled *The Silicon Web* because without the basic element silicon there would likely be no computers, optical fibers, or Internet. Silicon is to communication technology as carbon is to life—it is the basic building block. Fortunately, silicon is the second most abundant element on Earth, following oxygen.

Let us consider what makes up the Internet. **Figure 1.1** shows many of the components making up the Internet. The main component of the Internet is the computer, several of which are shown in the figure. Computers create, store, and process information in discrete, or ***digital***, form.[1] They do this using their internal circuits. Attached to the computer on the far left are a ***magnetic hard drive*** and an optical ***compact disc***—used for storing large amounts of information or data, whereas smaller amounts are stored and processed using ***integrated circuits*** inside of the computer. Also shown in the figure is a portable or laptop computer. Laptop users often use removable ***"flash" memory*** devices, which store digital information within an internal integrated circuit. Integrated circuits and magnetic, optical, and electronic storage devices operate according to the properties and behaviors of ***atoms*** and subatomic particles, called ***electrons, protons***, and ***neutrons***. The discovery of subatomic particles and their properties was one of the most important in the history of physics.

Individual pieces of information, called ***bits***, are shown in the figure as digital pulses traveling on a metal wire between a computer and a network node, where the bits are sent over long distances either through more metal wires (e.g., on telephone lines) or through ***optical fibers***. Before being transmitted on fibers, the bits must first be transformed from electrical form into optical, or light, form using ***lasers***. After a light

---

[1.] Terms in **bold italics** are defined in the Glossary at the end of the book and listed at the end of each chaper.

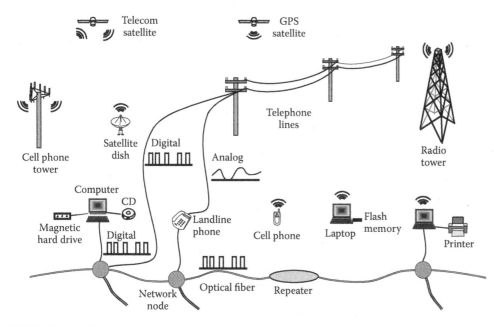

FIGURE 1.1 The Internet and its components.

signal from a laser has traveled a long distance, it loses energy, and optical amplifiers called *repeaters* are placed in the fiber to boost it back to its original strength. Voice telephone calls originating from typical household (land) phones are transmitted as *analog* electrical signals, rather than as digital signals. This harkens back to the telephone technology of the twentieth century, which is rapidly being replaced by all-digital technology. Mobile phones and laptop computers communicate with the network through digital wireless radio transmissions. These radio signals travel through the air in the form of electromagnetic waves. Satellites receive and transmit streams of digital data bits. Electrical, optical, and radio signals all share a common origin—called electromagnetism. The discovery of electromagnetism was another of the most important in the history of physics.

1.2 A ZOOMED-IN LOOK INSIDE A COMPUTER

If we open a computer, we see the components shown in **Figure 1.2**. The system board shown in **Figure 1.3** holds electronic memory chips and processing circuits, including the main processor chip, shown enlarged in **Figure 1.4**. A highly magnified image of the inside of an electronic memory chip is shown in **Figure 1.5**, where groups of transistors and other microscopic electronic devices become visible. **Figure 1.6** shows a transistor, in a highly magnified image, just one millionth of a meter (micrometers) on a side. The operation of a transistor depends on the properties of the atoms it is composed of and the subatomic particles called electrons that move through it. In a computer, all data are processed by moving electrons. Only recently have scientists learned how to acquire images showing individual atoms, using an instrument called an electron microscope. **Figure 1.7** is an extremely magnified image revealing many individual iron atoms on the surface of a gallium-arsenic crystal. The area of this image is about seven billionths of a meter (nanometers) on a side. It is difficult to acquire images on scales smaller than the size of atoms, but physicists have developed math-

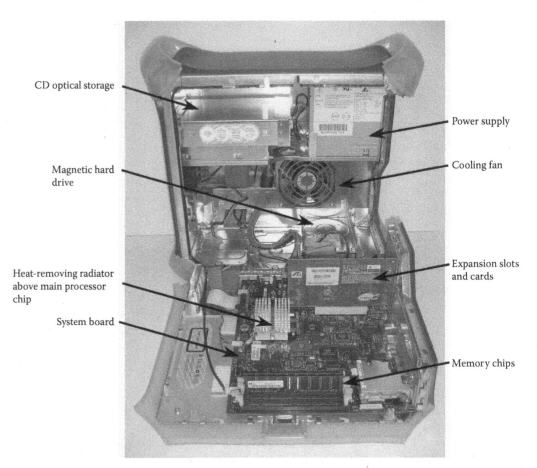

FIGURE 1.2   The main components inside an opened Apple Macintosh G3 computer. The area of this image is about 0.5×0.5 m.

ematical theories showing the most likely internal structure of atoms. **Figure 1.8** shows a computer simulation of the inside of a single atom. The area in this image is about 0.2 nanometers on a side. At this scale, electrons appear to move within an atom in the form of waves, or clouds, rather than as minute particles; that is, on this scale electrons appear to be "fuzzy."

FIGURE 1.3   System board in Apple Macintosh G3 computer. The area of this image is about 0.2×0.2 m.

Processor chip

Circuit boards

FIGURE 1.4 Motorola PowerPC processor chip with heat sink removed. Image area is about 5×5 cm (1 cm = 0.01 m).

FIGURE 1.5 Magnified photograph of integrated circuits in a Hynix Dynamic Random Access DRAM memory chip. Image area is about 2×2 mm (1 mm = 0.001m). (Courtesy of Jerry Gleason, University of Oregon.)

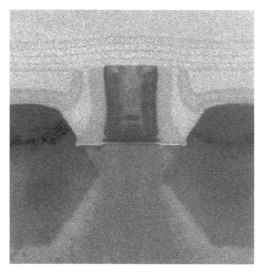

FIGURE 1.6 A cross-section of a state-of-the-art transistor used in Intel's 45 nm generation of processors. The thickness of a single transistor's switching "gate" (the rectangular region in the photo's center) is about 45 nm. The area of this image is about 150×150 nm (1 nm = 0.000000001 m.) (Courtesy of Intel Corp.)

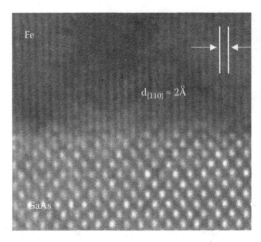

FIGURE 1.7 Electron microscope image of iron atoms (above) on a gallium-arsenic surface (below). Individual atoms can be seen in the lower part of the picture, which measures about 7×7 nm. (Courtesy of Achim Trampert, Paul-Drude-Institut, Berlin.)

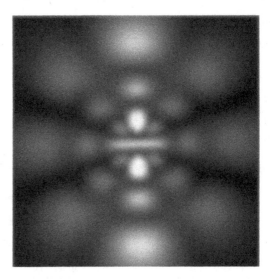

FIGURE 1.8 Computer simulation of the electron cloud surrounding a single atom. Size of image is about 0.2×0.2 nm. (Courtesy of Dauger Research, http://daugerresearch.com)

1.3 TIMELINE OF GREAT DISCOVERIES AND INVENTIONS IN PHYSICS AND COMPUTER AND COMMUNICATION TECHNOLOGIES

A fascinating way to visualize the development of a field of study is using a timeline, as shown in **Table 1.1**. In the left column are major advancements in physics and other sciences, and in the right column are major advances in computing and communication technologies. Some of the listed events deserve special mention. Mechanized printing was developed in stages in China (510), Korea (1403), and Germany (1455), giving an automated way to record, store, and reproduce information. In the early 1600s Galileo and Francis Bacon more or less invented the European form of the modern scientific method, although the word scientist was not used until 1834. In the mid-1600s Isaac Newton invented the modern mathematical approach and methods, including calculus, upon which all of physics theory is now based. He also introduced his three famous laws of motion and the effects of gravity.

TABLE 1.1

Timeline of Physics and Computer and Communication Technologies

Science/Physics	Technology/Computers
130 – Ptolemy writes about light, optics, reflection, and refraction	
	510 – Woodblock book printing invented by Chinese
517 – Philoponus shows that all objects fall with the same acceleration	
830 – Arabian mathematician Al-Khworizmi develops algebra	
1015 – Arabian physicist Alhazen explains how lenses, parabolic mirrors, and vision work	
1086 – First description of a magnetic compass by Chinese scientist Shen Kua	
1250 – Roger Bacon applies geometry to the study of optics and emphasizes the use of lenses for magnification	
	1403 – Movable type is cast in bronze by Koreans
	1455 – Printing of Gutenberg Bible, the first book to be printed with movable metal type
1480 – Reflection of light studied by Leonardo da Vinci	
	1494 – Leonardo da Vinci suggests the idea of a pendulum clock
1543 – Nicolaus Copernicus promotes the heliocentric Solar System model	
1572 – Books on optics by Alhazen and Witelo are translated into Latin by Freidrich Risner for the European scientific community	
	1593 – Galileo Galilei invents a water thermometer
1602 – Galileo studies the pendulum and determines that its swing period is independent of the amplitude	
1609 – Galileo Galilei establishes the principle of falling bodies descending to Earth at the same speed	
1620 – Scientific method developed by Francis Bacon	
1621 – Snell's law of refraction of light	
	1622 – First slide rule invented by Edmund Gunter
1633 – Galileo Galilei's published belief in Nicolaus Copernicus's solar system model condemned by the Roman Catholic Inquisition	1642 – First mechanical adding machine invented by Blaise Pascal
1645 – Female astronomer Marie Cunitz provided simplified versions of Johannes Kepler's planetary theories and laws	
	1659 – Robert Boyle devises an air pump
1665 – Wave theory of light published by Francesco Grimaldi	
1665 – Isaac Newton's law of universal gravitation	
1666 – Isaac Newton studies the separation of colors of white light by a prism	

(Continued)

TABLE 1.1 (*Continued*)

Science/Physics	Technology/Computers
1676 – Speed of light estimated at 140,000 miles per second by Ole Roemer	
1687 – *Principia* published; Isaac Newton's great work includes his three laws of motion	
	1712 – Thomas Savery and Thomas Newcomen build a practical working engine used in mining
1728 – Speed of light newly estimated by Bradley to be 183,000 miles per second	
1752 – Benjamin Franklin performs his kite experiments and shows that lightning is a form of electricity	
	1757 – James Watt begins work on the steam engine
1768 – Leonhard Euler suggests that the wavelength of light determines its color	1775 – Industrial Revolution in England starts
1776 – British colonies in America declare independence	
1787 – Silicon is discovered by Antoine Lavoisier	
1789 – Coulomb's law of electrostatic force	1792 – Semaphore (optical telegraph) system set up between Paris and Lille
1800 – The voltaic cell invented by Alessandro Volta; it is a prototype battery	
1801 – Thomas Young discovers interference of light	1801 – Joseph Jacquard invents the Jacquard loom, which used punch cards to create patterns in fabric; Jacquard loom precursor of modern computers
1803 – John Dalton develops the first useful atomic theory of matter	1811 – Members of the Luddite movement destroy factory machinery that threaten to eliminate their jobs
1816 – Fresnel explains the refraction of light	
1819 – Hans Orsted discovers electromagnetism	
1824 – Berzelius identifies silicon as a chemical element	
1826 – Ampere publishes electrodynamic theory	
1827 – Ohm's law of electrical resistance established	1830 – Joseph Henry builds first electromagnetic motor
1831 – Faraday discovers electromagnetic induction	1833 – Gauss invents the electric telegraph
1834 – William Whewell coins the term "scientist"	
1836 – The first reliable source of electric current is produced, the Daniell Cell, using copper and zinc	1837 – Electric motor first used in practical applications
	1838 – Samuel Morse makes the first demonstration of Morse Code for telegraphy
1842 – Principle of conservation of energy put forward by Julius Mayer	1840 – Charles Babbage designs (but is unable to build) the first general-purpose programmable mechanical computer
1843 – Joule describes the mechanical equivalent of heat	1843 – Ada Lovelace writes first computer program and article on computer programming
1847 – Hermann von Helmholtz proposes the Law of Conservation of Energy, the first law of thermodynamics	

TABLE 1.1   *(Continued)*

Science/Physics	Technology/Computers
1848 – Kelvin develops his temperature scale	
1849 – French physicist Armand Fizeau measures the speed of light	
1850 – Foucault observes that light travels slower in water than in air	1852 – Submarine telegraph cable is successfully laid under the English Channel
1854 – Development of theory of logic by George Boole	1858 – First telegraph cable spans the Atlantic Ocean but breaks after a few days
	1859 – Plucker invents the cathode ray tube
1865 – James Clerk Maxwell proves that electromagnetic waves travel at the speed of light	1866 – Telegraph cable laid beneath the Atlantic Ocean, providing high-speed communication between the United Kingdom and United States
1869 – The first Periodic Table of Elements is formulated and published by Mendeleev	
1874 – Kelvin puts forward the second law of thermodynamics	
1874 – Stoney suggests that electricity is composed of individual negative particles—electrons	
1887 – Hertz predicts the existence of radio waves and successfully detects them a year later	1889 – First direct dial telephone invented by Strowger
1897 – J. J. Thomson discovers that electrons are negatively charged particles with small mass	1896 – Marconi submits a full specification for the first wireless system of telegraphy using Hertzian waves
1898 – Marie and Pierre Curie discover radium and polonium radioactive isotopes	
1898 – Wien identifies a positive particle, later known as the proton	
1900 – Max Planck puts forward quantum theory	1901 – First transatlantic radio signal sent
1904 – Albert Einstein explains the photoelectric effect	1904 – Vacuum tube diode invented by Fleming
1905 – Einstein proves the existence of atoms by analyzing the observed motions of pollen grains in water	
1905 – Einstein proposes wave-particle duality of light and introduces the idea of discrete portions of energy, later called photons	1907 – First regular radio broadcasts of music (from New York)
1905 – Lise Meitner becomes the first woman to be awarded a PhD in physics (at the University of Vienna)	1910 – First movie picture with sound developed by Thomas Edison
1911 – Ernest Rutherford discovers that atoms contain a nucleus and orbiting electrons	
1914 – Niels Bohr proposes that electrons travel around the nucleus in fixed energy levels	
1916 – Using the photoelectric effect, Robert Millikan measures the Planck constant	1919 – Switching systems and rotary-dial telephones introduced in United States

(Continued)

TABLE 1.1   (*Continued*)

Science/Physics	Technology/Computers
1923 – Arthur Compton confirms the particle nature of light	1923 – Vladimir K. Zworykin applies for a patent on his iconoscope cathode ray tube used in televisions
1925 – Discovery of the exclusion principle for electrons by Wolfgang Pauli	1924 – Rice and Kellogg develop the loudspeaker
1926 – Wave mechanics introduced by Schroedinger	1925 – John Logie Baird successfully transmits the first recognizable television image
1927 – Lester Germer and Clinton Joseph Davisson perform experiments that confirm that matter can be wave-like	1927 – Quartz clock invented by Warren Marrison
	1928 – John Logie Baird uses wireless methods to transmit a picture across the Atlantic
	1928 – Philo Farnsworth develops the first working television sets
	1931 – Invention of electric guitar by the Rickenbacker Company based on electromagnetic pickups
	1933 – Invention of frequency modulation (FM) radio by Edwin Howard Armstrong
	1934 – Begun invents the magnetic tape recorder
	1942 – First digital computer is built by Atanasoff and Berry
	1945 – Electronic Numerical Integrator and Computer (ENIAC) built—funded by the US Army
1947 – John Bardeen, Walter Brattain, and William Shockley invent the transistor	1952 – Charles Townes and Russian scientists Prokhorov and Basov independently invent the maser, the forerunner to the laser
1948 – Claude Shannon develops information theory	1953 – First musical synthesizer invented at RCA
	1954 – Invention of the transistor radio
	1956 – First transatlantic telephone cable
	1958 – Jack Kilby, using germanium, invents the integrated circuit, the cornerstone of the modern electronics industry; Robert Noyce invents the silicon-based integrated circuit
	1960 – Theodore Maiman develops the first laser, using a ruby crystal
	1960 – The concept of packet switching for efficient data communication is developed
	1961 – First demonstration of use of optical fiber to transmit laser light signals
1963 – Maria Goeppert-Mayer becomes the second woman to be awarded a physics Nobel Prize (Marie Curie was the first)	1961 – Silicon chips are developed
1966 – Design for low-loss silica glass optical fiber for communication proposed by Charles Kao	1966 – ARPANET project (early version of Internet) begins
	1968 – RCA develops liquid crystal display
	1969 – Hoff develops the microprocessor
	1970 – Silicon chips used for the first time in computer memory applications

TABLE 1.1 (*Continued*)

Science/Physics	Technology/Computers
	1970 – Corning develops low-loss silicon-dioxide optical fibers
	1971 – First microprocessor introduced by Intel
	1972 – First e-mail message sent
	1973 – First portable cell phone call made
	1975 – First digital recording equipment introduced
	1975 – Transmission of television signals using optical fibers achieved
	1975 – First personal computer
	1977 – Telephone companies perform fiber-optic communication trials
	1981 – Global Positioning System (GPS) system becomes operational
	1983 – First compact disc (CD) player launched by Sony and Phillips
	1983 – Internet is formed
	1986 – Fiber optic cable across the English Channel
	1986 – First use of the word Internet
	1988 – First transatlantic fiber-optic cable
	1991 – World Wide Web software developed
	2005 – YouTube launched

Dates are in Common Era (CE). Sources listed at chapter's end.

In the 1700s the advent of coal mining to support the energy needs of the Industrial Revolution in Europe inspired engineers to invent steam engines, which turned out to play an important role in the development of the science of heat, or thermodynamics. This led to the formulation of a new concept called *energy*. Jacquard invented a fabric-weaving loom that used holes punched in paper cards to automate the weaving of complicated, often beautiful, patterns.

Since ancient times, philosophers and scientists have studied light, probably because we can see it and sunlight is essential for life on Earth. In 1015 the Arabian scientist Alhazen studied vision and how light behaves when reflecting from mirrors or passing through prisms. In 1665 Grimaldi explained light as a wave phenomenon. In 1792 French engineers invented methods to transmit messages between cities using light signals. As early as 1676 the speed of light was estimated from astronomical observations.

In the 1800s scientists discovered and explored the properties of electricity and magnetism. This led quickly to the invention of electric motors, radio, and the telegraph, including telegraph cables spanning large bodies of water. This anticipated the development of the world-spanning Internet.

In 1900 the "quantum nature" of the physical world was first exposed. Planck and Einstein explained that light comes in little bundles of energy, called quanta or *photons*. Later, in 1960, this concept proved crucial in the invention of lasers, which are now used for transmitting signal along optical fibers. In the early 1900s physicists gained a deeper understanding of the nature of atoms and subatomic particles called electrons, protons, and neutrons. Later, in 1947, this understanding proved to be crucial in the invention of the transistor, which is the basis of all modern computer circuits.

Early non-transistor-based computers were built in the 1940s. Silicon—the building block of information technology—was discovered in 1787. Almost 200 years later, the quantum properties of silicon scientists discovered and determined it to be the most useful element for making electronic circuits. In addition, glass made of purified silicon dioxide was found to be an excellent material for making optical fibers for communication systems based on laser light.

By 1966 all of the separate parts for the Internet had become known or at least dreamed of, and the first rudimentary long-distance computer network, the ARPANET, was constructed. Since then, Yahoo! Google! iTunes! Napster! YouTube!

1.4 THE METHODS AND SIGNIFICANCE OF SCIENCE

Science is a process by which repeated observations of natural events allow people to discover regularities in these events and reasonable explanations for them. In science, we have specific meanings for the terms *observation*, *rule*, *law*, *hypothesis*, *model*, and *theory*. The relationship between these concepts is diagrammed in **Figure 1.9**. Experimental observation can mean the outcome of a controlled, repeatable, man-made experiment or test. Alternatively, observation could mean carefully watching an event in its natural setting, such as using a telescope to watch a comet passing by Earth.

Rules and laws are based directly on experimental observations. By "rule" we mean an apparent regularity or pattern of occurrences (not an instruction that some object must obey). For example, in nineteenth-century France, André Ampère did many experiments in which he arranged wires, batteries, and compasses in various patterns and recorded what happened. He summarized those observations in several rules. A complete list of all of his observations might be useful to you, but it would be cumbersome to look through this complete list every time you wanted to predict the result of some experiment you were about to carry out. Also, if the list did not contain the particular experiment that you were interested in, then you would be out of luck. Ampère's rules are more useful, because they summarize regularity in the behavior of the world's phenomena. Using rules to summarize the outcomes of many experiments is more efficient than making big lists.

If a rule proves to have exceptionally wide applicability, it is called a law of nature, or simply a law.

A *law* is a statement describing how a specific aspect of the natural world behaves, which appears to be universal in its applicability. Laws describe rather than explain.

By universal we mean general; that is, we only call it a law if it applies everywhere and has never been observed to fail. Laws are generalizations that are based on experimental observations. A law is a statement of perceived fact about the universal

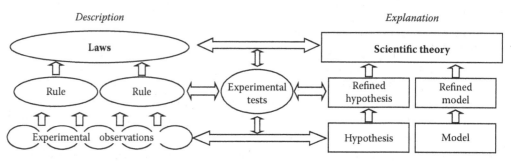

FIGURE 1.9 The development of laws and theories.

nature of natural things. A law describes what happens, but not why or how it happens. For example, it is a law (fact) of nature that all objects attract each other, as for example a stone is attracted by the Earth. This attraction we call gravitation.

We would like to go beyond rules and laws and try to explain how or why events occur. This is done using hypotheses, models, and theories. A first step in this process is making a hypothesis, which is a testable, educated guess about how or why something occurs.

A *hypothesis* is a preliminary conjecture or statement about the natural world that can be tested by experiments.

An example of a hypothesis is "During the next year, every Thursday will be warmer than the Wednesday just before it." Clearly this statement can be subjected to experimental testing (which would likely disprove the hypothesis). An example of a statement that is not a hypothesis is "During the following year, more Christians will go to heaven than will atheists." As far as is known by scientists, this could not be tested by physical experiments. Therefore, such a statement could be defended as a belief, but it cannot be called a scientific hypothesis. To confirm or disprove a hypothesis, you must carry out experiments.

By using successfully tested hypotheses, you can construct a model, that is, a complex scheme for reasoning and making predictions about experiments that have not yet been carried out. Then you should carry out these experiments to test the model.

A *model* is a tentative scheme for explaining and predicting how certain physical events occur.

Once constructed, a model enables us to carry out careful logical reasoning while taking into account many factors. Model building, as a type of reasoning, is different from simple description or pure observation. A model is usually tentative or temporary and might be used only for a specific case or example. We talk about "building" models because it is similar to, for example, building a toy train from small metal parts. The model train is not the actual train, but it has many features in common with the train and helps us understand how a real train behaves. Similarly, a physics model is not the actual object of interest, but in many ways it is similar to that. For example, the model of the solar system put forth by Copernicus in 1543 contains certain rules that the planets "should follow." Although the real planets may not follow these rules precisely, the overall behavior is convincing, and therefore his model is useful.

The culmination of hypothesis testing and model building is the construction of a scientific theory.

A *scientific theory* is a substantiated system of reasoning, based on a collection of well-tested hypotheses, models, and laws, which explains or predicts many physical phenomena.

A system of reasoning cannot be called a scientific theory until it has been subjected to rigorous testing by comparing its predictions with experimental results. Notice that the word theory is used differently in science than it is often used in everyday language. Many people might say, "That idea is just a theory," meaning that it is only unsubstantiated speculation. In science, theory is reserved only for those ideas that have been well tested and found to be successful in all cases tested.

To summarize:

A rule is an apparent regularity of occurrences.
A law is a universally valid description (not explanation) of some natural behavior.
A hypothesis is a testable statement or prediction about how a certain phenomenon will behave under certain conditions.
A model is a tentative explanation of how or why certain phenomena occur and is capable of predicting new phenomena.
A scientific theory is a widely accepted explanation for a large class of phenomena and is capable of predicting new phenomena.

Rules arise from experimental observations. Rules become laws if they are found to be universal. Models arise from tested hypotheses. Models and hypotheses turn into theories if they are found to be successful in explaining and predicting phenomena.

The process of science involves critical thinking, that is, an honest examination of all factors with an open mind to the possibility that previous scientific ideas might contain flaws. This does not mean that scientists believe that the now-accepted ideas are necessarily flawed. Nevertheless, by using skepticism, they hope to improve upon the current ideas to make them even more precise. See, for example, the opening quote for this chapter by Diane Grayson.

Physics is the scientific process by which repeated observations of physical events in the world allow people to discover rules about these events. Physical events are those that involve matter, energy, and the interactions between them. The field of physics includes acoustics, astrophysics, atomic physics, cosmology, electromagnetism, elementary particle physics, fluid dynamics, geophysics, mechanics, nuclear physics, optical physics, quantum physics, solid-state physics, thermodynamics, and others. These subjects cover a wide range of phenomena, over a huge range of size scales. For example, astrophysics studies objects the size of red super giant stars (1,000,000,000,000 meters in diameter), whereas nuclear physics studies objects the size of protons (1/1,000,000,000,000,000 of a meter).

A crucial aspect of physics is that observations, or experiments, are conducted in a quantitative manner, that is, in a way that yields numerical values. For example, if two people observe an apple falling from a tree, one might say, "The apple looked red, and it fell pretty fast." A second person might use a sophisticated instrument to record the apple falling, and state that the light from the apple contained 70% of its spectrum in the wavelength range of 600–700 nanometers and accelerated at a rate 9.8 meters-per-second-per-second. Both answers can be put to a test by a third person, but the more precise the answer is, the more carefully it can be checked and the more useful it is. Nobel laureate Robert Laughlin writes, "The existence of universal quantities that can be measured with certainty is the anchor of physical science." For example, the speed of light in a vacuum is known to be 299,792,458 meters per second. This is a very high precision of about 1 part in 300,000,000. If anyone, anywhere, measures the speed of light in a vacuum using proper methods, he or she will find this same result.

Because of the importance of the numerical values that arise from measurements, the natural language of physics is mathematics. That is not to say that ordinary language and concepts are not important—they are. Still, physicists have found that by using mathematics they can express the theories and laws of physics in the most precise manner possible. This, in turn, allows the most precise tests of the concepts to be made. Only through such precise experimental tests can a provisional theory of physics be verified or disproved. For example, after Ampère and others made their experimental observations of electrical and magnetic phenomena, James Maxwell found that a set of four particular mathematical equations could account for and describe perfectly every one of them. This is quite remarkable.

1.5   THE RELATION OF SCIENCE AND INFORMATION TECHNOLOGY

Engineering and technology are comprised of the principles, methods, materials, and hardware that allow people to achieve practical goals. Technology is often a direct consequence of the knowledge learned through the study of science. It is hard to imagine any one of mathematics, science, or engineering without the other two. Mathematics is a kind of highly structured language that can be used to discuss and analyze questions in science and engineering. Information science is the science of how information is created, stored, transmitted, and processed. Information technology arises through the desire to build practical systems to create, store, transmit, and process information.

Engineering has various aspects. One aspect attempts to use the predictive rules discovered by science to design and build useful devices, for example, compact disc (CD) players. The CD player was invented as a culmination of 100 years of research into many aspects of physics, followed by years of ingenious systems development by engineers. Many of these ingredients are studied in this book.

A different aspect of engineering is equally important—the empirical, exploratory approach to invention. In this approach, progress is made by trial and error, guided by the inventor's intuition. The mathematical laws of physics are not used as an essential guide. In fact, the devices invented using this approach often stimulate scientists to make advances in basic physics. An example of this is the invention of refrigerators. At the time of their invention, the scientific laws of heat (and cold) transfer between the inside and outside of a box were not fully known, but inventors did not let that stop them from building extraordinary contraptions that worked anyway. For example, the earliest room refrigerator (air conditioner) was built in 1844 by a medical doctor, John Gorrie, for keeping patients comfortable in his hospital in Apalachicola, Florida. The practical developments leading to this invention were not strongly dependent on the theoretical development of the science of thermodynamics, which was going on in parallel in Europe by scientists such as Julius Robert von Mayer. In 1842 von Mayer discovered that heat energy and mechanical energy are equivalent; that is, they can be converted into one another. This led to the law of the conservation of energy—one of the most important laws in all of physics.

A very influential mathematician who worked on information technology was Claude Shannon, a scientist at Bell Laboratories. In 1987 he wrote:

> I do what comes naturally, and usefulness is not my main goal. I like to solve new problems all the time. I keep asking myself, "How would you do this? Is it possible to make a machine do that? Can you prove this theorem?" These are my kind of problems. *Not* because I am going to do something useful.

This is a remarkable statement, coming from the mathematician who arguably did the most to bring the world into the Information Age. Given the incredible practical, economic, and social revolutions that communication technology has brought us, the man who laid its theoretical foundations was driven not by making something practical but by solving problems that interested him. The goal of the best scientists and technologists such as Shannon and David DiVincenzo, quoted at the beginning of the chapter, is to solve problems that occur in areas that they find interesting. What is interesting to a person is largely a matter of taste, but, as with music or art, taste can be cultivated. So, read on!

SUMMARY AND LOOK FORWARD

The global information network is made primarily of silicon, an element found in sand. Silicon is the foundation of many of the transmission and processing components making up the Internet in both its glassy form (as optical fibers) and in its crystalline form (as semiconductor crystals). Optical fibers, light detectors, optical and electrical switches, logic and processing circuits, and random-access memory all have silicon at their core. In addition, there are non-silicon-based components, such as lasers, CDs, magnetic storage media, electrical wires and cables, radio, and satellite systems, which serve as essential support elements for the network. Together these form a kind of delicate, tenuous, interconnected nervous system—a silicon web—transmitting information between centers of physical activity.

On the time scale of human history, most of the developments described above occurred very recently. Just how recently can be seen by considering that when my grandparents were born, the existence of atoms had not yet been confirmed conclusively. Although atoms had been postulated in ancient times, it was only in 1905 that Albert Einstein proved their existence by analyzing the observed motions of tiny pollen grains in water. The fact that science and technology have progressed so quickly since then is remarkable.

A scientific theory is not mere speculation based on wishful thinking. A set of ideas must be tested and confirmed by experimental observations before scientists call it a theory. If it fails a test, or if there are no experiments that can be done to test the theory, then it must be discarded or placed on the shelf labeled, "Untested Theories—Do Not Use." Over time, as experiments became more precise, the known laws and theories of nature were subject to revision. This occurred rarely and usually only for phenomena existing at the boundaries of our experience, such as the very small or the very large, where we need to use sophisticated instruments to make novel observations. Such revision of our basic theories is ongoing.

SOCIAL IMPACTS: SCIENCE AND TECHNOLOGY

Small differences of yesterday can have suddenly shocking consequences tomorrow.

Nicholas Negroponte
(1995)

Societal change is primarily driven by emerging technologies, and emerging technologies are never legislated into existence. In a newly apparent way, everything is out of control.

William Gibson
(1999)

Many chapters in this text include a brief essay on a social impact of science and technology. These essays should stimulate you to think about the impact of the science you are learning. Several general ideas to ponder [1]:

1. Scientific and technological developments have real and direct effects on every person's life. They usually include both desirable and undesirable effects or side effects.
2. A considerable gap exists between scientific knowledge and public understanding of it.

3. Scientific thought and knowledge can be used to support different positions. It is normal for scientists and technologists to disagree among themselves, although they may invoke the same scientific theories and data.
4. Scientists and engineers are human and therefore bring uncertainty and the proclivity for error that humans bring to most activities.

If you would like to learn more about this subject, type "Social Impacts of Science and Technology" into an online search engine and you will obtain millions of hits. Or, go to a library and ask a reference librarian for assistance. Also, see the suggested reading list below.

REFERENCES

1. Adapted from *Science: A Curriculum Guide for the Secondary Level Physics*, Copyright by Saskatchewan Education, Canada, http://www.sasked.gov.sk.ca/docs/physics/scilphyd.html.

SUGGESTED READING

General Physics Sources

For historical accounts of physics and its development:
Asimov, I. *The History of Physics*. New York: Walker, 1984.
Motz, L., and J. H. Weaver. *The Story of Physics*. New York: Avon Books, 1989.

An easy-to-read, conceptual introduction to the main topics of physics:
Hewitt, Paul G. *Conceptual Physics,* 9th ed. San Francisco: Addison Wesley, 2002.

Many university-hosted websites provide tutorials on physics. They tend to change, so links might not be reliable. You can use a search engine to locate the *Physics 2000* website at the University of Colorado (http://www.colorado.edu), which is intended for non-scientists and students. Another website covering physics at a level somewhat higher than the present course is *HyperPhysics*, which is maintained by Georgia State University (http://gsu.edu).
    The *MacTutor History of Mathematics* archive at the University of St. Andrews in Scotland (http://www.st-andrews.ac.uk) presents excellent historical summaries of physicists and mathematicians.

    For discussion of what constitutes a scientific theory, law, hypothesis, or model:
Working Group on Teaching Evolution, National Academy of Sciences. *Teaching About Evolution and the Nature of Science*. Washington, DC: National Academy Press, 1998). Available free online at http://www.nap.edu.

Technology and Computer Sources

A history of the personal computer revolution:
Freiberger, Paul, and Michael Swaine. *Fire in the Valley: The Making of the Personal Computer*, 2nd ed. New York: McGraw-Hill, 1999.

A popular account of radio and the Internet:
Naughton, J. *A Brief History of the Future*. Woodstock, NY: Overlook Press, 2000.

An account of how the Jacquard loom initiated the age of programmable machines and computers:
Essinger, James. *Jacquard's Web*. Oxford, U.K.: Oxford Press, 2004.

Some technology companies, whose scientists played major roles in the development of computers, offer historical and technical summaries of the science behind the Internet. You can search these companies' websites using terms such as *history, museum, research, silicon, labs,* and *information theory*:
Alcatel-Lucent Bell Laboratories, http://www.bell-labs.com
Intel, http://www.intel.com
International Business Machines, http://www.ibm.com
Texas Instruments, http://www.ti.com

Essays on Social Impacts

Wiener, Norbert. *The Human Use of Human Beings*. New York: Avon Books, 1950, 1967.
Negroponte, Nicholas. *Being Digital*. New York: Knopf, 1995.
Calcutt, Andrew. *White Noise*. New York: St. Martin's, 1999.
Kleinman, Daniel Lee. *Science and Technology in Society*. Oxford, U.K.: Blackwell, 2005.

References for Timeline

"Science Engineering and Technology Timeline," http://www.intute.ac.uk/sciences/timeline.html.
Marshall, Jim. "A Short History of Computing," Pomona College, http://maya.cs.depaul.edu/~classes/it130/history.
Carlson, Bob, Angela Burgess, and Christine Miller. "Timeline of Computing History," http://www.computer.org/portal/cms_docs_ieeecs/ieeecs/about/history/timeline.pdf.

Other Sources

The first quotation on page 1 is used with permission from: Grayson, Dianne, "Rethinking the Content of Physics Courses," Physics Today (Febuary 2006, pg. 31) Copyright American Institute of Physics.

KEY TERMS

Analog
Atom
Bits
Compact disc
Digital
Electromagnetism
Electron
Energy
Hypothesis
Integrated circuit
Laser
Law
Magnetic hard drive
Model
Neutron
Observation
Optical fiber
Photon
Proton
Repeater
Rule
Theory
Transistor

EXERCISES AND PROBLEMS

Exercises

E1.1 Explain what is wrong with the following request: "Write a statement offering a hypothesis about the number of angels that can dance on the head of a pin."

E1.2 Write a statement (80–120 words) offering a hypothesis about some subject, not necessarily a scientific one. Explain why your idea satisfies the requirements to be called a hypothesis.

E1.3 Give two examples each of theories, laws, and models other than those mentioned in the chapter.

E1.4 Write a statement (200–300 words) offering a speculation about which came first—science or technology?

E1.5

(a) How many years intervened between the invention of movable type and the first movie picture with sound?
(b) How many years intervened between the first motion picture with sound and the first e-mail?
(c) How many years intervened between the first proof that electromagnetic waves travel at the speed of light and the sending of the first transatlantic radio signal?
(d) How many years intervened between the introduction of the quantum concept and the first operation of a laser?
(e) Find and state one more interesting time interval between related events in the given timeline. Often there is a delay between a scientific discovery and its application.

E1.6 Look up the meaning of Murphy's Law (not in the text) and debate whether it qualifies as a law. Give pro and con arguments.

E1.7 Could there be a science of physics that uses only human language but no numbers or math? If not, why? If so, how would it be similar to and different from physics as it is currently practiced?

2

Mathematics: The Language of Science and Technology

[T]he mathematical formulation of the physicist's often crude experience leads in an uncanny number of cases to an amazingly accurate description of a large class of phenomena. This shows that the mathematical language has more to commend it than being the only language which we can speak; it shows that it is, in a very real sense, the correct language.

Eugene Wigner
(1960)

Albert Einstein writing equations on a blackboard in 1931. He was a master of expressing the description of physical phenomena in mathematical terms. (With permission of the Huntington Library, San Marino, CA.)

Claude Shannon of Bell Laboratories, who founded the theory of information and introduced the term "bit" in 1948. His mathematical work gave the basis for digital data storage, data compression, and communication. (With permission of Lucent Technologies Inc./Bell Labs.)

HOW TO USE THIS CHAPTER: This chapter reviews the mathematics concepts and methods that will be used in the text. Readers may choose to study this entire chapter before continuing on to the rest of the text or, instead, go directly to Chapter 3 and refer back to the relevant sections when needed. In particular, Sections 2.2 through 2.6 are necessary background for the main physics in Chapters 3, 4, 5, and 7 through 10. Sections 2.7 and 2.8 are helpful for understanding the computer science examples in those chapters and in Chapters 6, 8, 11, and 16. Section 2.9 is helpful background for Chapter 14 on lasers.

2.1   THE UTILITY OF MATHEMATICS IN SCIENCE AND TECHNOLOGY

An important part of learning about physics is to learn how to use the language of mathematics to describe physical phenomena. To paraphrase Hungarian physicist Eugene Wigner, mathematics is an extremely effective language for describing nature and its behavior. In our discussions of the inner workings of computers, lasers, and optical fibers, we will encounter many situations that beg for a mathematical description. Such a description allows us to summarize many possible phenomena by a single compact equation. It also allows us to predict the behavior of a physical device before we ever build and test it. This allows proper design of practical devices such as lasers, fibers, compact disk players, etc. Throughout this text, we will study examples in which we can describe physical phenomena using mathematics, but with a minimum amount of mathematical detail.

In this chapter, we will review several mathematical techniques and tools that will be used in the rest of the text. One indispensable mathematics technique is the use of scientific notation for representing very large or very small numbers. For example, it is easier to write 10^6 than to write 1,000,000. A convenient tool for discussing physical phenomena is the method of prefixes to represent large or small quantities. For example, it is easier to say kilometer (km) than to say 1000 meters (m). Another common tool is the use of graphs to represent a set of data or numbers. In addition, the concepts of *digital* and *analog* quantities are critical to understanding computers and the Internet.

Because this text integrates discussions of information technology with the study of physics, it is important to appreciate the way in which computers represent *information* in terms of numbers. In this chapter, we will study the concept of binary numbers, which is the language that computers use. In the *binary number system*, only two digits are used: 0 and 1. Each *digit* is called a *bit*, and 8 bits make 1 *byte*. We can then talk about kilobytes and megabytes as measures of the amount of storage space, or memory, in a computer. An important question is how much information can we store in a particular-sized memory, say 80 megabytes? To answer this, we need to understand how to count or measure information. The mathematics behind the theory of information was discovered by Claude Shannon, one of the founders of the Information Age. His theories guide scientists and engineers in understanding and designing information-handling systems such as the Internet.

2.2   GRAPHS

Graphs are used for representing data (numbers) that change as time progresses. A simple example is given by parents who record the height of their growing daughter by marking her height on a wall. Let us say that they record her height on the first of every January. Each time they record it, she moves to the right by one step, so that the marks make a pattern on the wall as shown in **Figure 2.1**. The recorded heights, measured

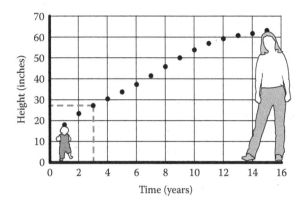

FIGURE 2.1 A child's height (measured in units of inches) graphed versus time (measured in units of years).

in inches (to the nearest 0.1 inch), are 17.9, 23.4, 27.2, 30.4, 33.7, 37.3, 41.4, 45.8, 50.0, 53.8, 56.9, 59.2, 60.7, 61.7, and 63.1.

The axis labeled *time* is called the horizontal axis, and the axis labeled *height* is called the vertical axis. The example shown by the dashed lines can be read as, "At the age of 3 years the child measured 27.2 inches in height." As a second example, ask the question, "At what age did she measure 50.0 inches?" The answer is 9 years.

If instead the parents had recorded her height every 6 months, the data might look like that graphed in **Figure 2.2**, with twice as many points. If the parents recorded her height every day, there would be thousands of data points, packed so closely together that they nearly make a continuous curve, as in **Figure 2.3**. This illustrates that height versus time is a smooth curve—we say that height is a continuous or analog variable. The girl's height changes continuously as she gets older.

Time is also a continuous variable. Between any two points on the time axis (e.g., between 8 and 10 years) there are an infinite number of points. By this, we mean that any time interval (say 1 year) can be divided into arbitrarily smaller intervals with no end to the process. Years can be divided into months, months into days, hours, seconds, milliseconds, microseconds, nanoseconds, etc. In practice, it is not possible to measure a continuous variable at every possible time, but the concept of continuous variables is useful nevertheless.

In the examples shown so far, the horizontal axis represents time and the vertical axis represents height. Of course, these can represent other quantities. For example, say

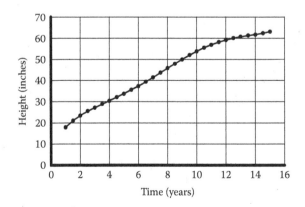

FIGURE 2.2 A child's height graphed versus time, with one data point every 0.5 years.

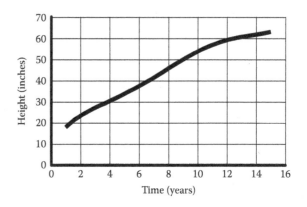

FIGURE 2.3   A child's height graphed versus time, represented as a continuous curve.

we want to describe the temperature of the air in a long hallway containing a heater at one end and an open window at the other end. We could make a graph, with the horizontal axis representing distance along the hallway and the vertical axis representing temperature. The temperature is a continuously changing quantity, so a smooth curve would again be appropriate in a graph of temperature versus position.

2.3   PRECISION AND SIGNIFICANT DIGITS

In the example in the previous section, the heights of the growing child were measured and given to the nearest 0.1 inch. The first few measurements were 17.9, 23.4, 27.2, and 30.4 inches. Here we say that 0.1 inch is the ***precision*** of the measurements. The precision of a measurement refers to the fineness with which the measuring process can distinguish between two nearly equal values of a continuous quantity.

Instead of measuring to a precision of 0.1 inch, the parents could have been less careful and measured only to a precision of 1 inch. In that case, the first four measurements would have been recorded as 18, 23, 27, and 30 inches. The original numbers have been rounded off to the nearest inch.

Or, if the parents wanted to be far more precise, they might have tried to measure to a precision of 0.01 inch (one hundredth of an inch). This would not be a very useful effort, because a person's height is not meaningful at this level of precision. A person's height can change depending on posture or even hairstyle. It is important in any measurement situation to decide to what level of precision a number should be recorded.

Figure 2.4 shows several rulers that are used for measuring distance. The first one has a mark at each 1-meter distance. (A meter is roughly the length of a 6-foot-tall person's leg.) Such a coarse ruler would not be very useful unless it were hundreds of meters long, in which case it could be used for measuring, for example, a large distance between two trees in a park. The second ruler shown has each meter divided into ten equal parts, providing higher precision. The third ruler, shown up close, provides still higher precision, with each meter divided into 100 equal parts, each having a length of 1 centimeter (cm). The third ruler allows you to measure distance with higher precision than you could using the other rulers.

Relative precision describes the precision of a number compared to the value of the number itself. That is, it is the ratio of the precision and the number itself:

$$\text{relative precision of a number} = \frac{\text{precision of the number}}{\text{value of the number}}$$

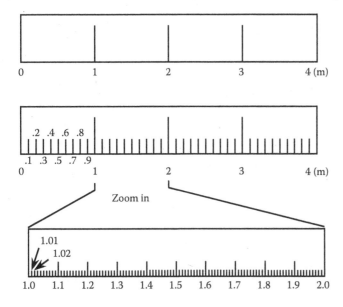

FIGURE 2.4   Three rulers having different levels of precision: 1 m, 1/10 m, and 1/100 m.

For example, if we are given the number 24.8, then the precision is 0.1 and the relative precision is 0.1 ÷ 24.8 = 0.004. We can state this relative precision as 4 parts in 1,000.

A digit is one of the symbols making up a number. For example, in the number 78.3, each symbol 7, 8, and 3 is called a digit. ***Significant digits*** are those digits in a number that actually convey useful information. For example, if an item in a store regularly costs $2.99, and it is marked down by one-third, you could calculate its price as $2.99 × (2/3) = $1.9933333..., where the digits repeat forever. Because we are talking about money, it makes sense to round the number to $1.99, a number having three significant digits.

For the purpose of counting the number of significant digits, it is irrelevant where the decimal point (.) is located. For example, the number 1.99 and the number 19.9 both contain three significant digits.

Sometimes the zero digit (0) can be a significant digit. For example, the zero in 705 is significant. On the other hand, the zero in 450 may or may not be significant, depending on the intent of the person writing it. If he meant 450, and not 460 nor 430, then it is not significant. But if he meant to say "really" 450, and not 451 nor 449, then the zero is significant. In this case, he could indicate that the zero is significant by adding a decimal point, and write "450." to represent the number.

2.4   LARGE AND SMALL NUMBERS AND SCIENTIFIC NOTATION

When working with numbers that are very large or very small, it is not convenient to write them out in long hand. For example, the distance from the Earth to the Sun is about 150,000,000,000 meters. Instead of writing all of these zeros, we could say that the number equals 150 followed by nine zeros, or 150 × 1,000,000,000, or 150 times 1 billion. A short-hand way to write 1 billion is 10^9, or ten raised to the ninth power. This means $10 \times 10 \times 10 \times 10 \times 10 \times 10 \times 10 \times 10 \times 10$. The following table summarizes the powers of ten. The powers (numbers in superscript) are called *exponents*.

QUICK QUESTION 2.1

Say that a long jumper clears a distance of 19.56 feet, and a second jumper clears eight-ninths of this distance. Calculate the value of the second distance, using the same number of significant digits as the first.

REAL-WORLD EXAMPLE 2.1: PRECISION OF DISPLAY PIXELS

Computer screens display images by lighting up many separate pixels (picture elements), which are small box-shaped regions arranged in a rectangular array. For example, a monitor with $1,680 \times 1,050$ resolution has $1,680 \times 1,050 = 1,764,000$ image pixels. Consider a gray-level monitor for simplicity.

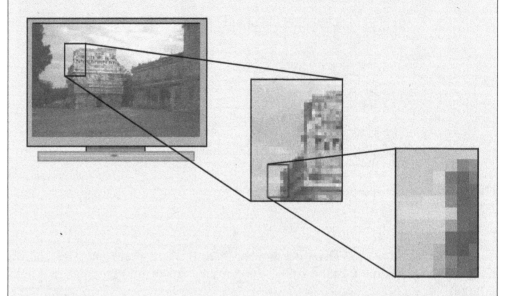

FIGURE 2.5 Each pixel can show a different brightness. For example, a computer might use a different whole number between 1 and 4,096 to represent each of 4,096 distinct brightness levels. The precision in this case is 1, whereas the relative precision is 1 in 4,096. A screen with higher brightness resolution might use 524,288 distinct levels, each represented by a different whole number. The relative precision in this case is 1 part in 524,288.

Powers-of-Ten Notation			
1	$= 10^0$	one	one
10	$= 10^1$	ten	ten
100	$= 10^2$	ten-squared	hundred
1,000	$= 10^3$	ten-cubed	thousand
10,000	$= 10^4$	ten to the fourth power	ten thousand
100,000	$= 10^5$	ten to the fifth power	hundred thousand
1,000,000	$= 10^6$	ten to the sixth power	million
1,000,000,000	$= 10^9$	ten to the ninth power	billion
1 (and n zeros)	$= 10^n$	ten to the nth power	

In scientific notation, we write a number with one digit to the left of the decimal and one or more digits to the right, depending on what level of precision is wanted. For example, the Sun-to-Earth distance is written as 1.5×10^{11} meters.

The form for writing a number in scientific notation is shown schematically by:

$$\Box.\Box\Box \times 10^{\pm\Box}$$

Each empty box indicates a place where a whole number goes. Any number of boxes can be used to the right of the decimal point.

To multiply two numbers containing exponents, add the exponents. For example,

$$100,000 \times 1000 = 10^5 \times 10^3 = 10^{5+3} = 10^8 = 100,000,000$$

To divide two numbers, subtract the exponents. For examples:

$$\frac{100,000}{1,000} = \frac{10^5}{10^3} = 10^{5-3} = 10^2$$

$$\frac{1,000}{1,000} = \frac{10^3}{10^3} = 10^0 = 1$$

Next, consider the example:

$$\frac{1,000}{100,000} = \frac{10^3}{10^5} = 10^{3-5} = 10^{-2} = \frac{1}{100} = 0.01$$

How did we know that $10^{-2} = 1/10^2$? It is a rule that for any exponent n, we have $10^{-n} = 1/10^n$. This rule is a result of the subtracting-exponents-for-division rule. For example:

$$10^{-2} = 10^{0-2} = \frac{10^0}{10^2} = \frac{1}{10^2}$$

The following table summarizes the negative powers of ten.

Negative Powers-of-Ten Notation		
0.1	$= 10^{-1}$	tenth
0.01	$= 10^{-2}$	hundredth
0.001	$= 10^{-3}$	thousandth
0.0001	$= 10^{-4}$	ten to the minus 4 power
0.00001	$= 10^{-5}$	ten to the minus 5 power
0.(n - 1 zeros)1	$= 10^{-n}$	ten to the minus n power

To multiply two numbers, break up the numbers, multiply, and recombine them. For example:

$$200,000 \times 6,000 = 2 \times (100,000) \times 6 \times (1,000) = 2 \times 6 \times 10^5 \times 10^3 = 12 \times 10^8$$

To write this in scientific notation, with one digit to the left of the decimal point, write:

$$12 \times 10^8 = 1.2 \times 10^1 \times 10^8 = 1.2 \times 10^9$$

To divide two numbers, break up the numbers, divide, and recombine them. For example:

$$\frac{200,000}{6,000} = \frac{2 \times 10^5}{6 \times 10^3} = \frac{2}{6} 10^{5-3} \approx 0.333 \times 10^2 = 33.3 = 3.33 \times 10^1$$

(The symbol "$\approx$" means "approximately equal to.")

Example 2.1

Say the budget deficit of the United States in a certain year is $300 billion. How many dollars is that per citizen? Assuming the population is 297,000,000, divide:

$$\frac{300 \text{ billion}}{297,000,000} = \frac{300 \times 10^9}{297 \times 10^6} = 1.01 \times 10^{9-6} = 1.01 \times 10^3 = 1,010$$

Example 2.2

How many data pits, each representing data, are on the surface of a compact disc? A CD has an area of about 100 square centimeters. Each pit on the surface occupies an area of about 10^{-8} square centimeters. The ratio of these areas gives the number of pits:

$$\frac{100}{10^{-8}} = \frac{10^2}{10^{-8}} = 10^{2-(-8)} = 10^{10}$$

Example 2.3

How many toothpicks can you get from one giant tree? Say one toothpick has a volume of 30 cubic millimeters (e.g., 1 mm × 1 mm × 30 mm) and a giant tree has a volume of 10^{11} cubic millimeters (e.g., 1000 mm × 1000 mm × 100,000 mm). The ratio of these volumes gives the number of toothpicks:

$$\frac{10^{11}}{30} = \frac{1 \times 10^{11}}{3 \times 10^1} = \frac{1}{3} \times 10^{11-1} \approx 0.33 \times 10^{10} = 3.3 \times 10^9,$$

which is over 3 billion toothpicks. Notice that the answer has been rounded to two digits because a more precise answer would not be meaningful given the rough nature of this estimate.

THINK AGAIN

If you are not told how many decimal places (digits) to include in an answer to a calculation, how can you decide this? Usually, the information given in the statement of the problem will guide you. If you are not sure, a good rule of thumb is to write the answer in scientific notation using three digits, one before the decimal and two after.

2.5 UNITS FOR PHYSICAL QUANTITIES

When dealing with quantities, or amounts, of a certain substance, such as water, or something more abstract such as wealth, it is helpful to have shorthand names for various amounts of the thing of interest; for example, 1 ton of bricks, 1 gallon of water, 6 kilometers of distance, 2 kilowatts of power, 5 amps of current, 4 dollars of money. The terms ton, gallon, kilometer, kilowatt, amp, and dollar are examples of *units*.

A *unit* is a fixed quantity (of length, volume, time, etc.) used as a standard of measurement.

Often there are many ways to express the same quantity by using different units—4 dollars is the same as 400 cents, and 6 kilometers is the same as 6000 meters. To explain the distance between Seattle and Boston, you could say 3046 miles, or 4902 km, or 4,902,000 meters. (This follows from 1 mi = 1,609.344 km, and keeping four significant digits.) The numbers used are different, but the distance is the same in each case.

2.5.1   Metric System Units

The units used throughout this book are metric units (meters for length, kilograms for mass, seconds for time, newtons for force, etc.). They are part of a standardized system called the Système International, or SI for short, which was adopted by the international science and engineering community in 1960.

Instead of always writing or saying the powers-of-ten notations for large or small numbers, people like to use prefixes to represent these. You are probably familiar with some of these terms, such as *kilo* in kilometer, and *mega* in megabyte, megawatt, or megabuck. You might have heard the term nanotechnology. This refers to technology constructed at the nanometer (10^{-9} m) scale. The following table gives the prefixes, along with their values and their abbreviations.

					Abbreviations					
10^{12}	10^{9}	10^{6}	10^{3}	10^{-2}	10^{-3}	10^{-6}	10^{-9}	10^{-12}	10^{-15}	10^{-18}
T	G	M	k	c	m	μ	n	p	f	a
tera	giga	mega	kilo	centi	milli	micro	nano	pico	femto	atto

We measure length, position, or distance in the metric units of millimeters (mm), meters (m), or kilometers (km). Of course, we could also use inches (in), feet (ft), and miles (mi), but it is more difficult to calculate with these than with metric units.

	Length	
1 kilometer	= 1 km	= 10^3 m = 0.621 mile
1 meter	= 1 m	= 39.37 inch (about 1 yard)
1 centimeter	= 1 cm	= 0.01 m = (1/100) m = 10^{-2} m = 0.394 inch
1 millimeter	= 1 mm	= 0.001 m = 10^{-3} m (thickness of a dime)
1 micrometer	= 1 μm	= 10^{-6} m (size of bacterium)
1 nanometer	= 1 nm	= 10^{-9} m (size of large molecule)

We measure volume in the metric units of cubic meters (m^3), or more commonly, liters (ℓ), which is the volume of a cube whose edges are 10 cm long.

	Volume	
1 kiloliter	= 1 kℓ	= 1 cubic m = 1 m^3
1 liter	= 1 ℓ	= 1,000 cubic cm = 10^3 cm^3 (1.06 U.S. quart)
1 milliliter	= 1 mℓ	= 0.001 ℓ = 10^{-3} ℓ = 1 cm^3 (volume of a thimble)

We measure time in units of seconds, abbreviated sec or simply s.

	Time	
1 second	= 1 sec	(about one heartbeat)
1 millisecond	= 1 msec	= 0.001 sec = 10^{-3} sec (flap of housefly wing)
1 microsecond	= 1 μsec	= 10^{-6} sec (high-speed strobe light flash)
1 nanosecond	= 1 nsec	= 10^{-9} sec (time for light to travel one foot)
1 picosecond	= 1 psec	= 10^{-12} sec (time for one vibration of a molecule)

Units for other quantities will be introduced as needed throughtout the book.

An important part of learning about science and technology is learning how to calculate physical quantities in real-life situations. This is useful in everything from estimating your home heating bill to understanding how computer circuits work. Here we will study a systematic method for calculating quantities involving units.

Consider a car traveling with speed equal to S. The formula for the distance (D) traveled in a certain time (t) by the car is: $D = S \cdot t$, where the dot means "times." Common sense tells us that if the car's speed is 100 kilometers per hour (100 km/hr), and the car travels for 2 hours, the distance it covers is 200 km. That is,

$$D = (100 \text{ km/hr}) \cdot (2 \text{ hr}) = 200 \text{ km}$$

There are several equivalent ways to write 100 kilometers per hour:

$$100 \text{ km per hr} = 100 \text{ km/hr} = 100 \frac{\text{km}}{\text{hr}} = \frac{100 \text{ km}}{\text{hr}}$$

The word "per" acts like division ($\div$) for the units.

Let us analyze the distance calculation in more detail:

$$D = S \cdot t = (100 \text{ km per hr}) \cdot (2\text{hr}) = \frac{100 \text{ km}}{\cancel{\text{hr}}} \cdot 2 \cancel{\text{hr}} = 200 \text{ km}$$

The hr unit appears both in the numerator and the denominator and it therefore cancels, just as in ordinary fractions involving numbers or variables; for example,

$$\left(\frac{a}{b} \right) \cdot b = a$$

Notice also that we must use values of the speed and time that are compatible, so that the unit of time (hours) will cancel properly, as in the above example. If instead we were to give the time in minutes (120 min) and put this into the formula, we would get

$$D = S \cdot t = \frac{100 \text{ km}}{\text{hr}} \cdot 120 \text{ min} = 1200 \text{ km} \frac{\text{min}}{\text{hr}}$$

Written this way, the minutes and hours units do not cancel, and we get a result that is hard to interpret. To remedy this, we need to use the following method.

2.5.2 The Method of Conversion Factors

In the preceding example, we were left with the awkward units min/hr, which we want to eliminate. We know that there are 60 minutes per hour; this means that there are 1/60 hour per minute. Another way to say this is that 60 minutes equals 1 hour, or 60 min = 1 hr. This means that 60 min divided by 1 hr equals 1.

$$\left(\frac{60 \text{ min}}{1 \text{ hr}} \right)_{CF} = 1$$

In this equation, the "1" on the right-hand side has no units. This quantity is called a *conversion factor*, and we write "CF" near the bottom of the bracket to remind us. It is also true that 1 hr divided by 60 minutes equals 1:

$$\left(\frac{1 \text{ hr}}{60 \text{ min}} \right)_{CF} = 1$$

We can multiply any quantity by any conversion factor without changing the quantity's value. Consider a simple example: What does 3 hours equal in minutes?

$$3 \text{ hr} = 3 \text{ hr} \cdot (1) = 3 \text{ hr} \cdot \left(\frac{60 \text{ min}}{1 \text{ hr}} \right)_{CF} = 3 \text{ hr} \cdot \frac{60 \text{ min}}{\text{hr}} = 180 \text{ min}$$

The numerical value (3) has been changed (to 180), but the quantity of time has not changed. We can also illustrate this method in the opposite direction:

$$180 \text{ min} = 180 \text{ min} \cdot (1) = 180 \text{ min} \cdot \left(\frac{1 \text{ hr}}{60 \text{ min}} \right)_{CF} = 180 \text{ min} \cdot \frac{1 \text{ hr}}{60 \text{ min}} = 3 \text{ hr}$$

When doing such a calculation on paper, it is most efficient not to write every step shown above. First write "180 min," then next to it write the conversion factor, then cancel units, then evaluate the numerical part, and write the answer. The final product looks like:

$$180 \text{ min} \cdot \left(\frac{1 \text{ hr}}{60 \text{ min}} \right)_{CF} = 3 \text{ hr}$$

It is good to always put brackets around conversion factors to remind you of the method.

Now we can apply the conversion factor method to the above example for the distance traveled in 120 minutes by a car moving with speed 100 kilometers per hour:

$$D = \frac{100 \text{ km}}{\text{hr}} \cdot 120 \text{ min} = \frac{100 \text{ km}}{\text{hr}} \cdot 120 \text{ min} \cdot \left(\frac{1 \text{ hr}}{60 \text{ min}} \right)_{CF} = 200 \text{ km}$$

We multiplied by the conversion factor (1 hr/60 min), canceled the minutes units with each other, and then canceled the hour units with each other, leaving kilometers.

There is another way to do such a calculation that some people find easier to understand, although it takes more steps. The speed (*S*) is given as 100 kilometers per hour, and the time (*t*) traveled is 120 minutes. The formula for distance is $D = S \cdot t$. The problem here is that the units are not consistently given in the speed and in the time. First, convert the time from minutes to units of hours before inserting it into the formula. This is done by multiplying 120 min by the proper conversion factor to give

$$120 \text{ min} = 120 \text{ min} \cdot \left(\frac{1 \text{ hr}}{60 \text{ min}} \right)_{CF} = 2 \text{ hr}$$

After converting the time into units of hours, insert it into the distance formula:

$$D = S \cdot t = \frac{100 \text{ km}}{\text{hr}} \cdot 2 \text{ hr} = 200 \text{ km}$$

Finally, note that you can also divide by conversion factors instead of multiplying by them. Because a conversion factor equals 1, either multiplication or division by 1 will leave the original quantity unchanged, except for its units. For example,

$$180 \text{ min} = \frac{180 \text{ min}}{\left(\dfrac{60 \text{ min}}{1 \text{ hr}} \right)_{CF}} = 180 \text{ min} \cdot \left(\frac{1 \text{ hr}}{60 \text{ min}} \right)_{CF} = 3 \text{ hr}$$

QUICK QUESTION 2.2

Use a conversion factor method to convert 3700 meters into kilometers. Recall that there are 1000 meters in a kilometer.

Dividing by a fraction is equivalent to multiplying by the fraction's inverse. For example,

$$\frac{a}{\left(\frac{b}{c}\right)} = a \cdot \left(\frac{c}{b}\right),$$

and therefore,

$$\frac{a}{\left(\frac{a}{c}\right)} = \not{a} \cdot \left(\frac{c}{\not{a}}\right) = c$$

IN-DEPTH LOOK 2.1: USING CONVERSION FACTORS

We can make up a general example of using conversion factors by using arbitrary names for the units. For fun, I will call these *whatnot* and *whosis*. Let us say we begin with a value of 125 *whatnot*. How many *whosis* does this equal? To answer, we need to know how many *whosis* make up a single *whatnot*, or vice versa. Let us use an example in which 1 *whosis* is equivalent to 25 *whatnot*. So, we can write

1 *whosis* = 25 *whatnot*, or 1 *whatnot* = (1/25) *whosis*

That is, 1 *whatnot* equals one twenty-fifth of a *whosis*. Because 1/25 = 0.04, we could also write this as 1 *whatnot* = 0.04 *whosis*. We can now create a conversion factor by noting that

$$\left(\frac{1\,whosis}{25\,whatnot}\right)_{CF} = 1$$

We can use this conversion factor to convert our starting number:

$$125\,whatnot \times \left(\frac{1\,whosis}{25\,whatnot}\right)_{CF} = 5\,whosis$$

We can also have an example that goes in the opposite direction. If we start out with 3 *whosis*, how many *whatnots* is this? We multiply by the other conversion factor:

$$3\,whosis \times \left(\frac{25\,whatnot}{1\,whosis}\right)_{CF} = 75\,whatnot$$

Now that you have struggled through this example using the silly terms *whatnot* and *whosis*, go back and reread it substituting *cents* for *whatnot*, and *quarters* for *whosis*. You will see that it makes a lot of sense. What we have argued is that 25 *cents* is the same amount of money as one *quarter*, and therefore (1 *quarter*/25 *cents*) = 1.

2.5.3   Guidelines for Calculating with Units

The following set of guidelines helps when calculating with units.

Guideline 1. Before inserting a number into a formula, convert it to a number having the proper units, so that the units being dealt with will cancel. For example, use $t = D/S$ to calculate time from distance and speed, where speed is equal to 0.4 km/sec. If you are given that the distance is 800 meters (800 m), you must convert this distance into kilometers before inserting it into the formula. This is done by:

$$800 \text{ m} \left(\frac{1 \text{ km}}{1000 \text{ m}} \right)_{CF} = 0.8 \text{ km}$$

Guideline 2. After inserting numbers into a formula, the unit symbols may be cancelled as if they were ordinary numbers. To continue the example, then

$$t = \frac{0.8 \text{ km}}{0.4 \text{ km/sec}} = 2.0 \text{ sec}$$

Guideline 3. Always write the units explicitly at every step of the calculation, including in the final answer, to avoid making mistakes or giving meaningless numbers as answers. For example, if someone asks you how far is it from your hometown to Antarctica, it is meaningless to say "six thousand," unless you also say what units this is being expressed in.

Guideline 4. After calculating (by hand or using a calculator) a final number, you need to decide how many significant digits to write when giving your answer to a problem. This determines what *precision* you will use in stating your answer. For example, the answer 2.0 seconds in the above problem is given using two digits: 2 and 0. This means that you believe the answer is not as large as 2.1 seconds, but is not as small as 1.9 seconds. Strictly speaking, it means that the answer is somewhere between 1.95 seconds and 2.05 seconds. If the data you used to calculate your answer are not actually known to this degree of precision, you should perhaps report your answer using only one digit. This implies a lower precision of your answer. For example, the answer 2.0 seconds could be reported instead as 2 seconds, which means that the answer is somewhere between 1.5 seconds and 2.5 seconds.

QUICK QUESTION 2.3

Use the following conversion factors to calculate the number of days in one century: 1 century = 100 years; 1 year = 52 weeks; 1 week = 7 days. Now repeat, using 1 year = 12 months; 1 month = 4 weeks; 1 week = 7 days. Again repeat, using 1 year = 365 days. Explain why each answer is different, though "correct," given the precision that each calculation is using.

2.6 PROPORTIONALITY

The simplest relation between two mathematical quantities is that of direct proportionality, meaning that if one quantity increases by a certain multiplying factor, the other increases by the same multiplying factor. For example, if you have a telephone billing plan that charges strictly by the amount of time you spend on the phone, then the cost of a call is proportional to the time connected. If the cost rate is 0.1 cents per second, then we can express the total cost by the equation

$$\text{total cost} = (0.1 \text{ cents per second}) \times (\text{time connected})$$

We say that the total cost is proportional to the time connected, and symbolize this by

$$\text{total cost} \propto \text{time connected}$$

Say that you talk for 100 seconds. This would cost 10 cents. If you double the time connected, the cost would double to 20 cents, etc.

Another kind of proportionality is inverse proportionality, meaning that if one quantity is increased by multiplying with a certain factor, the other decreases by dividing by the same factor. For example, in a long concert hall, the loudness of the music might be inversely proportional to your distance from the stage. If you double your distance, the loudness would be cut in half. That is,

$$\text{perceived loudness} \propto \frac{\text{loudness at 1 foot from stage}}{\text{distance in feet from stage}}$$

2.7   BINARY NUMBERS

If the base 2 is used, the resulting units may be called binary digits, or more briefly *bits*.

Claude Shannon
(*The Bell System Technical Journal*, 1948)

Computers use a system of numbers called the *binary number system*, which is different from the common **decimal number system** that people usually use. You have probably heard the term *bit* used to specify, for example, the speed of downloading data from the Internet. Using a DSL (or digital subscriber link) cable, you can achieve (in 2009) a speed of around 20 megabits per second (20 Mbps). In this section, we will discuss how computers use bits to represent numbers.

The concept of numbers comes from counting. How do you count? You can count to 10 on your fingers. (In fact, the word digit means "finger," and digital means "of, or like, a finger.") To keep track of the fact that you have passed 10 once, you could make a mark in the dirt with your right foot. Each time you pass 10 again, make another mark with your right foot. (You are counting the tens.) When you pass 100 (ten tens), you could make a mark with your left foot. (You are counting the hundreds.) For counting the tens we write a number to the left of the number we use for counting the ones. For example, 23 means "two tens and three ones" ($2 \times 10 + 3 \times 1$), and 385 means "three hundreds and eight tens and five ones" ($3 \times 100 + 8 \times 10 + 5 \times 1$). The numbers 0 through 9 are called digits. We call this the decimal number system because there are ten distinct symbols (0, 1, 2, 3, 4, 5, 6, 7, 8, 9). The origin of the word decimal is the Latin "deci," which means *ten*.

We call the values 1, 10, 100, 1000, etc., the **place values**. These are the values corresponding to each place or position in the number. They can also be represented equivalently as 10^0, 10^1, 10^2, 10^3, etc. In this system, the number 10 is called the base, and the decimal number system is also called the base-ten system.

For example, the decimal number 2385 can be broken down as shown in the following:

The Decimal Number 2385				
Place:	4	3	2	1
	10^3	10^2	10^1	10^0
Place value:	1,000	100	10	1
	thousands	hundreds	tens	ones
Decimal digit:	2	3	8	5

Computers use the binary number system. Instead of using the place values 1, 10, 100, 1,000, etc., as we do in the decimal number system, in the binary system we use as place values 1, 2, 4, 8, 16, etc. This is also called the base-two system. This is how you would have to count if you had only two fingers total. You could count only to two before having to resort to making marks with your feet to keep track of how many times you counted to two. An equivalent way to write the binary place values 1, 2, 4, 8, 16, etc., is 2^0, 2^1, 2^2, 2^3, 2^4, etc., as summarized in **Table 2.1**.

TABLE 2.1
Powers of Two (Note: $2^0 = 1$)

2^1	2^2	2^3	2^4	2^5	2^6	2^7	2^8	2^9	2^{10}
2	4	8	16	32	64	128	256	512	1024
2^{11}	2^{12}	2^{13}	2^{14}	2^{15}	2^{16}	2^{17}	2^{18}	2^{19}	2^{20}
2048	4096	8192	16384	32768	65536	131072	262144	524288	1048576

Because we use the numbers one and zero so often in this discussion, it helps to have a name for them.

A binary digit (0 or 1) is called a *bit*.

For example, the decimal number 14 requires four bits for its representation. The first time the word *bit* was used in this sense was in Claude Shannon's landmark paper, quoted at the beginning of this section. A group of eight bits is called a *byte*.

2.7.1 Converting from Decimal to Binary

A method for converting a decimal number to binary is to work from left to right (from largest place to smallest place). For example, let us convert the decimal number 327 (three-hundred twenty-seven) into its equivalent binary form. The calculation is summarized in the following table:

The Decimal Number 327 Converted into Binary								
Place	9	8	7	6	5	4	3	2	1
Place value	2^8	2^7	2^6	2^5	2^4	2^3	2^2	2^1	2^0
	256	128	64	32	16	8	4	2	1
Binary digit	1	0	1	0	0	0	1	1	1

The first step is to find the largest *power of 2* that the number (327) contains. As the table reminds us, 2^8 equals 256 (decimal). This is the largest power of 2 contained in the decimal number 327. We record this fact by placing a 1 in the far left column of the table. Then we subtract 256 from the starting number, and see what is remaining: $327 - 256 = 71$. The remainder, 71, contains zero 128s, so place a 0 below the 128 in the table. The remainder is still 71, which contains one 64, so place a 1 under the 64, and subtract 64: $71 - 64 = 7$. This remainder, 7, contains zero 32s, zero 16s, and zero 8s, so put 0s in those places. The remainder, 7, contains one 4, one 2, and one 1, so place 1s in those places. The binary representation of the decimal number 327 is therefore 101000111. This contains nine bits, so we say it is a 9-bit number.

Table 2.2 gives the binary versions for the decimal numbers between 1 and 16. Notice that zeros to the left of the left-most 1 do not carry any significance. For example, in binary, 0011 is the same as 011, which is the same as 11. This is also true for the decimal number system; for example, 0513 is the same as 513.

2.7.2 Converting from Binary to Decimal

It is easy to convert a binary number into decimal. Just note which places contain ones, and add these place values together. For example, the binary number 101101 equals in decimal:

$$1 \times 2^5 + 0 \times 2^4 + 1 \times 2^3 + 1 \times 2^2 + 0 \times 2^1 + 1 \times 2^0 =$$

$$1 \times 32 + 0 \times 16 + 1 \times 8 + 1 \times 4 + 0 \times 2 + 1 \times 1 =$$

$$32 + 0 + 8 + 4 + 0 + 1 = 45 \text{ (decimal)}$$

QUICK QUESTION 2.4

Convert the decimal number 211 into binary.

QUICK QUESTION 2.5

Convert the binary number 1110101 into decimal.

TABLE 2.2
List of Decimal Numbers 0 through 16 and Their Binary Equivalents

Decimal	Binary	2^4 16	2^3 8	2^2 4	2^1 2	2^0 1
0	00000	0	0	0	0	0
1	00001	0	0	0	0	1
2	00010	0	0	0	1	0
3	00011	0	0	0	1	1
4	00100	0	0	1	0	0
5	00101	0	0	1	0	1
6	00110	0	0	1	1	0
7	00111	0	0	1	1	1
8	01000	0	1	0	0	0
9	01001	0	1	0	0	1
10	01010	0	1	0	1	0
11	01011	0	1	0	1	1
12	01100	0	1	1	0	0
13	01101	0	1	1	0	1
14	01110	0	1	1	1	0
15	01111	0	1	1	1	1
16	10000	1	0	0	0	0
17						
18						
19						

This illustrates counting to decimal 16, which in binary is 10000. This means $1 \times 2^4 + 0 \times 2^3 + 0 \times 2^2 + 0 \times 2^1 + 0 \times 2^0$. Try filling in the empty cells.

QUICK QUESTION 2.6

What is the largest decimal number that can be represented using one byte?

What is the decimal equivalent of the binary number 100000000 (1, followed by eight 0s)?

REAL-WORLD EXAMPLE 2.2: ANALOG AND DIGITAL VARIABLES

Continuous variables are called analog, whereas discrete variables, including binary numbers, are called digital. Physicists and engineers typically use analog variables. This is because physics deals with physical quantities such as time and distance, which are continuous. Physicists have developed the mathematics of continuous variables for this purpose. Although in practice we cannot measure any variable with infinitely high precision, there is no theoretical limit to how high the precision could be.

In contrast, computers operate on a set of principles and rules that require them to treat numbers in digital form. Each *cell* in a computer's memory can store only one of two possible numbers: zero or one; that is, one binary digit, or bit. The only way to represent and store numbers with higher precision is to allocate more memory cells to each number location. For example, if a memory uses three cells to store each number, as shown in **Figure 2.6,** it could store one of eight different binary numbers in each location: 000 through 111 (which equals 7 in decimal). But, if the memory used four cells to store each number, then each location could store one of 16 different binary numbers: 0000 through 1111 (which equals 15 in decimal). This would allow higher relative precision, because a measurement range could be divided into finer intervals. When we write a number on a computer, it always has a finite (i.e., limited) number of digits, or binary place values, thereby limiting the precision.

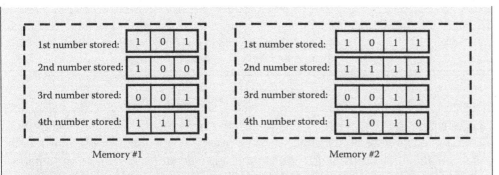

FIGURE 2.6 Two computer memories, each having space to store only four numbers. Each small square box is a memory cell and can hold 1 bit. The memory on the left uses three cells as binary place-value holders (each holding a bit) to store each number. The memory on the right uses four cells as binary place-value holders to store each number.

2.8   THE CONCEPT OF INFORMATION

Information is what you don't already know.

Neil Gershenfeld

When you send an e-mail to a friend, you are sending information. A one-page message contains a certain amount of information, whereas a two-page message contains roughly twice that amount. In the 1940s, long before the advent of the Internet, Claude Shannon asked himself the question, "How can the amount of information in a message be quantified?" By "quantify" we mean assign a numerical value to the amount of information. Or, to turn it around, how much information can be transmitted in a message of a given length? The answer to this question has far-reaching consequences for information technology. In 1948, Shannon published the landmark paper, "A Mathematical Theory of Communication," that revolutionized scientists' understanding of the concept of information, especially in the context of communication systems.

A gain of information decreases your uncertainty. If you are uncertain about the outcome of an event, then you lack some information. If someone tells you the outcome, then you gain information. In the case of two equally likely outcomes—say the result of flipping a coin—then the amount of information that is gained is one bit. This meaning of bit is slightly different from the usage in the previous sections in which bit meant simply a binary digit (0 or 1); however, these two meanings are closely related.

A *bit* is the smallest amount, or quantity, of information.

How can we quantify the amount of information in more complex situations? Say that someone has three coins, each with equal likelihood for heads or tails when tossed. If you do not know the outcomes of one toss for each coin, then it seems intuitively clear that you lack three bits of information—heads (H) or tails (T) for each. It is interesting that there are eight possible combinations of outcomes: HHH, HHT, HTH, HTT, THH, THT, TTH, and TTT. This is the same as the number of different numbers that can be represented by using just three binary digits: 000, 001, 010, 011, 100, 101, 110, and 111. Clearly, a strong connection exists between information and binary counting. A quantitative definition for information content can now be stated.

The amount of information, or ***information content***, in a message (or any other set of data) equals the minimum number of binary digits needed to faithfully represent the message (without having any knowledge in advance, and without just getting lucky).

Note that this definition does not say anything about the *meaning* of the message or data. Scientists have not learned how to quantify meaning. For example, the result of a coin toss could have no meaning whatsoever (e.g., if it is used only to illustrate the equal likelihood of heads and tails). On the other hand, it could have an important meaning (e.g., if it determines which football team gets to possess the ball first). In either case, the amount of information equals one bit.

To quote from Shannon's 1948 paper:

> The fundamental problem of communication is that of reproducing at one point either exactly or approximately a message selected at another point. Frequently the messages have meaning; that is they refer to … some system with certain physical or conceptual entities. These semantic aspects of communication are irrelevant to the engineering problem. The significant aspect is that the actual message is one selected from a set of possible messages.

A simple way to picture Shannon's definition of information is by imagining you are playing the game "20 Questions." Let us say that at a party a friend tells you that she is thinking of some object, and you must try to figure out what that object is by asking *yes* or *no* questions. "Is it larger than a bread basket? Is it made of plant material?" After answers are given to 20 such questions, you have gained 20 bits of information, which is usually sufficient information for you to correctly infer (make an informed guess about) the object's identity.

How many different combinations of yes/no answers could be given to 20 questions? For each question, there are two possible answers, so the total number of combinations of answers equals the number 2 multiplied by itself 20 times, or $2^{20} = 1,048,576$. This equals over 1 million possible combinations, so it is not surprising that 20 answers is usually sufficient to determine the identity of the object.

In a more general case, if we ask a question that has a number (N) of possible answers, then the amount of information (I) that we gain from its answer is equal to the exponent of 2 that is needed to produce the number N. That is, $2^I = N$. For example, if we play the game "Three Questions," we could combine our three questions into one combined question. Perhaps we ask, "Tell me if it is larger than an elephant, and tell me if it is an animal, and tell me if it would sink in water." There are eight possible combined answers: "no, no, no," or "no, no, yes," or "no, yes, yes," etc. Therefore, $N = 8$, so the amount of information is $I = 3$ bits.

An example of a question that has many possible answers (only one of which is correct) is illustrated in **Figure 2.7**. Say there are 128 identical stones lying in a field. One of the stones has a diamond under it that you would like to find. Your friend Bob knows which stone the diamond is under, but he refuses to tell you directly which one to turn over. How many answers to yes/no questions would it take you to find the diamond? You might try one of two strategies. The first would be to try to get lucky and guess directly which stone covers the diamond. If you guess correctly, then you obtain the diamond after only one answer, or one bit of information; however, your chances of success are very small (1 in 128). You would probably have to guess many individual stones before guessing the correct one. If this experiment were repeated many times, then on average you would need to make $128 \div 2 = 64$ guesses, or gain 64 bits of information, in order to find the diamond.

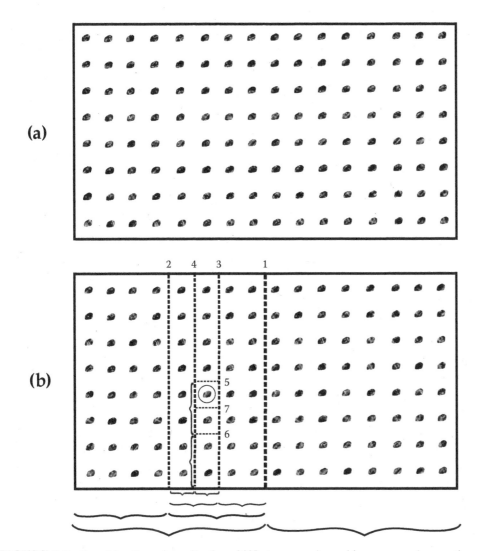

FIGURE 2.7 Searching through a collection of 128 stones requires asking seven yes/no questions.

There is a more efficient strategy, illustrated in Figure 2.7b. Divide the field into left and right areas, each containing one-half of the stones, and ask Bob, "Is the correct stone in the left half?" After hearing the answer, discard the side of the field that you now know does not contain the stone. Next, divide the remaining stones into two equal groups, ask the same question again, and discard the group of stones not hiding the diamond. Repeat this seven times, and you will be left with just one stone, under which will be the diamond. Mathematicians can prove that for this example there is no better strategy than the one that requires seven questions. For this reason, we say that the amount of information that must be gained in this case equals 7 bits. The reason that seven such yes/no questions are needed is that the total number of possible answers to the question, "Which stone hides the diamond?" is 128, which equals 2^7.

Another way to understand this diamond-finding example is to think about what type of message Bob could write on a piece of paper to convey to you where the diamond is hidden. A straightforward way to do this is to agree beforehand on a method of labeling each stone by a whole number between and including 0 and 127. There are 128 such numbers. Then Bob just has to write the number identifying the special stone on the paper and hand it to you. If you and Bob agree that he will write the number in binary, then the number will be between (and including) 0000000 and 1111111. For example, the special stone might be the one identified as 1100101. Clearly Bob will need to specify the value (0 or 1) of each of 7 bits in order to convey the needed information.

What if the number of possible outcomes or answers does not equal exactly a power of 2, that is, does not equal 2^n for some value of n? In this case, we can specify the amount of information to be within a range somewhere between two whole numbers. For example, if there are 135 stones in the field containing the diamond, then it will take on average slightly more than seven questions to unearth the proper location. Sometimes it will take seven questions, and sometimes eight questions. But, it will not take more than eight questions, which would be enough to sort through 256 stones. For this case, we can say that the amount of information is between 7 and 8 bits.

How much information can be transmitted in a message of a given length? If we can assume that the message is written in the most efficient way, then the answer is simple. Write the message as a single list of binary digits; that is, ones and zeros. Then the amount of information equals the number of binary digits, or bits, in the list. The question of how to ensure that the message is written in the most efficient way is subtle and will be discussed in a later chapter.

2.8.1   Bits, Bytes, and Other Units

The basic unit of information is the bit. One byte is defined as 8 bits and is abbreviated as B. For example, you might ask a salesperson how much memory a particular memory device has, and the response might be "1,000 bytes." This is the same as 8,000 bits. When the number of bits is much larger, we use other units—the kilobyte (kB), megabyte (MB), gigabyte (GB), and terabyte (TB). Recall that according to the standard metric system definitions,[1] the prefix k means 10^3, M means 10^6, G means 10^9, and T means 10^{12}. In common computer science usage, however, these symbols are often "misused" to mean 2^{10}, 2^{20}, 2^{30}, and 2^{40}, respectively. This usage arose out of the desire to have slang names for these quantities, and because of the near correspondence between the values: $10^3 = 1,000$ whereas $2^{10} = 1024$; $10^6 = 1,000,000$ whereas $2^{20} = 1,048,576$; $10^9 = 1,000,000,000$ whereas $2^{30} = 1,073,741,824$; etc. Throughout this text, we will use the standard base-ten definitions of k, M, G, and T, except where otherwise noted. For example, when we write GB, we mean 10^9 B or 10^9 bytes.

THINK AGAIN

Given a hard drive that can store 40,000,000,000 bytes, some computer sellers might state that it can store 40 GB, whereas another seller using a different definition for G might state that the same hard drive stores 37.25 GB.

THINK AGAIN

The word *bit* is used here in two different ways. Bit can mean a binary digit, 0 or 1. Bit can also mean the basic unit of information.

---

[1] The international standards for the physical sciences and for commerce are set by the International System of Units, abbreviated SI units from the French name Système International d'Unités. Although this system does not mention bits and bytes, it is clear on the meanings of the prefixes k, M, G, and T. Because of the potential confusion, a set of new binary prefixes for bits and bytes was introduced in 1998 by the International Electrochemical Commission (IEC). In this system, 1 kilobyte (1 kB) equals 1,000 bytes, whereas 1 kibibyte (1 KiB) equals $2^{10} = 1,024$ bytes. Likewise, mebi (Mi; 2^{20}), gibi (Gi; 2^{30}), and tebi (Ti; 2^{40}). The adoption of these prefixes has been limited.

TABLE 2.3
Information Units

Unit	Abbreviation	Physical Science	Computer Science
1 byte	1 B	8 bits	8 bits
1 kilobyte	1 kB	10^3 bytes	1024 bytes
1 megabyte	1 MB	10^6 bytes	$1024 \cdot 1024$ bytes
1 gigabyte	1 GB	10^9 bytes	$1024 \cdot 1024 \cdot 1024$ bytes
1 terabyte	1 TB	10^{12} bytes	$1024 \cdot 1024 \cdot 1024 \cdot 1024$ bytes

2.9 EXPONENTIAL GROWTH

In the study of lasers in Chapter 14 we will encounter the mathematical concept of exponential growth. A good example of exponential growth is the growth of a population. Consider an empty planet having near-infinite resources to support a population of rabbits. Every pair of rabbits produces two healthy offspring each year. The number of rabbit pairs doubles each year. After 1 year, there will be two pairs of rabbits, after 2 years, four pairs, after 3 years, eight pairs. The sequence is: 1, 2, 4, 8, 16..., or 1, 2, 2^2, 2^3, 2^4, ...2^n. That is, after n years, there will be 2^n pairs of rabbits. These numbers are graphed in **Figure 2.8**. After 10 years, there will be $2^{10} = 1,024$ pairs of rabbits. As Figure 2.8a shows, this appears to be a reasonably gradual increase over time.

The interesting feature of exponential growth is seen in Figure 2.8b. After only 5 times longer—50 years—the number of rabbit pairs is ridiculously large: $2^{50} = 1.2 \times 10^{15}$. If each rabbit pair occupied 1 square meter of area, they would more than cover the entire surface of an Earth-sized planet. (The surface area of Earth equals about 0.5×10^{15} m².) Notice that we did not account for rabbits dying, but if a rabbit's life expectancy were 5 or 10 years, then including this in our estimate would make little difference.

Exponential growth sneaks up on you. Early on, it seems that the growth is slow; then it appears to "explode," as seen in Figure 2.8b.

Exponential growth occurs when the growth during each succeeding period builds on the growth of the previous period. For every passing of a fixed time interval (such as 1 year), the quantity is multiplied by the same constant factor.

A good way to graph exponential growth is to plot the exponent of ten of the number on the vertical axis, instead of the number itself. **Figure 2.9** plots the data from

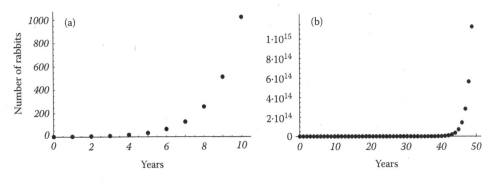

FIGURE 2.8   Exponential growth, illustrated by a population of rabbits, in which the number of pairs of rabbits doubles every year.

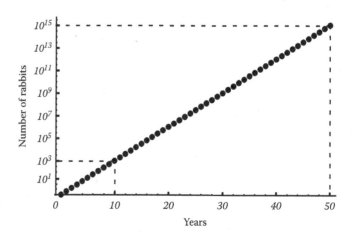

FIGURE 2.9 Exponential growth, plotted as the exponent of the number on the vertical axis (a log plot).

Figure 2.8b in this manner. In this kind of plot, called a log plot, the data appear as a straight line, instead of as sharply curving up. The same information is simply being displayed in a different manner. For example, the graph shows that at a time equaling 10 years, the exponent of ten equals about 3. This means the number of rabbits equals about 10^3, or 1,000, in agreement with the data in Figure 2.8a. Likewise, at time equaling 50, the exponent of ten equals about 15, so the number equals about 10^{15}, in agreement with the data in Figure 2.8b. The slope, or steepness, of the line is a reflection of the growth rate.

SUMMARY AND LOOK FORWARD

Mathematics is a necessary tool for doing physics properly. This is because nature appears to have many regular, predictable behaviors. This means we have a chance to understand these regular behaviors and use our knowledge to predict future behaviors and properties of parts of nature. That is what physics is about.

In this chapter, we discussed the fact that physics is usually explained in terms of continuous or analog variables, whereas computers use discrete or digital numbers. Any physically measured quantity (such as length, speed, etc.) can be represented by a digital number if enough digits are used. Therefore, computers, which store numbers as binary, can be used to solve problems in physics and engineering.

The more digits—decimal or binary—that are used to represent a number, the higher the precision of that number. In solving physics problems, we must decide the degree of precision we need. There is no sense in using a much higher precision than is needed. When numbers become very large or very small, we use scientific notation.

When dealing with physical quantities, units must always be specified. It makes no sense to tell someone you will pay him "1,000" to mow your lawn. The unit you probably meant to specify is *cents*. Physicists like to use nicknames for familiar quantities, such as kilo for 10^3. Calculations involving units are made simple by using conversion factors. This is akin to converting currency from, say, dollars to euros.

Number systems are founded on counting. We discussed the base-ten and base-two counting systems and how to convert between them. Computers use the base-two or binary counting system. This is appropriate, because for storing numbers computers use memory cells that can store only one of two possible symbols, called 1 and 0. A binary digit or bit is a number that can take on only one of two possible values, 1 and 0. This means that each memory cell can store 1 bit. The amount of information that can be stored in a computer memory equals the number of memory cells. This equals the number of bits that can be stored.

In Shannon's 1948 paper, he explained a mathematical way to analyze information traveling in communication systems such as telephone lines. As we will discuss in later chapters, Shannon's theory answered two main questions: How strongly can a message (or other data) be compressed without losing information? How much information (data) can a communication channel (such as a wire) transmit each second without errors occurring? These are important questions in the context of sending compressed data files, such as text, music, or video, across the Internet.

In the next several chapters, we will study the areas of physics called mechanics, heat, and electromagnetism. These so-called classical areas of physics were discovered and explored during the period leading up to around 1900. They provide a background for understanding how machines can be used to store and retrieve numbers, as well as to calculate using these stored numbers.

SOCIAL IMPACTS: THE EXPONENTIAL CHANGE OF NEARLY EVERYTHING

The exponential growth of computing is a marvelous quantitative example of the exponentially growing returns from an evolutionary process.

Ray Kurzweil

As technology advances, we get the feeling that its rate of advancing is increasing. By the rate of advancing, we mean the amount that some quantity changes in a constant time period. For example, a car moving at a steady speed of 60 miles per hour has a constant rate of advancing its position. In the first hour, it travels 60 miles. In the second hour, it travels 60 miles farther, etc. It would seem strange if in the first hour a car traveled 60 miles, in the second hour traveled 120 miles, and in the third hour traveled 240 miles, etc., doubling the distance covered each hour without limit. But, this is precisely how many trends in technology seem to be evolving—they double in capacity every 18 to 24 months or so. For example, the amount of computer memory in one chip doubles approximately every 18 months. This regular doubling behavior is an example of exponential growth, discussed in Section 2.9.

The human population on Earth has been growing roughly exponentially throughout history. Exponential growth of a quantity means that the rate of growth is proportional to the quantity's present value. From 1750 to 1927 the population grew at a rate of about 0.55% per year. Then from 1927 to 2000 it grew at a rate of about 1.5% per year. Since 2000, the annual growth rate has dropped slightly—to 1.3%. In very early times, from 1000 to 1750, the annual growth rate was much smaller—about 0.1%—because of high mortality in early life. **Figure 2.10** shows the data, along with a smooth line connecting the data points. If we continue the smooth exponential curve to the year 2050, it would predict a population of 13 billion. Thomas Malthus, in 1798, predicted that population would outrun food supply.

Until the year 2000, there was no indication of any slowdown of the growth rate. Whenever public health improved, as a result of new scientific information and education, the growth rate increased, due to a lowering of the death rate in the younger age groups. It seemed that we might be heading toward the scenario discussed above, with no room for all of the people on Earth. Since 2000, a slight decrease of growth rate—to about 1.3%—has been seen. This leads to a prediction of around 9 billion people in 2050, a number that is perhaps sustainable.

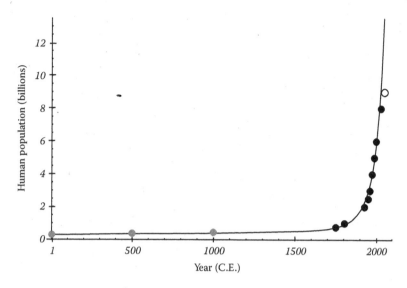

FIGURE 2.10 Human population over time. The lighter dots at 500 and 1,000 show rough estimates of early populations. The open circle shows a slowing down from the earlier growth rate.

Moore's Law

In the mid-1970s, Gordon Moore, the founder of Intel and one of the inventors of the integrated computer circuit, noticed that the number of transistors that could be placed onto a single circuit was doubling every 2 years. He speculated that this doubling would go on far into the future. He was correct, as **Figure 2.11** shows. This behavior is called Moore's Law, although it is not a law—it is only a description of a continuing, remarkable trend. A doubling every 2 years corresponds to a yearly growth rate of about 44%, as seen by $(1.44)^2 = 2.07$.

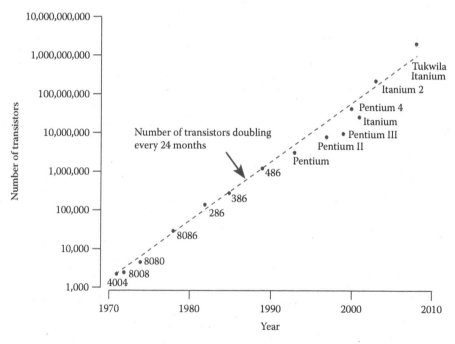

FIGURE 2.11 The number of transistors fitting on a single integrated circuit increases exponentially, doubling about every 2 years.

Other aspects of computer technology also double on a regular basis. The capacity of computer memory (DRAM and hard disk) increases at 60% per year, doubling every 18 months. The speed of computer memory writing and reading increases about 10% per year. The number of bytes of memory that can be purchased for a dollar increases at about 25% per year.

Why are such observations about exponential growth in technology relevant to thinking about society and its development? They raise several questions:

1. If human-made technology's capacity is growing exponentially, are other human-related activities also growing exponentially?
2. If the amount of information available to humans and to society is growing exponentially, can the human intellect and emotional capacity keep up? Is the exponential growth of information a blessing or a curse?
3. If human-made information technology is exponentially improved to help us manage the exponentially growing quantity of information, will there come a time when these artificial technologies become in certain ways "superior" to the human mind, or even sentient?
4. How will the exponential growth phase of technology, information, and human population end? Must it end? Most likely, yes. This idea is embodied as Barlett's Law.

Barlett's Law: All exponential growth must end.

University of Denver engineering professor James B. Calvert named this concept, or prediction, after Albert A. Barlett in 1999. Its basis is that when the numbers become too ridiculously high, something must break. Virtually all climate scientists now conclude that human activity has altered the Earth's atmosphere and climate, with potentially catastrophic consequences within 100 years or so. Ironically, this potential disaster has been facilitated by the spectacular success of science and technology in increasing energy use and in prolonging human health.

Supplemental Reading on Social Impacts

For a review of exponential growth:
Hewitt, P. *Conceptual Physics*. Boston: Addison Wesley, 2005.

A book that focuses on the technological predictions, but less on the environmental consequences:
Kurzweil, Ray. *The Singularity Is Near: When Humans Transcend Biology*. New York: Viking, 2005.

A skeptical view that argues that "the exponential model as a universal explanation for and predictor of technological change is at best an approximation and at worst a delusion":
Seidensticker, Bob. *Future Hype: The Myths of Technology Change*. San Francisco: Berrett-Koehler Publishers, 2006.

Bostrom, Nick. "Welcome to a World of Exponential Change". London: Demos, 2006, http://ieet.org/index.php/IEET/more/398 or http://www.demos.co.uk/BH3_pdf_media_public.aspx.

Kurzweil, R. "The Law of Accelerating Returns," http://www.kurzweilai.net/meme/frame.html?main=/articles/art0134.html, or http://www.kurzweilai.net/articles/art0134.html?printable=1 and http://www.kurzweilai.net/brain/frame.html?startThought=Exponential%20Growth.

A skeptical view is given in:
Tuomi, Ilkka. "The Lives and Death of Moore's Law," First Monday: Peer-Reviewed Journal on the Internet, http://www.firstmonday.org, Volume 7, 2002.

SUGGESTED READING

See the general technology references given at the end of Chapter 1.

For a popular-level in-depth discussion of information in physics:

Lloyd, Seth. *Programming the Universe*. New York: Knopf, 2006.

Lucent Bell Laboratories website (http://www.bell-labs.com) has useful information on technology history, information theory, and Claude Shannon.

Shannon, Claude. "A Mathematical Theory of Communication." *Bell System Technical Journal* 27 (1948): 379–423, 623–656. Reprinted in the following reference.

A fascinating biography of Shannon, as well as his original papers, can be found in:

Claude E. Shannon: Collected Papers. Edited by N. J. A. Sloane and Aaron D. Wyner. New York: Wiley-IEEE Press, 1993. http://www.research.att.com/~njas/doc/shannon.html.

KEY TERMS

Analog
Binary number system
Bit
Byte
Cell
Conversion factors
Decimal number system
Digit
Digital
Direct proportionality
Information
Information content
Place value
Power of 2
Precision
Relative precision
Significant digits
Unit

ANSWERS TO QUICK QUESTIONS

Q2.1 To four significant digits, $19.56 \times 8/9 = 17.386666... = 17.39$ ft

Q2.2

$$3700 \text{ m} = 3700 \text{ m} \cdot \left(\frac{1 \text{ km}}{1000 \text{ m}} \right)_{CF} = 3.7 \text{ km}$$

Q2.3

$$1 \text{ century} = 1 \text{ century} \times \left(\frac{100 \text{ years}}{1 \text{ century}} \right)_{CF} \left(\frac{52 \text{ weeks}}{1 \text{ year}} \right)_{CF} \left(\frac{7 \text{ days}}{1 \text{ week}} \right)_{CF} = 36400 \text{ days}$$

$$1 \text{ century} = 1 \text{ century} \times \left(\frac{100 \text{ years}}{1 \text{ century}} \right)_{CF} \left(\frac{12 \text{ months}}{1 \text{ year}} \right)_{CF} \left(\frac{4 \text{ weeks}}{1 \text{ month}} \right)_{CF} \left(\frac{7 \text{ days}}{1 \text{ week}} \right)_{CF} = 33600 \text{ day}$$

$$1 \text{ century} = 1 \text{ century} \times \left(\frac{100 \text{ years}}{1 \text{ century}} \right)_{CF} \left(\frac{365 \text{ days}}{1 \text{ year}} \right)_{CF} = 36500 \text{ days}$$

Because the conversion factors being used are not exact, the calculated answer depends on which combination of factors you use. Note that even the last one is subject to error in the case of leap years.

Q2.4 Convert the base-ten number 211 into binary. 211 =

$$128 + 64 + 0 + 16 + 0 + 0 + 2 + 1 =$$

$$1 \times 2^7 + 1 \times 2^6 + 0 \times 2^5 + 1 \times 2^4 + 0 \times 2^3 + 0 \times 2^2 + 1 \times 2^1 + 1 \times 2^0 =$$

$$11010011 \text{ (binary)}$$

Q2.5 Convert the binary number 1110101 into base-ten.

$$1 \times 2^6 + 1 \times 2^5 + 1 \times 2^4 + 0 \times 2^3 + 1 \times 2^2 + 0 \times 2^1 + 1 \times 2^0 =$$

$$64 + 32 + 16 + 0 + 4 + 0 + 1 = 117 \text{ (base-ten)}$$

Q2.6 One byte contains 8 bits. The largest 8-bit number is 11111111. This equals, in base-ten, $128 + 64 + 32 + 16 + 8 + 4 + 2 + 1 = 255$.

The base-ten equivalent of 100000000 is 256.

EXERCISES AND PROBLEMS

Exercises

E2.1 If the child's height in Figure 2.1 is recorded using units of meters rather than inches, the yearly numbers would be (to the nearest centimeter):

0.45, 0.59, 0.69, 0.77, 0.86, 0.95, 1.05, 1.16, 1.27, 1.37, 1.45, 1.5, 1.54, 1.57, 1.60

Use graph paper, carefully drawn lines on paper, or a computer to plot these numbers. Draw by hand a smooth curve through the points. Use your graph to estimate as precisely as you can the child's height at the following ages: 4.5 yr, 7 yr and 3 months, and 16 yr. Give your answers in meters, millimeters, and inches.

E2.2 Discuss the differences between a smoothly running electric clock that shows time by the angle of a rotating arrow (hand), and a specially made sand clock, which shows time by counting the number of sand grains that have fallen through a small opening in an upper chamber. Assume that the grains fall through at a constant rate (grains per second).

E2.3 The 2005 Mini Cooper automobile is listed as having the following specifications: fuel tank, 50 L; length, 143.1 in.; width, 66.5 in.; height, 55.4 in.; weight, 2,370 lb; cargo capacity (rear seat down), 23.7 ft³.

(a) Convert these into gallons, meters, kilograms, and cubic meters, respectively. Find the needed conversion factors in this or other texts, or online, and state your source. Show your work.
(b) Explain why you would or would not find it useful (reasonable) if a dealer listed:
    i.   The length as 143 in.?
    ii. The length as 143.137 in.?
    iii. The length as 4 m?
    iv. The length as 3.63574 m?

E2.4 Between 1889 and 1960, the meter was defined to be the length of a standard bar made of platinum-iridium metal, kept in Paris in a constant-temperature room.

(a) Imagine that to transfer this length standard from Paris to another city without moving the original bar, a copy was made to a precision of 0.1 mm. What is the relative precision of this copy?
(b) In 1983, the current definition of the standard meter was adopted: "The meter is the length of the path traveled by light in a vacuum during a time interval of 1/299,792,458 of a second." Given that it is possible to measure time precisely enough to use this new standard, what is the relative precision of this new standard? How much more precise is this standard than was the old metal bar standard? (http://physics.nist.gov/cuu/Units/meter.html)

E2.5 Say that you want to know the thickness of a piece of paper, but you have only a single 8.5 × 11 in. sheet of this paper and a 30-cm ruler with 1-mm precision. Can you think of a clever way to do this? (It is okay to destroy the paper if need be.) Estimate the best precision and relative precision you could achieve. Answer the same question if you had 100 identical sheets.

E2.6 Perform the following multiplications or divisions and write each answer in scientific notation. Example: $2,000 \times 70 = 1.4 \times 10^5$

(a) 1/(5,000)   (b) $10^6/10^3$ (c) $10^4/10^{-6}$
(d) $1,000,000 \times 0.001$ (e) 1/(10,000,000)   (f) $10^6 \times 10^{-7}$
(g) $2,500 \times 40$ (h) $1,200 \times 500$ (i) $(1.3 \times 10^2)(9.3 \times 10^3)$
(j) $(1.2 \times 10^3) \div (9.6 \times 10^3)$ (k) 0.0021×9 (l) $(0.003) \div (3 \times 10^{-2})$

E2.7 Perform the following multiplications or divisions and write the answer in scientific notation, keeping (rounding to) only three significant digits. Example: $2,100 \times 73.5 = 1.5435 \times 10^5 \approx 1.54 \times 10^5$.

(a) 1/(5,107)   (b) $(7.765 \times 10^6)/(5.555 \times 10^3)$ (c) $1,900,000 \times 0.0017$
(d) 1/(30,000,000)   (e) $2,511 \times 423$
(f) $(1.3 \times 10^2)(9.3 \times 10^3)$ (g) $(1.2 \times 10^3) \div (9.6105 \times 10^3)$

E2.8 Fully write out the conversion factor method for solving the following. (Useful data are given in the tables in Section 2.5.) How many:

(a) millimeters in a kilometer?
(b) centimeters in 40 m?
(c) kilometers in a centimeter?
(d) nanoseconds in a microsecond?
(e) picoseconds in 0.5 nsec?
(f) bacteria can you line up in a centimeter?
(g) vibrations of a typical molecule in 1 sec?
(h) flaps of a housefly's wing in half a second?

E2.9 We discussed how you could count by making marks in the dirt with your feet. Make marks (on paper, not in dirt) illustrating this counting technique for the following case. Label your drawing carefully, indicating which groups of marks correspond to which place values.

(a) the decimal number 7,128, counted in decimal

(b) the binary number 1011, counted in binary

E2.10 What is the largest decimal number that can be represented by (a) 1 byte? (b) 2 bytes? (c) 3 bytes?

Problems

P2.1 For each question, write out in full detail the conversion factor method. Give (round) answers to 3 significant digits.

(a) How long, in minutes, would it take to drive 1,550 mi if traveling with a speed 67.0 mi/hr?

(b) How long, in seconds, would it take to drive 1,550 mi if traveling with a speed 67.0 mi/hr?

(c) How long, in seconds, would it take a bullet to travel 0.500 km if traveling with a speed 896 m/sec?

(d) How long, in milliseconds, would it take a surface spot near the outer edge of a compact disk under the reading laser to travel once around the disk (distance = 37.7 cm), if traveling with the standard speed 1.25 m/sec?

P2.2 As of this writing, 1 dollar = 0.831 euro. How many euros are in 1 dollar? Use the method of conversion factors to convert: 1.75 dollars into euros; 1,546 dollars into euros; 0.50 dollars into euros; 4.25 euros into dollars; 3,999 euros into dollars.

P2.3 Find the decimal (base-ten) form of the following five-bit binary numbers: 01010, 10111, 10001, 01110, 11010, 11111.

P2.4 Find the binary form of the following decimal (base-ten) numbers: 17, 31, 45, 53, 62, 64, 67, 99, 128, 129, 130. Determine how many bits are necessary to represent each individual decimal number.

P2.5 Magazine subscription post cards, inserted into magazines, sometimes use a printed bar code for identification. An example is shown below. Assuming the simplest coding scheme (each bar can have one of two shapes—long or short), how many distinct (unique) codes can be represented by this particular bar code format? Compare your answer with the number of humans on Earth.

P2.6 DNA (deoxyribonucleic acid) is made up of two long molecular strands, connected by "linkages" (like rungs on a long, twisted ladder). There are only four types of linkages: AT, TA, CG, and GC, where the letters stand for combinations of the four molecules adenine (A), cytosine (C), guanine (G), and thymine (T).

(a) Consider a small section of a DNA molecule containing three linkages. How many distinct variants of this three-linkage section could occur?

(b) Answer the same for a five-linkage section.

(c) Answer the same for a 150-linkage section. Compare the answer to the estimated number of protons in the known universe (10^{80}).

P2.7 In the discussions above, we have always considered cases in which the different outcomes of each event were equally likely; for example, flipping a fair coin. Let us consider a case in which the two possible outcomes are not equally likely. Say that the NFL champion team agrees to aggressively play a football game against a middle-school football team. Your friend tells you the outcome, and you say, "Well, duh!" Estimate how much information your friend's message contained. Explain.

P2.8 Have a friend think of a particular number (call it N) between 1 and 256 (including possibly 1 and 256) and write it down. Your task is to determine the chosen number by asking a series of yes/no questions. Before doing this exercise, write down how many questions you guess it will take if you adopt the best strategy. Next, using your best strategy, begin asking questions, writing down each question and answer (and turn these in) until you determine the number. If you repeat this more than once, give the results of each trial and explain what you learned from each trial. Finally, explain what is the best strategy and why. How much information (in bits) does your friend initially have that you do not?

P2.9 A general rule is that to find the total amount of information present when combining two sets of information, add the amounts of information in the two sets. Consider the 20-questions game discussed in the chapter, modified so each player can ask only ten questions. Alice gains 10 bits of information after asking her set of questions. The number of different possible combinations of answers in this case equals $2^{10} = 1,024$. Bob gains 10 bits of information after asking his set of questions. If Bob and Alice combine their information, how much information will they have in total? How many different possible combinations of answers are there in this combined case?

P2.10 Say that there is a field with 131,072 identical stones lying on it. Your friend Bob knows under which stone is a diamond.

(a) Using the best strategy, how many bits of information would you need to get from Bob to find the diamond? Explain.

(b) What if, instead, the number of stones is 150,000? Estimate the amount of information you would need to get from Bob to find the diamond; that is, state what two whole numbers the amount of information falls between.

P2.11 Estimate the amount of information in one page of English text containing 50 lines of text with about 100 characters per line. (A space is considered a character.) Each character in the English alphabet requires 8 bits for its representation. For example, the capital letter A is represented by the group of 8 bits: (01000000), and the capital letter B is represented by the 8 bits: (01000010). Give the answer in bits, bytes, and kilobytes.

Mechanics: Energy Enables Information Technology

[R]ational mechanics will be the science of motion resulting from any forces whatsoever, and of the forces required to produce any motion ...

Isaac Newton
(The Principia, 1687)

Tony Hawk goes for the rail at the skate park in Springfield, Oregon. (Courtesy of Willamalane Park and Recreation District.)

Portrait of Isaac Newton by Godfrey Kneller (1642—1727), the founder of modern physics. (With permission of the Trustees of the Portsmouth Estate.)

3.1 FROM LOOMS TO COMPUTERS

In this chapter, we begin building a bridge between the abstract ideas of binary numbers and information on the one hand, and the real physical devices that are needed for computer technology on the other. A computer is a programmable machine—a device that can take in instructions and perform the corresponding operations. How does a computer receive, store, and manipulate bits of information? The earliest programmable machine was the Jacquard loom, which Joseph-Marie Jacquard in Napoleonic France designed so it could weave any intricate pattern into a silk fabric. That is, Jacquard in

FIGURE 3.1 A reproduction of the original Jacquard loom perfected by Joseph-Marie Jacquard in 1804, showing the punched "programming" cards. This model was built by students of Professor Marjorie Senechal at Smith College. (Photograph by Stan Sherer.)

1801 found a way to convey instructions to the loom for what pattern it should weave. The instructions were "written" in the form of holes punched into cards, which were then "read" by the machine. The Jacquard loom performed mechanical steps, but it did not calculate or compute. The 1800s were a remarkable period during which science and technology went hand in hand in the advancement of both. Jacquard's advance in programmable machines rightly stands as a major achievement of the time.

The earliest machines that calculated or computed were mechanical. They used physical objects, such as gears and other mechanisms, to record and process information. Modern computers use silicon-based electronic circuits. In later chapters, we will explore the physical basis of such semiconductor circuits. The communication networks that make up the Internet send information between computers on the network.

Before we study electronic circuits and communication networks, we will discuss the concepts of *force* and *energy*. These are at the heart of Isaac Newton's great contributions to physics in the 1680s and formed the basis of the Industrial Revolution of the 1800s. Newton was perhaps the most influential physicist of all time. He, along with Galileo before him, changed the subject from a qualitative one based on philosophers' ideas of how the world "should" be, to a quantitative system of mathematical laws based on how the world is observed to be. We call this system Newtonian classical mechanics. One of his amazing achievements, when he was 24 years old, was the invention of calculus to solve physics problems. Later he was a professor at Cambridge University in England during a time of great technical and scientific progress.

This chapter introduces the concepts of force, *acceleration*, and energy. This sets the stage for our study of matter, electricity, and magnetism in the following two chapters. Following those discussions, we will begin exploring how a computer performs useful tasks by using energy to manipulate information.

3.2   SPEED, ACCELERATION, AND FORCE

The important concept of energy is closely tied to the concept of force—an influence that can change the direction of motion or the speed of an object. An example is a wooden bat imparting a sudden force to a baseball when it hits the ball. Let us begin by studying the motion of objects, a subject called mechanics. The major concepts in mechanics are speed, force, mass, acceleration, and energy.

3.2.1   Description of Motion

How do we describe an object's motion? One way is to make an extensive list of its locations or positions at many instants in time. We can also use the concept of *distance*, which means the difference between two positions. We can also give its speed at every instant. *Speed* means the rate of change of position or distance with time. Speed is measured in meters per second, abbreviated m/sec. The distance (D) traveled in a certain time interval (t) by an object moving with a constant speed (S) equals the speed multiplied by the time interval. That is, distance is in direct proportion to the time. As a formula:

$$Distance = speed \times time\ interval$$
$$D = S \times t$$

Consider an object traveling with a constant, steady speed $S = 50$ m/sec. In **Figure 3.2a**, the two graphs show the speed and the distance traveled by the object as time advances.

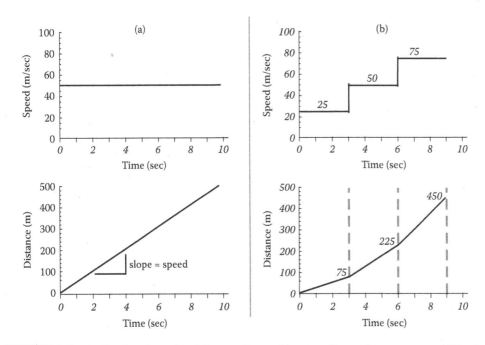

FIGURE 3.2   (a) Graphs of speed and distance for an object traveling with constant speed 50 m/sec. (b) An object's speed is steady for 3 sec, then increases suddenly every 3 sec thereafter.

The slope of the line showing distance versus time equals the speed. By *slope* we mean how steep the line is, or mathematically, the distance divided by the time. The higher the speed, the steeper the slope of the line.

In Figure 3.2b, the two graphs show what happens if an object travels with speed equal to 25 m/sec for 3 seconds, then its speed is suddenly increased to 50 m/sec and it travels with this higher speed for the next 3 seconds, then its speed is suddenly to increased to 75 m/sec, and it travels with this speed for the next 3 seconds. During the first 3 seconds, the object travels a distance 75 m. During the next 3 seconds, the object travels a distance 150 m, putting it at a total distance of 225 m. During the third 3 seconds, the object travels a distance 225 m, putting it at a total distance of 450 m. Notice how the slope of the distance graph keeps getting steeper, meaning that the object covers a greater distance during each successive 3-second time interval. This "curving up" of the distance graph is a characteristic in cases when the speed is increased during the motion.

QUICK QUESTION 3.1

Consider an electron (elementary particle of electric charge) traveling in a television tube from the "gun" that sends electrons from the rear of the television tube to the front screen where the image is formed. These electrons typically travel with a speed of $S = 2 \times 10^7$ m/sec. What distance would such an electron travel in a time of 1 microsecond (1 μsec)?

When an object's speed is increasing in time, we say the object is *accelerating,* or that it is experiencing acceleration. For motion in a straight line, **acceleration** means the rate of change of an object's speed. This is why the units of acceleration are m/sec^2, which means meters per second of speed increase per second.

Consider a jet-powered racing car, as in **Figure 3.3**, which starts accelerating at time equal to zero ($t = 0$); that is, the time when you start a stopwatch. Say the jet car accelerates with a steady acceleration equal to 10 m/sec^2. As illustrated in the upper graph and in the table in **Figure 3.4**, the acceleration is constant, meaning in this example the acceleration does not change in time. When acceleration is constant, the car's speed increases at a steady rate, and if we graph it versus time, as in the middle graph, we see a straight line. When an object undergoes a constant acceleration (a) for a certain time interval (t), the speed changes by an amount equal to the acceleration multiplied by the interval. As a formula:

$$Change\ of\ Speed = acceleration \times time\ interval$$
$$\text{or}$$
$$\Delta S = a \times t$$

We use the symbol ΔS to represent the change of speed. The Greek letter Δ, pronounced "delta," reminds us of the word *difference,* which means *change.* Notice again that if the acceleration has a negative value, then the change of speed is negative, meaning the object loses speed; that is, it slows. The constant of proportionality, a, in this equation is the acceleration.

FIGURE 3.3 The North American Eagle accelerating under jet engine power. This jet car is designed to exceed land speeds over 760 mi/hr, or 339 m/sec. (Courtesy of Tim Finley.)

The lower graph shows the distance traveled by the car as time goes on. At first, the speed is small, so in the first 1 second a distance of only 5 meters is covered. During the final 1-second interval, the speed is almost 100 m/sec and the car covers a distance of 95 meters. This behavior is indicated by the "curving up" of the distance-versus-time curve shown in the lower figure, similar to that also seen in Figure 3.2b.[1] At the end of 10 seconds, the car's speed is 100 m/sec, which is about 219 miles per hour.

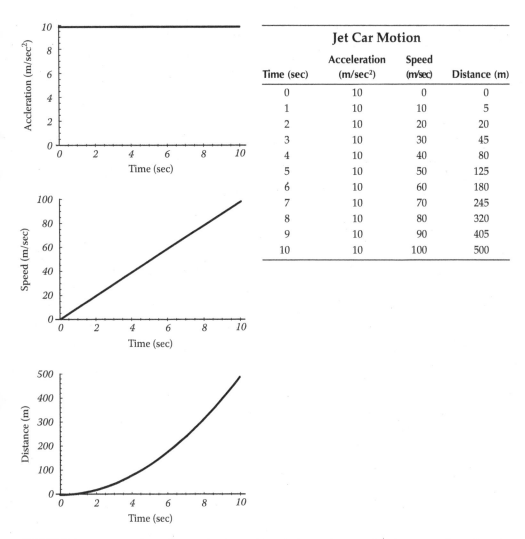

Jet Car Motion

Time (sec)	Acceleration (m/sec²)	Speed (m/sec)	Distance (m)
0	10	0	0
1	10	10	5
2	10	20	20
3	10	30	45
4	10	40	80
5	10	50	125
6	10	60	180
7	10	70	245
8	10	80	320
9	10	90	405
10	10	100	500

FIGURE 3.4   A steady acceleration leads to a linearly increasing speed and an upward-curving distance-versus-time graph.

THINK AGAIN

For motion in a straight line, speed tells how fast you are changing your position. Acceleration tells how fast you are changing your speed.

[1] Distance is given by $x = (1/2) \cdot a \cdot t^2$, assuming a constant acceleration, as discussed in In-Depth Look 3.1.

If the object is initially moving with some speed *S(initial)*, and then experiences a steady acceleration (*a*) for a time interval (*t*), its new speed after this time will equal the initial speed plus the change of speed. That is,

$$New\ Speed = Initial\ Speed + Change\ of\ Speed$$
$$S(new) = S(initial) + a \times t$$

In the next example, in **Figure 3.5**, the jet car starts accelerating as in the previous case, but then at time = 4 seconds its engine quits, and the acceleration suddenly drops to zero. The car coasts at a steady speed of 50 m/sec for a while (until ***friction*** between the wheels and the ground begins to slow it). The distance graph starts out with an upward curve as the car accelerates, but after time = 5 seconds increases linearly when the speed becomes essentially constant, reaching 375 meters after 10 seconds.

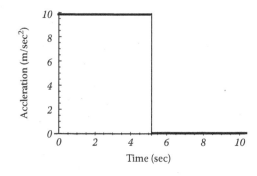

Time (sec)	Acceleration (m/sec²)	Speed (m/sec)	Distance (m)
0	10	0	0
1	10	10	5
2	10	20	20
3	10	30	45
4	10	40	80
5	5	50	125
6	0	50	175
7	0	50	225
8	0	50	275
9	0	50	325
10	0	50	375

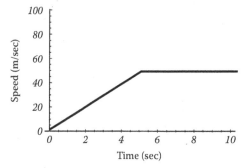

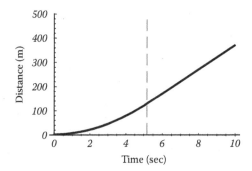

FIGURE 3.5   A starting acceleration that suddenly decreases to zero leads to an increasing speed followed by a constant speed and a linearly increasing distance graph.

Acceleration can also have a negative value, referring to a decreasing of the speed with time—putting the brakes on. This is called ***deceleration***. **Figure 3.6** shows a period of negative acceleration between time = 3 sec and 6 sec. In this example, the duration of deceleration is such that the speed is brought back to zero, after which the car stops moving, so the distance becomes constant after time = 6 sec.

THINK AGAIN

If the deceleration in Figure 3.6 were continued beyond the time 6 seconds, then the car would begin moving backwards; that is, toward smaller distances. In this case, it would be tempting to say that the speed becomes negative, but the conventional language of physics requires us to use the word velocity instead of the word *speed* (i.e., velocity means the same as speed, except that velocity can be either positive or negative).

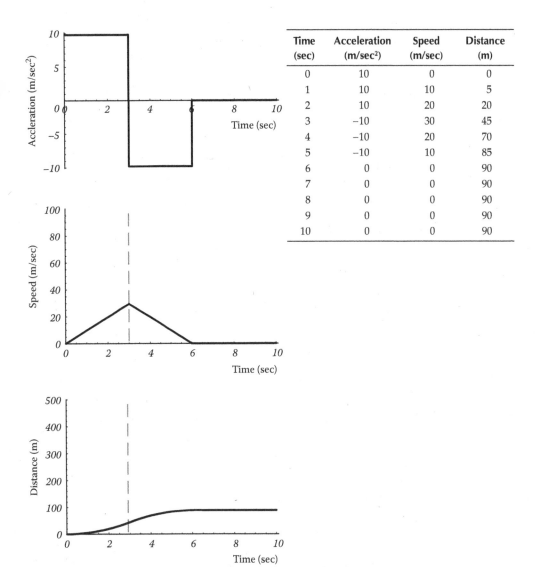

Time (sec)	Acceleration (m/sec²)	Speed (m/sec)	Distance (m)
0	10	0	0
1	10	10	5
2	10	20	20
3	−10	30	45
4	−10	20	70
5	−10	10	85
6	0	0	90
7	0	0	90
8	0	0	90
9	0	0	90
10	0	0	90

FIGURE 3.6   An initial steady acceleration that suddenly reverses sign from plus to minus at time = 3 sec, beginning a period of deceleration. This initially leads to an increasing speed followed by a speed that decreases to zero and a flattened distance-versus-time graph as the car comes to a halt at time = 6 sec.

QUICK QUESTION 3.2

Draw the speed and distance graphs corresponding to the acceleration graph to the right. There is no need to get the numerical values exactly right—draw the shapes to the best of your ability.

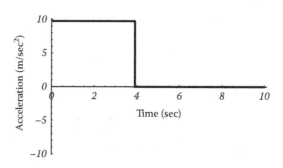

IN-DEPTH LOOK 3.1: DISTANCE TRAVELED UNDER CONSTANT ACCELERATION

If an object begins at rest and then undergoes a constant acceleration for a time interval equal to t, how far will it travel in that time? This is the situation described in Figure 3.4 for the jet car. In that example, the acceleration has a constant value $a = 10$ m/sec². Starting from zero speed, after a time (t) of 10 seconds, the speed will equal:

$$final\ speed = S = a \cdot t = 10\ m/sec^2 \cdot 10\ sec = 100\ m/sec$$

Because the speed increases linearly in time, during the earlier half of this period the speed is smaller than 50 m/sec and for the latter half the speed is greater than 50 m/sec. The average speed during this complete time interval equals 50 m/sec, or one-half of the final speed. We can write this relation as

$$average\ speed = \frac{final\ speed}{2} = \frac{a \cdot t}{2}$$

The total distance traveled during the 10-second time interval equals the average speed times the time interval:

$$distance = (average\ speed) \cdot t = (50\ m/sec) \cdot 10\ sec = 500\ m$$

We can give a general result for the distance an object travels if it begins at rest and then experiences a constant acceleration equal to a for time equal to t. The distance traveled is calculated by combining the two formulas given above, that is:

$$distance = (average\ speed) \cdot (t) = \left(\frac{1}{2}a \cdot t\right) \cdot (t) = \frac{1}{2}a \cdot t^2$$

REAL-WORLD EXAMPLE 3.1: SEEK TIME OF A HARD-DRIVE HEAD

A hard disk drive (see **Figure 3.7**) stores large amounts of information in small magnetic regions on a metal-coated disk. The data are read from the disk surface by a tiny magnetic pickup head that moves or "flies" just above the surface of the rapidly spinning disk. The data regions are arranged in concentric circular *tracks*. Typical fast drives can access any track on the disk in about 8 milliseconds, called the *seek time*. Of that time, about 5 milliseconds is the time it takes the head to move from one track to the other, and 3 milliseconds is the time it needs to stabilize after it stops at the proper track. For a 3.5-inch diameter disk, with a used storage area about 1 inch wide, the typical distance that the head might move between tracks is about one-third of an inch. How large an acceleration must be applied to the head to accelerate it from a standstill to move it a distance of one-third of an inch (or about 8 mm) in a time of 5 milliseconds?

A good method to get the head moving and then stop it is to apply acceleration similar to that shown in **Figure 3.8**. For the first half of the 5-millisecond travel time, apply a

FIGURE 3.7 In a hard disk drive, an arm rotates rapidly to move the read head to different tracks.

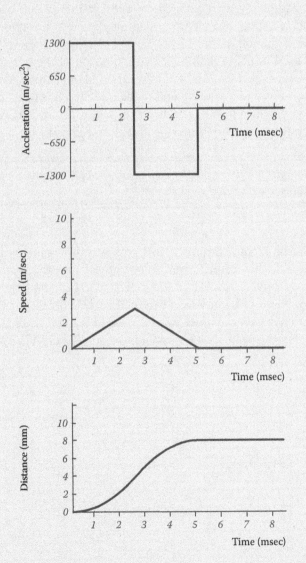

FIGURE 3.8 The acceleration, speed, and distance of a hard drive read head as it makes a typical move between tracks.

steady acceleration. Then for the second half, apply a deceleration with a strength equal to the acceleration applied in the first half. For the head to move 4 millimeters during the first 2.5 milliseconds, it must accelerate with a value $a = 1300$ m/sec^2 during that time interval. You can verify this if you wish by using the formula given in In-Depth Look 3.1. At the end of this interval, the speed will equal:

$$Speed = S = a \cdot t = 1300 \text{ m/sec}^2 \cdot 0.0025 \text{ sec} = 3.25 \text{ m/sec}$$

Then, during the following 2.5-millisecond interval, subject the head to a deceleration of −1300m/sec^2. During that time, the head will move another 4 millimeters, and then come to rest. This means that during the 5-millisecond time interval, the head moves a total of 8 millimeters.

The acceleration (1,300 m/sec^2) applied to the head is huge! Recall that 1 g ("gee") force, provided by gravity at Earth's surface, creates an acceleration of only 9.8 m/sec^2. The hard drive head is being accelerated 130 times more strongly than the acceleration caused by gravity. We say the head experiences 130 g, or 130 gees. A human could not survive such accelerations.

In the above examples, the jet car's engine creates a force that increases its speed, that is, causes acceleration. Acceleration can also refer to the rate of change of an object's *direction* of motion, rather than its speed. We use the word "direction" in the same sense it is used in map reading—as an indication of the bearing of an object's motion—north, east, northwest, up, down, etc. For example, consider a hockey puck sliding across the surface of a frozen pond in the direction north, as in **Figure 3.9**. If a strong wind suddenly gusts toward the direction southwest, it could force the puck to make a change of direction while traveling at the same speed.

A *force* is an influence that can cause a change of speed and/or direction of motion of a material object; that is, force causes acceleration.

In ancient times, long before Newton, it was observed, for example, that if a ball is rolled across a flat horizontal surface it will gradually slow and come to a stop. This led people to conclude that a force needs to be applied steadily to an object to keep it moving. That was wrong! A constant force is needed to keep an object moving only if friction is present. Now, with humans experiencing space travel, we are more familiar with the idea that no steadily applied force is needed to keep an object moving in the vacuum of space. For example, if an astronaut is outside of a craft in orbit around Earth,

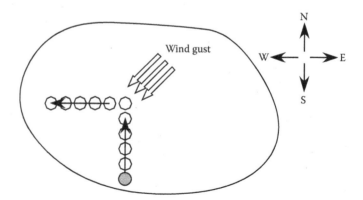

FIGURE 3.9 Bird's-eye view of a frozen pond with a hockey puck initially sliding in the direction north. A sudden gust of wind forces the puck to begin sliding west. After the wind dies down, the puck continues to slide in the west direction.

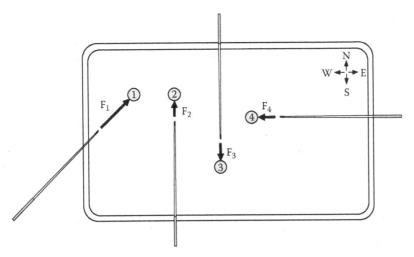

FIGURE 3.10 Force vectors, illustrated by cues striking balls on a billiards table. The direction the arrow points indicates the direction of the force, and the length of the drawn arrow represents the strength of the force.

and if she gently lets go of a wrench, the wrench will stay next to her, as if floating. Then, if she gives the wrench a sudden push (applies a force), the wrench will begin moving away from her at a constant speed. It will keep moving with a constant speed, in a straight line, until some other force acts on it. For example, the Earth's gravity will act weakly on it, likely causing it to move in an orbit around the Earth. If the wrench were in deep space, far from any planet, it would move in nearly a straight line for a very long time. In the case of the ball rolling on the grass, the grass exerts a force—the force of *friction*—on the ball, opposing its forward motion and slowing it down. We will hear more about friction later.

3.2.2   Force Vectors

A force on an object accelerates it in a particular direction. The direction could be up or down, left or right, or forward or backward. For example, when a billiards cue strikes a ball straight on, the ball feels a force in the same direction that the cue is moving, as shown in **Figure 3.10**. In physics, we represent the direction of a force by an arrow, called a *force vector*. The direction the arrow points indicates the direction of the force, and the length of the drawn arrow represents the strength of the force. ("Longer means stronger.") In Figure 3.8, the 1-ball feels a large force in the northeast (NE) direction, the 2-ball feels a small force in the north (N) direction, the 3-ball feels a small force in the south (S) direction, and the 4-ball feels a small force in the west (W) direction.

IN-DEPTH LOOK 3.2: NET FORCE VECTORS

If two forces act simultaneously on the same object, then the forces add to give a resultant force, or *net force*. For example, in **Figure 3.11**, if both forces $F3$ and $F4$, which are equal in strength, act on a ball at the same time, the resultant force is in the southwest (SW) direction. And, if both forces $F1$ and $F2$ act on one ball at the same time, the resultant force is in the north-northeast (NNE) direction. In each case, the net force is found by redrawing each force vector extending beyond the object (shown in gray), forming a square or other parallelogram with the force vectors as two sides. Then the net force is an arrow connecting the two opposite corners of the parallelogram.

For example, because the forces $F3$ and $F4$ are equal in strength and at a right angle, the net force is at 45 degrees to both, and has strength somewhat larger than either. (Using the Pythagorean Theorem for relating the sides of a right triangle, it is not hard to show that the net force vector has length equal to $\sqrt{2} \approx 1.4$ times the length of each individual force vector.)

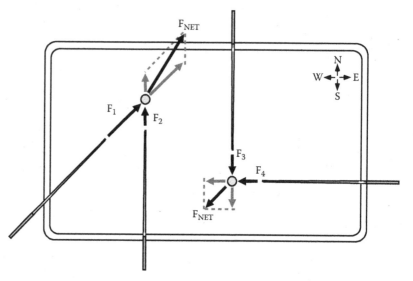

FIGURE 3.11   When two forces act together on one object, the net force is given by a new vector found using a parallelogram.

REAL-WORLD EXAMPLE 3.2: ACCELERATION IN CATHODE-RAY TUBES
(Note: the calculations in this box are more advanced.)

Older televisions and computer monitors use cathode-ray tubes, or CRTs, for their screens. *Electrons* (elementary particles of electric charge) travel in a CRT from the "gun" at the rear of the television tube to the phosphor front screen where the image is formed.

After being accelerated by an *anode,* these electrons typically travel with a speed $S = 2 \times 10^7$ m/sec (see Quick Question 3.1). If the distance from the gun to the screen is 0.4 m, it takes only 2×10^{-8} sec or 20 nanoseconds to travel the distance, as seen by:

$$t = \frac{D}{S} = \frac{0.4\text{m}}{2 \times 10^7 \text{m/sec}} = 2 \times 10^{-8}\text{ sec}$$

$$= 2 \times 10^{-8}\text{ sec}\left(\frac{10^9 \text{n sec}}{\text{sec}}\right) = 20\ \text{n sec}$$

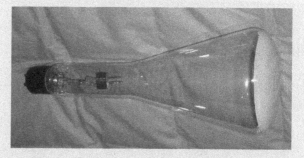

FIGURE 3.12   Old-style CRT, Western Electric 224-B (1927). (Courtesy of Early Television Museum, Hilliard, OH.)

All electrons are initially aimed toward the center of the viewing screen, which lights up at a tiny spot wherever electrons hit. To create an image, the electrons must be deflected from the center to a chosen spot on the screen. This is done by deflection plates, which create electric forces and will be explained in a later chapter. Each electron feels the force created by the deflection plates for only a brief time while passing between the deflection plates.

In the example shown, the force, shown as a bold arrow at the deflection plate, acts in the "up" direction, which is perpendicular to the initial "forward" direction of travel. How strong must the deflection force be to make the electron hit the screen 0.1 meters above the center spot? During the 20 nanoseconds of time the electron is traveling from the deflector to the screen, it must travel 0.1 meters in the "up" direction. This means that its up speed must be

$$S = \frac{D}{t} = \frac{0.1\,\text{m}}{20 \times 10^{-9}\,\text{sec}} = 5 \times 10^6\,\text{m/sec};$$

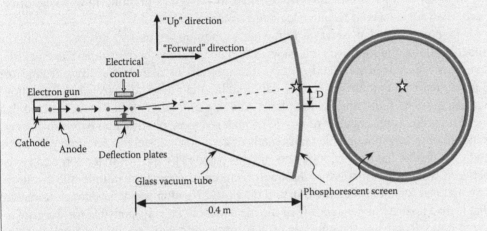

FIGURE 3.13   Electron acceleration in CRT.

that is, 5 million m/sec. The upward acceleration of the electron takes place only during the time interval when the electron is between the two deflection plates. This very short acceleration interval (t') is about 1 nanosecond in duration. The amount of upward acceleration (a) provided by the plates during that time must be truly enormous to achieve such a fast upward speed (S). Recall that $S = a \cdot t$, so:

$$a = \frac{S}{t'} = \frac{5 \times 10^6\,\text{m/sec}}{1 \times 10^{-9}\,\text{sec}} = 5 \times 10^{15}\,\text{m/sec}^2$$

Compare this to the acceleration provided by gravity near the Earth's surface, which equals 9.8 m/sec². The upward or sideways acceleration experienced by the electrons in the CRT is about 5×10^{14} times higher! The forces used to create such huge accelerations are electrical in nature and are the subject of Chapter 5.

3.3   PRINCIPLES OF MECHANICS

Isaac Newton was the first to realize that the motions of objects obey a particular set of rules. A physical rule that is never observed to be violated is called a law, as we discussed in Chapter 1. Newton was the first to state these laws as quantitative (mathematical) rules that were capable of not only summarizing what was happening in a performed experiment, but also of predicting the outcome of experiments not yet carried out. This was revolutionary in the evolution of science.

Newton recognized three important principles of mechanics, often called Newton's Laws of Motion. The first is:

MECHANICS PRINCIPLE (I)

Newton's first law: An object's speed and direction remain constant unless an external force, which is not balanced by other forces, acts on the object.

This implies, for example, that if there is no friction present, no force is needed to keep an already moving object in motion.

Any change in an object's speed or direction requires a force. Simply maintaining the speed and direction does not require a force. In particular, an *unbalanced* force is required to change the object's speed or direction, because two forces could be applied to an object in a way that their influences balance or cancel each other. For example, if the designer of a jet car mistakenly specified two engines pointing in opposite directions, their forces would balance, leaving zero acceleration.

It is remarkable that Newton, following a tradition begun by Galileo, was able to understand this nonintuitive fact. Newton had no access to human experience in outer space, as we do, but he could observe the motions of planets traveling around the Sun. He realized that any object (we can imagine the astronaut's wrench) would move according to the same rules as the planets moving around the Sun. He recognized that these rules are a consequence of gravity, which is a force of attraction between any two material objects. Although Newton's mathematical formulas could accurately describe and even predict the motion of planets, he was puzzled by the fact that a force can exist between two objects that are so far away from each other; for example, the attractive gravitational force between the Sun and the Earth. Newton was wise enough to realize that he had no deep understanding of the mechanism that is responsible for the gravitational force. Regarding the question of how gravitation was transmitted from object to object across the void (the vacuum of space), he said, "I make no hypotheses."

Newton also recognized that the same amount of force being applied to two different objects could result in different amounts of acceleration if the two objects have different internal properties. For example, a force acting on a tennis ball will lead to a larger acceleration than would the same force acting on a bowling ball. Newton used the word *mass* to indicate the reluctance of an object to accelerate when subjected to a force. He formulated his most important principle as follows:

MECHANICS PRINCIPLE (II)

Newton's second law: Acceleration of an object is proportional to the force applied to it, and inversely proportional to the object's mass. In equation form, acceleration equals force divided by mass:

$$\text{acceleration} = \frac{\text{force}}{\text{mass}} \quad \text{or} \quad a = \frac{F}{M}$$

Newton's second law can also be written in a different form if we multiply both sides of the equation by M:

$$\text{force} = \text{mass} \times \text{acceleration}$$
$$F = M \times a$$

Mass refers to the amount of "stuff" that makes up an object. Nowadays we know that this stuff is atoms (made of protons, electrons, and neutrons), but in Newton's day,

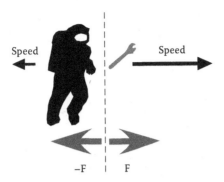

Speed   Speed

−F   F

FIGURE 3.14   A free-"floating" astronaut exerts a force equal to F on a wrench, causing it to accelerate. The wrench exerts an equal and opposite force on the astronaut, causing her to accelerate in the opposite direction, but to a much smaller speed than the wrench.

scientists simply thought of the amount of "material." In honor of Sir Isaac Newton, we measure the strength of forces in terms of units (amounts) called newtons, abbreviated N. One newton is defined as the amount of force needed to accelerate an object with mass 1 kg at a rate of 1 m/sec². Newton's first law is actually a consequence of his second law: if the force is zero, then the acceleration is zero. The reason the first law is stated as a law is for emphasis.

As an example, imagine our astronaut in orbit floating freely and holding a wrench, as in **Figure 3.14**. She places the wrench in her open palm and pushes it away toward space. Say she exerts a force equal to 1 N. During the push, the wrench will experience an acceleration equal to 1 N divided by the mass of the wrench. If the wrench has mass equal to 0.5 kilograms, the acceleration will be:

$$\text{acceleration} = \frac{1 \text{ N}}{0.5 \text{kg}} = 2\text{m} / \sec^2$$

After the push is finished, the wrench will feel no further acceleration (except perhaps from gravity, but let us ignore that here).

Note that for this example calculation to give the correct answer, we had to express the mass in kilograms (1 kg = 1,000 g), and the acceleration in m/sec². We will discuss the matter of units in more detail later in this chapter.

REAL-WORLD EXAMPLE 3.3: FORCE ON A HARD-DRIVE HEAD

In Real-World Example 3.1, we found that the read/write head in a hard disk drive experiences a huge acceleration when moving from point to point above the disk. How much force is needed to achieve such a high acceleration of the head? To answer this, we need to know the head's mass. An estimated value is 1×10^{-5} kg, a small mass. We find the required force using Newton's second law:

$F = M \cdot a$
  $= 1 \times 10^{-5}$ kg $\cdot$ 1,300 m/sec² $= 0.013$ kg m/sec² $= 0.013$ N

We conclude that the force applied to the head during the first 2.5 milliseconds equals 0.013 N. The force applied during the second 2.5 milliseconds is the negative of that. This is a rather small force, but when applied to such a small mass, it results in a very high acceleration.

The value 0.013 N is the force applied to the head alone. The motor must also accelerate the metal arm that holds the head, and this weighs about 10 times more than the head. This requires an actual applied force about 10 times higher than our estimate for the head alone. This force is exerted in the form of a rotational force, called a *torque*, around the rotation axis of the arm swivel axis.

MECHANICS PRINCIPLE (III)

Forces always come in pairs. This is described by Newton's third law:

Newton's third law: When one object exerts a force on a second object, the second object also exerts a force, equal in strength and opposite in direction, back on the first object.

For example, if you lean steeply with your hand against a rigid wall, your hand exerts a force on the wall, yet the wall does not accelerate. This is because the wall is strong enough to oppose the applied force of your hand. This means that the wall bends by an imperceptible amount, and in doing so it acts like a stiff spring that exerts a force back on your hand. The fact that neither the wall nor your hand are accelerating proves that the two forces involved are equal in strength and opposite in direction.

Considering again our example in Figure 3.14, this law says that when the astronaut pushes away the wrench with a force of 1 N, she will feel a pushing force back from the wrench, also equal to 1 N. But, because her mass is larger than the wrench, say 100 kilograms including her spacesuit, her acceleration will be much less the wrench's:

$$acceleration = \frac{-1 \text{ N}}{100 \text{ kg}} = -0.01 \text{ m/sec}^2$$

Her acceleration is denoted as negative, because she accelerates in the opposite direction than does the wrench.

QUICK QUESTION 3.3

Explain how Newton's third law applies to the acceleration of a jet car, keeping in mind that the purpose of the jet engine is to eject gas out of the back end at high speed.

3.3.1 Gravity's Force

The force of gravity on an object near the Earth's surface equals 9.8 N for every kilogram of mass in the object (i.e., 9.8 N/kg). Consider a skateboarder, with a mass of 34 kilograms, who suddenly drops over the lip of a large ramp. The downward force of gravity on him is

34 kg × 9.8 N/kg = 333 N (about 75 lb of force)

One newton equals the force that Earth's gravity (at sea level) exerts on an object with mass equal to 0.102 kilograms. One newton of force is equivalent to 0.225 pounds, or about one-quarter of a pound. That is, to hold up a quarter-pound hamburger, you need to exert a force of 1 N in the upward direction. It is a fun fact—but only a coincidence—that a medium sized apple weighs about a quarter pound. This means that the force of gravity on that proverbial apple that Isaac Newton watched falling from a tree equaled about 1 N, in today's terminology.

We can express gravity's force acting on an object with mass M by the equation:

force of gravity $= M \times 9.8\text{N/kg}$

For example, a bowling ball with a mass of 4 kilograms would experience a force of gravity equal to:

force of gravity $= 4\text{kg} \times 9.8\text{N/kg} = 39.2\text{N}$

Gravity is the special case of the attraction between the Earth and all other objects. The force of gravity on an object near the Earth is proportional to the amount of mass in

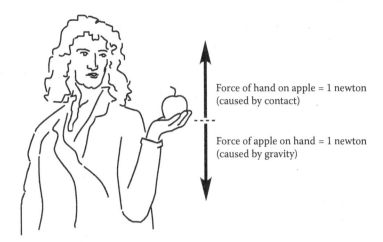

Force of hand on apple = 1 newton
(caused by contact)

Force of apple on hand = 1 newton
(caused by gravity)

FIGURE 3.15   Newton holding an apple up against gravity.

the object. This leads to the fact, made famous by Galileo, that all objects feel the same acceleration due to Earth's gravity, with a value equal to 9.8 m/sec^2. This can also be stated as 9.8 N per kilogram, or 9.8 N/kg, because 1 N/kg equals 1 m/sec^2. It follows that the force exerted on a mass of 1 kilogram near the Earth's surface is 9.8 N. We can sum up much of our discussion so far using a drawing of Newton simply holding a 0.102-kilogram apple, shown in **Figure 3.15**. Earth's gravity creates a downward force of 1 N on the apple. The apple, by virtue of its contact with the hand, exerts a downward force of 1 N on the hand. Newton, using his muscles, creates an upward force of 1 N, which is exerted on the apple. At the point of contact between the apple and the hand, these two forces are equal in strength and opposite in direction. Therefore, the acceleration is zero.

3.4   THE PHYSICS OF ENERGY

There is an intuitive feeling that one will not be able to get something for nothing. It therefore seems proper and orderly to suppose that the universe possesses a fixed amount of something or other and that, while this may be distributed among different bodies of the universe in various ways, the total amount may neither be increased nor decreased.

Isaac Asimov
(*The History of Physics*)

The concept of *energy* gives us a deeper understanding of the motion of objects. If, at the end of a long day, you feel the need for an energy boost, there are two ways you could get one. You could drink a cup of coffee or you could eat a so-called "energy bar." These would have quite different effects, from a physics perspective. The coffee would stimulate your body to a higher level of activity, requiring a more rapid conversion of stored *chemical energy* into physical activity for a certain time. After a while, this would decrease the energy stored in your body. In contrast, the energy bar, which is loaded with sugar, would add to the energy content of your body, allowing you to perform more vigorous physical and mental tasks or the same tasks for a longer time. What is energy?

Energy is an intangible quantity in nature that enables one object to cause the motion of another object.

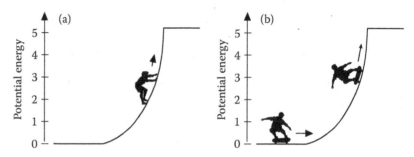

FIGURE 3.16   The top lip of a skateboarding ramp represents higher gravitational potential energy than the bottom of the ramp, which is indicated as zero energy.

Energy comes in distinct forms: kinetic, potential, chemical, and thermal, among others. As an example of **kinetic energy** and **potential energy**, consider a skateboarder at the bottom of a skateboard ramp, as in **Figure 3.16**. If he wants to get to the top of the ramp so he can drop in and begin his ride, he first needs a boost of energy to do so. This energy boost allows him to overcome gravity—the force of the Earth pulling him toward the ground. He could get that boost in several ways–he could use his muscles, converting stored chemical energy (Remember that candy he ate earlier?) into **mechanical energy**, which allows him to climb up the ramp. Another skater decides to use a different technique to reach the top of the ramp to start her ride. She backs up and accelerates by the familiar foot-on-ground pushing motion. If, before reaching the base of the ramp, she has gained sufficient kinetic energy, then when she reaches the base she will smoothly sail up the ramp, all the while slowing down, and land gracefully on her feet at the top (and catch her board in her hand if she is good). In this example she has first converted stored chemical muscle energy into kinetic energy, then—by rolling up the ramp—has converted that kinetic energy into potential energy. Energy is measured in joules or kilojoules. One kilojoule is the amount of energy required to raise a 34-kilogram skater to a height of 3 meters against gravity (on Earth at sea level).

Once at the top, the skaters have expended some chemical energy from their muscles, and have gained in stored **gravitational energy**. We say that the gravitational energy is stored as potential energy. It has the potential to be released at any time, causing an object's motion. This stored energy can be released suddenly by the skater dropping in from the edge of the ramp, after which he would quickly gain kinetic energy—defined as the energy associated with motion.

Now our skateboarders are at the top, ready to drop in. Say that this ramp is a half pipe with equal heights at both sides, as in **Figure 3.17**. Our first skater knows that if he

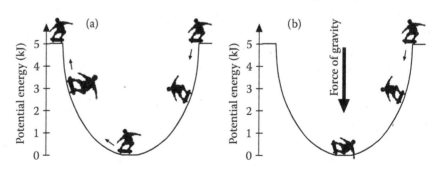

FIGURE 3.17   (a) The skater drops in from the right lip and rolls, without friction, ending up at the lip on the left, with zero speed. (b) The skater falls and, because of friction, ends up at rest at the bottom of the ramp

drops in suddenly and heads straight down the ramp, he will reach a maximum speed at the bottom (lowest point) and, if the ramp is very smooth with no friction (and if he has recently cleaned and oiled his bearings), he will sail exactly up to the lip of the ramp at the other side, where he can simply step onto it and again be at rest. He will not overshoot the lip and sail 20 feet beyond it, and he will not fail to reach the lip entirely.

We are assuming that this skateboard exhibition takes place on Earth, where gravity is present. (Try to imagine skateboarding on a ramp in outer space, where there is little gravity.) This means that there is always a downward force of gravity, pushing a skateboarder toward the center of the Earth. Therefore, when the first skateboarder slowly climbs up the ramp, his muscles are exerting themselves against the force of gravity. He is working against gravity. Physicists call this kind of exertion doing work.

A useful way to think of *work* is that it is an action, involving motion, that you would be willing to pay money to someone for performing, such as pushing you up the ramp so you do not have to tire yourself by climbing. Another example would be that you are willing to pay the power utility company to provide you with electricity—the amount you pay each month is proportional to the amount of electrical energy you use to enable your household devices to perform work for you. Washing machines wash; light bulbs light; CD players play. A final example would be that you are willing to pay for a lift ticket at a ski resort because, after the lift has done enough work to enable you to reach the top, you have acquired sufficient potential energy that you can then ski down the hill and have all kinds of fun. (Fun is not a technical physics term, although maybe it should be.) When we say that work is an action that you would be willing to pay for, we mean by the word *action* that there is a force acting on a moving object through a distance. Therefore, if a person stands still and holds your heavy suitcase off the ground for 1 minute, he or she might be expending chemical muscle energy, but that does not mean the person is doing work in the technical sense.

Work is defined as follows:

Work is the process of applying a force over a certain distance.

More precisely, work is the amount of energy transferred when moving an object through a certain distance by applying a force in the direction of motion. When you do work on an object, you transfer energy from yourself to it. In terms of an equation, work equals the strength of the applied force multiplied by the distance the object moves.

$Work = Force \times Distance\ Moved$
$W = F \times D$

Using this concept of work, we can turn the definition of energy around and say:

Energy is the capacity of a physical system to do work.

Force doing work is illustrated in **Figure 3.18**. The astronaut braces her back against the space shuttle and exerts a steady force, say 1 N, on a ball over a certain distance, say 0.5 meters. In Figure 3.18a, the ball is a heavy bowling ball, so after the force is exerted, and the work is done, this ball ends up moving at a fairly slow speed. In Figure 3.18b, the ball is a light ping-pong ball, which is easier to accelerate. So, after the force is exerted over the same distance, and the same work is done, this ball ends up moving at a higher speed. In both cases, the amount of work done is the same (*force* × *distance*). Therefore

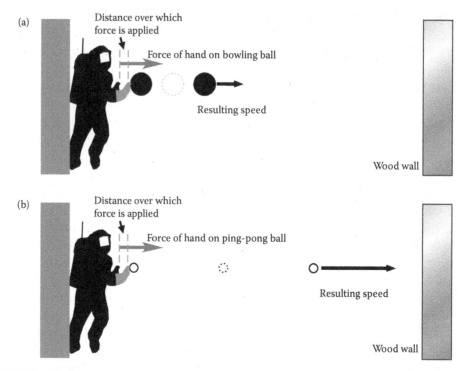

FIGURE 3.18   By applying the same force over the same distance, an astronaut does equal amounts of work on a bowling ball and on a ping-pong ball.

the amount of kinetic energy given to the balls is the same.[2] That is, the amount of work done on an object does not depend on the size of the object's mass—only on the force and the distance through which it is pushed.

Let us try to understand this counterintuitive conclusion in the case of the two balls. We can get a feeling for how much kinetic energy a moving object possesses by thinking about how much damage the object could do if it were to run head-on into a fixed wall made of a material such as wood covered with a thick, soft fabric so that the object does not bounce from the surface. Both balls could break the wooden wall. It should seem reasonable to you that the fast-moving ping-pong ball could do as much damage to this wall as could the slow-moving bowling ball. This is true, because they have the same amount of kinetic energy before they hit the wall.

A second example shows how the potential energy of an object is increased by doing work on it. In **Figure 3.19a**, a wheeled cart is slowly pulled up a ramp against the force of gravity by a steady, applied force equal to 5 N. The cart gains a height of 1 meter. The length of the ramp is 3 meters, so the work done on the cart equals

$$Work = 5\ N \times 3\ m = 15\ J$$

Therefore, the potential energy gained by the cart is 15 joules. In Figure 3.19b, the same cart is pulled straight up, gaining the same height, 1 meter. The steady force required to pull the cart straight up against gravity equals 15 N, larger than in the case of pulling it up the ramp.

$$Work = 5\ N \times 1\ m = 15\ J$$

In both cases, the potential energy given to the cart depends only on the height to which it is raised, in both cases 1 meter. That is, after being raised to this height,

---

[2] The formula for the amount of kinetic energy possessed by an object having mass M, moving with speed S is: $(1/2)MS^2$, but we will not use this formula.

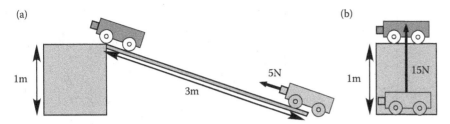

FIGURE 3.19   The amount of work that needs to be done to raise a cart by a height of 1 m is the same if we pull it up an incline or raise it straight up.

the cart has the potential to do a fixed amount of work (or damage) on some other object.

To summarize, changes of energy (increases or decreases) involve either having work done on an object or work done by that object. Forces are needed for this to occur.

3.5   FRICTION AND THERMAL ENERGY

If, as in Figure 3.17b, the skateboarder falls off her board just after dropping in, then the *friction* between her clothes or skin and the ramp surface will cause her to slow down, and she will not sail up to the edge of the ramp lip at the far side of the half pipe. What happened to her stored potential energy, which had been converted to kinetic energy as she dropped in? The answer is that her clothes and skin (and the ramp) heated up. All of the stored potential energy was converted into ***thermal energy*** (see the next section). *Friction* is the process by which kinetic energy is converted into thermal energy when two surfaces rub together.

What causes friction? If you push two surfaces together and try to slide them past each other, there will be some resisting force, which is friction. To understand it, first realize that no surface is perfectly smooth. Some roughness will always be seen under a powerful enough microscope, as illustrated in **Figure 3.20**. Think of two pieces of sand paper being rubbed together. Friction between them is caused by the tiny protrusions on the two surfaces bumping into each other and possibly sticking together temporarily. Friction causes a force between two objects when their surfaces are in contact and they are moving relative to each other. The friction force always opposes the motion of one object relative to the other. Typically, the force provided by friction increases for the higher speed of one surface relative to the other. When an object falls through the atmosphere, air resistance causes a type of friction on the object.

Thermal energy is the microscopic kinetic and potential energies associated with the random motions of the particles, called atoms, making up a solid, liquid, or gas. Thermal energy is also called internal energy. The warmer a substance is, the faster its internal atoms move on average. ***Temperature*** is the term we use to designate the

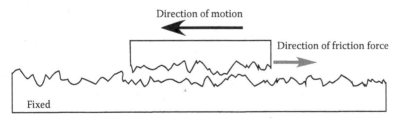

FIGURE 3.20   Two surfaces sliding while in contact, viewed under high magnification. The friction force is in the opposite direction from the motion.

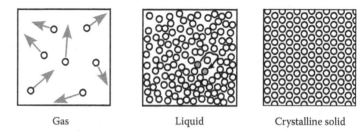

Gas   Liquid   Crystalline solid

FIGURE 3.21   The behavior of atoms in a gas, liquid, or crystalline solid (a crystal).

"hotness" of an object, as discussed in the next chapter. The behavior of atoms is different, depending on whether they are in a gas, a liquid, or a solid (see **Figure 3.21**).

- In a gas, such as air, the atoms move in straight lines until they hit a wall or other object, or another atom in the gas. Then the atom will change its direction of motion and its speed. You can visualize a gas as being many frictionless billiard balls on a frictionless pool table, so that after the balls are put into motion, they keep moving and colliding forever. The higher the temperature, the higher their speed.
- In a liquid, the atoms are in continuous physical contact with other atoms, like shoppers in a very crowded store, all jostling one another as they slowly move randomly from place to place. The higher the temperature, the faster they jostle and move.
- In a solid, atoms are fixed in location. Because of thermal energy, the atoms in a solid jiggle rapidly in place, but do not go anywhere, because their neighbors refuse to yield their positions.

QUICK QUESTION 3.4

Think of other examples in which thermal energy can be used to do work.

Automobile engines use thermal energy to do work. When the air-fuel mixture inside an engine cylinder is compressed, the sparkplug fires, the mixture burns and pushes the piston outward in the cylinder, the piston cranks turn the crankshaft, the crankshaft rotation is transmitted to the wheels by the driveshaft, and the automobile moves forward. In such a system, and any other system, there are definite limits to the efficiency with which thermal energy can be converted into work. These limits are discussed in Chapter 4.

To summarize the two general types of energy:

- Mechanical potential energy is the energy that is stored in objects by virtue of their positions. This stored energy can potentially be released later and cause some work to be done. Examples of potential energy include gravitational potential energy gained when you climb (or someone pushes you up) a ramp, chemical energy stored in your muscles, and electrical energy stored in a battery.
- Kinetic energy is the energy associated with a moving object. An example of kinetic energy is a moving skateboard and rider.

3.6   THE CONSTANCY OF ENERGY

This leads us to the most important principle about energy—its constancy or conservation.

MECHANICS PRINCIPLE (IV)

Energy conservation principle: Energy cannot be created or destroyed; it just gets converted between various forms. This means that the total amount of energy in any closed system does not change, but is constant (i.e., conserved).

The word *closed* means that a system cannot exchange energy with other systems. For example, imagine a closed room whose walls are very thick and made of materials that allow no heat, light, radio, or any other form of energy to enter or exit. In this room, there are various stores of supplies, such as firewood, oil, and food, and there are apparatuses such as steam engines, water wheels, water pumps, and electrical generators. A person in this room could perform many tasks involving energy—move objects, boil water, drive steam-powered generators, power light bulbs, etc. But, all of these tasks involve only the conversion of energy from one form to another—not the creation or destruction of energy.

THINK AGAIN

We use the word *conservation* in the technical sense meaning *constancy*, not in the sense of *sparing use*, as we would mean regarding saving costs on electricity by turning lights off.

THINK AGAIN

A seeming exception to the energy conservation rule occurs in nuclear reactions—for example, in nuclear power generating stations, where some of the energy stored as mass of the fuel atoms is converted into thermal energy. This is not really an exception, because mass can be thought of as just another way to store energy, which can be released in such reactions.

In terms of the skateboarder dropping into the half pipe—the amount of gravitational potential energy that he gained when climbing to the top of one side of the ramp is exactly the same as that needed to reach the top of the other side of the ramp (if the sides are of equal height). When he drops in from one side, he accelerates because of the force of gravity. Assuming little or no friction is present, when he reaches the lowest point on the ramp, all of his earlier stored potential energy has now been converted into kinetic energy (he is going fast). This gives him just the right amount of kinetic energy to allow him to reach just to the lip of the other side of the half pipe. This is conservation of energy in action!

Another example is when gasoline is burned in a car engine, energy stored in chemical bonds is converted into mechanical (kinetic) energy that propels the car. Also, when sunlight is absorbed by a photocell, light energy (a form of electromagnetic energy) is converted into electrical energy.

3.7   UNITS FOR MECHANICS

In Chapter 2, Section 2.5.1, we discussed the metric units for distance, time, and speed, and how to use *conversion factors* to change between different units, for example, meters or kilometers. Here we will summarize the units for mass, force, and energy.

3.7.1   Units for Mass

The standard *unit* for mass is kilograms (kg), not grams (g) as you might expect.

Mass		
1 kilogram	= 1 kg	= mass for which Earth's gravity force at sea level equals about 2.2 lb
1 gram	= 1 g	= 10^{-3} kg (1 paper clip)
1 milligram	= 1 mg	= 10^{-6} kg (mosquito)
1 microgram	= 1 μg	= 10^{-9} kg (small grain of sand)

3.7.2   Units for Acceleration

Recall that the unit of acceleration is meters per second squared, or m/sec². This can also be written as $m \cdot sec^{-2}$, where the dot (·) means multiplication. There is no special name for this unit.

3.7.3   Units for Force

We measure the strength of forces in units called *newtons*, abbreviated N. One newton is defined as the amount of force needed to accelerate an object with a mass of 1 kilogram at a rate 1 m/sec². That is,
$1N = 1kg \cdot m/sec^2$, or $1N = 1kg \cdot m \cdot sec^{-2}$

For example, to accelerate a 4-kilogram bowling ball at a rate 3 m/sec², you would need to apply a force equal to $F = m \cdot a = 4kg \cdot 3m/sec^2 = 12kg \cdot m/sec^2 = 12N$. Alternatively, we could ask how much acceleration would we achieve for this bowling ball if we applied a given amount of force; for example, 20 N. The acceleration in this case would be

$$acceleration = a = \frac{20N}{4kg} = \frac{20kg \cdot m/sec^2}{4kg} = 5m/sec^2$$

THINK AGAIN

The Earth exerts a stronger gravitational force on a bowling ball than it does on a tennis ball. You might think that this means the bowling ball would accelerate faster than the tennis ball. Is that correct?

3.7.4   Units for Energy

How much energy is needed to raise a skater from the bottom of a ramp to its lip, if the lip is 3 meters high? Or, how much energy is used to run electric lights for 3 hours of night skating? To answer questions of this type, we need some terminology that refers to specific amounts of energy.

The basic unit of energy is the *joule* (pronounced "jool," and abbreviated J), named after English brewer James Joule, whose hobby was doing physics experiments. Joule carried out careful experiments in the 1840s that first showed that the various forms of energy—mechanical, electrical, and thermal—can be changed one into another. His work is discussed in detail in Chapter 4. The joule can be defined in several different, but equivalent, ways:

1. One joule is the amount of mechanical energy needed to generate a force equal to 1 N and to use it to push an object a distance of 1 meter. This means 1 J = 1 N · 1 m, or J = N · m.
2. One joule is the amount of thermal energy needed to raise the temperature of 1/(4,000) liter of water by about one degree centigrade (the more precise number is 0.96°C). (Note that 1 L equals about 1 quart, and a change of 1°C equals a change of 1.8°F.)
3. One joule is the amount of electrical energy needed to operate a 1-watt light bulb for 1 second (or a 100-W bulb for one-hundredth of a second).

One kilojoule (abbreviated kJ) equals 1,000 J. In terms of our skateboarder, if his mass is 34 kilograms (corresponding to about 75 lb), then, as we discussed earlier, the downward force of gravity is 333 N. Therefore, to push him up to a height of 3 meters requires $333 \times 3 \approx 1{,}000$ J = 1 kJ of work, which gets stored as gravitational potential energy. This example gives a good feeling for the amount of energy contained in 1 kJ.

3.8   GRAPHING ENERGY

In previous figures showing a skater on a ramp, the drawing of a ramp surface can be thought of as a graph of potential energy plotted versus the horizontal position of the skater. The graphs in **Figure 3.22** are constructed by drawing a scale in the vertical

QUICK QUESTION 3.6

If our skateboarder's mass is 50 kilograms (corresponding to about 110 lb), then how many newtons does the downward force of gravity equal? In this case, how many joules of energy are required to push him up to a height of 2 m?

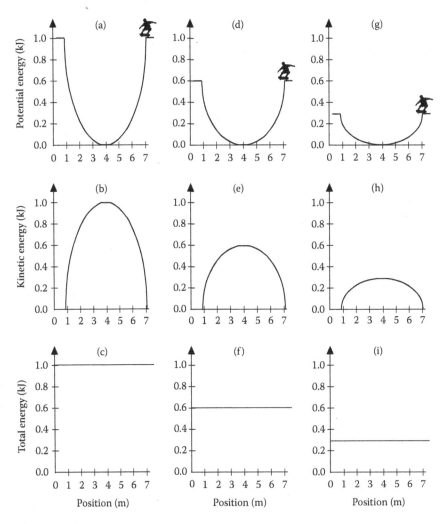

FIGURE 3.22 Graphs of potential, kinetic, and total energy for a skater dropping into a ramp. The three cases, (a, d, and g) illustrate different starting heights, and therefore different amounts of energy.

direction (as if a ruler is standing on end) indicating the potential energy of the skater. The horizontal direction (axis) of the graph is the position of the skater measured in meters from some chosen point on the left, called zero. Graphs like these will prove to be useful in later chapters when we discuss the behavior of electrons in semiconductor computer circuits. We use skaters here to make the examples more familiar.

As shown in Figure 3.22a, the skater, who weighs 75 pounds (meaning he has a mass of 34 kg), begins at a horizontal distance indicated as 7 meters on the scale. This places him at the top of the ramp (which is 3 m high) with a potential energy equal to 1 kJ relative to the lowest point on the ramp. When he moves from right to left in horizontal position, he goes down the ramp, losing potential energy, until he reaches the bottom of the ramp, where his potential energy is a minimum. We assign a value of zero to this minimum potential energy.

The choice of the value zero to designate the minimum potential energy is arbitrary. That is, the only important quantity is the change of potential energy in going from one height to another. For example, going from a height of 1 meter above the ground to a height of 4 meters above the ground involves the same change of potential energy as going from a height of 5 meters above the ground to a height of 8 meters above the ground. It is common to choose the minimum energy in a given scenario to be zero.

In addition to graphing the potential energy, it is useful to graph kinetic energy versus horizontal position. As shown in Figure 3.22b, when the skater moves from right to left in his horizontal position he picks up kinetic energy, which increases to a maximum value of 1 kJ when he is at the bottom of the ramp (position 4 m), where his potential energy equals a minimum. Then, as he moves up the left side of the ramp, his kinetic energy decreases as he slows down, and eventually it reaches zero again when he reaches the lip, where he comes to rest. The total energy (potential plus kinetic) is constant during his ride down and back up the ramp. In Figure 3.22c, this total energy is graphed as a flat, horizontal line at 1 kJ.

Figures 3.22 d–f correspond to the same sequence of events in the case that the skater begins at a lower level, where his starting potential energy is 0.6 kJ. Likewise, if he starts at an even lower level, 0.3 kJ, then the sequence is given by Figures 3.22 g–i.

3.9 POWER

We have seen that energy is something (a resource) that is conserved in the universe. It can be converted from one form to another, and it can be delivered from one place to another. The question we are concerned with here is, "How fast are we converting (or delivering) the energy?" Consider the example of electric house heating. The power company, through power lines, provides a supply of energy that you can use by plugging in electric space heaters to wall outlets. When you "use" it, this means you are converting electrical energy into thermal energy. If you plug in five large heaters, you are clearly using (converting) more energy per hour than if you plug in one small heater. We say you are using more **power**.

Power equals energy delivered or converted per second. A power of one **watt** (abbreviated W) equals one joule of energy delivered per second. (1W = 1 J/sec)

The energy could be delivered by an electric current, by a light beam, or by some other means, such as a gasoline engine. The unit of power is the **watt** (W), named after James Watt, a Scottish instrument maker whose steam engine helped to spur the Industrial Revolution in the late 1700s.

Conversely, consider a device that is continuously "using" a constant amount of power P. After a period of time t, the total amount of energy that will have been used is proportional to P and proportional to t. The total energy E delivered by a source with power P in a time interval of duration t is equal to the product of the power and the duration. As a formula:

$$Energy = Power \times time$$
$$E = P \times t$$

That is,

$$joules = watts \cdot seconds$$

For example, a typical incandescent light bulb uses 100 W when it is lit. If you leave it on for 100 seconds, the energy used during that time is $E = 100$ W $\cdot$ 100 sec $= 10,000$ (J/sec) $\cdot$ sec $= 10^4$ J. If you leave this bulb on for 100 hours, the energy used is E $= 100$ W $\cdot$ 100 hr $= 100$ W $\cdot$ 360,000 sec $= 3.6 \times 10^7$ W $\times$ sec $= 3.6 \times 10^7$ J.

Another way to express energy is in units of kilowatt·hours. One kilowatt·hour equals the amount of energy delivered by a 1-kilowatt source in 1 hour. That is,

$$1 \text{ kW} \cdot \text{hr} = 1,000 \text{ W} \cdot 3,600 \text{ sec} = 3,600,000 \text{ J} = 3.6 \times 10^6 \text{ J}$$

The energy used in 100 hours by the bulb in the above example can be expressed as

$$E = 100 \text{ W} \cdot 100 \text{ hr} = 10,000 \text{ W} \cdot \text{hr} = 10 \text{ kW} \cdot \text{hr}.$$

When the bulb uses this electrical energy, where does this energy go? About 5% of the electrical energy is converted into visible light, and about 95% is converted into thermal energy, including invisible infrared light and warming of the light fixture and the air around it.

3.9.1   An Example on Power Costs

Consider the problem of figuring out how many kilowatts (a unit of power) your household used on average last January to heat your house using electrical energy (or, if you live in a hot climate, to air condition your house last July). Say your electric bill was $30, and you note on the bill that the rate you pay is 7 cents per kilowatt·hour. To solve this problem, analyze the units. We can express 7 cents per kilowatt·hour equivalently as $0.07/(kilowatt·hour). We know January has 24 $\cdot$ 31 $=$ 744 hours. The answer we are looking for has units of kilowatts. We know the problem is one that can be solved by simple multiplication or division. How many ways can we combine $30 with $0.07/(kilowatt·hour) and 744 hours to obtain units of kilowatts? Only one way! Namely,

$$\frac{30 \text{ dollars}}{(744 \text{ hours}) \cdot (0.07 \text{dollars}/\text{kilowatt} \cdot \text{hours})} = 0.58 \text{ kilowatts}$$

Notice how the units cancelled in this calculation.

We can extend this example to answer how many houses of your size could be supplied power by a 3-megawatt power plant? A megawatt, or MW, equals 1 million W, or 10^6 W. A kilowatt, or kW, is 1,000 W, or 10^3 W. So, a 3-MW plant produces 3,000 kW. We can write the single-house power usage as 0.58 kW per house, or 0.58 kW/house.

We need to combine this with 3,000 kW to obtain an answer having houses as the units. There is only one way to do this and have the units cancel properly:

$$\frac{3,000 \text{ kilowatt}}{0.58 \text{ kilowatt/house}} = 5,172 \text{ houses}$$

QUICK QUESTION 3.7

Thermal energy is generated every time a computer performs an operation on a data bit, such as reading it from or writing it to memory. A typical desktop personal computer uses around 100 W of power when carrying out intensive processing, such as running a video game. If you play such a game for 90 minutes, how much energy (in joules) will it use?

THINK AGAIN

There is no such thing as "watts per hour," or "watts per second." Watt is a unit of power, which is the instantaneous rate of using energy, not an amount of something.

REAL-WORLD EXAMPLE 3.4: MOTION SENSORS IN LAPTOPS

Devices like laptop computers and portable music players contain hard drives for data storage. The read head in such drives travels, or "flies," at a height less than 0.1 micrometer (10^{-7}m) above the surface of the magnetic disk on a cushion of air. If the device is shaken, dropped, or moved too quickly, the head can crash into the disk surface, destroying both. Many devices have an ingenious technique for protecting against such head crashes. A motion sensor called an accelerometer is built into the unit to detect any sudden movements, and if such a movement occurs, the head is automatically lifted a safe distance from the disk surface until the device again becomes stable.

An *accelerometer* consists of an object, called a test mass, suspended above a platform (called the body) by thin supports with flexible joints called flexures, shown in **Figure 3.23**. The body is attached firmly to the case of the computer unit. If the case suddenly lurches in some direction, the supports flex, and the test mass remains momentarily at its original position. The flexures bend like soft springs rather than applying a strong force to the test mass. Eventually, the flexures, which have been compressed, push back and apply a force to the test mass, making it move.

As this lagging motion of the test mass takes place, an electronic sensor measures and records the position of the test mass relative to the body of the accelerometer. In this way, sudden accelerations of the body are detected. Notice that such a sensor does not detect the speed of the body. For example, if the laptop computer is in an airplane that is moving

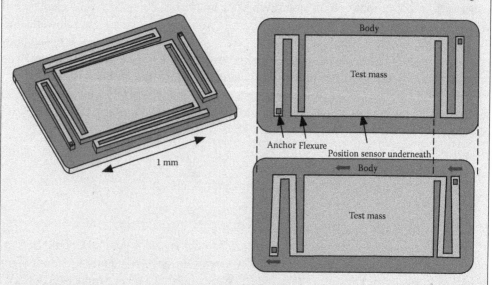

FIGURE 3.23 Micro-machined silicon accelerometer.

fast, the test mass moves along with the body and has no motion relative to the body. But, when the plane decelerates for landing, the body will move relative to the test mass, and this will be detected.

A remarkable fact about modern accelerometers is that their body and test mass are made entirely from a single tiny piece of silicon. This piece of silicon is less than 1 millimeter in length. Special chemical processes—the same that are used for making computer circuits—are used to "micro-machine" the moving parts. It is as if Michelangelo could carve the *David* statue out of a single grain of sand. This type of microdevice is an example of MEMS (micro-electrical-mechanical systems).

QUICK QUESTION 3.8

Why is the often-used phrase "power consumption" not valid?

SUMMARY AND LOOK FORWARD

The principles of mechanics are summarized as follows:

i. Acceleration of an object (change in its speed or direction) requires the application of a force. This implies that if there is no force applied, there can be no acceleration. If there is no friction, then no force is required to keep a moving object moving.

ii. Acceleration is proportional to the force applied, divided by the mass of the object.

iii. Energy is the potential to move an object through a distance by the continuous application of a force, that is, the possibility to do work.

iv. The total amount of energy in any closed system is constant. Kinetic energy is the energy associated with a moving object. Potential energy is the energy that is stored in an object, or arrangement of objects, depending on their positions. Thermal energy is the kinetic and potential energies associated with the microscopic, random motions of atoms and electrons in an object.

What do force, acceleration, energy, and power have to do with computers and the Internet? Mechanical components, such as hard drives, play an important role. Power levels in computers set the upper bounds for processor speed. When electrical power is converted within a processor to run the logic circuits, thermal energy is always generated. If the thermal energy is not removed efficiently, the heating of the components will cause data errors and other malfunctions. Power management is a large part of processor design.

Computers and networks move information around by the use of energy. Wireless devices, such as cell phones, send data or information through the air in the form of electric and magnetic waves, which carry energy from the sender to the receiver. Wired networks transmit information across long distances by delivering energy from the sender to the receiver as electrical pulses in metal wires or as light pulses in optical fibers. In computers, logic is performed by moving electrons (elementary particles of electric charge) around; this requires forces, which accelerate the electrons. The very structure of semiconductor crystals, which make up the circuit components (e.g., transistors), depends on the forces between atoms in the crystal. Finally, the way that electrons move through the semiconductor crucially depends on the electrons' energy in different places in the crystal. The movement of the electrons in the semiconductor generates thermal energy, which must be removed through a "heat sink" and a fan.

In the following chapters, we will study the properties of matter, electricity, and magnetism, because those are the basis of the physics behind computers and the Internet. In later chapters, we will study the makeup of atoms and how they form semiconductor devices such as computer processors and lasers.

SOCIAL IMPACTS: SCIENTIFIC THOUGHT AND METHODS HAVE ARGUABLY CHANGED THE COURSE OF HUMAN HISTORY MORE THAN ANYTHING ELSE[3]

And new Philosophy cals all in doubt,
The Element of fire is quite put out;
The Sunne is lost, and th'earth, and no mans wit
Can well direct him where to looke for it.
And freely men confesse that this world's spent,
When in the Planets, and the Firmament
They seeke so many new; they see that this
Is crumbled out againe to his Atomis

John Donne
(*An Anatomy of the World: The First Anniversary,* 1611)

A thousand years ago, there were no accurate maps of any continent, much less of the whole world. In the European world view, the Earth was the center of the universe, and the Sun revolved around it on a Celestial Sphere, as depicted in **Figure 3.24**. The dominant mode of intellectual thought was dictated by the Church. Now, using a handheld Global Positioning System (GPS) receiver we can instantly find our location to within a couple of meters anywhere on Earth. The Earth revolves around the Sun, and thought has been freed, for better or for worse.

Humans by nature love to explore—both geographically and mentally. Although geographical exploration once took months by ship and horseback, Google Earth now allows us to virtually fly above the globe, revealing any region in startling detail. Surfing the Web allows a mental and social exploration of the present state of humanity.

How did we arrive at our present state of humanity and human thought? Let us try to list those arenas of human activity that have shaped what it means to be human, while recognizing that any list would likely reveal biases of the person making the list. In no particular order:

agriculture	art	weapons and warfare
development of tools	music	business and shopping
writing, recording	literature and theater	global communication and travel
printing	philosophy	exploration
mathematics	social interactions and family	sports
science	religion and spirituality	videogames
technology	democracy and politics	others…

Not all activities that are of great interest to people are unique to being human, for example, sex and eating. On the other hand, philosophy, religion, spirituality, and science, among others, appear to be uniquely human.

Is there a single idea or thought that most influenced and enabled the human transition in the previous millennium? In 1999, on the eve of the millennium, novelist Richard Powers wrote, "It lies beyond all reasonable doubt that no single idea has had a more profound or ubiquitous impact on what the human race has become, or what it has worked upon the face of the planet, than the vesting of authority in experiment."

To make his point, Powers traces the beginnings of the experimental method back to Abu Ali al-Hasan Ibn al-Haytham,[4] born around the year 965 in Basra, in what is now Iraq. al-Haytham, whose portrait is shown in **Figure 3.25**, made important discoveries about optics—the study of light, lenses, prisms, etc. More importantly, al-Haytham

---

[3] Inspired by Richard Powers. "Eyes Wide Open" *New York Times Magazine Special Millennium Issue*, no.1, 1999. http://www.nytimes.com/library/magazine/millennium/m1/powers.html.
[4] Abu Ali al-Hasan Ibn al-Haitham, known in the West as Alhazen.

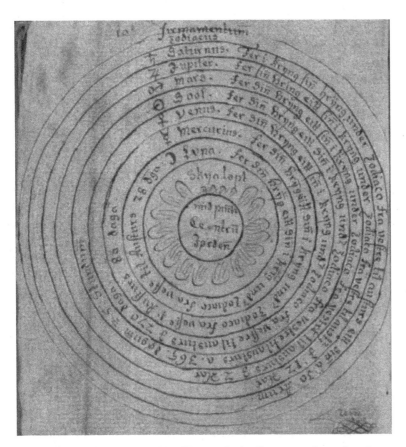

FIGURE 3.24   The geocentric world view. (From an Icelandic manuscript, written between 1747 and 1752. Courtesy of the Árni Magnússon Institute, Iceland.)

FIGURE 3.25   Portrait of Ibn al-Haithem (or al-Haytham) from an Iraqi 10000 Dinar note. (Courtesy of Jacob Bourjaily.)

demonstrated the power of direct observation in resolving natural philosophical arguments by the following example concerning the nature of human vision. Some prior philosophers—Euclid and Ptolemy—had believed (without evidence) that when a man saw an object with his eyes, light traveled from his eye to the object. Others—Aristotle and atomists—had believed (without evidence) that light traveled from the object to the eye. al-Haytham demonstrated that the latter was correct by instructing students to stare at the Sun, whereupon the students' eyes became dazzled, momentarily losing the power of vision. Because looking at a dim candle flame does not blind the eye, it is clear that light travels from objects to the eye, and not the reverse.

Although to us this seems simple and obvious, in those days it was radical. Then wisdom was handed down from the top; bottom-up thinking was not encouraged. The minds of the time, even of the very bright, were not in the habit of believing that a mere person's observation could dictate how the world was made up. This seemed the province for only God. al-Haytham's writings and ideas directly influenced Western natural philosophers such as Roger Bacon (c. 1220–1292), an early proponent of the experimental method. The now-modern practice of looking at and believing "the data" led, in the seventeenth-century Age of Reason, to a wrenching, growing doubt in the supreme authority of the Church(es) (see the excerpt from John Donne's 1611 poem above). The Western modern world is a direct consequence of the application of the scientific ways of thought. (Although it may seem Eurocentric to focus here on so-called Western thought and achievements, the fact is that in Europe modern science developed so rapidly since the time of Newton that it eclipsed progress in other parts of the world. Happily, much of the world is now caught up or is rapidly catching up.)

In present times, a variant of this discussion continues in some parts of the culture. Highly publicized campaigns carry forward the claim that the scientific method is improperly treading on matters best reserved for religion, such as the veracity of Darwin's evolution theory. In a 2006 essay, Murray Peshkin, a U.S. physicist, clarifies that "a proposition is not a scientific theory at all unless it's falsifiable in principle. Absent a possible experiment, science does not know the meaning of the proposition." [1] A current controversy exists over groups of intellectuals called creationists or advocates of intelligent design, who argue that evolution could not possibly be a process unguided by a God's active intervention. Peshkin goes on:

> Science and religion have different assumptions, different rules of inference, and different definitions of truth or reality. … Scientists may have opinions about religion, but they cannot honestly invoke the authority of science on the other side of the fence. Similarly, creationists and advocates of intelligent design should not pretend to be conducting a scientific argument.

Peshkin points out that a scientific theory is one that is well established by many telling experiments. It is not merely a hypothesis or a speculation, as the word *theory* is often used in common language. He explains how all scientific theories are subject to improvements based upon new experimental observations. We could say simply that a scientific theory is the "best we can do" logically, given the experimental data.

It is important to understand the distinction between "revealed truth" and "discovered truth." Although both may play a valid role, history is filled with examples of a discovered truth overturning the reigning interpretation of revealed truth. The classic example is Galileo, and his claim that the Earth revolves around the Sun, versus the Inquisition. His research and teaching on that subject led the Catholic Church to decree in 1616 that the Sun-centered Copernican view was "false and erroneous," a decree that was not officially rescinded by the Church until 1992.

Questions to Ponder

1. If science is only about observing the world with your own eyes, something all humans have done from time immemorial, why is that a radical idea in the history of thought?
2. What other great ideas can compete with the experimental method for being the greatest thought of the past 1,000 years? The past 5,000 years?
3. Recently, some scholars have questioned the validity of the dominant scientific view or method, on the grounds that it is a product of a particular culture and not free of bias. Although perhaps true, does this criticism undercut the scientific enterprise? Is the reverse also true—that respected sources of revealed truth are also interpreted culturally?
4. A recent poll for *Seed Magazine* found that more Americans are interested in science than in sports or religion. [2] The subjects that were found to be of higher interest than science are: current events, music, television, and technology. Surprised?

Terms to Research

creationism, empiricism, Galileo, geocentric view, intelligent design, inquisition, paradigm, relativism, Roger Bacon, Thomas Kuhn

REFERENCES

1. Peshkin, Murray. "Addressing the Public about Science and Religion."*Physics Today,* July 2006, 46–47.
2. Seed Media Group, http://www.seedmediagroup.com.

SOURCES

Sagan, Carl. *Cosmos*. New York: Random House, 1980.

SUGGESTED READING

See the general physics references given at the end of Chapter 1.

A highly readable account of how the Jacquard loom began the age of programmable machines and computers:
Essinger, James. *Jacquard's Web*. Oxford, U.K.: Oxford, 2004.

KEY TERMS

Acceleration
Accelerometer
Atoms
Chemical energy
Conversion factor
Distance
Electron
Energy
Force
Force vector

Friction
Gravitational energy
Joule
Kinetic energy
Mass
Mechanical energy
Net force
Neutron
Newton's first law
Newton's second law
Newton's third law
Potential energy
Power
Proton
Speed
Temperature
Theory
Thermal energy
Unit
Watt
Work

ANSWERS TO QUICK QUESTIONS

Q3.1 In a time interval of 1 μsec, the electron would travel a distance
$D = S \cdot t = (2 \times 10^7 \text{ m/sec}) \cdot (1 \, \mu \text{ sec}) = (2 \times 10^7 \text{ m/sec}) \cdot (1 \times 10^{-6} \text{sec}) = 2 \times 10^1 \text{ m} = 20 \text{ m}$

Q3.2

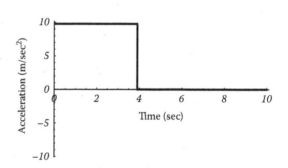

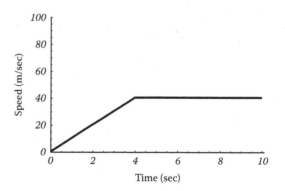

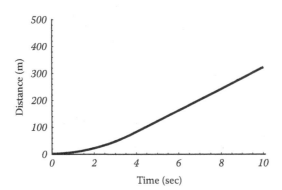

Q3.3 A jet engine takes in air, heats it, and accelerates it at high speed out the back. The engine exerts force on the air to accelerate this air. According to Newton's third law, the air therefore imparts an equal and opposite force to the engine, accelerating the car forward.

Q3.4 Other examples in which thermal energy is used to do work include a steam-powered turbine such as found in nuclear energy plants, a rising hot-air balloon, and an egg being jostled by boiling water.

Q3.5 When she kicks the ball, a soccer player converts stored chemical energy in muscles into kinetic energy in her leg and foot, which is transferred to the ball through direct contact. During the time that the foot and ball are in contact, the foot does work on the skin of the ball, temporarily compressing the ball to a smaller size, and storing potential energy in the compressed air. As the compressed ball expands against the foot, the potential energy in the compressed air gets converted into kinetic energy of the moving ball. As the ball rises up against the force of gravity, its kinetic energy decreases. Friction between the ball and the air also causes the ball to lose kinetic energy. It slows down as it reaches its peak height, where much of the ball's kinetic energy is now stored in gravitational potential energy. Then gravity does work on the ball, pulling it toward Earth and converting its potential energy back into kinetic energy.

Q3.6 The force of gravity on a 50-kg skateboarder equals (9.8 N per kg) · 50 kg = 490 N. To push him up to a height of 2 m require $Work = Force · Distance = 490 \text{ N} · 2 \text{ m} = 980 \text{ J}$

Q3.7 100 W × 90 min = 100 J/sec × 90 min × (60 sec/min) = 540,000 J = 540 kJ

Q3.8 The phrase, "power consumption," is not valid because energy is conserved. Therefore, energy and power cannot be "consumed," a word that seems to imply that the energy is destroyed. A better phrase would be "power conversion" or "power usage."

EXERCISES AND PROBLEMS

Exercises

E3.1 From the data and graphs given in Figure 3.5, estimate the acceleration, speed, and distance of the jet car at time = 2.5 sec.

E3.2 If a jet car experienced acceleration as shown below (and no friction), sketch graphs of its speed and distance. There is no need to be numerically accurate for graphing the distance.

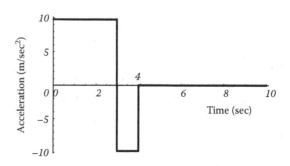

E3.3 If the jet car experienced the same accelerating forces as shown in the top graph in Figure 3.5, except that, starting at time = 6 sec, the car started dragging a tail pipe, causing a lot of friction between the car and the road, how would the three graphs be modified after time = 6 sec?

E3.4 In Figure 3.9, a wind gust changes the north-to-south speed of a puck, and the east-to-west speed.

 (a) Using a drawing, give an example of a wind gust that would change only the north-to-south speed.
 (b) Give an example of a wind gust that would change only the east-to-west speed.

E3.5 Two astronauts, Alice and Bob, are floating freely in space, with negligible gravity forces exerted on them. Bob's mass is 160 kg and Alice's mass is 80 kg. They place their palms against each other's palms and push away from each other with a steady force for a 1-sec interval. After this push, Bob is moving away from his original position at a speed 1.0 m/sec.

 (a) Compare the force that Alice feels from Bob relative to the force Bob feels from Alice.
 (b) What is Alice's final speed? *Hint*: Compare Alice's acceleration to Bob's speed?

E3.6 Three smooth rocks with equal mass sit on smooth ice, where friction is so little that we can ignore it. A certain force, F_0, is applied to one of the rocks, resulting in the rock accelerating at a rate 0.5 m/sec^2. If the three rocks are tightly tied together by a string to make a single object with three times the mass of one rock, and the same force F_0 is applied to one of the rocks, what is the acceleration?

E3.7 Gravity's force on an object near Earth's surface is proportional to the object's mass. Consider two identical blocks of metal. If the two are dropped at the same time, they obviously accelerate at the same rate and hit the ground at the same time. Now consider the two blocks being glued together and dropped as a single object. Compare the acceleration rate in this case to the case when they are separate (ignore air friction). Explain.

E3.8 Gravity's force on an object is proportional to the object's mass ($F_{GRAVITY} = C \cdot M$, where C is some constant), whereas the object's acceleration is proportional to the force applied to it ($a = F_{GRAVITY} / M$). Therefore, in the absence of friction or air resistance, all objects accelerate at the same rate, equal $a = C = 9.8$ m/sec². Discuss what would happen if you dropped four objects from the top of a tower (say the Leaning Tower of Pisa) at the same time: (1) a stone ball with a diameter of 4 mm and a mass 0.1 g, (2) a plastic ball with a diameter of 10 mm and a mass of 0.1 g, (3) a plastic ball with a diameter of 1 mm and a mass of 0.01 g, and (4) a feather with a mass of 0.1 g. Consider two separate cases: (a) the objects fall through air, or (b) the objects fall through a perfect vacuum.

E3.9 Refer to Figure 3.16. During the time the skater is initially moving horizontally to the right at nearly constant speed, what forces, including gravity, are acting on the skater and his board? Using arrows, indicate the direction of each force.

E3.10 Refer to Real World Example 3.4 on motion sensors. Draw a picture of a laptop on a person's lap in a seat in an airplane (a schematic is fine). Draw the picture for the two cases below, showing the relative positions of the test mass and the body of the accelerometer in each:

  (a) The airplane is in steady forward motion with constant speed.
  (b) The airplane is in deceleration.

E3.11 The graph below shows the height versus horizontal position of a large skate-boarding ramp. Assume that there is essentially no friction in the following cases:

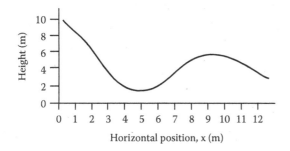

  (a) If a ball is released at a height of 10 m (horizontal position $x = 0$), how far will it roll in the horizontal distance?
  (b) If a ball is released at a height of 4 m (horizontal position $x = 3$ m), how far will it roll?
  (c) For the case that the ball is released at a height of 4 m, make three graphs of the type in shown in Figure 3.22, one for kinetic energy, one for potential energy, and one for total energy.

E3.12 Only differences of potential energies are meaningful, not the values of energies themselves. For example, in the figure, the moving skater starts at the level on the left, which we call zero potential energy for reference, then he rolls up to the top of the ramp, where the potential energy is 1 kJ. The difference, or change, of potential energy is +1 kJ. Then he goes down a second ramp and bottoms out at 2 kJ lower than the maximum at the top of the ramp. This means the energy at the bottom of the second ramp equals -1 kJ, that is, a negative number. Assume there is no friction. Carefully redraw this figure, and draw below it two more figures, one graphing kinetic energy and one graphing total (potential plus kinetic) energy.

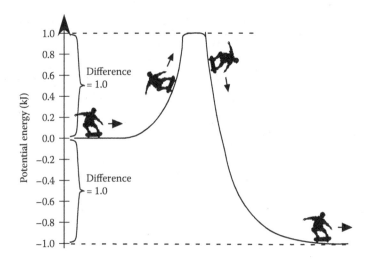

E3.13 Explain the nature of thermal energy and why it is considered a form of energy. For your discussion, think of and use at least one example of conversion of thermal energy into another form that is not mentioned in this chapter.

E3.14 A simple analogy that is helpful for understanding the units of energy and power is the following: The hens in an egg factory produce eggs at a rate of 3,600,000 eggs per hour, called the "production rate." Say that 3,600,000 eggs just fit into one delivery truck. Call this many eggs one "truckload" of eggs. Therefore, the factory produces one truckload per hour. In 24 hr of time, the factory produces 24 truckloads.

(a) How many eggs are produced in 24 hr?
(b) Develop a close analogy of the above scenario in terms of a power generating station, using the following correspondence: 1 egg ↔ 1 J of energy. What is the energy-related correspondence to the egg production rate? What are the energy-related units for this quantity?
(c) What is the energy-related correspondence to one truckload of eggs? What are the energy-related units for this quantity? *Hint*: Consider watts and kilowatt-hours.

E3.15 Use the method of conversion factors to perform conversions. *Hint*: One of the given examples is impossible.

(a) How many centimeters in 3.2 m?   (b) How many centimeters in 5.8 km?
(c) How many millimeters in 0.5 km?   (d) How many meters in 2.0 mm?
(e) How many millimeters in 2.35 m?   (f) How many seconds in 1 mm?
(g) How many seconds in 3.6 hr?   (h) How many seconds in 2 msec?
(i) How many meters per second in
    4.4 mi/hr?

E3.16 To convince yourself that using decimal numbers and metric units really is easier than using fractions and English units (now used almost exclusively by Americans), do the following conversions, which refer to very similar problems:

Hint: 1 mi = 5,280 ft.

(a) How many millimeters are there in 4.890449 km?
(b) How many one-sixteenth inches are there in a distance equal to 3 mi, 204 ft, and 9 3/8 inches?

E3.17 Roughly estimate how many toothpicks can be made from one 50-m tall Douglas fir tree? List all of your assumptions.

(This website gives many fun estimates of numbers of things: http://hypertextbook. com/facts/index-topics.shtml.)

E3.18 In Real World Example 3.2 about CRTs, the diagram shows the method for deflecting the electron beam only in the up-down direction. Design and describe a system that would allow creating two-dimensional (up-down and left-right) pictures on the screen.

Problems

P3.1 A motorcycle starts from rest and at time = 0 begins moving with constant acceleration equal to 1 m/sec^2. After 9 sec, the acceleration suddenly drops to zero. Assume there is no friction between track and motorcycle and no air resistance.

 (a) At time = 9 sec, what is the cycle's speed?
 (b) Make graphs of the acceleration and speed.
 (c) During the time interval between 9 and 14 sec, how far does the cycle travel?

P3.2 In the example in Figure 3.14, the astronaut, whose mass including her suit equals 100 kg, exerts a constant force of 10 N for a time 0.5 sec on the wrench, whose mass is 0.5 kg. What is the final speed of the wrench and her final speed in the opposite direction?

P3.3 (a) If a hard disk head (see Real World Example 3.1) is moving with a steady speed of 2 m/sec, how long will it take to move between two tracks separated by 13 mm?

(b) If a hard disk head starts at rest and then undergoes a constant acceleration of 1,100 m/sec^2 for a time interval of 3 msec, how fast will it be going at the end of this interval?

(c) (Optional—more quantitative) In P3.3b, how far will the head have traveled?

P3.4 (a) If electrons in a CRT (See Real World Example 3.2) are accelerated to a speed of 1.4×10^7m/sec, how long would it take them to travel 10 cm in a small television tube?

(b) The acceleration of the electrons is provided by an attractive force applied on the electrons, pulling them from the cathode toward a hole in the anode. Assume the acceleration has a constant value a and is applied for a time interval of 3 nsec. What value of a is required to achieve an electron speed of 1.4×10^7 m/sec?

P3.5 (a) Consider a skater on a horizontal cement surface initially moving to the right at nearly constant speed. She then runs head-on into a pile of four soft feather mattresses stacked front-to-back vertically against a brick wall, which compress and stop her. What forces, including gravity and the force of contact with the ground, act on the skater during these events? Give the direction of each force and describe the accelerations they cause. Also, explain the main forms of energy present during these events and how they convert from one to the other.

(b) Consider a skater on a horizontal surface initially moving to the right at nearly constant speed. She then runs head-on into a pile of four mattress box springs stacked front-to-back vertically against a brick wall, which compress and stop her, then spring back and send her in the opposite direction. What forces, including gravity, act on the skater during these events? Give the direction of each force, and describe the accelerations they cause. Also explain the main forms of energy present during these events and how they convert from one to the other.

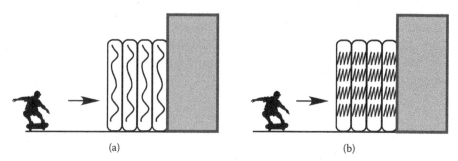

(a)   (b)

P3.6 Complete each diagram by drawing the net force vector:

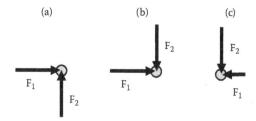

P3.7 (a) Draw the speed and distance graphs corresponding to the acceleration graph shown in case (a) in the figure below. Label the time axes as in the acceleration graph. For each speed graph, carefully determine the proper numerical values of speed at each time. For each distance graph, there is no need to get the numerical values exactly right—make estimates to the best of your ability.

(b) Do the same for case (b). *Hint*: Negative acceleration means deceleration, and when speed apparently becomes negative, we give it a different name—velocity. Negative speed means moving in the opposite (backward) direction.

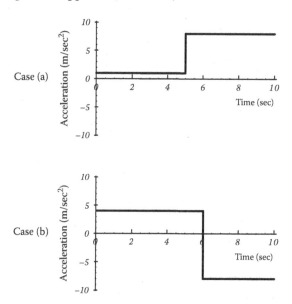

P3.8 Draw the acceleration and distance graphs corresponding to the speed graph shown below. Label the time axes as in the acceleration graph. For the speed graph, carefully determine the proper numerical values of speed at each time. For the distance graph, there is no need to get the numerical values exactly right—make estimates to the best of your ability. *Hint*: Negative speed (or velocity) means moving in the opposite (backward) direction.

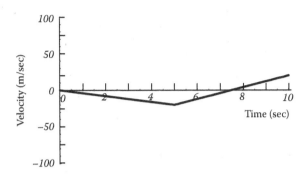

P3.9 A possible way to gather and transport energy for the future automobile transportation system would be to construct thousands of wind turbines in a remote area near the Aleutian Islands between Alaska and Russia. The wind could turn windmills, which could generate electricity to power a chemical factory, the purpose of which is to break ocean water into molecules of hydrogen and molecules of oxygen. The hydrogen gas could be transported in pipelines to the U.S. mainland, where it could be used in fuel cells to power cars. Make a list of all of the sources, forms, and transformations of energy that occur in this scenario.

P3.10 (a) McDonald's has sold millions of hamburgers over the years. If we assume each burger to be 0.5-in. thick and incompressible (they do not collapse when pressed together), roughly how many burgers would it take to reach the moon if they were stacked on top of each other? The moon is 3.84×10^8 m from the Earth.

(b) The Sun is about 1.5×10^{11} m from the Earth. The speed of light is 3×10^8 m/s. How many seconds does it take light from the Sun to reach the Earth? Use the method of conversion factors to convert your answer into minutes.

P3.11 We measure the area of a large room. Say that we pace it off and find that it is 15 paces by 26 paces, where one pace is equal to 80 cm (or 0.8 m). Say that we want to stack a lot of bricks in this room, and that each brick has a length of 15 cm, a width of 10 cm, and a height of 7 cm.

(a) If you line up bricks end-to-end (lengthwise), how many bricks could you fit in a single line along the floor in the 26-pace direction of the room?
(b) If you tile the whole room floor with bricks (one layer only), how many bricks will fit?
(c) If you add more layers or bricks to this first layer, so the whole stack of bricks is about one story, or 4 m high, how many bricks will fit in total?

P3.12 Perform the following calculations. Write your answers in scientific notation and keep track of the units. Write your answers in the units that are specified in brackets, [ ].

Example: Speed of sound in air: (20,580m)/(1 hr), units [km/sec]

$$\frac{20{,}580m}{hr} = \frac{20{,}580 \; \cancel{m}}{\cancel{hr}} \left(\frac{1 \; km}{1000 \; \cancel{m}} \right)_{CF} \left(\frac{1 \; \cancel{hr}}{3{,}600 \; sec} \right)_{CF} = 0.005717 \; km \,/\, sec$$

$$= 5.717 \times 10^{-3} km \,/\, sec$$

(a) Speed of sound in water: $(7.5 \times 10^8$ mm$)/(5 \times 10^2$ msec), units [m/sec]

(b) Processing rate of a Pentium 4 processor: $(32$ bits$) \times (3 \times 10^9$/sec), units [bytes/sec]

(c) Population density of Alaska and New York:
Alaska: $(6.269 \times 10^5$ people$)/(570{,}000$ mi$^2)$, units [people/mi^2]
New York: $(1.898 \times 10^7$ people$)/(47{,}000$ mi$^2)$, units [people/mi^2]

(d) Average number of atoms making a human: $(70$ kg$)/(2 \times 10^{-23}$ g/atom$)$, units [atoms]

(e) National debt per person: $(7.025 \times 10^{12}$ dollars$)/(293$ Mpeople$)$, units [dollars/person]

P3.13 Refer to Real-World Example 3.4 on motion sensors. The accelerometer body and test mass are at a standstill, and then the body begins accelerating with a constant acceleration of 2 gee. (1 gee equals 9.8 m/sec^2.)

(a) After a time of 1 sec, what is the relative speed of the body and the test mass?

(b) (Optional—more quantitative) After a time of 1 sec, how far has the body moved relative to the test mass, assuming the test mass is stationary during this time? *Hint*: See In-Depth Look 3.1.

P3.14 (a) The electrical energy used by households is usually measured in units of kilowatt · hours (kW · hr). One kilowatt · hour equals the amount of energy used in 1 hr if the power is 1kW. Recall that 1J = 1 W · sec. Use this fact, and conversion factors, to determine how many joules of energy are contained in 1 kW · hr.

(b) Suppose that by mistake you leave five 60-W light bulbs turned on at your house for 24 hr. Say that the cost of electricity is 3 cents per kW · hr. How many Joules are used by the five light bulbs over the 24 hours? From this, calculate the cost (in dollars) for this mistake.

(c) If the typical household uses, on average, 2 kW of electrical power constantly (this is equivalent to leaving two toasters running continuously), what would the electrical bill be each month? Use the estimate that 1 kW · hr costs three cents.

(d) Again assuming the typical household continuously uses 2 kW of electrical power, determine the number of homes that can be served by a power plant putting out 100 MW of electrical power.

P3.15 An incandescent filament light bulb converts about 5% of the electrical power it uses into visible light. The rest goes into heat (i.e., thermal energy). A fluorescent bulb is typically 4 times more efficient than an incandescent bulb. For example, a fluorescent bulb using 15 W of electrical power makes as much light as an incandescent bulb using 60 W of power. If electrical energy costs 9 cents/kW · hr, how much money would you save by using one of these fluorescent bulbs instead of one incandescent for 1 yr?

P3.16 In 2005 and 2006 Microsoft, Yahoo, and Google constructed large facilities called server farms along the Columbia River between Oregon and Washington, where cooling water and electrical power are relatively low cost. Each farm contains a very large number of computers, or servers, for processing the huge numbers of Internet "hits" that each company serves. Although details are secret, it has been estimated that a typical server farm continuously uses around 50 MW of electricity.

(a) If all of this power were to be used to run servers, each using 200 W, how many servers would be present?

(b) If Microsoft pays only 2 cents/kW · hr instead of the national average of around 8 cents (in 2006), how much would did they save in 1 yr?

4

Matter and Heat: Cooling Computers Is Required by the Physics of Computation

Motion is the very essence of what has hitherto been called matter.

Lord Kelvin
(Sir William Thomson)

The bright dots are individual silicon atoms sitting on the surface of a silicon crystal. (Courtesy of Franz Himpsel, University of Wisconsin–Madison.)

Ludwig Boltzmann, shown at age 24, was a strong proponent of the atomic nature of matter. He died in 1905, without knowing that in the same year Albert Einstein published a paper proving the existence of atoms. (With permission of the Austrian Central Library for Physics, Vienna University.)

4.1   FROM STEAM ENGINES TO COMPUTERS

Computer chips are made of semiconductor crystals—a regular arrangement of atoms. In later chapters we will discuss how electricity moves through such crystals. In this chapter we discuss the different forms of matter—gas, liquid, and solid—and how their properties, such as pressure and temperature, can be understood. We discuss heat, which means the transfer of **_thermal energy_** from one place to another. Understanding heat is important for understanding how computer chips are cooled. All machines obey certain physical laws related to heat, called the laws of thermodynamics. These tell what the limits are to extracting useful work from heat. A computer is a machine that moves bits of information around, so the laws of thermodynamics ultimately limit its performance. It turns out that cooling computers is required by the physics of computation.

4.2   MATTER AND ATOMS

The everyday objects and materials around us are made of matter. In ancient times, philosophers did not know if matter was a continuous substance or was made of **_atoms_**— tiny lumps of mass that determine a material's properties. We now have indisputable evidence that matter is made of atoms. Until recently, scientists had to rely on indirect evidence of atoms' existence. Still, that evidence was overwhelming, and after 1905 or so, all scientists accepted atoms as being real. Since the 1980s, physicists have been able to construct detailed maps (like a topological map used for hiking in the mountains) of the surfaces of crystals, in which each small bump represents a single atom. The image on the first page of this chapter shows such a map of a silicon crystal that has been specially prepared by making tiny grooves where extra silicon atoms can rest. Images such as this are not made using a microscope, but by slowly passing a metal needle across the surface and measuring how much electric current passes through the needle to the surface at any location. The data are then assembled into a topological map of the surface.

Since 1980, we have been able see single atoms and to record their images with a video camera. **Figure 4.1** shows such an image of a single cadmium ion (an atom with one of its electrons stripped away), taken with an ultraviolet-sensitive camera. The distance between the two needle tips is about 0.1 millimeters, and the small spot between the tips is the image of ultraviolet (UV) light reflected from a single cadmium atom when a laser beam is shined on it. The spot of light in the image is about 3,000 nanometers across, an area much larger than the actual size of the single atom (under 1 nm). No camera lens is capable of resolving an image much better than this, so the image

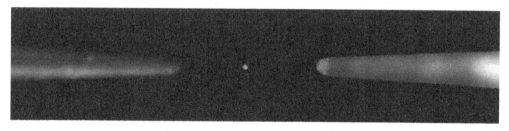

FIGURE 4.1   A single electrically charged cadmium atom photographed in UV light. It is held by electric and magnetic forces between the tips of two sharp needles. (Courtesy of Christopher Monroe, University of Maryland.)

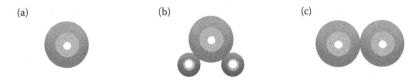

FIGURE 4.2 (a) A single oxygen atom represented as a more or less spherical object. (b) A water molecule, H_2O, consists of two hydrogen atoms bonded to one oxygen atom. (c) An oxygen molecule consists of two bonded oxygen atoms.

is blurred. Nevertheless, it would have seemed astounding to those early philosophers that individual atoms can indeed be seen directly.

A *molecule* is a structure made of two or more atoms held together by electrical forces. For example, water (H_2O) is made of two hydrogen atoms and one oxygen atom, as in **Figure 4.2**. Typical sizes of molecules are 1–10 nanometers. A molecule is the smallest entity possessing the properties of the substance that it forms. For example, water is wet, but if all of the molecules of water are split into their hydrogen and oxygen atom constituents, the resulting substances are not wet.

4.3 GASES, LIQUIDS, AND SOLIDS

The four basic types of substances are *gases*, *liquids*, *noncrystalline solids*, and *crystals*. Examples of gases are helium, steam, and air (a mixture of oxygen and nitrogen molecules). Examples of liquids are water, methyl (rubbing) alcohol, and mercury. Examples of noncrystalline solids are glass, plastic, and rubber. Examples of crystals are diamond, quartz, and silicon. These are illustrated in **Figure 4.3**. In a gas, the atoms or molecules are free to move in nearly straight lines until they encounter another atom or the wall of a container, where they bounce off. In a liquid, the atoms or molecules are tightly packed, so they are in nearly continuous contact, but they are free to gradually move in a jostling motion. In a noncrystalline solid, such as glass, the atoms are packed tightly and randomly. They jiggle, but cannot move far. In a crystal, the atoms are held rigidly in a regular pattern.

Figure 4.4 shows microscopic, three-dimensional views of a crystal and a liquid. A crystal can be represented by a simple model in which the atoms are held near each other by connecting springs. A real solid has forces between neighboring atoms that act in much the same way as springs—if two atoms get too far apart they feel an attraction to one another, whereas if they get too close they feel repulsion. The atoms are always vibrating—moving toward and away from the neighboring atoms. Such vibrations

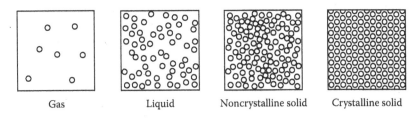

| Gas | Liquid | Noncrystalline solid | Crystalline solid |

FIGURE 4.3 The four basic types of substances and their atomic arrangements. Each circle represents one atom or molecule. There are many different ways to arrange atoms in crystals in regular patterns. The figure on the right shows only one such way.

(a)   (b)

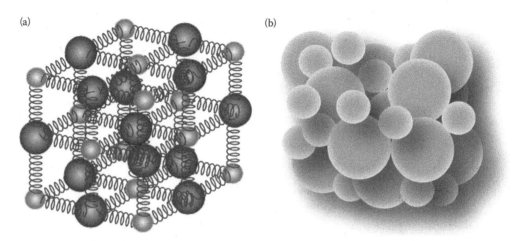

FIGURE 4.4   Models of (a) a crystal and (b) a liquid. In a crystal, the forces between atoms are imagined as springs that hold atoms in an orderly arrangement while allowing them to vibrate within. In a liquid, atoms are tightly packed together and can slide from one place to another. This causes a jumbled, changing arrangement of atoms in the liquid.

contain both kinetic and *potential energy*. The motion of an atom represents *kinetic energy*. The compression or stretching of a spring represents potential energy; that is, a compressed or stretched spring stores *energy*, which can be used later to do work. A liquid can be described by a model in which the atoms are like compressible rubber balls that are tightly packed together. There are no fixed springs connecting them, allowing the atoms to move from place to place in the liquid, all the while vibrating toward and away from their neighbors.

THINK AGAIN

Not all substances fall neatly into one of the categories: gas, liquid, or solid. Consider pudding, Silly Putty, or quicksand.

IN-DEPTH LOOK 4.1: SIZE AND NUMBERS OF ATOMS

How many atoms are there in a tiny salt crystal, 0.1 millimeter on a side? The size (diameter) of atoms ranges from about 0.1 nanometers to about 0.5 nanometers, depending on the type. The volume of a single atom is therefore about

$$(0.1 \text{ nm})^3 = (1 \times 10^{-10} \text{ m})^3 = 1 \times 10^{-30} \text{ m}^3$$

The volume of the salt crystal is

$$(0.1 \text{ m})^3 = (1 \times 10^{-4} \text{ m})^3 = 1 \times 10^{-12} \text{ m}^3$$

So, the number of atoms in the salt crystal equals about

$$number\ of\ atoms = \frac{volume\ of\ crystal}{volume\ of\ atom} = \frac{1 \times 10^{-12}\,m^3}{1 \times 10^{-30}\,m^3} = 10^{-12+30} = 10^{18}$$

This is an enormous number—a billion-billion. Think about that the next time you salt some french fries.

The typical speed of a molecule in air at room temperature is about 500 meters per second (m/sec). This is comparable to the speed of a sound impulse (about 345 m/sec) from, say, a thunderclap, but slower than the fastest speeding bullet (up to 1,500 m/sec). While moving rather fast, a typical atom in a gas does not travel far before striking another atom, deflecting or reversing its path. For example, this is why it takes a significant time for an odor, which is carried by molecules that we can smell, to travel from one side of an air-filled room to another.

In a liquid, the atoms are continually colliding, like a mob of people in a crowded store on a sale day. Imagine you are in that mob, and it is so packed with people that you cannot go in the direction you want to. So, you decide to just let the crowd buffet you and see where you end up. First, you are pushed to the east 3 feet, then to the north 4 feet, then east 2 feet, then south 5 feet, etc. After a long while, you end up some 30 feet from where you started, but at an unpredictable location in the store. This kind of random drifting motion from being pushed in many small steps is called *diffusion*.

Figure 4.5 illustrates diffusion. The four pictures shown can be thought of as frames in a movie—snapshots showing the locations of all of the atoms (people in a mob), including one color-coded in black (you). At the starting time ($t = 0$), the black-coded atom starts near the center of the container of liquid. As time progresses in 1-sec intervals, all of the atoms are seen to jostle and move, and the black-coded atom gets pushed to an unpredictable location.

Motion of this type was first seen in 1827 by Robert Brown, a botanist, who saw under a microscope that tiny grains of pollen (dust, or soot particles) in still water were moving around randomly, as if doing a wild, unpredictable dance. We now call motion of this type Brownian motion. Clearly, the particles were not propelling themselves; they must have been pushed by some agents too small to be seen using a microscope. A detailed mathematical analysis by Albert Einstein, published in 1905, showed conclusively that the tiny invisible agents must be atoms or molecules making up the water itself. Remarkably, until Einstein published this analysis, many scientists were still skeptical of the existence of atoms. So, it was only in the twentieth century—not so long ago—that the existence of atoms was fully accepted.

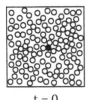

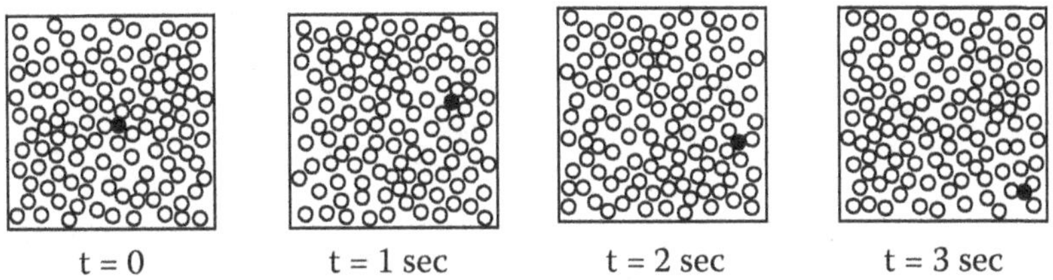

t = 0   t = 1 sec   t = 2 sec   t = 3 sec

FIGURE 4.5 The atom labeled by the dark shading is pushed in a random manner by many collisions with other atoms. After a while, it happens to diffuse to the lower right corner of the container. Later, it may diffuse to any location.

QUICK QUESTION 4.1

A plastic, helium-filled balloon, with no flaws that would cause leakage, will fairly soon lose its helium and deflate. Explain this in terms of the concepts discussed above. *Hint*: Helium forms one of the smallest possible molecules. Oxygen molecules are much larger. Would an oxygen-filled balloon deflate faster or slower than a helium-filled balloon?

REAL-WORLD EXAMPLE 4.1: GROWING SILICON CRYSTALS FOR COMPUTER CHIPS

A nearly perfect silicon (Si) crystal is needed to begin the making of the electronic circuit in a computer chip. These are called integrated circuits, or ICs. Crystals for making ICs are grown by a natural process in which atoms find their way from a liquid onto the surface of a crystal, which constantly increases in size. You might be familiar with crystal growth from grade-school days, where a common experiment is to grow sugar crystals from a highly concentrated solution of sugar in water. A string is placed into the solution and small crystals grow onto the surface of the string. To make ICs, we need to grow large Si crystals, up to 16 inches in diameter. The surface of a crystal of this size can hold hundreds of ICs, which are then separated by cutting the crystal.

The common method to grow large, pure Si crystals is to start with clean sand (silicon dioxide, SiO_2). The sand is purified by chemical reactions and then melted at a high temperature in a furnace to produce a liquid of SiO_2 that is pure to a few parts per billion. A large crystal is grown from this hot liquid by starting from a small Si crystal, called a seed crystal. This seed is grown by other techniques. As in **Figure 4.6**, the seed is lowered into the melted SiO_2, and the seed begins to grow larger by the growth of new material onto its surface. The atoms that attach themselves to the seed crystal do so in a way that maintains the perfect regularity of the seed crystal structure. As the seed crystal grows larger, it is slowly pulled away from the liquid SiO_2, while still maintaining contact. A typical growth rate is a few millimeters per minute. This leads to a long, cylinder-shaped Si crystal. This method of "pulling" a crystal is called the Czochralski method, after its inventor. Jan Czochralski, a Polish chemist, discovered the method in 1916 when he accidentally dipped his pen into a crucible of molten tin rather than his inkwell. He pulled his pen out to discover that a thin thread of solidified metal was hanging from it.

After the crystal is grown and cooled, a diamond saw is used to slice it into thin wafers 0.5 to 0.7 mm thick. Each wafer surface is then polished to be extremely flat and smooth, as shown in **Figure 4.7**.

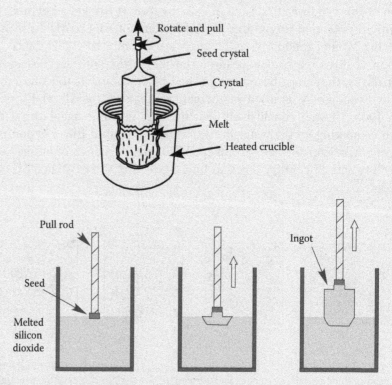

FIGURE 4.6   The method for growing silicon crystals.

FIGURE 4.7   Silicon crystals (ingots), and wafers sliced from an ingot and polished. (Courtesy of SUMCO Corporation, Japan.)

4.4   PRESSURE IN A GAS

If you are in a very tall elevator that begins to descend rapidly, sometimes your ears will "pop," or even feel pain. This happens because the air pressure changes with altitude. What, precisely, is *pressure*?

Pressure is the steady net force exerted on a single surface by a gas, liquid, or solid, divided by the area of the surface. As an equation:

$$pressure = \frac{net\ force}{area}, \quad \text{or} \quad p = \frac{F_{NET}}{A}$$

Net force means the total *force* exerted by all of the atoms or molecules striking just one side of an object. **Figure 4.8** shows the eardrum, with surface area equal to A (about 10 mm^2). When the molecules in the air (primarily oxygen and nitrogen) impinge on the eardrum, each momentarily exerts a force on its surface. The number of atoms that hit the eardrum's surface each second is a very large number, far greater than can be illustrated in the figure. As many molecules bounce from the surface, a nearly steady force is exerted on it. This means that any gas, liquid, or solid exerts a pressure on a surface that it contacts. Pressure is force per unit area.

The units of pressure are newtons (N) per square meter (m^2). That is,

$$pressure = \frac{net\ force}{area} = \frac{newtons}{square\ meter} = \frac{N}{m^2}$$

A force of 1 newton acting on an area of 1 square meter creates a pressure that is called 1 pascal, which is given the symbol Pa. That is,

$$1\ Pa = 1\ N/m^2$$

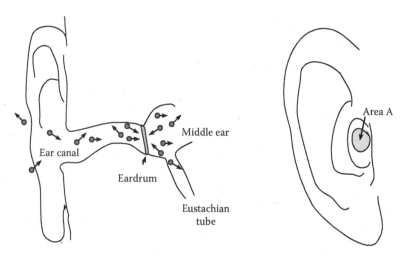

FIGURE 4.8 Rapidly moving air molecules in the ear canal create pressure on the outer surface of the eardrum. Air molecules in the middle ear create pressure on the inner surface of the ear drum. The actual number of air molecules is far greater than shown.

A pressure of 1×10^5 N/m², or 1×10^5 Pa, is called 1 **bar**, a unit for pressure used by the National Weather Service when giving the so-called barometric pressure. One bar is a force of 100,000 N acting on an area of 1 m². On a typical day, the pressure of the air equals roughly 1 bar. Air pressure at sea level typically does not vary more than a few percent. Air pressure decreases at higher altitudes, and at the summit of Mt. Everest (8,848 m) it is about one-third of that at sea level. Because the air is less dense, fewer atoms collide with the surface each second.

If we know the pressure p, we can calculate the force on a surface with area A by using the equation:

$$net\ force = pressure \times area, \quad \text{or} \quad F_{NET} = p \times A$$

For example, atmospheric pressure acting on your eardrum, with an area of about 1×10^{-5} m², creates a force on it equal to about:

$$net\ force = pressure \times area = (1 \times 10^5\,\text{Pa}) \times (1 \times 10^{-5}\,\text{m}^2) = (10^5\,\frac{\text{N}}{\text{m}^2}) \times (10^{-5}\,\text{m}^2) = 1\,\text{N}$$

There is a force on the outer surface of the eardrum from the air in the ear canal, and a force on the inner surface of the eardrum from the air in the middle ear. The total resulting force on the eardrum equals the difference of these two forces. In general, if both sides of an object are exposed to the same gas or liquid under the same conditions, then the forces on the two sides will cancel, and the total force on the object will be zero.

Next, consider the pressure of a gas that is held in a closed container. An example is air in an otherwise empty bottle whose cap is tightly sealed. The net force exerted on the inner surface of the bottle is proportional to the number of molecules that bounce from this surface per second. For any gas, that number is proportional to the total number of atoms, N, in the container divided by the total volume, V, of the container. We can express this relation as an equation:

$$pressure \propto \frac{number\ of\ atoms}{volume}, \quad \text{or} \quad p \propto \frac{N}{V} \text{ (for a gas)}$$

Empty a glass water bottle at the top of a 5,000-meter-high mountain, where the atmospheric pressure is about 0.5 bar, as in **Figure 4.9a**. Then tightly seal its cap and

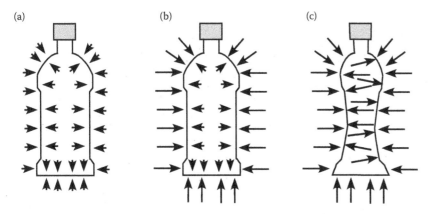

FIGURE 4.9 (a) A bottle is sealed at the top of a mountain, where the atmospheric pressure is about 0.5 bar, and in (b) and (c) is brought down to sea level, where atmospheric pressure is about 1 bar. The length of each drawn arrow indicates the amount of pressure at a point of the surface. If the bottle is rigid glass, as in (b), it maintains its shape. If the bottle is flexible plastic, as in (c), it becomes compressed.

pack it down to sea level, where the atmospheric pressure is 1 bar. The bottle is rigid, and its volume remains constant, as in Figure 4.9b, so the pressure in the bottle stays constant (assuming the temperature also remains constant). At sea level, there is a difference between the pressures inside of the bottle (low) and outside of the bottle (high).

Next, empty a plastic water bottle at the top of a 5,000-meter-high mountain, then tightly seal its cap and pack it down to sea level. You will observe that the bottle gets compressed, meaning its volume becomes smaller. Because the plastic is flexible, the bottle compresses to an extent such that the air pressure inside and outside of the bottle becomes nearly equal. The number of air molecules in the bottle did not change while coming down the mountain, and the bottle volume at the mountaintop is about twice its value at sea level (again assuming the temperature remains constant).

4.5 PRESSURE IN A LIQUID

Exerting a force on the surface of a solid or liquid will tend to compress it, as shown in **Figure 4.10**. Although gases are highly compressible, as in Figure 4.9, liquids are not. In a liquid, the atoms are already very close together, and cannot be pushed much closer by an outside force. Water, for example, is a nearly incompressible liquid. The consequence of applying an external force to a liquid is to increase its internal pressure without changing its volume significantly.

As compression occurs, the pressure inside the material and at its surface increases. In the case of a liquid, the increase of pressure is felt equally at all locations in the liquid. This effect is called Pascal's principle.

Pascal's principle: Pressure applied to one region of an enclosed liquid or gas increases the pressure at every location in the liquid or gas by an amount that is equal to the applied pressure.

As a consequence, pressure applied externally to the surface of an enclosed liquid or gas increases the pressure at every location in the liquid or gas by an amount equal to the applied pressure.

These properties of liquids are illustrated by the experiment shown in **Figure 4.11**, in which you apply an external force to a sliding piston inserted in one end of a sealed glass container full of water. Two small holes allow water to escape. The rate of escape and the angle of each resulting stream of water indicate how much pressure is in the

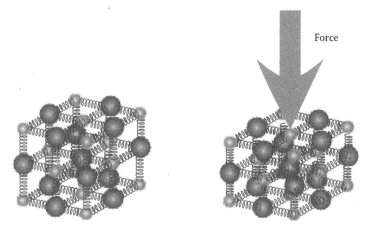

FIGURE 4.10 A model of a crystal, with atoms connected by springs, before and after an applied force compresses it.

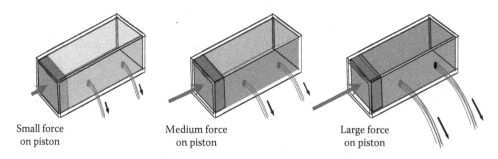

Small force
on piston

Medium force
on piston

Large force
on piston

FIGURE 4.11 A sealed glass container containing water and having a piston that exerts pressure on the water.

water immediately behind each hole. First, note that for each example shown, the two streams are identical, indicating equal pressure behind each hole, although the water is being compressed only from the end of the container. This illustrates Pascal's principle.

Second, notice that as the force applied to the piston is increased, the water pressure increases, as indicated by the darker color and by the angle of the streams. Also notice that the piston moves only slightly, because water is nearly incompressible.

Pascal's principle allows us to understand the transfer of pressure from one container to another. In **Figure 4.12**, a hollow pipe connects two containers, and a compression force is applied only to the container on the left. The pressures in both containers are equal, as long as there is not too much flow (gallons per second) through the connecting pipe.

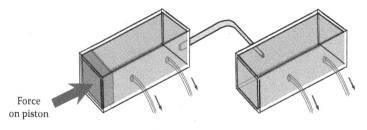

Force
on piston

FIGURE 4.12 Two sealed containers connected by a hollow pipe. External force applied to a piston in one container creates increased and equal pressure in both containers.

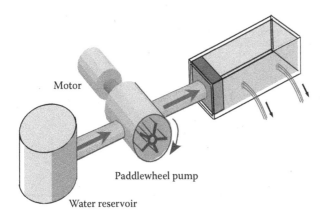

Motor

Paddlewheel pump

Water reservoir

FIGURE 4.13   A pump increases the pressure of water in a container.

4.6   PUMPS, CURRENT, AND RESISTANCE

An alternative way to increase the pressure of a liquid is by using a pump. A pump can be made with a spinning paddlewheel driven by a motor, as in **Figure 4.13**. The blades of the wheel impart forces to the water, pushing it into the container. The harder the pump pushes, the greater is the pressure in the container.

For liquids, we define **current** as the rate of flow, measured in liters (ℓ) per second. (1 ℓ equals 1.06 quarts.) The amount of current through a pipe depends on the pressure at the entrance to the pipe and on the internal diameter of the pipe. In **Figure 4.14**, a pump forces water to flow around a loop of hollow pipe. If the pipe diameter is large, the pressure is nearly equal everywhere in the pipe. If we partially fill the pipe with obstructions, as in part Figure 4.14b, they hinder or resist the flow of current. We say that the pipe now offers **resistance** to the flow. This lessens the water current in the pipe. The pressure drops from high (indicated by dark shading) at the pumped side of the obstructed region to low (light shading) at the other end of the obstructed region. A region containing resistance to flow is called a **resistor**.

The current leaving the resistor region in Figure 4.14b is equal to the current entering the resistor region. The pressure (P_2) at the resistor output is smaller than the pressure (P_1) at the resistor input. Some of the microscopic, random, kinetic energy of the water's molecules has been converted into thermal energy by friction at the obstructions, heating up the water and the pipe. The water with lower pressure has less ability to do work than does the high-pressure water.

P_1 P_2

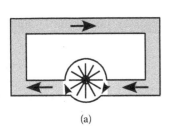

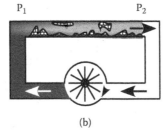

(a)   (b)

FIGURE 4.14   A pump forces a liquid around a loop of pipe. Arrows indicate the liquid flow. In (a) the pressure is nearly equal everywhere in the pipe. In (b) the pressure decreases as the liquid flows past the obstructions. Higher pressure is indicated by darker shading.

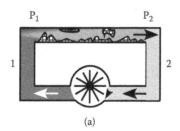
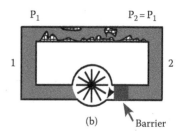

(a)   (b)   Barrier

FIGURE 4.15 In (a) the pressure decreases as the liquid flows past the obstructions. In (b) there is no flow of water and no drop of pressure.

THINK AGAIN

Earlier, we said that the current exiting the resistor region is equal to the current entering the resistor region. How do we know this is true?

A resistor, such as the pipe containing obstructions in **Figure 4.15a**, hinders the transfer of pressure from one location to the location at the other side of the resistor region. The decrease of pressure is proportional to the amount of resistance in the resistor and proportional to the amount of current. If there is no current, then there is no pressure drop. For example, in Figure 4.15b, a solid barrier is placed in the pipe at the entrance to the pump, preventing a steady flow of water. Pressure builds up at the barrier, equalizing the pressure everywhere. On the other hand, when the barrier is not present, as in Figure 4.15a, the pressure at location 2 does not build up to equal the pressure at location 1 because the pump removes water from location 2 faster than it can be resupplied by flow through the resistor.

The decrease of pressure when current passes through a resistor is described by Ohm's law for liquids.[1]

Ohm's law for liquids: The decrease of liquid pressure across a resistor region is approximately proportional to the amount of resistance in the resistor and proportional to the amount of current.

To express Ohm's law as an equation, use P_1 to denote the pressure of the liquid in the region entering the resistor region and P_2 to denote the pressure of the liquid leaving the resistor region. Denote the liquid current by the symbol I, whose units are liters per second (ℓ/sec). Then,

$$P_1 - P_2 = R \times I,$$

where the proportionality constant, R, is called the resistance of the resistor region.

As an example, assume that the current is 2 ℓ/sec and the resistance equals 0.3 bar · (sec/ℓ). (Recall that 1 bar = 1×10^5 N/m².) Then the change of pressure that takes place in going through the resistor equals

$$P_1 - P_2 = R \times I = (0.3\,\text{bar} \cdot \frac{\text{sec}}{\ell}) \cdot (2\frac{\ell}{\text{sec}}) = 0.6\,\text{bar}$$

Therefore, if the pressure of the liquid at the entrance to the resistor region is 2 bar, then the pressure exiting the resistor region equals 2 bar – 0.6 bar = 1.4 bar. Notice that if

---

[1] The more precise law for fluid flow is called Poiseuille's law, which describes smooth flow in pipes.

we were to double the current, the pressure change would double, because the pressure change is proportional to the current.

Another way to use Ohm's law for liquids is to predict the liquid current if you know the change of pressure. Say that the water pump can generate 10 bar of pressure at its output, and we know from previous measurements that the resistance of the resistor region is 0.3 bar · (sec/ℓ). If the input side of the pump is open to the atmosphere (1 bar) (i.e., the pressure there equals 1 bar), then we predict the current to be:

$$I = \frac{P_1 - P_2}{R} = \frac{10\,\text{bar} - 1\,\text{bar}}{0.3\,\text{bar} \cdot \text{sec}/\ell} = \frac{9\,\text{bar}}{0.3\,\text{bar} \cdot \text{sec}/\ell} = 30\,\ell/\text{sec}$$

QUICK QUESTION 4.2

Say that a water pump on the Space Shuttle in orbit can generate 4 bar of pressure at its output, which is connected through a pipe, with resistance equal to 0.4 bar · (sec/ℓ), into outer space, where the pressure is nearly zero. What is the flow current?

REAL-WORLD EXAMPLE 4.2: A WATER-PRESSURE-OPERATED COMPUTER

Although it would not really be practical, it should be possible to construct a simple computer that is operated by water. A computer is a device that processes digital bit values (1 or 0) in a way that represents logical reasoning. By combining many such simple operations, complex tasks are performed. Consider the water-operated valve in **Figure 4.16**. It consists of a control pipe and a source pipe that enter a junction containing a movable plug. If the water pressure in the control pipe is lower than the pressure in the source pipe, as in Figure 4.16a, then the valve is in the CLOSED position. A metal spring holds the movable plug, blocking any flow from the source to the drain. If the water pressure in the control pipe is higher than the pressure in the source pipe, as in Figure 4.16b, then the water in the control pipe pushes the plug down into the body of the junction, creating an open channel for water to flow through. The valve is in the OPEN position.

Figure 4.17 shows a device that processes bits, 1 or 0 (for review of binary digits, or bits, see Section 2.7). Pumps create pressure in four pipes. Two of these provide a steady pressure called the source pressure. The other two pipes provide pressure for controlling two valves connected to the source pipes. Two manual ON/OFF switches allow you to enter bit values 1 or 0 at each control junction. Think of moving these switches like punching buttons on a calculator; this is how you enter the data. When a switch is in the OFF position (representing a 0 bit value), low pressure appears in the attached control pipe, and the valve is closed. If you move a switch to the ON position (representing a 1), high pressure appears in the control pipe, and the valve opens, passing water and pressure from the source into the answer chamber. The pressure in this chamber determines the answer to the calculation.

One switch is called the "upper bit," and the other is the "lower bit." If you set both bit switches to OFF (both bit values equal 0), then the answer chamber will have low pressure, which we say represents a 0-bit value. If you set either bit to ON, representing a bit value of 1, the answer chamber will experience an increase of pressure to the same pressure as the source, representing a 1-bit value for the answer. If one switch is already

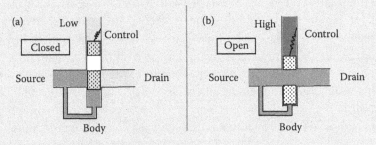

FIGURE 4.16 Water-controlled valve operation, with higher pressure indicated by darker shading.

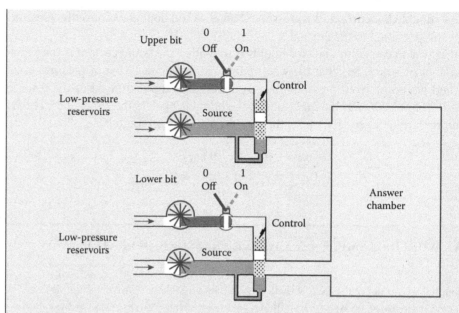

FIGURE 4.17 A water-pressure-operated computer.

ON, with the other OFF, and then you turn the other one ON, the pressure in the answer chamber does not change, because both source pressures are the same.

In this way, the water device computes the operation called the "OR operation." It simply says that if either the upper bit or the lower bit (or both) equals 1, then the answer equals 1. You can think of this as a computer for answering the question, "Should I carry an umbrella today?" Let us represent a YES answer by a bit value of 1 and a NO answer by bit value of 0. Let us represent the answer to the question "Is it cloudy outside?" by a bit value 1 (YES) or 0 (NO) at the upper bit. Let us also represent the answer to the question "Is it raining outside?" by a bit value 1 (YES) or 0 (NO) at the lower bit. Then our "computer" gives the following advice: If it is cloudy or if it is raining (or both), then YES, "you should carry your umbrella." Strange as it may sound, the only thing computers can do is to process bit values in ways similar to the manner illustrated here. By processing millions of bits, complex tasks can be carried out. It turns out that the water model given here is a good analogy for the operation of electronic transistors, which are the basis of real computers.

4.7 TEMPERATURE

A way to make a cork pop out of a bottle is to heat the gas inside.[2] Say we start with a corked bottle containing air at the same temperature and pressure as the atmosphere surrounding the bottle. Then, as shown in **Figure 4.18**, we gently heat the bottle and the air inside it, but not the surrounding air. As the gas heats, the speed of atoms inside the bottle increases, as indicated by longer arrows. Now each atom, when it bounces from the cork inside the bottle, exerts a greater force on the cork than does an atom of the outside air, which is traveling slower. Therefore, the pressure of the air inside of the bottle is greater than the pressure of the air outside of the bottle. The cork begins to move upward.

When we heat an object, its temperature increases. What is temperature?

---

[2] This could be dangerous, since the glass walls might shatter before the cork pops out.

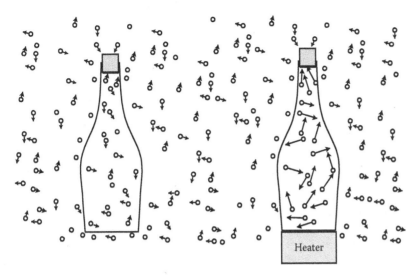

FIGURE 4.18 On the left, the gas inside and outside of the bottle has equal temperature and pressure. The net force on the cork is zero. On the right, the gas in the bottle has been heated, so the atoms move faster. Both the temperature and pressure increase inside. The net force on the cork now tends to push the cork out of the bottle.

Temperature is a measure of the internal kinetic-plus-potential energy of randomly moving atoms in a gas, liquid, or solid.

To state this mathematically, we need to decide what temperature scale we will use. The scale used internationally is the Celsius scale, in which temperature is given in degrees Celsius, denoted by °C. The Celsius scale is related to the Fahrenheit (°F) scale by the formula

$$T_F = (9/5)T_C + 32°F$$

This means that we multiply the temperature given in Celsius by 9/5 and add 32°F. For example, 0°C is the same as $T_F + (9/5)0 + 32°F = 32°F$, which is the temperature at which water freezes. As another example, 100°C is the same as $T_F + (9/5)100 + 32°F = 212°F$, which is the temperature at which water boils.

Water freezes at 0°C, but temperature can go far lower, of course. How low can temperature go? There is a lowest possible temperature, called **absolute zero**. On the Celsius scale, it equals –273.15°C. In experiments, gases can be cooled to within one-millionth of a degree of absolute zero, but it is not possible to go below absolute zero. This is the temperature at which the kinetic energy of atoms is zero. It would make no sense to try to go lower than that.

Scientists prefer to use another temperature scale, in which temperature equals zero at absolute zero. This is called the Kelvin scale, or degrees Kelvin, denoted K. The scale is named in honor of Lord Kelvin (Sir William Thomson). If the temperature using this scale is denoted by T_K, then T_K equals zero at absolute zero temperature. In the Kelvin scale, an increase of 1K is the same increase as 1°C. Therefore, when the Celsius temperature is 0, the Kelvin temperature T_K equals 273.15K. Water freezes at this temperature. Water boils at the Kelvin temperature $T_K = 273.15 + 100 = 373.15$K. In general, the Kelvin scale (K) is related to the Celsius (°C) scale by $T_K = 273.15°C + T_C$, or $T_C = 273.15°C - T_K$.

Table 4.1 gives some important temperatures in all three scales for comparison, and **Figure 4.19** compares the three temperature scales using the imagery of rulers or thermometers.

QUICK QUESTION 4.3

A laptop computer processor chip heats up to around 60°C when working hard. What is this temperature in Kelvin? in Fahrenheit? First use the picture of the rulers to estimate your answer, then calculate them using the given formulas.

TABLE 4.1
Typical Temperatures

Kelvin	Celsius	Fahrenheit	Notes
1811	1538	2800	Iron melts
1687	1414	2577	Silicon melts
373.15	100	212	Water boils*
323.15	50	122	Typical running computer chip
310.0	36.8	98.2	Average human body temperature
293.15	20	68	Room temperature
273.15	0	32	Water freezes/ice melts*
90.2	−183.0	−297	Oxygen becomes liquid
0	−273.15	−459.67	Absolute zero

* At standard pressure, 1.013 bar (1.03×10^5 N/m²), also called 1 atmosphere.

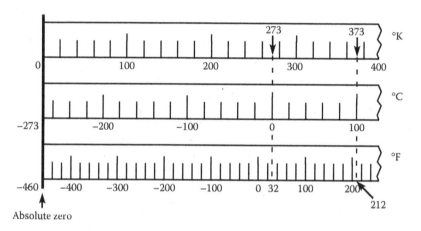

FIGURE 4.19 Temperatures shown on three different scales: Kelvin, Celsius, and Fahrenheit. The separation between tic marks on a scale indicates a change of 20° on that scale.

The average kinetic energy associated with the movement of atoms or molecules from place to place in a gas is proportional to the temperature of the gas in the Kelvin scale. If we denote the total kinetic energy of all atoms in the gas by KE, and temperature in the Kelvin scale by T_K, then we can write:

$$\text{kinetic energy} \propto \text{temperature}, \quad \text{or} \quad KE \propto T_K$$

We will not prove this proportionality relation. Instead, we could take it as a definition of temperature. An example of this relation is the case of a simple atomic gas, such as helium gas—most commonly found inside balloons at parties. By an atomic gas, we mean one that consists only of atoms that are not combined into molecules. In an atomic gas containing N atoms, one finds that the total kinetic energy is given by the formula:

$$KE = 1.5 \, N k_B T_K$$

The constant k_B is called **Boltzmann's constant**, after Ludwig Boltzmann, a physicist who in the latter 1800s developed the theory of random atomic motion and how it relates to temperature. The value of Boltzmann's constant is

$$k_B = 1.38 \times 10^{-23} \text{ J/K}$$

Here, J/K reads "joules per degree Kelvin."

As an example, consider a gas containing 10^{23} atoms (say, helium) at 273K, the temperature at which water freezes. The kinetic energy of this whole gas equals

$$KE = 1.5 \cdot 10^{23} \cdot \frac{1.38 \times 10^{-23} \text{ J}}{\cancel{K}} \cdot 273 \cancel{K} = 1.5 \cdot 1.38 \text{J} \cdot 273 = 565 \text{ J}$$

If this same gas were heated to the boiling point of water, its average kinetic energy would be

$$KE = 1.5 \cdot 10^{23} \cdot \frac{1.38 \times 10^{-23} \text{ J}}{\cancel{K}} \cdot 373 \cancel{K} = 772 \text{ J}$$

4.8 THE IDEAL GAS

Low-pressure gases behave simply, and can be described using a simple relation between pressure and temperature, called the ideal gas law. We have already discussed why pressure is proportional to the ratio of the total number of atoms to the volume of the enclosure; that is, $p \propto N/V$. We also know that pressure is proportional to the average force being exerted on the enclosure wall by atoms bouncing from it. Recall from Chapter 3 that a faster moving object, with the ability to exert a larger force when bouncing from a surface, has a larger kinetic energy. It should not be surprising then that pressure is proportional to the average kinetic energy $(\overline{KE})$ of the atoms in the gas. Because $\overline{KE}$ is proportional to temperature in the Kelvin scale, we infer that pressure also is proportional to temperature; that is, $p \propto T_K$.

These two proportionality relations are both contained in the relation called the ideal-gas law, which reads

$$p = \frac{N}{V} k_B T_K$$

This relation captures many different behaviors of gases. If the number of atoms N is increased, holding all else constant, the pressure goes up. If the enclosure volume V is increased, holding all else constant, the pressure goes down. If the temperature T_K is increased, holding all else constant, the pressure goes up. If both the number of atoms N and the volume V are doubled, holding all else constant, the pressure does not change, etc.

Let us continue the example above, involving a gas containing 10^{23} atoms (e.g., helium). At room temperature (20°C), this gas contained in a bottle with volume 1 liter, or 10^{-3} cubic meters, would have a pressure equal to

$$p = \frac{N}{V} k_B T_K = \frac{10^{23}}{10^{-3} \text{m}^3} 1.38 \times 10^{-23} \frac{\text{J}}{\cancel{K}} \cdot 293 \cancel{K} = 404,000 \text{ J/m}^3$$

Recalling from Chapter 3 that $1 \text{ J} = 1 \text{ N} \times 1 \text{ m}$, we can express this pressure as

$$p = 404,000 \frac{\text{J}}{\text{m}^3} \left(\frac{1 \text{N} \cdot \text{m}}{1 \text{J}} \right)_{CF} = 404,000 \frac{\text{N}}{\text{m}^2}$$

This is the same as 0.404 bar of pressure, which is about 40% of the pressure of the air around us at sea level.

4.9   HEAT AND THERMAL ENERGY TRANSFER

As we discussed briefly in Chapter 3, **thermal energy** is the microscopic kinetic and potential energies associated with the random jiggling motions of the atoms and electrons internal to a solid, liquid, or gas. As more thermal energy is transferred to a substance, the temperature of the substance becomes higher and the atoms in it vibrate or move faster. The unit of thermal energy is joules (J).

Before 1900 or so, it was thought that perhaps thermal energy was a substance that moved from one object to another. That was wrong. Although thermal energy is not a substance, we can visualize it as migrating from one region or object to another. If you jump into a cold lake, you feel the effects of thermal energy leaving your body. The rapidly jiggling molecules in your skin are hitting the slower moving water molecules, making the water molecules move faster, thereby transferring energy into the water and removing energy from your body.

The amount of thermal energy in an object influences the condition or state of the object. (Here we will use the word *object* to refer to a solid, a liquid, or a gas.) As we transfer thermal energy to an object, its temperature rises proportionally. That is,

(change in temperature) ∝ (change in thermal energy).

Let us denote the thermal energy in a given object by Q. Also denote the change of temperature by ΔT_K, and denote the amount of change of thermal energy in the object by ΔQ. The proportionality is

$$\Delta T_K = \frac{\Delta Q}{C}, \quad \text{or} \quad \Delta Q = C \times \Delta T_K$$

The constant of proportionality C is called the **heat capacity**. (It is not to be confused with "Celsius," denoted by °C.) The word *capacity* is meant to suggest how much thermal energy an object can receive and store while incurring a certain temperature rise. For example, it seems sensible that a 50-gallon drum of water can receive a large amount of thermal energy and incur only a small temperature rise, whereas a thimbleful of water cannot receive this large of an amount of thermal energy without incurring a large temperature rise. The water-filled drum has a larger heat capacity than the water-filled thimble. The heat capacity depends on the amount of matter (mass) that is in an object and what material it is made of. For a gas, its value also depends on the situation, for example, whether the gas is held at constant volume or at constant pressure while the thermal energy is being transferred to it.

Heat capacity is the amount of thermal energy transferred to a certain object when its temperature is raised by 1K (or 1°C). The heat capacity of an object is proportional to the amount of mass M in the object ($C = C_0 \times M$); the constant C_0 is called the specific heat capacity characterizing that particular object and has units joule per kilogram per 1K, or J/(kg · K).

For liquid water, the specific heat capacity equals 4,184 J/kg · K. Therefore, 1 kilogram of water (which happens to have a volume of almost exactly 1 ℓ) has a heat capacity of 4,184 J/K. For a silicon crystal, important in computer circuits, the specific heat capacity equals 700 J/kg · K. For air kept at a constant pressure, the specific heat capacity equals around 1,020 J/kg · K.

If thermal energy is the microscopic kinetic and potential energies associated with the jiggling motions of the atoms in substance, then what do we mean by the word *heat*? In physics, the word *heat* refers to a process, not a quantity of something. For example, recall how we used the word *heat* in the discussions above:

"A way to make a cork pop out of a bottle is to <u>heat</u> the gas inside."

"The gas in the bottle has been <u>heated</u>, so the atoms move faster."

"A laptop computer processor chip <u>heats</u> up to around 60°C when working hard."

"Some of the microscopic energy of the water's molecules has been converted into thermal energy, <u>heating</u> up the water."

Heat is the process of transferring thermal energy from one object or place to another.

Heat does not flow. That would be a misuse of language. In addition, when we talk of the heat capacity of an object, we do not mean how much heat the object can contain (that makes no sense grammatically). We mean how much thermal energy can be transferred to the object, in connection with a certain change of the object's temperature.

THINK AGAIN

It might be tempting to think that thermal energy and temperature are the same concept, because the change of temperature is proportional to the change in thermal energy. This is not the case. To see this, compare 1 cup of water to 1 gallon (16 cups) of water, both initially at room temperature. Now put both of these onto identical stove burners for equal lengths of time, thereby increasing the thermal energy contained in each by the same amount. If the gallon of water becomes warm to the touch, the cup of water will become very hot. The temperatures are not equal, although the thermal energy increases are equal.

How fast does thermal energy transfer? That is, what is the rate of thermal energy transfer? Newton hypothesized that the rate of thermal energy transfer from one object to another is proportional to the difference of the temperatures of the two objects. This is correct, and we call this Newton's law of cooling.

Newton's law of cooling: The rate of thermal energy transfer (J/sec), from one object at temperature T_A to another object at temperature T_B, is proportional to the difference in temperature. That is,
(Rate of thermal energy flow from object A to object B) $\propto (T_A - T_B)$

This equation summarizes the obvious fact that if object A is hotter than object B (i.e., $T_A > T_B$), and they are put into contact, then thermal energy will flow from object A into object B. On the other hand, if T_A is smaller than T_B, then the rate of flow is negative, meaning thermal energy flows in the reverse direction—from object B to object A.

Figure 4.20 shows three blocks of iron, A, B, and C, in close thermal contact. Their temperatures are 400, 300, and 273°C. Because the blocks A and B are cube-shaped, the areas of the touching surfaces between any two blocks are equal. In Figure 4.20a, each block is held at a constant temperature by heaters or coolers, which are not shown in the figure. The length of each arrow indicates the rate of thermal energy flow. According to Newton's law of cooling, the thermal energy flow should be greatest between the two objects having the largest difference of temperatures; this is $T_A - T_C = 400°C - 273°C = 127°C$ between objects A and C.

In Figure 4.20b, the blocks are initially at the same temperatures as in Figure 4.20a, but after they are put into contact, the blocks are placed in a thermally insulating enclosure. Common sense tells us that thermal energy will flow between the three blocks until all are at the same temperature, at which time the flow of thermal energy ceases.

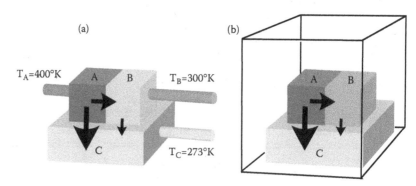

FIGURE 4.20 (a) Three iron blocks held at constant temperatures by external heaters or coolers (not shown). (b) The same three objects, initially having the temperatures indicated in part (a) and surrounded by a thermally insulating barrier.

As we discussed above, the change in the temperature of an object is proportional to the change of thermal energy in that object. In situations such as in Figure 4.20b, in which no external sources of heat are applied after two objects, A and B, are put in contact, Newton's law of cooling can be then stated as:

(Rate of decrease of temperature of object A) $\propto (T_A - T_B)$

According to this relation, if an object A starts out much hotter than its surroundings B, so that $T_A - T_B$ is large, it will begin cooling quite rapidly. Let us assume the surroundings stay at constant temperature. After some time, the temperature difference is smaller, so in the next few moments the cooling rate is smaller than it was previously. As the object cools, the rate of cooling slows. Eventually the temperature difference equals zero, at which point cooling stops.

4.9.1 Heating by Conduction, Convection, or Radiation

There are three ways in which thermal energy can flow—by conduction, convection, or radiation. If two objects are in physical contact (i.e., touching one another), then thermal energy can flow by conduction. ***Heat conduction*** is the direct transfer of thermal energy from one object to another by collisions of the atoms on the surfaces of the two objects that are touching. For example, **Figure 4.21** shows three objects in contact—a crystal, a piece of glass, and a gas. The crystal might be an ice cube inside a glass container, which is in a freezer compartment. Conduction of thermal energy

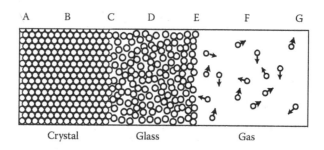

FIGURE 4.21 Thermal energy can transfer between different materials by collisions of atoms at the interfaces (C and E) between the materials. Thermal energy also transfers through a material of one type by collisions between atoms of the same type (for example at position B).

occurs between the crystal and glass at the boundary labeled C, where atoms of the two materials jostle and collide with each other, transferring internal kinetic and potential energy from one to the other. If a fast-moving atom in the glass collides with a slow-moving atom in the crystal, some kinetic energy will be transferred to the atom of the crystal, and vice versa. The same type of interaction occurs at the boundary between the glass and the gas. Atoms of these two substances randomly collide and exchange kinetic energy.

We can understand Newton's law of thermal-energy exchange by thinking about the microscopic picture in Figure 4.21. Assume, for example, that the temperature of the crystal is initially higher than that of the glass and the gas, which are at equal temperatures. Then there will initially be more high-energy atoms on the crystal side of the crystal-glass boundary than on the glass side. On average, the atoms of the crystal will give more kinetic energy to the atoms in the glass than vice versa. This will lead to a net transfer of kinetic energy from the crystal to the glass. Then the atoms in the glass at the boundary labeled C will jostle other atoms in the glass and pass some of their thermal energy to atoms in the glass at position D. During this process, atoms in the glass do not move large distances—they only vibrate back and forth because many surrounding atoms confine them. This conduction process continues until thermal-energy transfer reaches the boundary between the glass and gas at position E. Conduction of thermal energy now takes place between the atoms in the glass surface and the freely moving atoms in the gas. The moving gas atoms bounce from the warmer glass surface, typically gaining kinetic energy on each bounce. This heating of the gas by conduction occurs until the gas and the glass have equal temperatures—they are now in ***thermal equilibrium***.

Thermal equilibrium is the situation in which two or more objects that are capable of exchanging thermal energy have no net transfer of thermal energy between them. This occurs when they have equal temperatures, so that $(T_1 - T_2) = 0$.

Thermal equilibrium of the crystal, the glass, and the gas in Figure 4.21 will occur only after sufficient thermal energy has transferred between them to equalize their temperatures. As long as any two temperatures are unequal, thermal energy will continue to transfer, leading to a lessening of the temperature difference between two touching objects. Eventually, the temperature differences will decrease to zero, at which point thermal equilibrium is achieved. The microscopic makeup of a material determines its ability to conduct heat readily. This ability is summarized by a material's ***heat (or thermal) conductivity***.

Heat (or thermal) conductivity is the property of a material that indicates its ability to conduct thermal energy.

Typically, crystals have the highest heat conductivities, followed by noncrystalline solids, and then gases. **Table 4.2** lists the heat conductivities of several materials. Diamond has the highest heat conductivity of any substance, and is about 40,000 times more conductive than air. Metals, such as silver and aluminum, have high heat conductivities. Crystalline quartz and pure silica glass have the same chemical makeup, but different heat conductivities. As is typical, the crystalline form has the higher conductivity. An exception to this rule is that the crystalline form of water (i.e., ice) has a lower conductivity than the liquid form. Gases have the lowest conductivities, simply because they have far fewer atoms per cubic meter to partake in the transferring of thermal energy. The unit of heat conductivity is W/(m · °C), or watts per meter per degree Celsius.

TABLE 4.2
Typical Heat Conductivities of Substances

Material	Heat Conductivity	Comments
Diamond	1000 W/(m · °C)	The highest value of any substance
Silver	429	The highest of the common metals
Copper	401	Cooking pots
Aluminum	237	Cooking pots
Silicon	149	The main material in computer chips
Sodium	141	Can be a liquid above 98°C
Crystalline quartz (SiO$_2$)	6-11	Depending on direction in crystal
Pure silica glass (SiO$_2$)	1.6	Room temperature, purity 99.995%
Common glass	0.8	SiO$_2$ with metal impurities
Water (H$_2$0)	0.61	At 27°C
Ice (H$_2$0)	0.59	
Styrofoam	0.033	Polystyrene plastic containing air pockets
Air	0.026	78% nitrogen, 21% oxygen, at 27°C and 1 bar pressure

QUICK QUESTION 4.4

How does thermal underwear provide good insulation to lessen the transfer of thermal energy away from your body? Why does it cease to be effective when wet?

To understand by a simple argument why noncrystalline solids, such as glass, have lower heat conductivities than do crystals, consider the analogy discussed above in which all neighboring atoms in a solid are connected by tiny springs. Think of a mattress box spring with all of the soft padding taken out, in which each metal wire coil is connected to each neighboring coil by a springy loop of wire, as in **Figure 4.22**. In this model of a crystal, all of the springs are identical and are lined up perfectly in rows and columns. Any motion at one end of the crystal will be rapidly transmitted down the length of the crystal by the organized vibration of all of the coupled springs. In contrast, the noncrystalline solid is similar to the disorganized pattern of coils and loops shown in the bottom part of Figure 4.22. In such a pattern, the separations of coils and the lengths of connecting loops are random. This random pattern impedes the transfer of vibrations (thermal energy), because the coils will all vibrate at different speeds and the vibrations must follow a tortuous path from one end to the other.

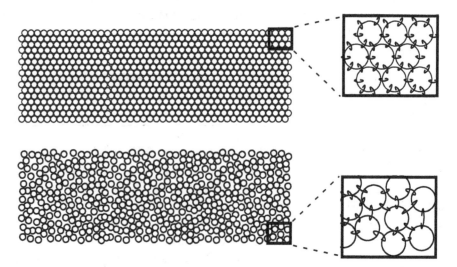

FIGURE 4.22 Model of heat conduction, in which each atom is thought of as a springy metal wire coil connected to each neighboring coil by a springy loop of wire. The inserts show magnified drawings of small regions of the arrangement of coils and loops, both for a regular crystal-like pattern and a noncrystal-like pattern.

In addition to conduction, thermal energy can also be transferred by convection or by radiation. Convection involves a liquid or gas flowing from one place to another, carrying thermal energy with it.

=====

Convection is the transfer of thermal energy from one location to another by currents in a gas or a liquid.

=====

The current responsible for convection can be either forced or it can be natural. An example of forced convection is cooling a computer chip by air that is blown by a fan inside a computer box. Through conduction, the air picks up some of the thermal energy from the chip and then carries this away by convection. The effectiveness of convection depends on the heat capacity of the coolant—air or a liquid. An example of natural, or free, convection is the cooling of a stovetop burner after it is switched off. If the air in the kitchen is initially still, with no air currents, the air heated at the surface of the burner will rise through the cooler air above it, because hot air is less dense, and therefore buoyant. The rising air creates a current that takes thermal energy away by convection.

Thermal energy can also transfer by radiation.

=====

Radiation is the heating or cooling of an object by light waves and infrared waves absorbed or emitted by the object.

=====

If an object is very hot, its atoms are jiggling very rapidly. This leads to the emission of light and other energy waves, which are electric and magnetic in nature. The following two chapters discuss the physics of radiation. For example, a red-hot stovetop burner glows red because it emits radiation in the form of light that can be seen by the eye. If you place your hand a few inches above the burner, you will feel the radiation as heat because your hand absorbs this radiation and heats up. Consider a stovetop burner in a spacecraft from which all of the air has accidentally leaked out. There can be no convection cooling of the burner after it is turned off. Instead, the burner radiates all of its thermal energy and cools to the background temperature (in this case, very cold!).

QUICK QUESTION 4.5

Several objects are at room temperature—a large diamond, an empty drinking glass, and a dry sponge. Using a glove to avoid feeling the objects with your hand, you pick up each in turn and place it to your cheek. How would the heat sensations differ? Why? *Hint*: Your cheek is warmer than each of the objects.

QUICK QUESTION 4.6

Outer space is a near-perfect vacuum, with very few atoms per cubic meter. Estimate the value of the heat conductivity of outer space. How does thermal energy travel from the Sun to Earth?

REAL-WORLD EXAMPLE 4.3: COOLING COMPUTER CHIPS

The electronic chips inside a computer generate a lot of heat when they run. A typical processor chip, the "thinking" part of a desktop computer, generates heat at a rate between 25 and 100 watts when performing complex tasks. The thermal energy produced must be removed from the chip, or it will fail to operate. Let us estimate how rapidly this can be done. Consider a chip that is 1 millimeter thick and 10×10 millimeters on its edges. In **Figure 4.23**, the chip is bonded to a block of highly conducting material such as copper. According to Newton's law of cooling, the greater the temperature difference between the chip and the block, the faster thermal energy will be removed from the chip by conduction.

The rate of heat conduction through a slab of material is given by the formula:

$$rate\ of\ heat\ conduction = \frac{(electrical\ conductivity) \times area}{thickness} \times (temperature\ difference)$$

We can use this to estimate the rate of heat conduction through the chip to the block. A chip is made primarily of silicon, which has a heat conductivity of about 150 W/(m · °C). The area of the chip is 10×10 millimeters = 10^{-4} m². Its thickness is

FIGURE 4.23   Computer chip mounted on a block of heat-conducting copper.

1 millimeter = 10^{-3} m. If the temperature difference between the chip and the block equals 5°C, then we calculate:

$$rate\ of\ heat\ conduction = \frac{(150W \cdot m/°C) \times 10^{-4} m^2}{10^{-3} m} \times (5°C) = 75W$$

This is a sufficient rate for cooling a typical chip, but to achieve this we need to keep the block cool. This is usually done by blowing cool room air across the block.

Some computer users like to modify or "mod" their computers so the chip runs faster than the safe rating stated by the manufacturer. This generates heat faster than normal, so extreme cooling is required. These users modify the cooling method, using either a radiator or a water-cooling system. A radiator, shown in **Figure 4.24**, is a collection of thin metal plates over which cool air passes. The metal block at the bottom of the radiator is in contact with the top of the chip to conduct thermal energy away. A special heat-conducting paste consisting of silver particles in oil is applied between the top of the chip and the bottom of the radiator block. The heat conductivity of the paste is at least 7 W/(m · °C). Because this paste layer is only 0.02 millimeters thick, it conducts heat rapidly. The increased surface area of the radiator helps speed up the transfer of thermal energy to the passing air. One liter of dry air has a heat capacity of about 1.3 J/°C. So, ideally, if the air carried away the maximum amount it could carry, and if it were heated by 10°C as it passed through the radiator, then 1 liter of air could carry away (1.3 J/°C) × 10°C = 13 J of thermal energy. A fan would need to blow a minimum of 5 liters of air through the radiator each second to remove 5 × 13 = 65 watts of power in the form of heat.

If radiator cooling is not good enough, you can attach a water-cooled block to the top of the chip. Cold water flows through hoses and past the chip, removing thermal energy very fast. One liter of water has a heat capacity of about 1,500 J/°C. So, if the water is heated by, say, 10°C as it passes through the heat exchanger, 1 liter of water could carry away a maximum of 15,000 J of thermal energy. In this case, it would be sufficient to pass only 0.01 liter water per second to carry away 150 J of thermal energy, assuming excellent heat transfer to the water from the chip. For more information on chip modding, carry out an Internet search for "cpu heatsink" or "overclocking."

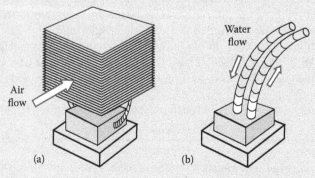

FIGURE 4.24   Air or water is used to cool chips when overclocking. (a) Chip-radiator cooling. (b) Water-block cooling.

Future chips are projected to generate over 1,000 watts of heat, and this cannot be easily removed using existing cooling techniques that are consumer friendly (water blocks are not). Chip manufacturers are therefore carrying out research to find new techniques.

4.10 PRINCIPLES OF THERMODYNAMICS: EXTRACTING WORK FROM HEAT

The scientific study of heat and thermal-energy transfer is called *thermodynamics*. The two big questions, which, from Newton's time, took over 150 years to answer, are: (1) If energy is the prime mover in the world, how does it move around from one form to another? and (2) How much of the energy that is "out there" can be harnessed by us to do work? The answers to these questions are called the first and second laws of thermodynamics. They are crucial for modern technology, which is largely concerned with harnessing energy to perform useful work. An obvious example is an automobile engine, which converts chemical energy to thermal energy to mechanical energy. Even a computer is a machine that does useful work: it pushes electrons around in a circuit to achieve computation, as we will see in later chapters.

The first law of thermodynamics is a specific form of the energy-conservation law, which we stated in Chapter 3. Recall the statement of that law—energy cannot be created or destroyed; it just gets converted between various forms. In the case of thermal-energy transfer, this can be stated specifically as:

First law of thermodynamics: The increase of the energy of a substance (gas, liquid, or solid) is equal to the thermal energy transferred to it, plus any work done on the substance.[3]

Long before Joule's studies, physicists understood that *mechanical energy* was conserved—kinetic and potential energies can be converted back and forth at will with an efficiency approaching 100%. In this case, only friction prevents perfect efficiency, and friction can be minimized to nearly zero in a very good machine.

THINK AGAIN

Notice that we do not say, "The increase of the energy of a substance is equal to the heat added to it, plus any work done on the substance," because this would imply that heat is like a substance that could reside in a body. Remember—heat is a physical process, like "combing your hair." You would not say that the combing now "resides in" your hair, but that your "hair is combed." Likewise—a substance "is heated." It does not "gain heat."

To say the first law in another way—work and heat refer to two different forms of energy transfer, and these two forms can be converted into one another within certain limitations. This concept is often called the equivalence between mechanical work and heat. James Joule (1818–1889) was the first to realize this relation in the 1840s. In fact, Joule was the first to fully develop and verify the concept that there exists a property

---

[3] Note to instructors: The convention that the work is done on the object is consistent with most European texts, but is opposite of many American texts. (Arons, A.B. *Teaching Introductory Physics*. New York: Wiley, 1997. p. 147.)

called energy that is conserved in nature. It is fitting, then, that we name the basic unit of energy after him.

Joule came to his realization of the equivalence between mechanical work and heat by carrying out precise experiments that simultaneously involved both work and heat. His most convincing experiment consisted of the apparatus shown in **Figure 4.25**. He argued to himself that it should be possible to observe a slight increase in temperature after vigorously stirring a liquid, such as water. The thermal energy, he thought, would be generated by the friction of the stirring object with the liquid, as well as the internal friction of the moving water itself. To verify his hypothesis, he first measured the temperature of water contained in a metal enclosure at room temperature. He then rotated, by using a hand crank, a specially constructed paddlewheel inside the water. He again measured the water's temperature, and found a slight increase. Being a good physicist, he thought to make a connection between the amount of mechanical work done and the amount of temperature increase. To do this, he attached a weight, with mass M, to a string, which was wound around the axis of the paddlewheel. He then used the known relation involving gravity's force and the distance that an object falls to calculate the amount of mechanical work done by gravity in turning the paddlewheel. From this, he was able to find a relation between mechanical work and heat.

In particular, as the mass falls a distance D under the force of gravity, the amount of work done equals gravity's force F times the distance D that the mass falls. This also equals the amount of work done by the paddle on the water. Recall that the force of gravity on an object at sea level, having mass M, equals $F = M \times 9.8$ N/kg. As we discussed earlier, as we transfer thermal energy in the amount Q to an object, its temperature rises proportionally; that is, $\Delta T_K = C_W \times Q$. The constant of proportionality, C_W, in this case is the heat capacity of water, a concept that Joule first introduced during his thinking about this experiment.

Let us imagine a specific example of a measurement that Joule might have done. Let us say that he used a weight with mass 100 kg that slowly descended through a distance of 10 meters. Then, the work done on the water would have been equal to

$$work = F \times D = 100 \text{ kg} \times 9.8 \text{ N/kg} \times 10 \text{ m} = 9{,}800 \text{ N} \cdot \text{m}$$

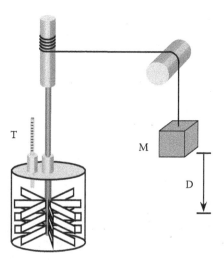

FIGURE 4.25   James Joule's 1849 apparatus for studying the equivalence of mechanical work and heat. As a weight with mass M falls a distance D under the force of gravity, a brass paddlewheel rotates in water contained in a copper enclosure, heating the water slightly. A thermometer measures the water's temperature T.

Let us also say that his container held 10 kilograms of water, and that he observed a temperature increase in the water of about $\Delta T_K = 0.2\text{K}$. From this result, he would have deduced that the heat capacity of 10 kilograms of water equals

$$C_W \text{(for 10 kg)} = \frac{9800\,\text{N} \cdot \text{m}}{0.2\,\text{K}} = \frac{9800\,\text{J}}{0.2\,\text{K}} = 49,000\,\text{J/K} \quad \text{(estimate)}$$

Joule found that if the amount of water was doubled, the heat capacity also doubled. If the amount of water was decreased by a factor of 10, to 1 kilogram, the heat capacity decreased by a factor of 10 to about 4,900 J/K. He thus would have deduced that the heat capacity per kilogram of water equals about

$$C_W \text{(for 1 kg)} = 4,900\text{J/K} \quad \text{(estimate)}$$

Joule's measurement techniques were not as accurate as modern ones. Later experiments, with more accurate measurements, showed the actual value to be

$$C_W \text{(for 1 kg)} = 4,184\ \text{J/K} \quad \text{(actual)}$$

We can state this value as $C_W = 4,184$ J/(kgK); that is, 4,184 Joules per kg per degree Kelvin. Using this more accurate value for C_W, we can calculate the expected temperature increase of 10 kilograms of water upon transferring 9.8 N · m of thermal energy to it:

$$\Delta T_K = \frac{\Delta Q}{C_W} = \frac{9.8\text{N} \cdot \text{m}}{10\,\text{kg} \cdot 4184\text{J/(kg} \cdot \text{K)}} = 2.34 \times 10^{-4}\ \text{K}$$

In modern terms, we say that 9.8 N · m of thermal energy equals 9.8 J of energy, because 1 joule (J) is the amount of mechanical energy needed to generate a force equal to 1 N and to use it to push an object a distance of 1 meter; that is, 1 J = 1 N · m.

Joule's experiments with this paddlewheel convinced the leading scientists of the day of the existence of energy as a conserved quantity. Joule was able to perform this experiment only because he had spent years inventing means to measure very small temperature changes. Of course, Joule did not use the units of meters, kilograms, and Joules, which were invented many years afterwards. He used feet, pounds, and calories (where 1 cal = 4.184 J).

4.10.1   The Second Law of Thermodynamics

There are crucial limitations on the extent to which we can convert heat into work. A common-sense example of such a limitation is shown in **Figure 4.26**. Considering only the law of energy conservation, a person with little common sense might think it should be possible for the mass M to spontaneously rise, thereby gaining gravitational potential energy as long as the water slightly cooled and lost thermal energy equal to the energy gained by the mass. Clearly, this kind of motion is never actually observed in nature. Why not? It is not ruled out by the first law of thermodynamics. There is another important principle of nature at work here—the second law of thermodynamics.

Consider what would happen if you grabbed the axle holding the paddlewheel and rotated it by hand to make the mass M rise a certain distance. The mass would gain potential energy, and the water would heat up slightly. The work done by your hand goes not only into raising the mass, but also heating the water. The point is that rotating

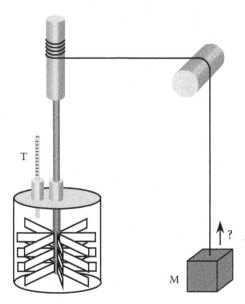

FIGURE 4.26 In this setup, which is similar to that in Figure 4.25, why does the mass *M* not spontaneously rise, cooling the water?

the wheel axle in either direction—clockwise or counterclockwise—heats up the water. It is not possible to cool the water by rotating the paddlewheel. It seems that it is easier to convert mechanical energy into thermal energy than vice versa.

Of course, it is possible, within limits, to convert thermal energy into useful work. A good example is the steam engine, invented in Britain in the 1700s for pumping water out of coal mines. Historically, the desire to understand how the steam engine works was a major driving force behind scientists' push to discover the laws of thermodynamics. Steam engines are used today to drive electric generators in power plants. The electrical power generated is used, for example, to power your computer. Earlier versions of the steam engine were greatly improved upon by Scottish inventor James Watt in 1769. This engine proved so important for the progress of science and engineering that the unit of power—the watt (W)—was named in his honor.

A schematic drawing of the steam engine invented by Watt is shown in **Figure 4.27**. A boiler containing water is heated by an external heat source, such as burning coal, creating steam. If valve 1 is open and valve 2 is closed, then the high-pressure steam pushes the piston to the right, doing work on any object connected to the piston. After the piston moves to the right, valve 1 is closed and valve 2 is opened. Hot steam enters the condenser, the inner walls of which are kept cool by flowing water from a cold source, such as a river. The steam condenses, rapidly decreasing the pressure in the piston. The external air pressure pushes the piston to the left. The valve settings are switched, and the cycle repeats.

It is not surprising that hot expanding steam can do work on a piston. What is not so obvious is whether all of the thermal energy stored in the steam can be converted into work in this manner. If an engine could do this perfectly, we would say that it has 100% efficiency. After much experimentation and theoretical reasoning, physicists came to a firm conclusion, summarized as the second law of thermodynamics.

Second law of thermodynamics: Any engine that converts thermal energy into useful work cannot be 100% efficient—it must exhaust some waste heat.

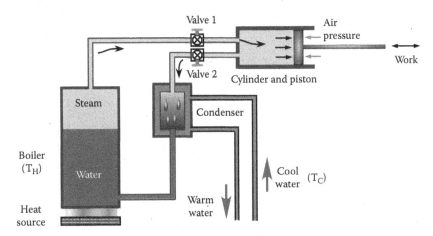

FIGURE 4.27 James Watt's steam engine. Water is boiled to create high-pressure steam. Opening valve 1 allows the steam to move a piston, doing work. Closing valve 1 and opening valve 2 sends the steam into the condenser, where it becomes liquid water, which greatly decreases the pressure in the cylinder, which allows the external air pressure to move the piston left. Wasted thermal energy is ejected through the warmed water leaving the condenser.

Waste heat refers to the thermal energy that was not converted into mechanical work; it is wasted. In the case of Watt's steam engine, the waste heat is removed by the warm water leaving the condenser. It requires a constant source of cold water for cooling, as well as a constant source of hot water, to run the engine. If the "cooling" water came from the same source as the "hot" steam entering the cylinder, the engine would not run. This is an unfortunate fact of life, which prevents us from using a hypothetical perpetual-motion steam engine to solve our energy needs.

How high can the efficiency of a steam engine be? By "efficiency," we mean the ratio of useful work W extracted to the total amount of thermal energy Q extracted from the hot region—the boiler.

$$efficiency = \frac{W}{Q}$$

To prove what the maximum possible efficiency is would take us far from the main goals of this text. Instead, let us make a simple, plausible argument in two parts. First, if the temperature of the hot region T_H (the boiler) equals the temperature of the cold region T_C (the cooling water), the engine should not work at all, meaning its efficiency is zero. Let us thus suppose that the engine's efficiency is proportional to the temperature difference, $T_H - T_C$. Second, if the temperature of the hot region T_H is much higher than the temperature of the cold region T_C, the engine should work with very high efficiency, nearly equal to 1. We can summarize these two ideas in a single equation:

$$maximum\ efficiency = \frac{T_H - T_C}{T_H} = 1 - \frac{T_C}{T_H}$$

Here the temperatures must be expressed in degrees Kelvin, relative to absolute zero. By using detailed calculations that are beyond the goals of this text, this formula can be proven accurate for any conceivable steam engine or any other engine that extracts work by the use of a hot region and a cold region. For example, if the steam has a temperature 300°C (or 573K) and cooling water has temperature 40°C (or 313K), then the engine's efficiency could not exceed:

$$maximum\ efficiency = 1 - \frac{313}{573} = 0.45\ \text{or } 45\%.$$

This means that 55% of the heat we generate is ejected into the cooling water as waste heat. Where does that waste heat go? It likely goes into a river or the atmosphere.

Most modern electrical power generators burn coal to generate steam to turn steam turbine engines. In a turbine, expanding steam passes by many small fins, which are like airplane propellers, causing the shaft on which they are mounted to rotate, which does work to create electrical power. Such turbines have an efficiency less than 50%, meaning that for every watt of heat power converted to mechanical work, more than 1 watt is discarded into the environment as waste heat.

The second law can be stated in a more general and powerful way.

Second law of thermodynamics (general statement): Within any closed, isolated system, the amount of unusable energy is increasing in time.

That is, the amount of usable energy is decreasing in time. By a "closed, isolated system" we mean a region that cannot take in or expel energy in any form; that is, it contains a constant amount of energy. By "unusable energy" we mean energy in a form that cannot be converted into mechanical work. The second law tells us something profound about how the universe is evolving—it is evolving from a more ordered (organized) condition toward a more disordered (disorganized) condition, at least on the large scale.

THINK AGAIN

You might hope that the waste heat could be captured and reused to drive some other engine, perhaps at a lower temperature. This is true, but there is a limit to how far this process can be continued. The formula for maximum efficiency given above assumes that the most clever and efficient process of heat use is being applied, and still there is a limit to the efficiency.

4.11   COOLING COMPUTERS IS REQUIRED BY THE PHYSICS OF COMPUTATION

Imagine a primitive computer—a steam-driven adding machine, shown in **Figure 4.28**. This machine is capable of performing the calculations: $0 + 0 = 0$, $1 + 0 = 1$, $0 + 1 = 1$, and $1 + 1 = 2$. The input data—the numbers to be added—are represented by the settings of the valves that connect the hot and cold regions (boiler and condenser) to the two pistons. In the case shown in the top figure, the valves are set so that each piston is connected to the cold region, so there is low pressure in each piston. This results in low pressure in the "adder piston," representing a zero (0) as the output of the calculation. This computes $0 + 0 = 0$.

In the second figure, the upper valve has been switched, allowing hot steam into the upper piston, causing it to move to the right and exerting a force on the adder piston. This compresses the gas in the adder piston, which increases its pressure to a medium value. This computes $1 + 0 = 1$. In the third figure, the lower valve has been switched, which drives the calculation $0 + 1 = 1$. In the fourth figure, both the upper and lower valves have been switched, which allows hot steam into both pistons, which causes them both to move to the right and exerts force on the adder piston. This strongly compresses the gas in the adder piston, which increases its pressure to a high value, which we interpret as the value 2. This computes $1 + 1 = 2$.

Let us say we have successfully calculated $1 + 0 = 1$, and now we wish to calculate some other sum, say $0 + 1 = 1$. To do this, we need to switch the valves to their opposite positions. When we do this, the lower piston returns to low pressure by exhausting hot steam into the cold condenser region. This is waste heat, as we discussed in connection with Figure 4.27. We can describe this in another way. Say that after we calculate the answer to any of the three problems: $1 + 0 = 1$, $0 + 1 = 1$, and $1 + 1 = 2$, we wish to return the pistons to their positions shown in the first (top) figure, with both pistons at low pressure. We call this operation "erasing the input data." In each of these cases, we must release some waste heat when one or both pistons return to low pressure. We see, then, that erasing the input data always creates some waste heat.

This waste heat must be removed from the computer, so it does not overheat and fail. This requires some kind of cooling method. A fan to blow air over the heated computer chips is the most common, although high-performance computers sometimes use water circulating in tubes for cooling.

Although this example is clearly not a serious proposal for building a computer, it allows us to understand an important physical principle about computers—Landauer's principle.

Landauer's principle: Any process that erases data in a computer must release some waste heat.

By erasing data, we mean returning all stored data (bit) values to 0. Real computers perform erasing operations many times every second as they transfer data into and

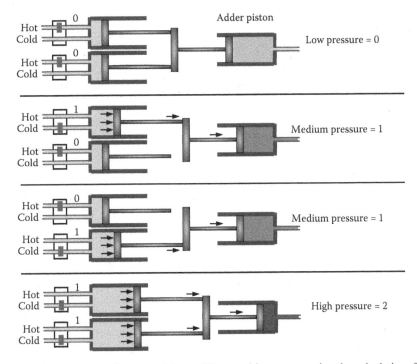

FIGURE 4.28   Four cases of a steam-driven adding machine, representing the calculations $0 + 0 = 0$, $1 + 0 = 1$, $0 + 1 = 1$, and $1 + 1 = 2$. Input data determine the positions of valves (dark rectangles), which may allow hot steam into each piston on the left, causing its expansion, which drives compression of the piston on the right. Higher pressure is indicated by darker shading.

out of their internal "adding machine." Rolf Landauer, a physicist at IBM, discovered this principle in 1961. Landauer's principle is true for any possible computer design, whether it uses mechanical or electrical methods to perform its calculations. Thus, we see that cooling computers is required by the physics of computation.

Present-day computers do not operate close to the maximum efficiency allowed by the laws of physics. The amount of thermal energy that they waste is far greater than the lower limit set by the second law of thermodynamics. In the future, however, as computers become more and more efficient, they might run into the limit set by the second law of thermodynamics.

THINK AGAIN

The waste heat generated in a computer is not necessarily due to the presence of a region that is hotter than room temperature, as you might think. For example, we could use nitrogen gas, with boiling temperature 77K (−196°C or −320°F), instead of steam to run the piston adding machine. The "hot" region could be at 0°C (the temperature at which water freezes) and the "cold" region could be at −196°C (cold indeed). The operating cycle would be the same as described in Figure 4.27, and the machine still would release waste heat.

As an example, I performed a test: My laptop has a feature that tells me the temperature of its processor chip. After turning off all programs and letting the computer "sleep" for a while, the temperature was 46°C. Then I started running a program that required many calculations each second. After a few minutes, the processor's temperature rose to 62°C, heating up my lap, at which point the internal fan started blowing air across the chip to cool it. It returned to less than 60°C.

SUMMARY AND LOOK FORWARD

Objects and substances are made of constituents called atoms. Even without describing the detailed nature of individual atoms, we are able to understand many properties of solids, liquids, and gases. The differences between solids, liquids, and gases are in the arrangement and motions of the atoms. Atoms in a substance are in constant motion, and their jiggling causes pressure—the force per area exerted on an object in contact with the substance. For a gas or liquid, a pressure applied externally to the surface of the substance increases the pressure at every location in the substance by an amount equal to the applied pressure. This is Pascal's principle.

The flow of a liquid through a pipe always meets some resistance to its flow, caused by friction between the liquid and the pipe walls or obstructing objects inside the pipe. The pressure in a flowing liquid decreases as it passes through the resistive region. The decrease of pressure is proportional to the current, or rate of flow, and to the amount of resistance in the resistive region. This is Ohm's law for liquids.

Temperature is a measure of the average kinetic energy of randomly moving atoms. In a gas, the average kinetic energy associated with the movement of atoms or molecules is proportional to the temperature in the Kelvin scale. The proportionality constant, k_B, is called Boltzmann's constant, equal to 1.38×10^{-23} J/K. Low-pressure gases can be described using the ideal gas law. This states that gas pressure is proportional to its temperature, as well as to the ratio of the total number of atoms to the volume of the enclosure, that is, $p = (N/V)k_B T_K$, where temperature, T_K, is measured in the Kelvin scale.

Thermal energy is the kinetic and potential energies associated with the random jiggling motions of the atoms in a substance. As we heat an object (i.e., transfer thermal energy to it), its temperature rises proportionally. The amount of heat (ΔQ) needed to raise the temperature of an object by amount ΔT_K is proportional to ΔT_K, with a constant of proportionality, C, called the heat capacity. That is, $\Delta Q = C \times \Delta T_K$.

There are three ways in which thermal energy can transfer: conduction, convection, or radiation. Heat conductivity is the property of a material that indicates its ability to transmit thermal energy by conduction. Thermal energy will transfer from a hotter object to a colder object. According to Newton's law of cooling, the rate of thermal-energy transfer (J/sec) from one object (A) at temperature T_A to another object (B) at temperature T_B is proportional to their temperature difference, $(T_A - T_B)$. This means that the rate at which the temperature of object A decreases is proportional to $(T_A - T_B)$. This leads to a smoothly decreasing temperature of object A as time passes until their temperatures are equal. When two objects have equal temperatures, they are said to be in thermal equilibrium, and there is no transfer of thermal energy between them.

Computer processors, or chips, presently generate about 50 watts of heat power, and to avoid malfunctions, this thermal energy must be removed by conduction, convection, or radiation. In the near future, chips will generate over 1000 watts of heat, and improved cooling techniques will be needed.

The general principles of heat and thermal-energy transfer are described by the laws of thermodynamics. The two basic principles are:

1. First law of thermodynamics: The increase of the energy of a substance is equal to the thermal energy transferred to it plus any work done on the substance.
2. Second law of thermodynamics
   Version 1: Any engine that converts thermal energy into useful work cannot be 100% efficient; it must exhaust some waste heat.
   Version 2: Within any closed, isolated system, the amount of unusable energy increases.

The first law of thermodynamics applies to a much wider range of situations than we have discussed in this chapter. Energy can be added to an object by different means—electrical and magnetic being the most important in the context of computers. Inside a computer chip, electrically charged particles (electrons) are pushed around by electrical forces created by batteries or other voltage sources, such as a wall outlet. Work is done on the electrons as they are pushed around. The second law of thermodynamics leads to the conclusion that any operating computer must generate some waste heat. It cannot be perfectly efficient. In addition to the expected heating caused by any machine because of friction, the second law says that there is an absolute minimum to the amount of heat that must be wasted and expelled by any computer. This limit might impact the future course of computer development in the next 20 years or so.

The next step in understanding computers and the Internet is to study electricity and magnetism. These are the basis of electronic circuits, which do all of the work in a computer. This work generates heat, as we pointed out in this chapter.

SOCIAL IMPACTS: THE INDUSTRIAL REVOLUTION AND THE INFORMATION REVOLUTION

In 1950—at the dawn of the Computer Age—founding father Norbert Wiener predicted that the oncoming computer revolution would be comparable in scope and impact to the Industrial Revolution of 1760–1830 [1]. The social problems in each are similar: whether the new technology would be used for the "ennoblement or degradation" of humans [2]. Wiener feared that the new information technology would make it difficult to prevent the "use of a human being in which less is demanded of him and less is attributed to him than his full status." He feared such improper, inhuman uses would be the result of the loss of control by individuals, and increased control by organizations, made possible by the increasing control of the flow of information by those organizations. He correctly identified digital information as a key commodity or resource, which would be subject to ownership and to laws.

What parallels are there between the present age and the Industrial Revolution that led Wiener to recognize so early this potential crisis? The Industrial Revolution began in England in the 1780s and quickly spread in Europe and North America. It developed from the maturing of the physics of Newton after his death in 1727. As scientists and engineers realized that Newton's physics had applications beyond predicting the motions of the planets, they invented new technologies, such as steam engines and automated weaving and other textile machines. Rapidly adopted by industry for manufacturing, the technology greatly impacted society, creating leisure time for some, but also causing a large population movement of workers from the countryside to cities. This development has been compared to the Neolithic Revolution, when mankind many thousands of years ago developed agriculture, allowing the first permanent human settlements.

The invention of accurate clocks, occurring between about 1710 and 1740, paved the way for precise navigation across the oceans—a boon for trade and military purposes—and set the stage for the invention of other precision machines. In 1769, Scottish inventor James Watt invented an improved steam engine, which replaced a large portion of the brutal human and horse-powered labor previously used as the source of energy for the pumping of water from coal mines and moving boats up and down rivers. Steamboat transportation on the Mississippi River and the Erie Canal made possible the more rapid settlement of the interior of the United States. In the textile weaving industry, automated spinning machines powered by steam engines allowed factories to produce far more products per week. Wealth increased and became more broadly distributed. The fact that steam engines were scarce, expensive, and large dictated that automated spinning machines—formerly household appliances—be brought together in large numbers at factories, where many such machines could be run through pulleys powered by one steam engine. The machines ran the people as much as the people ran the machines. The physical form of the technologies strongly influenced the social changes. Also, people became more confident in their ability to influence the world and nature through their understanding of science. There was a movement against reliance on religious ideas by some intellectuals.

Often the new, more efficient machines required less expertise to operate than did the earlier, simpler machines. Brutal child labor and other nightmares ensued, as society adjusted to the new situation. As automation improved, some people lost their jobs to machines. Between 1811 and 1814 a group of skilled English workers and crafts-people, whose jobs were at risk by the new automated machines, protested by writing threatening letters to factory owners and in some cases smashing the new machines in the factories with hammers. (See the illustration on the next page.) These men were called Luddites after their fictional leader, "General" Ned Ludd. Even today, people who are perceived as antitechnology are often called Luddites.

In a 1995 interview by Kevin Kelly for *Wired*, Kirkpatrick Sale, a self-proclaimed modern-day Luddite, and Kelly discussed whether nature could be a model for a more human-friendly technology. Kirkpatrick Sale said:

This is simply an attempt to use science and its technologies to manipulate nature. This is an attempt to make nature technological, so that humans can determine everything about nature. ...It might be possible for you in the language of technology to come up with something faster, but you can't come up with something smarter, because you don't have that in your language bank.

Kelly responded:

That's where I think you are fundamentally wrong. Because you are stuck on an old language of technology, and we are creating a new one. It is possible to make an improved, smarter, wiser, more organic technology that can serve us better. [3]

In 2000, Doc Searls, a senior editor of *Linux Journal*, wrote that the neo-Luddites had already won, but they did not know it:

The Industrial Age is over, and the Information Age is already well underway. Workers are walking away from rusty old industrial machines that are dying for a single transcendent reason: they mistook people for parts. Now those people are going off to ply their crafts as competent human components of better, smaller machines—or large machines with better, smaller, more craftlike—i.e. autonomous—parts. These are machines they run, not machines that run them." [4]

Perhaps the Computer Age—computers as machines—has been supplanted by the Internet Age—computers as portals. A portal allows you to walk through, at least virtually, to nearly any place on Earth. Although that is a wonderful thing, perhaps it carries its social price as well.

Questions to Ponder

1. Does new automation, including robotics, cause unemployment or not?
2. Who benefits and who, if anyone, is harmed by new automation and information technology, including online technology?
3. Is the present-day resistance to automation largely among intellectuals or does it include employed and/or unemployed workers?
4. Can there be a new language of technology, or just the same old language?

5. What types of technologies promote centralization? Decentralization?
6. Is science apolitical by nature?
7. If humans are part of nature, can humans do anything that is "unnatural?" When human beings process silicon into computer chips, build computers, fashion the Internet, and communicate with each other by keyboard, are they being any less natural than bees who extract nectar from flowers, build nests, communicate, and come together to manufacture honey?

Terms to Research

Industrial Revolution, Luddite, neo-Luddite, scientism, spinning jenny

REFERENCES

1. Wiener, Norbert. *The Human Use of Human Beings.* New York: Avon Books, 1950, 1967.
2. Kaplan, Abraham. "Sociology Learns the Language of Mathematics," in *The World of Mathematics,* edited by James R. Newman, 1294–1313. New York: Simon and Schuster, 1956.
3. Kelly, Kevin. "Interview with the Luddite." *Wired* June 1995, 3.06. http://www.wired.com/wired/archive/3.06/saleskelly_pr.html.
4. Searls, Doc. "Let Freedom Ping," *Linux For Suits, Linux Journal,* November 2000, http://www.searls.com/linuxforsuits/lfs_nov00.html.

SOURCES

Encyclopedia Britannica
Kleinman, Daniel L. *Science and Technology in Society.* Victoria, Australia: Blackwell, 2005.
Russell, Colin A. *Science and Social Change 1700–1900.* New York: St. Martin's Press, 1983.
Sagan, Carl. *Broca's Brain.* New York: Random House, 1979.

The image of silicon atoms on the opening page of this chapter is from the following source: Bennewitz R., J. N. Crain, A. Kirakosian, J.-L. Lin, J. L. McChesney, D. Y. Petrovykh, et al. "Atomic Scale Memory at a Silicon Surface." *Nanotechnology,* Volume 13, 2002, 499–502.

SUGGESTED READING

See the general physics references given at the end of Chapter 1.

For popular-level, but still detailed, discussions of heat and thermodynamics:
Goldstein, Martin, and Goldstein, Inge. *The Refrigerator and the Universe.* Cambridge, MA: Harvard University Press, 1993.
Lloyd, Seth. *Programming the Universe.* New York: Knopf, 2006.
In particular, for discussions of the minimum waste heat generated during computation, see Goldstein (p. 226) and Lloyd (p. 77).

For instructors:
Feynman, Richard. *Feynman Lectures on Computation.* Reading, MA: Addison Wesley, 1996, pp. 148–151.

KEY TERMS

Absolute zero
Atom
Boltzmann's constant
Convection
Crystal
Current
Diffusion

ANSWERS TO QUICK QUESTIONS

Q4.1 The helium atoms are able to diffuse, one by one, through the plastic walls of the balloon.

Q4.2 $\quad I = \dfrac{P_1 - P_2}{R} = \dfrac{4\,\text{bar} - 0\,\text{bar}}{0.4\,\text{bar} \cdot \text{sec}/\ell} = \dfrac{4\,\text{bar}}{0.4\,\text{bar} \cdot \text{sec}/\ell} = 10\,\ell\,/\,\text{sec}$

Q4.3 $\quad T_K = T_C + 273.15^\circ\text{C} = 60 + 273.15 = 333.15\,\text{K}$

$\quad\quad T_F = (9\,/\,5)\,T_C + 32^\circ\text{F} = (9\,/\,5)(60^\circ\text{C}) + 32^\circ\text{F} = 140^\circ\text{F}$

Q4.4 Thermal underwear has many tiny air pockets or cavities trapped between the fibers. Air has low heat conductivity. Filling the cavities with water increases the heat conductivity.

Q4.5 The diamond would feel coldest because it would conduct thermal energy most rapidly away from your cheek. The dry sponge would feel the least cold, because it is in large part air, with low heat conductivity.

Q4.6 Outer space has almost zero heat conductivity and no convection. Thermal energy travels from the Sun to Earth by radiation.

EXERCISES AND PROBLEMS

Exercises

E4.1 How many ping-pong balls will fit into a cubical box with 40-cm sides? The standard size for a table tennis (ping-pong) ball is 40 mm in diameter. Assume the balls are packed as shown here:

E4.2 On the basis of the pictures in Figure 4.3, would you expect a liquid or a solid to have a higher density (number of atoms per volume)? Assume the size of atoms to be the same in the two cases.

E4.3 On the basis of the pictures in Figure 4.3 and Figure 3.20 (Chapter 3), explain the fact that adding water between two rough surfaces (like your boot and the street) can decrease the friction.

E4.4 Pure silica glass—the type used for optical fibers—is composed exclusively of SiO_2. Each such molecule has one Si atom tightly bound by chemical bonding to two O atoms, illustrated as:

Assuming that these molecules always stay intact in this shape, make drawings, similar to those in Figure 4.3, showing your best guesses for the typical distributions of particles in an SiO_2 gas, liquid, and crystalline solid.

E4.5 Which of these are examples of diffusion? Explain.

(a) Liquid chlorine is added at one end of a swimming pool and later is found to be irritating to the eyes of swimmers at the other end.
(b) A large rock dropped into a lake sinks to the bottom.
(c) A rumor, after many retellings, reaches a person 300 miles from its source.
(d) Black soot emitted in a powerful industrial smokestack stream reaches an elevation of 20 ft. above the top of the smokestack.

E4.6 To lift a heavy object such as an automobile using a smaller force, engineers devised the hydraulic lift shown in the figure below. A smaller cylinder is connected through a tube to a larger cylinder. The surface area of the fluid in the smaller cylinder is A_1, and in the larger cylinder A_2. Assume that $A_2 = 4A_1$, and that the fluid is incompressible.

(a) Explain why the two fluid heights are the same, when both are exposed to the air, as in figure (a), where air pressure is indicated by arrows.
(b) In figure (b), a piston is pushed down into the smaller cylinder a distance $d = 1$ m. To what height, h, does the surface of the fluid in the larger cylinder rise?

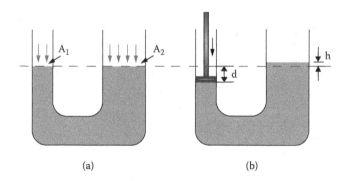

(a)   (b)

E4.7 To test one of his basic hypotheses about mechanical work and heat, while visiting the Swiss Alps, James Joule measured the temperature of the water at the top of a high waterfall and at the bottom. What do you think he found? Explain on the basis of the discussion in the chapter.

E4.8

(a) Explain why the water pressure near the bottom of a 10 m-high water tank, at the Earth's surface, is greater than near the top.

(b) Copy the side-view drawing below, which shows a water-filled tank with two small holes and two horizontal collection pans equal distances below each corresponding hole. To the best of your ability, draw the water streams going from each hole to the pans. Note: It will not be possible to precisely predict the trajectories of the water streams given what you have learned here. A principle called Bernouolli's principle is needed to do that.

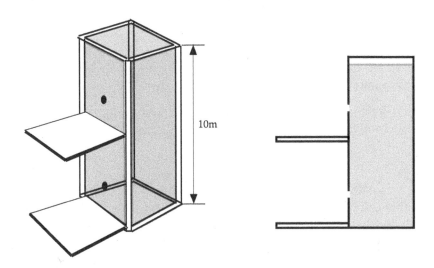

E4.9 A pump generates pressure to make water flow through a pipe containing obstructions, as shown.

(a) Of the locations labeled 1, 2, and 3, which corresponds to the highest pressure in the pipe? Which corresponds to the lowest? Explain in terms of water resistance.

(b) Three holes, small enough to allow only a small amount of water to leak, create small water spouts, like the one shown. Draw the three spouts, indicating roughly their relative heights.

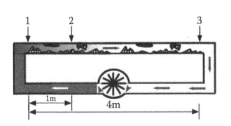
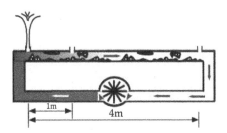

E4.10 Write a half-page description of how the paddlewheel in Joule's experiment heats up the water. What is going on at the microscopic, atomic level? Also explain why, after stopping the wheel, the thermal energy that was added to the water will not cause the wheel to begin turning. *Hint*: Which direction would it turn?

E4.11

(a) Butter melts at around 72°F. What is this in °C and in K?
(b) Tungsten melts at around 3,410°C. What is this in K and in °F?
(c) Dry ice, which is solid carbon dioxide (CO_2), melts at around −70°F if held at high pressure. What is this in °C and in K?

E4.12 Which sentence is correct? Explain.

(a) If you want to raise the temperature of a cup of water, transfer thermal energy to it.
(b) If you want to raise the thermal energy of a cup of water, transfer temperature to it.

E4.13 Give an example in which thermal energy is transferred to a substance but its temperature does not increase.

E4.14 You are outside on a cold day and you have two identical thermometers inside your warm coat. You quickly remove them and insert one into a mud puddle and the other into some dry dirt. Both substances are at 3°C, just above freezing.

(a) Which, if either, reaches equilibrium temperature more quickly?
(b) After reaching equilibrium, will both thermometers read the same?

E4.15

(a) A 1 ℓ container of milk and a 0.5 ℓ container of milk are both at room temperature (20°C). Do they contain the same amount of thermal energy? Explain.
(b) Do individual water molecules in the two containers of milk have the same kinetic energy, on average?

E4.16 A small plastic balloon is inflated slightly with helium and tied by string to a rock at the bottom of a large cooking pot that has a small amount of water at the bottom. The pot, with a lid that is slightly open, is placed on a stovetop and gradually heated, eventually warming the water to near boiling. Describe what happens to the size (volume) of the balloon as this experiment progresses, citing the physical principles you use. Assume that the balloon is not damaged during the experiment.

E4.17

(a) If you had a 1 g diamond and a 1 g piece of glass, both at 5°C, and you held one to your right cheek and the other to your left cheek, which would feel colder? Why?
(b) If, instead, you held them together, what would happen to their thermal energies and to their temperatures?

E4.18 If object A and object B are in thermal equilibrium with each other, and object C is in thermal equilibrium with object A, must objects B and C necessarily be in thermal equilibrium? Verify your answer using Newton's law of cooling.

E4.19 Imagine that you are a patent examiner (as was Albert Einstein). Consider a patent application for the following machine. A Joule paddlewheel setup, with a falling mass, is used to heat up water hot enough to run a steam engine, which is used to return the mass to its original height. This cycle is repeated without any external source of energy required. How do you rate this patent application? Why?

E4.20 Perhaps we could build a more economical steam engine by recycling the cooling water. In Figure 4.27, consider hooking the warm water outlet to the cool water inlet and using a pump to move the water around this loop. After a while, what happens? Why does the steam engine stop operating? What is the value of the efficiency after it stops operating?

Problems

P4.1 In E4.1, what fraction of the box is filled with air after the balls are packed in? Do not count the air that is inside the balls. *Hint*: The volume of a ball of radius r is $4\pi r^3$.

P4.2 A compressed or stretched spring stores energy, which can be used later to do work. Say that a 300 kg person falls from a 3 m-high balcony onto a mattress, sinks into the mattress, and then bounces back up 1 m.

(a) How much potential energy was stored in the mattress at the time the person came to rest?
(b) How much of that potential energy was released as the person bounced?
(c) Where did the unreleased potential energy go? That is, in what form is the energy stored afterward?

P4.3 A closed gas-filled container is separated into two equal-volume regions, each with volume V_0, by a thin, rigid divider wall. The number of gas atoms to the left of the divider is 3×10^{24}, and the number to the right is 6×10^{24}. The pressure in the left region is 5 bar. A thin, flat object is just to the left of the divider, as shown. Temperature is kept constant and uniform throughout by a large reservoir of water surrounding the container. The divider is pulled out, exposing the flat object to pressures from both sides. The object slides left or right like a piston, without gas leakage around it.

(a) Immediately following the removal of the divider, what is the net pressure on the object, and which way does it cause the object to move?
(b) After the object comes to rest, what are the volumes of the two regions, and what are the pressures?

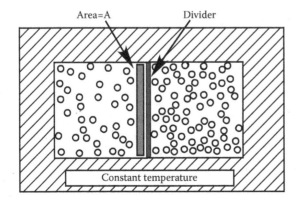

P4.4 As in E4.6, a hydraulic lift is shown in the figure. A smaller cylinder is connected through a tube to a larger cylinder. The surface area of the fluid in the smaller cylinder is A_1, and in the larger cylinder it is A_2. Assume that $A_2 = 4A_1$ and that the fluid is incompressible. A heavy object is placed on a piston in the larger cylinder, as in the figure below. A smaller force F can exert a larger force F' on the other piston. If $F = 100$ N, what is the value of F'? *Hint*: Recall how force relates to the pressure p, which is the same at the top of each cylinder, assuming we can ignore effects of gravity.

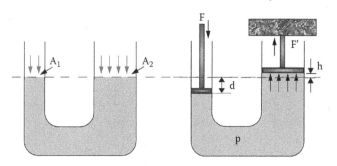

P4.5 The figure shows a hydraulic disk brake in an automobile. The small piston, which is pushed by the foot pedal, has a smaller diameter than does the brake piston. The brake fluid is a nearly incompressible liquid. The purpose of the design is to transfer a relatively small force from your foot, acting through a relatively large distance, into a larger force at the brake pad, acting through a smaller distance. Explain in words and simple equations how this works. What physics principle is behind this action? *Hint*: Consider the difference in the distance the small piston moves compared to the distance moved by the large piston.

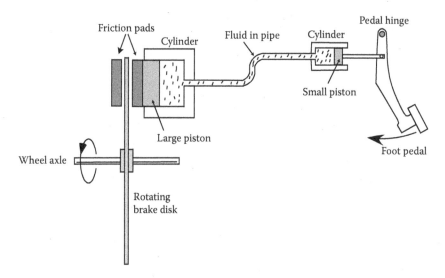

P4.6 Say that the water pump shown in E4.9 generates 12 bar of pressure at its output, and the pressure at the pump's input is 1 bar. The resistor part of the pipe has resistance equal to 1.2 bar · (sec/ℓ). Assume there is zero resistance in the parts of the pipe containing no obstructions.

(a) Without the small holes present, what is the current?

(b) Assuming the holes are small enough not to change the pressure and flow in the pipe, find the pressure at the location of the middle hole. *Hint*: Consider the resistor region to be separated into two distinct resistor regions, one on either side of the middle hole. Their resistances are 0.3 bar · (sec/ℓ) and 0.9 bar · (sec/ℓ). The currents through each resistor region are the same as that determined in (a).

P4.7 The number of atoms in a container with volume 5 ℓ is $n = 4 \times 10^{24}$, and the pressure is 7 bar. If the number of atoms is increased to $n = 9 \times 10^{24}$, and the container's volume is increased to 14 ℓ (while keeping the temperature constant), what is the new pressure?

P4.8 A gas contains 3×10^{24} atoms (e.g., helium) at room temperature.

(a) Calculate the average kinetic energy of this whole gas.
(b) What is the kinetic energy of one atom?
(c) If this gas is enclosed in a volume of 4 ℓ, what is the pressure?

P4.9 In 1845, James Joule predicted that the water at the bottom of Niagara Falls (about 50 m high) should be warmer than at the top by about one-fifth of a degree Fahrenheit. What is the equivalent in Celsius? Verify that his prediction was correct by considering a 1 kg quantity of water falling through gravity for a distance of 50 m (164 feet) and converting all of the gained energy into thermal energy. *Hint*: Recall that the increase of gravitational potential energy incurred by lifting an object with mass M through a height h is $M \times (9.8$ J/m $\cdot$ kg$) \times h$.

P4.10 A Joule-type paddlewheel is fitted with two weights, one with mass $M_1 = 150$ kg, and the other with a variable mass M_2. As they descend slowly under gravity, they both do work on the water. The container holds 5 kg of water.

(a) If only the first mass M_1 is attached, and it slowly descends a distance $d = 5$ m, how much work would be done, and what would be the temperature change of the water?
(b) If the second mass also equaled 150 kg, and it was also attached, answer the same question.
(c) If the second mass equaled 75 kg, and it was also attached, answer the same question.

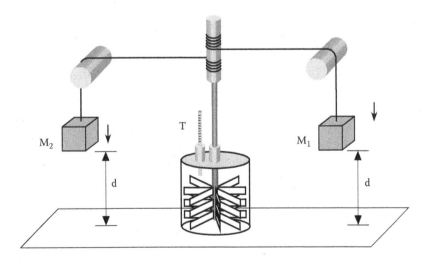

P4.11 (See Real-World Example 4.3.) Consider a bar-shaped substance with heat conductivity denoted by HC, cross-sectional area A, and length L. If the temperature difference between the two ends is $(T_1 - T_2)$, then the rate of thermal-energy flow from one end of the bar to the other end equals $(T_1 - T_2) \times HC \times A \div L$. For example, a square bar of silver, with widths $d = 0.1$ m, has cross-sectional area $A = d \times d = 0.01$ m^2 and length $L = 1$ m. With a temperature difference between the ends

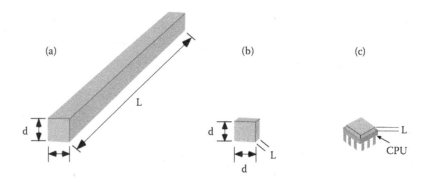

of 30°C, it would conduct thermal energy at a rate equal to 30°C × 429 W/(m · C) × 0.01 m² ÷ 1 m = 1.29 W.

(a) If the bar's widths were doubled to $d = 0.2$ m, how would the rate of heat conduction change?

(b) If the bar were made of aluminum, and its widths were $d = 0.01$ m and length $L = 0.001$ m, what would the rate of heat conduction be? *Hint*: See Table 4.2.

(c) A processor chip in a personal computer generates about 50 J of thermal energy per second, or 50 W. Say it runs at a temperature of 60°C and is cooled by contact with the aluminum thin "bar" described in part (b). The hot side of the aluminum bar is at 60°C, and the other side is kept at 35°C by flowing air. Is this arrangement sufficient to prevent the chip from overheating? Explain.

P4.12 This problem explores Newton's law of cooling. A cup of coffee is initially at 100°C temperature. It sits in contact with cool air at temperature 20°C, and gradually cools until it reaches 20°C. It is assumed that the quantity of air is so large that the air does not appreciably heat up, and that no new sources of heat are applied. How does the object's temperature change with time as it cools? First, notice that at the start, the temperature difference is 80°C, and as the object cools, the difference decreases. This means that the rate of cooling is largest at the start, and decreases as time goes on. Say that in the first minute the temperature decreases by 8°C, meaning that during that 1-min time interval, the rate of temperature change was about 8°C/min. This change is a certain fraction of the initial temperature difference of 80°C. This fraction is 8/80 = 0.1, or 10%. After this first minute, the temperature equals 80 − 8 = 72°C, as listed in **Table 4.3** below. This is a fraction 72/80 = 0.9 of the initial temperature. At the end of each subsequent minute, the temperature decreases to a value given by a factor 0.9 multiplied by its temperature at the start of that minute. This factor, 0.9, is the decay factor.

The table summarizes how the temperature changes in time. The numbers in the column on the far left give the temperature difference between the coffee and the air at the beginning of each 1-min period. The third column gives the temperature at the end of each 1-min period, the result of multiplying the initial temperature by the decay factor 0.9. The fourth column gives the degrees by which the temperature changes in the corresponding 1-min period. The amount of temperature change becomes smaller as the temperature difference becomes smaller, in accordance with Newton's law of cooling. The last column gives the temperature of the coffee at the beginning of each period. It changes from its initial value of 100°C to nearly 20°C after 30 min.

(a) Use a calculator to complete the table.

(b) On graph paper, make a careful plot of the "ending temperature difference" versus time, in minutes. Also make a plot of "actual temperature" versus time.

TABLE 4.3

Temperature Changes

Initial Temp. Difference	Decay Factor during One Minute	Ending Temp. Difference	Temp. Change	Actual Temp.
80.0	× 0.9 =	72.0	8.00	100.
72.0	× 0.9 =	64.8	7.20	92.0
64.8	× 0.9 =	58.3	6.48	84.8
58.3	× 0.9 =			
52.5	× 0.9 =			
47.2	× 0.9 =	42.5	4.72	67.2
42.5	× 0.9 =			
38.3	× 0.9 =			
34.4	× 0.9 =			
31.0	× 0.9 =			
27.9	× 0.9 =			
25.1	× 0.9 =			
22.6	× 0.9 =	20.3	2.26	42.6
20.3	× 0.9 =			
18.3	× 0.9 =			
16.5	× 0.9 =			
14.8	× 0.9 =			
13.3	× 0.9 =	12.0	1.33	33.3
12.0	× 0.9 =			
10.8	× 0.9 =			
9.73	× 0.9 =			
8.75	× 0.9 =			
7.88	× 0.9 =			
7.09	× 0.9 =			
6.38	× 0.9 =			
5.74	× 0.9 =			
5.17	× 0.9 =			
4.65	× 0.9 =			
4.19	× 0.9 =			
3.77	× 0.9 =			

The shapes of the curves in your plots are characteristic of exponential decay. They begin falling rapidly and become less steep as time goes on.

P4.13 Say you had $1,000 saved in a bank account, and you withdrew 5% (or a fraction, 1/20) of the remaining balance at the end of each month. How much remains after one withdrawal? Use a calculator to find out how much would remain after 1 year. Comment on the connection of this problem and Newton's law of cooling.

5

Electricity and Magnetism: The Workhorses of Information Technology

Nothing is too wonderful to be true if it be consistent with the laws of nature.

[W]hen a piece of metal is passed either before a [magnetic] pole, or between the opposite poles of a magnet, ... electrical currents are produced across the metal transverse to the direction of motion.

Michael Faraday
(1832 and 1849)

Portable music players use miniature magnetic hard disks to store digital data. (Courtesy of Apple, Inc.)

Michael Faraday, the discoverer of magnetic induction, upon which magnetic storage and many other technologies are based.

5.1   ELECTRICITY AND MAGNETISM ARE THE BASIS OF COMPUTERS AND THE INTERNET

One of the greatest scientific discoveries is that electricity, magnetism, and light are aspects of the same physics. Prior to the mid-1800s, these phenomena were thought to be unrelated. With the leading experimental studies of Michael Faraday and the theoretical ingenuity of James Clerk Maxwell, scientists learned that these phenomena are all connected to electric charges and their movement. This is an important example of one of the main goals of science—to bring a large number of phenomena under one roof of understanding.

Electricity and magnetism play essential roles in most aspects of computers and Internet technology. There are electronic circuits for fast data storage and processing, magnetic hard drives and tape for high-capacity data storage, and radio and light waves for wireless and optical fiber communication. Electronic circuits are at the heart of computers. Each circuit carries out basic operations called logic operations. The operation of such circuits is governed by the principles of electricity and magnetism. More broadly, the world's economy and the arts and entertainment industries run primarily on electricity and magnetism. From lighting in homes and studios to computers used in making music, art, and videos, they are basic to almost every activity.

5.2   ELECTRIC CHARGE

The key to understanding electricity is to learn the basic properties of *electric charge*—the property of particles that determines electrical forces between them. Charge comes in two types, called *positive* (+) and *negative* (–). These are also called *plus* and *minus*. Negative electric charge is inherent in *electrons*.

An *electron* is an elementary particle that carries a negative electric charge, denoted e.

By *elementary*, we mean that the electron is not made up of any smaller parts, as far as we know. If we consider e to be the *unit* for charge, then the electron's charge is equal to $-1e$. Let us call the amount of charge on an electron an *elementary unit of negative charge*. Electrons are tiny, and every common object contains very many electrons. For example, the ink making up the period at the end of this sentence contains roughly 10^{16} electrons. (The mass of an electron is very small: 0.00000000000000000 000000000000091 kg. In scientific notation, this is 9.1×10^{-31} kg.)

We know that like charges repel each other, and that opposite charges attract; that is, positive charge repels positive charge, negative repels negative, positive attracts negative, and negative attracts positive. How we know this is discussed in In-Depth Look 5.1.

IN-DEPTH LOOK 5.1: THE CONCEPT OF PLUS AND MINUS ELECTRIC CHARGE

How did scientists arrive at the concept that there are two, and only two, types of electric charge? Why are they called *plus* and *minus*? For thousands of years, people have observed that when amber is rubbed against cloth or fur, it begins to attract lightweight objects such as feathers. Most people probably considered this only a curiosity, or perhaps an irritant. Others began to take a more scientific interest in these behaviors. By the 1730s, it was noted that certain objects, after both had been separately rubbed, would repel rather than attract. It was also observed that after being rubbed, an object could be made to transfer its repelling or attracting property to another object, simply by connecting a metal wire between the two.

The simplest idea, or mental model, to explain these observations was evidently that some special substance, called electricity, was being transferred between the objects by the presence of the wire. And, to account for the fact that some rubbed objects repelled whereas others attracted, it was supposed that there were two types of electricity. More than two types could have been hypothesized, but because there is no virtue in making the explanation any more complicated than necessary, such an idea was dropped. In 1747, American scientist and statesman Benjamin Franklin named these two types of electricity plus and minus types. He also correctly surmised that only one of these two types is transferred between objects through a connecting wire, although he incorrectly guessed that it was the plus electricity that is transferred. We now know, through further experiments, that it is the minus electricity (electrons) that is transferred. Franklin also correctly stated that the amount of electricity is constant; that is, it is conserved. Franklin could have named the two types *A* and *B*, or anything else for that matter. He chose plus and minus because it allowed him to use the concept of adding and subtracting amounts of electricity, also called electric charge. He reasoned that objects are normally neutral, meaning that the sum of plus and minus charges added to zero. He surmised that after one type of charge departed an object, it then had more of the opposite type of charge left than the type that had departed.

Although scientists understand the properties of charge and the behaviors of charged objects, they do not know why there is charge. Science is a method for describing how the world is, not why it is.

In addition to electrons, which are negatively charged, there are particles called *protons*.

A *proton* is a particle that carries a positive (+) electric charge $+e$, equal in magnitude but opposite in sign to that of the electron.

Considering e to be the *unit* for charge, the proton's charge is equal to $+1e$.

As a familiar example of electrical forces, when you take your clothes out of the dryer, some objects—such as socks—are stuck tightly together. This arises because some socks have a surplus of electrons, making them negative, and some have a deficit of electrons, making them positive. Let us discuss electrical attracting and repelling forces in some detail, using objects other than socks.

When a plastic rod is rubbed with a fur cloth, electrons move from fur to plastic, and the plastic acquires a negative charge. Protons do not normally move between objects because they are tightly bound up in the structure of the materials. If two such charged plastic rods are brought near each other, they will repel; that is, they have a repulsive force between them. In another example, when a glass rod is rubbed with a silk cloth, electrons move from glass to silk, and the silk acquires negative charge. The glass rod is left with a positive charge, corresponding to a deficit of negative charge. If two such charged glass rods are brought near each other, they repel. On the other hand, if a glass rod, after being rubbed by silk, is brought near a plastic rod, after being rubbed by fur, the two rods attract. These observations are illustrated in **Figure 5.1**.

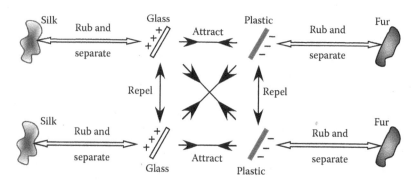

FIGURE 5.1 After being electrically charged from being rubbed by silk or fur, like objects (glass-glass or plastic-plastic) repel, whereas unlike objects (glass-plastic) attract. Minus signs show an excess of electrons; plus signs show a deficit of electrons. The black arrows indicate the forces between rods.

THINK AGAIN

There are two ways to make a formerly uncharged object become positively charged. You can add an excess of protons to the object. Or, you can remove electrons, causing a deficit of negative charge.

QUICK QUESTION 5.1

You have three objects, labeled A, B, and C. In an experiment, you find that A attracts C and that A repels B. Predict whether B and C will attract or repel. What are the possible explanations in terms of charge on each?

This leads to the first principle of electricity.

ELECTRICITY PRINCIPLE (I)

Electric charge is a property of particles that determines the electric force that acts between two particles. Like charges (+ + or − −) repel by electric force. Opposite charges (+ − or − +) attract by electric force.

During experiments of this type, it is always found that the total amount of charge shared between any number of objects is constant; that is, charge is never created nor destroyed, but only moved around. It is also observed that after being partially charged positively or negatively, an object can be made to transfer some of its charge to another object by connecting a metal wire between the two, as in **Figure 5.2**. When one object gains a certain amount of positive charge during the transferring process, the other object gains an equal amount of negative charge. If enough charge is transferred, both

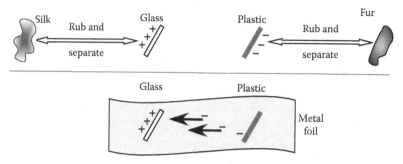

FIGURE 5.2 After glass and plastic rods are electrically charged, they are placed onto a piece of metal foil. Electrons move through the metal foil, from the plastic to the glass.

objects end up with neither a surplus nor a deficit of electrons. Each object has equal numbers of electrons and protons, and has zero net charge. We describe such objects as electrically **neutral**.

It was found that the total, or net amount, of charge on an object could be kept track of by simple addition. For example, say that a small piece of dust has an excess of 1,000 electrons. A second small piece of dust has a deficit of 200 electrons. If these two dust pieces are brought together to make a single, larger dust piece, its net charge is $-1000e + 200e = -800e$. Therefore, if an electron is brought near this combined object, it is repelled.

These facts are summarized by a principle:

ELECTRICITY PRINCIPLE (II)

Electric charge conservation and addition: The total amount of electric charge is conserved; that is, charge is never created nor destroyed. When various numbers of electrons and/or protons are placed together on an object, the values of the electric charges are summed to give the object's net charge.

Electrons are far more mobile than protons. That is, electrons can easily move between two objects whereas protons cannot. Therefore, we can understand that the net charge on an object equals the number of excess electrons or the deficit of electrons on the object.

5.3   ELECTRIC FORCES: COULOMB'S LAW

The force between two charged objects decreases as they are moved farther apart. This seems reasonable—you would not expect a plus-charged sock in Kansas City to strongly attract a minus-charged sock in Seattle. In 1785, the French physicist Charles de Coulomb carried out careful experiments to precisely determine how the force decreases as distance increases. He found a simple rule to explain all of his observations, which we now call Coulomb's law.

ELECTRICITY PRINCIPLE (III)

Coulomb's law: The strength of the electric force between two small (pointlike) charged objects is directly proportional to the product of their charge values and is inversely proportional to the square of the distance between them. In the form of an equation, this is:

$$F = \left(2.3 \times 10^{-28} \ \frac{\text{N} \cdot \text{m}^2}{e^2}\right) \times \frac{(\text{charge on first object}) \times (\text{charge on second object})}{(\text{distance between})^2}$$

$$F = \left(2.3 \times 10^{-28} \ \frac{\text{N} \cdot \text{m}^2}{e^2}\right) \times \frac{q \cdot q'}{d^2}$$

In the equation, the symbols q and q' equal the net charges on the two objects; that is, the excess or deficit number of electrons on each times the unit of charge e. The symbol d equals the distance separating the charges. The number $2.3 \times 10^{-28} \ \text{N} \cdot \text{m}^2 / e^2$ is the constant of proportionality.

Note that the formula for Coulomb's law includes the possibility for describing either positive or negative charges. If q is positive and q' is negative, then the product $q \times q'$ is negative, meaning the force pulls the two objects toward each other.

QUICK QUESTION 5.2

A small piece of dust has an excess of 100 electrons. A second small piece of dust has a deficit of 250 electrons. Do these two pieces repel or attract? If these two dust pieces are brought together to make a single, larger dust piece, what is its net charge? Is it positive, negative, or neutral? What force, if any, does a single electron feel if it is near this larger object?

For example, consider two objects, each charged with 66 billion (6.6×10^{10}) excess electrons, and separated from one another by 1 millimeter. Each will feel a repelling force equal to:

$$F = \left(2.3 \times 10^{-28} \; \frac{N \cdot m^2}{e^2} \right) \times \frac{(-66 \times 10^9 e) \times (-66 \times 10^9 e)}{(10^{-3} m)^2}$$

$$F = \left(2.3 \times 10^{-28} \; \frac{N \cdot m^2}{e^2} \right) \times \frac{4.36 \times 10^{21} \, e^2}{10^{-6} \, m^2} = 1.0 \; N$$

This is 1 newton (N) of force. If these objects are moved twice as far away, to 2 millimeters, they will feel a force one-fourth as strong, or 0.25 N. If the charge on only one of the objects is doubled, the force will double. If the charge on both objects is doubled, the force will quadruple, etc.

A second example involves quarter-pound hamburgers. Each burger contains roughly 10^{26} electrons. If we could transfer just one out of every 10^{15} of the electrons in one burger to the other burger, that would create $10^{26} \div 10^{15} = 10^{11}$ excess electrons on one burger and a deficit of the same number on the other. That would create a force slightly stronger than 1 N. Then the two burgers would attract each other by a force stronger than the force of the Earth's gravity on one burger, which is about 1 N. If we were to hold the upper burger up by hand, the lower burger would be held by it, and would not fall under the force of gravity, as in **Figure 5.3**. It seems remarkable that such a small fraction of the burgers' electrons can cause a force stronger than gravity.

For another example, one piece of dust has charge $2 \times 10^9 (-e)$, and another has a charge $3 \times 10^{11} (+e)$. If they are 2 millimeters apart, the electric force between them equals:

$$F = 2.3 \times 10^{-28} \; \frac{N \cdot m^2}{e^2} \times \frac{2 \times 10^9 (-e) \times 3 \times 10^{11} (+e)}{(0.002 \; m)^2} = -0.034 \; N$$

The minus sign indicates that this force is attractive, not repulsive.

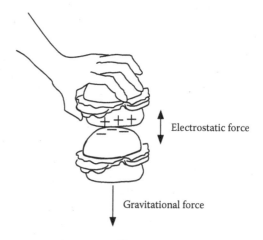

FIGURE 5.3 Transferring just one part in 10^{15} (one millionth of a billionth) of the electrons in one burger to another burger would create an electrostatic attractive force stronger than 1 N, which is strong enough to counteract the force of the Earth's gravity on the lower burger.

IN-DEPTH LOOK 5.2: THE DISCOVERY OF THE ELECTRON

How was the electron discovered? In the 1850s through the 1890s, several German and British physicists were studying a peculiar phenomenon—a glowing gas in a glass tube containing a low-pressure atmosphere. A glass tube, with two metal wires inserted through the walls, was partially evacuated of air. When the two wires were hooked to opposite sides of a **battery**, as illustrated in **Figure 5.4**, a glow was observed in the low-pressure air between the wires. This was the precursor of the modern-day fluorescent light bulb, but at that time, scientists did not know what caused the glow. They suspected that some kind of "rays" were emitted by the cathode, and they called these cathode rays. They did not know what the rays consisted of.

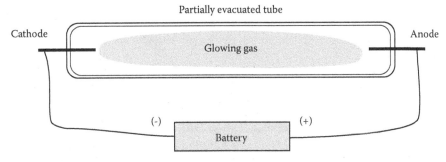

FIGURE 5.4 Experiment used in the nineteenth century to study electrically excited air.

In 1897, J. J. Thomson, a physics professor at Cambridge University in England, constructed a different design of tube, illustrated in **Figure 5.5**. He reasoned that if the cathode rays were charged particles, they should be susceptible to electric forces. To test this hypothesis, he added a second region to the tube where the rays would pass between two metal plates that were connected to a second battery. A battery charged the cathode negatively, creating the cathode rays, which were attracted toward the positively charged anode. With this design, the rays that were generated by the cathode could pass through a small hole in the anode and continue on, passing between the metal deflection plates. When the rays struck the far right end of the tube, which was coated on the inside with a special material called a phosphor, the material glowed at a small spot. Thomson found that when he applied a battery to the deflection plates, the position of the glowing spot moved. He made careful measurements of the deflection of the rays for various combinations of batteries.

Thomson said, "I can see no escape from the conclusion that [cathode rays] are charges of negative electricity carried by particles of matter." He also inferred that the particles were about 1,000 times lighter than a hydrogen atom and were basic constituents of atoms. Thomson was awarded the Nobel Prize in Physics in 1906 for these discoveries. Thomson's cathode ray tube (CRT) is the precursor of the television or TV tube and the computer monitor. These are discussed in Chapter 3 in Real-World Example 3.2. Remarkably, Thomson discovered the electron and invented the TV tube at the same time!

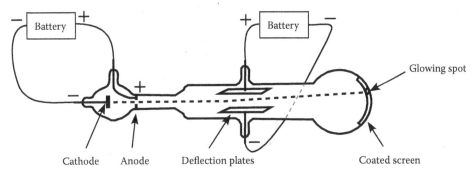

FIGURE 5.5 Experimental apparatus used by J.J. Thomson for discovering electrons.

5.4   ELECTRIC FIELDS

How does one charged object exert a force on another if the two are a distance apart? Physicists have discovered that this occurs indirectly, rather than directly. For example, if a positively charged dust particle comes into the vicinity of a positively charged glass ball, it will be repelled away from the ball. In a sense, they reach out and push on each other. The agent by which charged objects reach out is called an *electric field*.

An *electric field* is an (invisible) agent through which a charged object exerts a force on another charged object.

The electric field can exist in a medium such as air or water, or even in a *vacuum* (i.e., a region such as deep outer space, where there are virtually no atoms or other material medium).

An electric field at a particular location can be represented by an arrow, drawn pointing in the direction of the force on a proton at that location. The length of the arrow, called a *vector*, is proportional to the strength of the force. **Figure 5.6** shows an electric field vector as a big arrow at a specific location. The direction in which the arrow points indicates the direction of the *field*. At this location a proton, which carries an elementary unit of positive charge, would feel a force in the same direction as the direction of the field vector. On the other hand, if an electron, which carries negative charge, were placed at the same location, it would feel a force opposite to the direction of the field vector. The following rule holds:

A positive-charged object feels a force in the same direction that the electric field vector points and is proportional to the length of the field vector.

A negative-charged object feels a force opposite to the direction that the electric field vector points and is proportional to the length of the field vector.

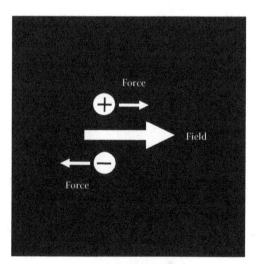

FIGURE 5.6   The larger arrow illustrates the electric field vector. A proton, having positive (+) charge, would feel a force (the small arrow) pointing in the same direction as the field vector. An electron, having negative (−) charge, would feel a force in the direction opposite to the direction of the field vector.

(a)   (b)

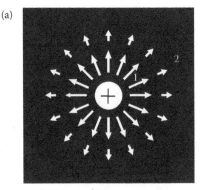

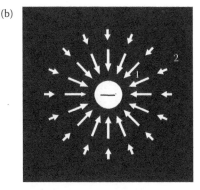

Electric field created by a plus charge   Electric field created by a minus charge

FIGURE 5.7 (a) A positively charged ball creates an electric field with a pattern shown in the left diagram. The field is strongest closest to the center, as indicated by longer arrows. (b) A negatively charged ball creates an electric field with a pattern shown on the right. In both cases, the arrows show the direction that a positively charged object would be pushed if placed at that location.

Consider the electric field that is created by a positively charged, small plastic ball. A broad view of the electric field created can be represented by a collection of vectors, as in **Figure 5.7a**. The arrows in Figure 5.7a indicate the direction and strength of the force that a proton, which carries positive charge, would feel if it were located near the ball. At the position labeled 1, the proton would feel a strong force away from the ball, along the line connecting the proton and the ball. At the position labeled 2, the proton would feel a weaker force in the direction away from the ball. On the other hand, if an electron, which has negative charge, were placed near the positively charged plastic ball in Figure 5.7a, it would feel an attractive force toward the ball. An electron feels a force in the direction opposite to the direction of the field vector at that point.

Figure 5.7b shows the electric field vectors that surround a negatively charged ball. If any positively charged object enters the ball's vicinity, it will be attracted toward the ball. A negatively charged object would be repelled from the ball. The drawings of the field in Figure 5.7 are in agreement with Coulomb's law. Namely, the field, and thus the resulting force, is stronger at locations that are closer to the central charged object.

An electric field is not a substance; it does not have mass. Yet it does contain energy. We call this electric-field energy, an additional type of energy not mentioned in our earlier discussions.

THINK AGAIN

It might seem that the concept of a field is not really needed. It might seem to be just an intermediary step in our reasoning about forces. However, electric fields have the ability to exert force on any particle or body that possesses an imbalance of plus or minus charges. Fields carry energy through solids, gases, and empty space. Fields are real. We will see in the following chapter that radio and light waves are best understood in terms of the concept of fields.

The word *field* might remind you of a field of wheat, with the stalks pointing in various directions, as in **Figure 5.8**. To make another fanciful analogy, let us say there

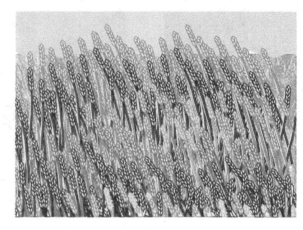

FIGURE 5.8   A collection, or field, of wheat stalks, each of which point in a different direction.

are two types of bugs that live in a wheat field. Assume a windless day, so each stalk is pointing in a fixed direction. One type, call them ladybugs, feel a strong affinity for the tip of a stalk and repulsion from the root of a stalk. The ladybug will move in the direction of the stalk pointing toward the tip. The other type of bug, call them gentleman bugs, feel a strong attraction to the root of a stalk and repulsion from the tip. Let us draw a vector to represent each wheat stalk, with the arrow pointing from the root to the tip, as in **Figure 5.9**. The length of each vector indicates the strength of the force a bug will feel. Looking at the picture of the wheat field, we can immediately predict in which direction either type of bug will feel a force. A ladybug feels a force in the direction of the vector she is closest to, whereas a gentleman bug feels a force opposite to the direction of the vector.

To summarize, a field can be represented by a pattern of field vectors with various lengths and directions at different points in space. Depending on the charge of an object placed at a particular location, the field vector at that point tells us the direction and strength of the force exerted on the object.

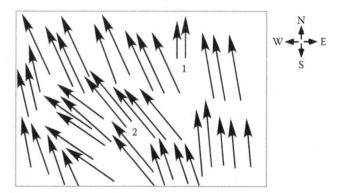

FIGURE 5.9   A field of vectors, each of which may point in a different direction and have a different length. The length of a vector indicates the strength of the field at that location. A ladybug at the location labeled 1 will feel a small force in the north direction, and a gentleman-bug there will feel a small force in the south direction. At the location labeled 2, a ladybug will feel a large force in the northwest direction, and a gentleman-bug will feel a large force in the southeast direction.

IN-DEPTH LOOK 5.3: ELECTRIC FIELD LINES

When describing electric fields surrounding charged objects, it is simpler to draw *electric field lines* rather than the more complicated pattern of field vectors, such as those in Figure 5.7. Electric field lines are lines or curves that indicate the direction of the electric force exerted on a plus-charged object located at that position. The field lines trace through a series of connected field vectors. For example, we can redraw Figure 5.7 showing the electric field around a plus or a minus more simply using field lines, as in **Figure 5.10**.

The electric field lines around a positive charge are drawn "leaving" the charge. The electric field lines around a negative charge are drawn "approaching" the charge. That is,

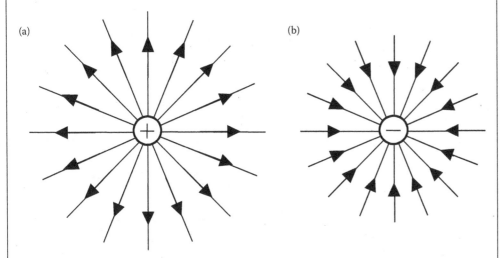

(a)   (b)

FIGURE 5.10   Electric field lines around a plus- or minus-charged object.

electric field lines begin on positive charges and end on negative charges. The field lines show the direction, but not the strength, of the field at each location.

If a positively charged ball is near a negatively charged ball, the field lines around them combine into a single pattern, as shown in **Figure 5.11**. If a small, positively charged object, such as a single proton or a dust speck with a deficit of electrons, is placed somewhere between these two balls, it will feel a force pushing it along the field line that it is on. This will push it toward the negatively charged ball, as expected. A small, negatively charged object will be pushed along the field line in the opposite direction.

If two flat plates of metal are charged oppositely, the field lines form nearly straight lines between them, as in **Figure 5.12**. If a small, positively charged object is placed somewhere between these two plates, it will feel a force pushing it along the field line that it is on. This will push it toward the negatively charged plate.

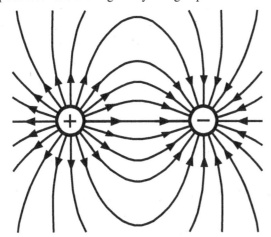

FIGURE 5.11   Electric field lines around two nearby plus- and minus-charged objects.

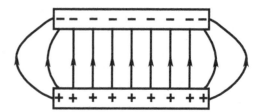

FIGURE 5.12   Electric field lines between two oppositely charged metal plates.

5.5   ELECTRIC CURRENT AND CONDUCTORS

Electric current is the flow of electric charge, as a river current is the flow of water. For example, when a copper wire is connected between the two ends of a flashlight battery, electrons flow from the minus end of the battery, through the wire, to the plus end of the battery. Electrons can flow in some materials and not in others. For example, a copper wire allows current to flow within it, but the plastic sheath around a wire is there to prevent current from flowing from the wire to another object, such as your hand, should you touch it. Typically, electrons can flow through a material, such as metal, but plus-charged objects cannot.

Electric current is defined to be in the direction in which positive charge (i.e., a deficit of electrons) flows. Therefore, in a circuit in which electrons are flowing, the current is in the direction opposite to that of the movement of the electrons.

For example, in Figure 5.2 the electrons were moving in the directions indicated by the solid arrows. Current, however, is defined to flow in the direction opposite of these arrows. It is helpful to refer to this as positive current to remind us which way positive charge is moving. If this definition of current seems awkward, blame Benjamin Franklin, who guessed wrongly that positive particles move in materials, when actually it is the negative particles that move.

Because very many electrons are moving in a typical electric current, we use a unit of charge called the *coulomb*, which equals the charge of 6.2×10^{18} protons. The symbol for the coulomb is C. Therefore,

$$1 \text{ C} = 6.2 \times 10^{18} e$$

Current is expressed in units of *amperes*, called *amps* for short. One ampere (A) is the amount of current present when 1 coulomb of positive charge (i.e., $6.2 \times 10^{18} e$) moves between two locations in 1 second. That is,

$$1 \text{ A} = \frac{1 \text{ C}}{1 \text{ sec}}$$

For example, if the heater wire in a toaster carries 10 A of current, then 10 coulombs pass through it each second.

An *electrical conductor* is a material, such as metal, through which electric current can flow. An *electrical insulator* is a material, such as glass or plastic, through which electric current cannot easily flow. A *semiconductor* is a material with properties in between those of conductors and insulators. The origins of the conducting or insulating properties of various materials will be discussed in Chapter 9. We can describe the degree to which a type of material can conduct electricity by its *electrical conductivity*.

Electrical conductivity indicates the ability of a material to transport (i.e., *conduct*) electric current.

QUICK QUESTION 5.3

A toaster draws 10 A of current when operating. In 2 seconds, how many electrons pass through the heater coil of this toaster?

A good conductor is said to have high conductivity, and a poor conductor has low conductivity. **Table 5.1** gives some values of electrical conductivity relative to the value for silver, which has the highest conductivity of any metal. For example, the conductivity of silver is at least 1 billion times greater than that of drinking water. Silver, copper, and aluminum are conductors. Glass, pure water, and plastics such as Teflon are insulators. Materials in between, such as silicon, are semiconductors.

The amount of current that will pass through a bar of material is proportional to the electrical conductivity of the material making up the bar. **Figure 5.13** illustrates that a 1 millimeter long wire made of silicon (a fair conductor) has the same conducting ability as an equal-diameter wire made of silver (a good conductor) that is 100 meters long. A bar made of Teflon would allow essentially zero current to flow.

A simple model for why some materials conduct readily whereas others do not is illustrated in **Figure 5.14**. In a material with high conductivity, electrons encounter few obstructions and pass through easily. In a material with low conductivity, electrons encounter many obstructions, which may be irregularities in the structure of the material, or may be due to the nature of the atoms or molecules making up the material. Electrons bounce from the obstructions, hindering their passage through the material.

This simple model is adequate for comparing the conductance of two similar materials, such as two metals. It does not explain why some materials, such as glass, are insulators, an understanding that comes only with knowledge of atomic structure.

TABLE 5.1
Electrical Conductivity of Materials (at 20°C)

Material	Electrical Conductivity (Relative to Silver*)	Comments
Silver	1.0	Highest conductivity of any metal
Copper	0.95	Most common for circuits
Aluminum	0.60	Low cost
Tungsten	0.30	Light bulb filament
Silicon (pure)	1×10^{-5}	The main material in computer chips
Sea water	1×10^{-7}	Contains ions (charged atoms)
Diamond	6×10^{-9}	High cost
Drinking water	1×10^{-9} to 1×10^{-11}	Depends on ion content
Deionized water	1×10^{-13}	Low ion content
Glass (SiO_2)	1.6×10^{-17} to 1.6×10^{-21}	Depends on ion content
PTFE	1.6×10^{-22}	Polytetrafluoroethylene, the plastic Teflon (Dupont)

* The electrical conductivity of silver is 6.3×10^{7} $(\text{ohm-m})^{-1}$, although we will not use these units here.

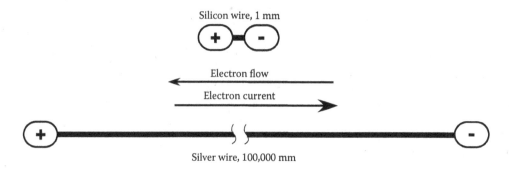

FIGURE 5.13 A silicon wire that is suddenly connected between two oppositely charged objects will allow electric current to flow between them, eventually making both neutral. A silver wire of the same diameter and 100,000 times longer will allow the same amount of current to flow as does the shorter silicon wire.

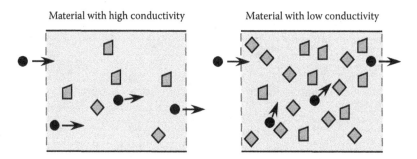

FIGURE 5.14   The number of obstructions in a material determines its electrical conductivity.

5.6   ELECTRICAL ENERGY AND VOLTAGE

We learned in Chapter 3 that accelerating an object through a distance by applying a force to the object increases its energy. The increased energy equals the work done on the object, and can be in the form of kinetic energy (energy of motion) or potential energy (energy of position).

Electric potential energy is the potential energy gained by a charged object if it is accelerated through an opposing electric force. If an object has positive electric potential energy, this means it is located at a position where it has the potential to be pushed to a new position by the existing electric field. It is measured in energy units, joules (J).

Consider two parallel metal plates, which are oppositely charged, as in **Figure 5.15**. If a single proton, which has charge +e, is placed between the plates, it will "feel" a force directed toward the negative plate. Let us say that the proton begins at the negative plate, and you use your hand to overcome the electric force and push the proton to the positive plate. You did physical work on the proton, although this is a very small amount of work,[1] because a proton has a very small charge. Now the proton, being at the plus plate, has the potential to be pushed by the electric field back to the minus plate if you release it; that is, the proton has electrical potential energy by virtue of its location. We specify a charge's potential energy using the concept of *voltage*.

Voltage is a measure of a charged object's change in electric potential energy that is associated with moving from one location to another.

---

[1] The work equals 0.16×10^{-18}J, or 0.16aJ, which is sometimes called one electron-volt.

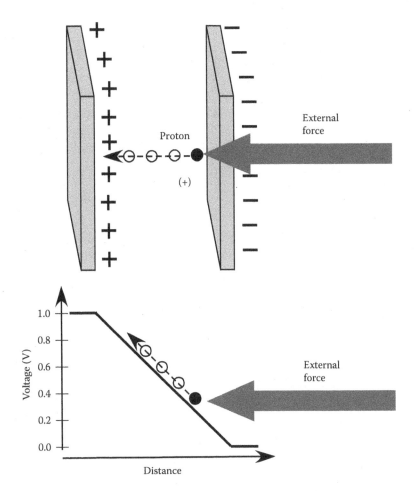

FIGURE 5.15   A proton between oppositely charge plates. An external force pushes the proton toward the plus plate, doing work on the proton and moving it to a location of higher voltage.

Voltage is not specified in terms of the amount of charge on an object, but in terms of the electrical environment around the object. We say that the positive plate on the left is at higher voltage than is the negative plate on the right. The voltage at an object depends only on its position in an electric field.

Consider a situation in which you have a small object—call it the "test object"—having 1 C ($6.2 \times 10^{18} e$) of positive charge on it, and you move it by hand from the negative plate to the positive plate in Figure 5.15. If it requires doing 1 J of work to move the test object from one plate to the other, then we say that the voltage between the plates equals one *volt*. The symbol for volt is V.

Volt: One joule of energy, or work, will move 1 C of charge through 1 V. That is,
$$1J = 1C \cdot 1V$$

The relationship between voltage, potential energy, and charge is:

$$\textit{electrical potential energy (J)} = \textit{charge (C)} \cdot \textit{voltage (V)}$$

$$\textit{voltage (V)} = \frac{\textit{electrical potential energy (J)}}{\textit{charge (C)}}$$

For example, pushing an object having a net charge of 3 C between two plates with a voltage of 9 V between them requires 3 C × 9 V = 27 J.

A simple analogy helps us understand this relationship of voltage, charge, and energy. Say that a certain hill is 20 feet high, and you carry a bucket of water to its top. This requires you to expend a certain amount of energy while increasing the potential energy of the water. If you were to carry two buckets up the same hill, it would require twice as much energy. If the hill were twice as high, it would require twice as much energy to carry up the same number of buckets. The height of the hill is analogous to voltage, whereas the amount of water is analogous to charge. To determine the potential energy, you must know both.

We can summarize the above discussion by the following principle:

ELECTRICITY PRINCIPLE (IV)

Electrical energy is required to move a charged object when its motion is being opposed by an electric force. The amount of required energy equals the number of coulombs of charge on the object multiplied by the voltage between the object's starting and ending locations. That is,

number of joules = (number of coulombs) × (voltage)

potential energy (J) = *charge* (C) × *voltage* (V)

THINK AGAIN

It is tempting to use the phrase "voltage difference," but this is redundant. Why is that so?

5.6.1 Voltage Sources—Batteries

A battery is a device that creates a voltage between two locations, which are called terminals. As discussed above, a charged object experiences a potential energy difference between these two locations that is equal to the charge (in coulombs) times the voltage difference. A simple way to think of a battery is to imagine two chambers with an impenetrable, nonconducting barrier separating them, as in **Figure 5.16a**. One chamber (the minus side) has an excess of electrons and the other chamber (the plus side) has a deficit of electrons. The barrier effectively blocks any electrical attraction directly through the barrier between charged particles in opposite sides. Yet a negatively charged test object placed near the battery minus terminal feels forces pushing it away from the minus terminal and pulling it toward the plus terminal.

Figure 5.16a is not a realistic drawing of a typical battery's shape. Figure 5.16b shows a more typical battery with conducting metal wires attached to terminals. The wire provides a conduit for electrons to move in. Electrons within the wire feel an attractive force along the direction of the wire that leads to the plus side. In the figure, an electron is symbolized by a minus sign (–). A region where there is a deficit of electrons is represented by a plus sign (+). One chamber (the minus side) has an excess of electrons and the other chamber (the plus side) has a deficit of electrons. When allowed to flow, some of the electrons from the minus side will rush out of this side because of the electric forces being exerted by the remaining electrons. If given a pathway, such as a metal wire, they will move along the path and enter the plus chamber to which they are attracted.

Figure 5.16b is a simple example of an *electric circuit*, meaning a group of connected objects through which electrons can flow. Electric current will flow steadily

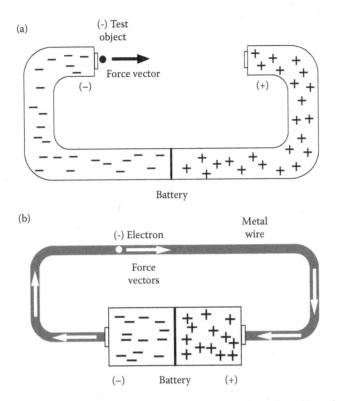

FIGURE 5.16   (a) Simple concept of a battery. (b) Metal conducting wires allow electrons to escape the minus side of the battery and move to its plus side.

only if there is an unbroken path of conducting material between two locations of unequal voltage.

THINK AGAIN

In circuits, only the minus-charged objects (electrons) move through wires. Plus-charged physical objects do not move through wires.

A battery provides positive potential energy to a plus-charged object and negative potential energy to a minus-charged object. A helpful analogy is a skater on a skateboard ramp, as in **Figure 5.17**, with the height providing positive potential energy. The battery has a wire connected between the plus end (B) and the minus end (A). If

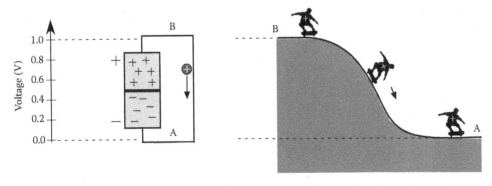

FIGURE 5.17   Skater analogy for motion of a plus charge in the presence of a voltage created by a battery.

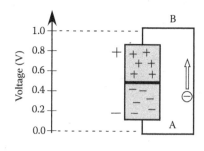
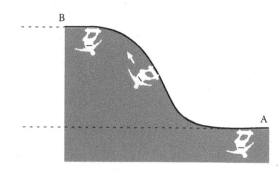

FIGURE 5.18 A pretend skater analogy for motion of a minus charge in the presence of a voltage created by a battery.

a plus-charged object is located at position B, it will be pushed toward the A position. This is analogous to the skater at the top of a ramp going downhill under the force of gravity. The skater had potential energy at the top of the ramp and kinetic energy at the bottom of the ramp.

In the case of gravity, there is only one type of skater—all skaters go downhill under the force of gravity. In contrast, in the case of electric forces, there are two types of "skater"—one plus and one minus charged. Whereas the plus objects go "downhill" with respect to the voltage difference, negatively charged objects go "uphill" with respect to the voltage. Although for this there is no physical analogy using gravity, we can make up a pretend analogy, as in **Figure 5.18**. A new type of skater, called here a negative skater, is drawn with a minus sign on her shirt and inverted compared with an ordinary skater. It is drawn inverted to remind us that we are talking only about a pretend analogy. This negative skater would move uphill naturally, without making any effort, just as an ordinary skater naturally moves downhill.

THINK AGAIN

In the case of electric forces there is a third type of object—neutral objects. What would be the motion of the analogous "neutral skater" on the ramp in this case?

If conducting wires are connected to the battery terminals, as in **Figure 5.19**, each wire is brought to the same voltage as the corresponding terminal. The connected plates now act as extensions of the battery terminals. If a 1-V battery is connected in this way to a pair of metal plates, as in Figure 5.19a, a voltage equal to 1 V is created between the plates. To move a test charge from the negative plate to the positive plate requires applying an external force. If the test object contains 6.2×10^{18} excess protons, or 1 C of charge, then to move it from the negative plate to the positive plate requires doing 1 J of work.

If you make the wires shorter, as in Figure 5.19b, the voltage between the plates still equals 1 V. Therefore, if you apply an external force to move the same test object from the minus terminal to the plus terminal, it still requires 1 J of work. The amount of work does not depend on what path you use in moving the charge. The voltage and the work depend only on voltages at the starting and ending locations of the charged object.

You would need nearly the same amount of energy to move the test object from the plus plate to the minus plate as is needed to move the object from the battery's plus side

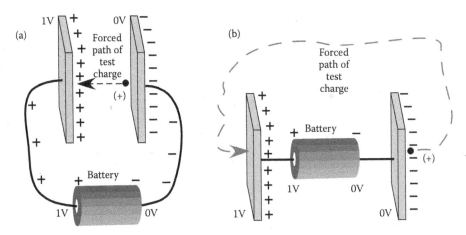

FIGURE 5.19   (a) A 1 V battery connected to a pair of metal plates creates a voltage equal to 1 V between the plates. If an outside agent (not shown) pushes a plus-charged object toward the plus plate, the agent does work on the object, increasing the object's energy. (b) The amount of work required in moving the object from the minus to the plus plate does not depend on the locations of the plates, nor on the path taken by the object when the agent pushes it.

to the battery's minus side. This means that the voltage is nearly equal everywhere on the plus wire and the plus plate. Also, the voltage is nearly equal everywhere on the minus wire and minus plate. The two voltages differ by the amount of voltage provided by the battery. We can summarize as:

Rule: For an object made with a material having very high electrical conductivity (such as copper), the voltage is nearly equal everywhere on the object.

5.6.2   Energy Stored in a Battery

In **Figure 5.20**, a positive test object is located near the plus plate. In this case, rather than requiring an outside agent to do work on the test object, the test object can do work on an outside agent. If you put your hand between the test object and the minus plate, the test object would exert a force on your hand toward the minus plate. If you allowed the test object to push your hand toward the minus plate, the test object would be doing work on your hand.

From this, we can see that a battery stores energy, which can be transferred to other objects, such as your hand. To recharge a spent battery, we use a battery charger to push many electrons into its minus end, and pull electrons out of its plus end. Clearly, this requires us to do work on the electric charges. The energy stored is electric potential energy, which has the potential (ability) to be released later to do work, such as running a computer or an electric motor.

The voltage provided by an outlet in a house is typically 120 or 240 V. Such large voltages are needed when large forces must be created, such as for driving the drum of a washing machine. In contrast, in a device like a computer, where large forces are not required, smaller voltages are used—12 V for a laptop battery.

5.6.3   Energy Stored in a Capacitor

The combination of two metal plates we have been discussing makes a device called a *capacitor*. The name arises from the capacity of the device to store charge. **Figure 5.21** shows a sequence of events, using a battery, a capacitor, and two electrical switches.

QUICK QUESTION 5.4

If the test object in Figure 5.20 carried a charge of plus 5 C, and the battery was 1.5 V, how much work (in joules) would the test object do as it pushed your hand toward the minus plate?

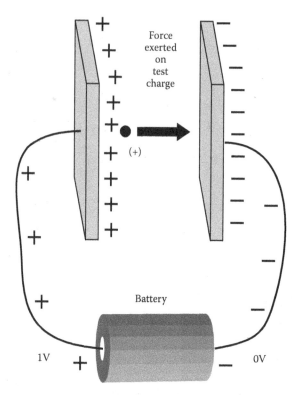

FIGURE 5.20   A 1 V battery connected to a pair of metal plates creates 1 V of voltage between the plates. If a positive test charge (the solid dot) is located near the plus plate, it will feel a force pushing it toward the minus plate.

QUICK QUESTION 5.5

In Figure 5.21b, charge is not flowing. Why not?

Charge can flow through a switch only when it is closed. If a switch is open, it prevents charge from moving through that part of the circuit. In Figure 5.21a, switch 1 is suddenly closed, causing electrons in the minus side of the battery to "feel" a force along the wire toward the plus end of the battery. These electrons can travel only as far as the plate on the right, where a minus charge builds up. Electrons originally on the left plate feel repulsion from the electrons now on the right plate and an attraction to the plus end of the battery, so they flow to the plus terminal of the battery. In this case, there is no closed-loop circuit, as the capacitor breaks the circuit. Therefore, charge cannot flow indefinitely. The current continues only until the voltage across the two plates equals the voltage of the battery. Then charge ceases to flow.

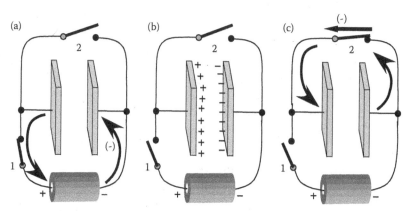

FIGURE 5.21   Capacitor circuit: (a) charging, (b) storing, and (c) discharging.

After this voltage equality is established at the two plates, switch 1 is opened, as in Figure 5.21b. Both switches are now open, and charge is prevented from moving. Charge, and therefore energy, is stored in the capacitor. To discharge the capacitor, switch 2 is closed, as in Figure 5.21c, allowing the minus charge to leave the right capacitor plate and flow to the left plate. This makes the plates neutral again.

A practical way to make a capacitor is to press a thin piece of paper or plastic, which acts as an insulator, between two smaller sheets of metal foil and then roll them up into a cylinder shape. Capacitors, in miniaturized form, are used in many electrical circuits, including those in electronic computer memory, as explained in Real-World Example 5.1.

REAL-WORLD EXAMPLE 5.1: CAPACITOR COMPUTER MEMORY

Data storage in computers is accomplished by means of capacitors. Electronic memory in a computer is random-access memory (**RAM**). This means that you can access the data in any individual memory location, or cell, without reading out the whole memory or some large block of it. This differs from, for example, compact-disk memory, which requires you to read out large blocks of data. Most of the memory in computers is dynamic RAM or DRAM, which can store a bit value for only about 1 millisecond. In DRAM, each bit value is represented by the amount of charge stored on a particular capacitor, as in Figure 5.21. The scheme for controlling the read-in, storing, and readout of charge is shown in **Figure 5.22**. The region enclosed by a dashed line is one cell of memory. Inside the memory cell is a capacitor and a switch, controlled by current passing from outside

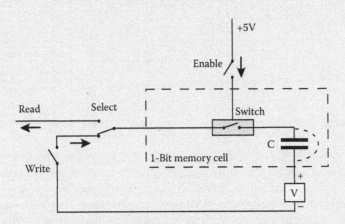

FIGURE 5.22 Capacitor-based memory cell.

of the chip into the chip on the wire shown. The switch is a transistor device, whose operation will be examined in later chapters. For now, you can think of it as a mechanical switch.

When the *Enable* switch is open, as shown in the figure, the capacitor is cut off from the rest of the world, and whatever charge is on the capacitor is stored there for a while. When the *Enable* switch is closed, then charge can be transferred to or from the capacitor, writing or reading a bit value. To write a bit value onto the capacitor, set the *Select* switch to *Write*, as shown. Then the position of the *Write* switch determines whether 1 is written (*Write* switch closed) or 0 is written (*Write* switch open). If, instead, you wish to read the bit value stored on the capacitor, set the *Select* switch to *Read*. Then any charge that is on the capacitor will flow into the *Read* circuit (not shown). Charge will not remain on a capacitor forever. There is a small, unintended leakage of charge between the plates, indicated in the figure by the dashed, curved line.

In a typical DRAM memory chip, millions or billions of cells each store one bit. **Figure 5.23** shows how this is done. Cells are arranged in rows, labeled Row 1, 2, 3, etc. This arrangement is called a memory array. There is a collection of *Enable* switches, arranged at the tops of columns labeled A, B, C, etc., in the figure. Each *Enable* switch controls one whole column of cells: when that switch is closed, every cell in the column below it is enabled. In the configuration shown, the *Enable* switch labeled "B" is closed. The other *Enable* switches are open. To select which individual cell will be accessed, a *Row Select* switch is used to connect the data *Read* or *Write* lines to only one row of cells. In the example shown, Row 1 is selected. Therefore, only one cell is both enabled and selected; this cell is indicated by a dashed oval. Now data can be read or written at this individual cell, leaving all other cells unaffected. To address other cells, we would use different combinations of *Enable* and *Row Select* switch settings.

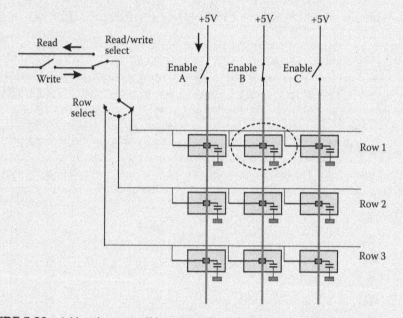

FIGURE 5.23   Addressing one cell in a dynamic random-access memory (DRAM) array.

In a semiconductor DRAM cell, electric charge does not stay on the capacitor forever. It "leaks" from one capacitor plate to the other, because the insulator material between them (a form of silicon) is not a perfect insulator. Every material has a nonzero conductivity, meaning it will conduct current at least weakly. Because of charge leakage, after 1-millisecond passes, each data bit value in the memory has to be rewritten. Imagine writing a 100-page novel on paper using disappearing ink, every letter of which has to be rewritten every thousandth of a second! DRAM is used because it is cheap to make in large quantities. A 500-megabyte DRAM chip might cost $40. That is 4 billion capacitors, at 0.000001 cents each.

5.7   RESISTORS, CONDUCTORS, AND OHM'S LAW

Recall from Section 5.5 that electrical conductivity is a material's ability to conduct electric current. A term that is inversely related to conductivity is resistance. ***Electrical resistance*** is the ability of a wire or bar of any material to resist the flow of charge through it. Therefore a bar made of a material with low conductivity has high resistance.

Resistance is analogous to friction. For example, in Figure 5.13 the short silicon wire and the long silver wire have the same electrical resistance. This is because silver has higher conductivity than does silicon. Electrical conductivity is a property of only the material making up the wire, whereas resistance is a property of the wire as a whole, which depends also on the dimensions of the wire. For two bars having the same dimensions, the one made of the material with the highest conductivity has the lowest resistance. The unit for electrical resistance is the **ohm**, abbreviated by the Greek letter omega, Ω. A typical copper wire, 1 ft. long, has a resistance less than 1 Ω.

A *resistor* is a piece of some material manufactured in a controlled way to determine its electrical resistance. Resistors are used in electronic circuits, including those in computers. A typical resistor in a circuit might be made of graphite (powdered black carbon), thin films of metal, or long, thin wires. In **Figure 5.24**, points A and B are held at unequal voltages by the battery, and a wire connects each of these points to the ends of a resistor. Current flows through the resistor. As shown in Figure 5.24b, an incandescent light bulb contains a type of resistor that emits light when heated by current passing through it. It is helpful in thinking about resistors to imagine that each resistor is a light bulb.

The amount of current in a circuit is given by a rule called *Ohm's law*, named after Georg Ohm.

Ohm's law: The current in a circuit is proportional to the voltage across the circuit, and inversely proportional to the resistance of the circuit. That is,

$$current \ (A) = \frac{voltage \ (V)}{resistance \ (\Omega)}$$

If a resistor is made so that a 1 V voltage across it leads to a 1 A current through it, then we say this resistor has a resistance equal to 1Ω. This defines the unit ohm (Ω). A resistor in a circuit might have a resistance anywhere from 50 Ω to 1 million Ω.

For the example in Figure 5.24, the resistor (or light bulb) has resistance equal to 100Ω, and the battery is 12 V. Then the current flowing through the resistor will be:

$$current \ (A) = \frac{12 \ V}{100 \ \Omega} = 0.12 \ A$$

Usually the wires used in a circuit, such as in Figure 5.24, have very low resistance compared to the resistors used in the circuit. In this case, the voltage is nearly constant everywhere on the wire segment labeled A. Likewise, the voltage is nearly constant everywhere on the segment labeled B, although the voltage on B is not equal to that on A.

When current flows through a resistor, the voltage (a measure of potential energy) decreases from one end of the resistor to the other. If we know the resistance of a circuit

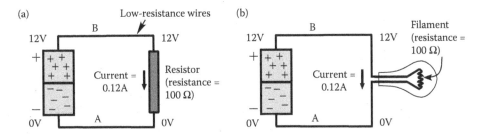

FIGURE 5.24 (a) A resistor connected in a closed circuit to a battery allows a certain amount of current to flow around the circuit. (b) A light bulb is a type of resistor that emits light when heated by current passing through it.

and the current in it, we can calculate the decrease of voltage from one end of the resistor to the other side by rearranging the Ohm's law formula to:

$$voltage\ decrease\ (V) = current\ (A) \times resistance\ (\Omega)$$

For example, if the current is 3 A and the resistance is 100 Ω, then the corresponding voltage change is: *voltage* = 3 A × 100 Ω = 300 V. This corresponds to the fact that electrons lose potential energy as they pass through a resistor in a circuit.

The pressure in water is analogous to a circuit's voltage. A water pump that increases the pressure is analogous to a battery that increases voltage. In Chapter 4, we discussed Ohm's law for liquids, which states that the decrease of pressure is proportional to the amount of resistance in the resistor and proportional to the amount of liquid current flowing. Using Figure 4.14 in Chapter 4, we recognized that many small obstructions in a pipe cause a resistance to the liquid flow.

THINK AGAIN

A few words about the use of words: Voltage does not "flow." Only charged particles flow through a circuit. Often people, including this author, say "current flows" when they actually mean "charged particles flow," or for short, "charge flows." Current is equivalent to the flow of charge.

It is important to understand that as electrons pass through a resistor they do not gain nor lose kinetic energy. They neither speed up nor slow. To see this, note that the current exiting the resistor equals the current entering it. Charge is conserved, and it does not build up anywhere in the circuit, except in the battery. This means that the current is equal everywhere in the circuit. The current in the wire labeled *A* equals the current in the wire labeled *B* in the circuit in Figure 5.17. This means that the speed of the electrons is the same in both wires. Electrons do lose potential energy when moving through a resistive wire, corresponding to a decrease in voltage. That occurs as the electrons momentarily lose energy when they bounce off obstructions within the resistor, and are then reaccelerated by the voltage produced by the battery. This lost energy goes into heating the resistor. If the resistor heats enough, it will glow.

An analogy for this situation is a skater rolling downhill on a very bumpy, rocky road, as in **Figure 5.25**. (In a real resistor, it is the minus charges that move. The drawing is easier to understand, however, viewed as plus charges.) If the road is bumpy

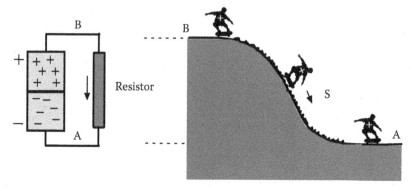

FIGURE 5.25 Skater analogy for motion of a charge through a resistor. The speed (S) of the charged particle remains constant as it moves through the resistor.

enough, the skater will roll straight downhill without gaining or losing speed. The bumps take away the skater's energy as quickly as gravity gives energy to the skater. This analogy for how a resistor works is close to the actual physics inside a bar of resistive material. The material is made of atoms, which act like energy absorbers for electrons. As electrons bounce off the atoms, they change direction rapidly in a random manner, while eventually making their way "uphill" toward higher voltage, which for electrons is lower potential energy.

5.8 ELECTRICAL POWER

Recall from Chapter 3 that power equals energy delivered (or converted) per second. A power of 1 watt, abbreviated W, equals 1 J of energy delivered per second. (1 W = 1 J/sec) The total energy E delivered by a source with power P in a time interval of duration t is equal to the product of the power and the duration. As a formula, this is:

$$Energy = Power \times time$$
$$E = P \times t$$

That is, joules = watts · seconds.

When electric charge flows through a resistor under the influence of a battery, the resistor heats up. Potential energy of the electrons is converted to kinetic energy, which is then partially lost to thermal energy in the resistor. The amount of energy converted per second is the power lost in the resistor. It is related to the current and the resistance by the formula:[2]

$$Power = current \times voltage$$

If the power is high enough the resistor will begin to glow and emit light. This is how an incandescent light bulb creates light. The bulb filament is made of a metal with high resistance, so a lot of thermal energy is transferred from the flowing charges to the filament. As mentioned in Chapter 3, only about 5% of this electrical power is converted into visible light. For example, a typical household light bulb operates at 120 V. If a current of 0.5 A passes through it, the power converted into heat equals:

$$Power = 0.5 \text{ A} \times 120 \text{ V} = 60 \text{ W}$$

According to Ohm's law, the resistance of this bulb filament equals:

$$resistance \ (\Omega) = \frac{voltage \ decrease \ (V)}{current \ (A)}$$
$$= \frac{120 \text{ V}}{0.5 \text{ A}} = 240 \ \Omega$$

5.9 MAGNETISM

Magnetism refers to phenomena involving magnets and magnetic forces. Ancient peoples knew about lodestones—naturally occurring ore containing magnetic iron that attracts iron objects. For us perhaps the most familiar example of magnetism is the use of a magnetic compass to determine the directions pointing to the two magnetic poles

---

[2] Current = charge per second, and charge × voltage = energy. Therefore, current × voltage = energy per second, or power.

of the Earth. This use for navigation (perhaps originating in China) has been known since the eleventh century. During the nineteenth and twentieth centuries, it was discovered that magnetism is intimately related to electricity. Magnetic effects are crucial for computers and Internet technology. From magnetic memory to electromagnetic relays (switches), magnetism is the physical basis of many essential devices.

A bar magnet is a rod of an iron-containing metal that has been magnetized; that is, it has been made magnetic by exposure to a strong magnetic force. This process of making a piece of metal into a permanent magnet is called *magnetization*.

The results of experiments yield the following principle:

MAGNETISM PRINCIPLE (I)

Every magnet has two poles, called north (N) and south (S). Like magnetic poles repel, and unlike magnetic poles attract.

You can verify this by using three bar magnets, whose pole identities are unmarked. Label them A, B, and C. Suspend each magnet by string so it is free to rotate, as in **Figure 5.26**. First mark one end (it does not matter which) of magnet A with a white sticker and the other with a black sticker. Now bring magnet B close to magnet A and you will observe that magnet B rotates so that one end is closest to the black-marked end of magnet A. Place a white sticker on that end of magnet B, and place a black sticker on the other end of magnet B, as shown in the figure. Now remove magnet B, and bring magnet C near magnet A and perform the same marking of magnet C. Finally, bring magnets B and C together, and you will find that their like-marked ends repel and their unlike-marked ends attract. This shows that there is a consistent behavior of any pair of magnets, and that there are only two types of ends. Any number of additional magnets can be marked this way and compared to all others with the same conclusion—there are only two types of poles. This categorizing of magnetic poles is reminiscent of the categorizing of electric charges into two types, but the underlying physics is quite different, as we will see. Historically, the end of a magnet that is attracted toward the Earth's magnetic north pole was called the north (N) pole of the magnet.

A compass needle is a small bar magnet, having its own N and S poles. It feels forces exerted by the magnetic fields of the Earth and any nearby magnet. The magnetic field is best visualized by drawing *magnetic field lines*, as **Figure 5.27**.

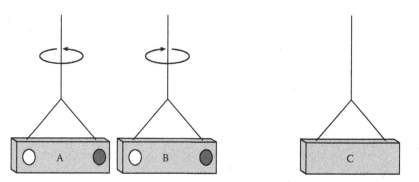

FIGURE 5.26 By observing repelling and attracting behavior between the ends of three suspended, unmarked bar magnets, we can determine that there are two types of poles.

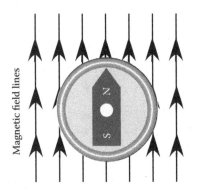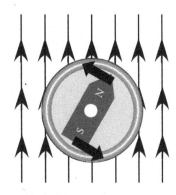

FIGURE 5.27 A compass consists of a small bar magnet with poles labeled N and S and supported at its center so it can freely rotate. If the needle is not parallel with the nearby magnetic field lines, it will experience a force that tends to rotate it so it becomes parallel with the lines.

Magnetic field lines are imaginary lines that indicate the directions of a magnetic field. An arrow can be placed on each line to indicate the direction of the field at a location. The lines are drawn closer together where the field is stronger.

A compass needle (or any bar magnet) will feel forces from the magnetic field that tend to turn the needle to be parallel with the nearby magnetic field lines. This rule is summarized as:

Magnetism Principle (II)

A freely rotating compass needle always aligns itself with the direction of the magnetic field at the location of the compass.

The space around a bar magnet is occupied by a magnetic field. The form of the magnetic field is visualized by drawing the magnetic field lines, as in **Figure 5.28.** Magnetic field lines always form closed loops. Figure 5.28a shows the outline of a bar magnet and its magnetic field lines, all of which form closed loops when followed through the body of the magnet. Some of the loops are not fully drawn where they exceed the boundaries of the drawing. Four compasses are shown at various locations around the magnet. Each compass needle has been rotated by the magnetic force so that its needle is parallel to the nearest field line, and its N pole points along the field line leading to the S pole of the bar magnet. Figure 5.28b is a photograph of thousands of magnetic iron particles (each shaped like a tiny toothpick) that were placed on a piece of paper just above a bar magnet. Each iron particle acts like a compass and has rotated to be parallel to the nearest field line.

5.10   ELECTROMAGNETISM

In the 1800s, scientists discovered that electricity and magnetism are very closely related. In fact, it is not possible to deeply understand either without studying them together. The behaviors and properties of electricity and magnetism together are called *electromagnetism*.

(a)

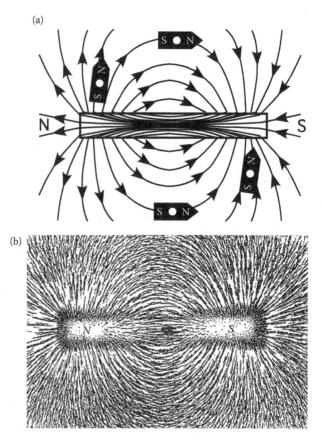

(b)

FIGURE 5.28 Magnetic field lines surrounding a bar magnet. (a) The lines are drawn with compasses aligned along them. (b) The field lines are visualized by many tiny iron particles, which have aligned themselves along the field lines. (Photograph from Practical Physics, Macmillan and Co., New York, 1914.)

5.10.1   Electric Current Creates Magnetic Field

Hans Christian Oersted, a Danish physics teacher, first observed in 1820 that an electric current in a wire produces a magnetic field in the region surrounding it. He used a battery to create an electric current in a straight piece of metal wire, and when he brought a compass near the current-carrying wire, he saw that the compass needle deflected. When he turned off the electric current, the needle returned to its normal direction, pointing north in the Earth's magnetic field. The compass needle reacted to the magnetic field that was created by the electric current.

Figure 5.29 depicts a copper wire passing through a small hole in a horizontal platform, on which a compass needle is located. When the switch is open, no current is present in the wire. The compass needle points N, as dictated by the Earth's weak magnetic field. After the switch is closed, current is present in the wire. The wire remains electrically neutral, so a nearby charged particle (at rest) would feel no force.[3] Nevertheless, the magnetic compass needle experiences a force that causes it to rotate so it is parallel to an imaginary circle whose center is pierced by the wire. Regardless of the location of the compass on the platform, its needle feels forces causing it to rotate so that it is parallel to one of the circles. The circular lines drawn around the wire are *magnetic field lines*.

---

[3] A magnet exerts magnetic forces on an electrically charged particle only if the object is moving in a direction perpendicular to a magnetic field line. This force is in a direction perpendicular to the direction the object is moving and perpendicular to the magnetic field line.

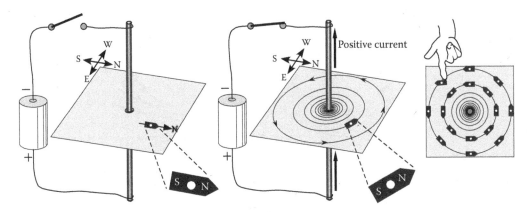

FIGURE 5.29 Oersted's experiment. A steady electric current in a wire causes magnetic forces on a compass needle (shown also in expanded view). The top view shows many possible locations for the compass, the alignments of which map out circular magnetic field lines.

With the current flowing in the wire, if you push with your finger on the needle's end, you do work on the needle as you rotate it around the pivot at its center. After you release it, the needle will rotate back to be parallel with the field line. This shows that the needle can store magnetic potential energy, which can later be released to cause the needle to realign itself with the field.

After further experiments were carried out, it was found that all magnetic field lines are in the form of closed loops. Magnetic field lines have no beginning or ending, unlike electric field lines, which begin and end on electric charges.

In 1820, André Marie Ampère, a French physicist, proposed that electric current (comprised of moving electrons) is the source of all magnetic forces and all magnetization. Thousands of experiments have since verified that Ampère was correct. His idea is summarized in an important principle:

ELECTROMAGNETISM PRINCIPLE (I)

Ampère's law: All magnetism is caused by charges that are moving. That is, magnetic forces are interactions between electric currents in two objects. The strength of the magnetic field along a certain field line is proportional to the amount of electric current passing perpendicularly through the loop made by the closed field line.

For the situation shown in Figure 5.29, the magnetic field is equally strong everywhere along any circular field line, but this is not the case if the current-carrying wire is curved rather than straight.

You might wonder what creates the magnetic field of a bar magnet. According to Ampère's law, moving electric charge is required, but if so, where are these electrons in the case of a bar magnet? As we will discuss in Chapter 9, every material is made of atoms, which have moving electrons inside of them. In certain cases, all of the electrons in the atoms within a bar magnet move in a coordinated way, creating the magnetic field around the bar.

A setup used for creating a magnetic force by using electric current is a coil of insulated metal wire, as shown in **Figure 5.30**. When such a coil, called a solenoid, is connected to a battery so that electric current passes through the coiled wire, a magnetic field is created. The magnetic field lines are shown. Rather than encircling each wire individually, the field lines encircle groups of wires making up the coil. The field lines pass through the hollow "core" region made by the coil, and each line eventually closes on itself. The field is strongest inside of the core region.

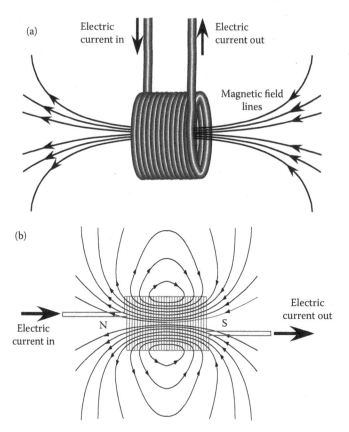

FIGURE 5.30 (a) A wire coil (solenoid) carrying electric current. (b) Magnetic field lines around the coil.

This arrangement is an ***electromagnet***, used in everything from pinball machines to stereo speakers. If a permanent bar magnet is placed near one of the coil ends, it will feel a force, as in **Figure 5.31**. It will be either repelled from or attracted to the interior of the coil, depending on the direction that the current flows in the wire coil.

THINK AGAIN

Do not think that there is such a thing as magnetic charge. After careful searching, scientists have never found this type of charge. To explain all electric and magnetic phenomena that we know about, electric charge is sufficient.

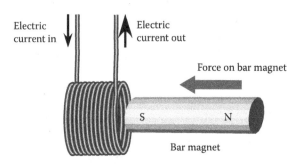

FIGURE 5.31 Electromagnet. A current in a coiled wire creates a magnetic force on a permanent bar magnet.

REAL-WORLD EXAMPLE 5.2: THE TELEGRAPH, PRECURSOR TO THE INTERNET

The precursor to the Internet was the telegraph system. Around 20 years after 1820, when Oersted observed that an electric current in a wire produces magnetic forces in the region surrounding it, many scientists worked on schemes to use this phenomenon for long-distance communication. The idea is simple: a push switch and a battery in New York can be used to activate an electromagnet in Minneapolis if a pair of long wires connects them. As in **Figure 5.32**, the electromagnet pushes a bar magnet down onto a moving strip of paper to record the dots and dashes, which constitute the Morse code. This two-value code was the precursor of the modern use of binary digits, or bit values, discussed in Chapter 2. In later chapters, we will study how bit values now are transmitted over the Internet.

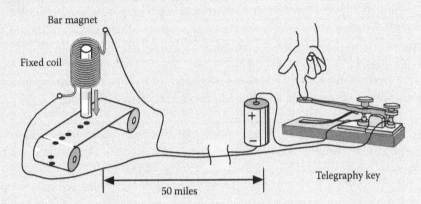

FIGURE 5.32 Telegraph.

Transmitting voltage and current over very long distances is not easy because resistance in the wires decreases the voltage, according to Ohm's law. The resistance of a wire is proportional to its length. Therefore, the voltage decreases in proportion to the length of the wire. For example, a 1-meter-long copper wire with diameter equal to 1 millimeter has resistance of about 0.02 Ω. A 50-mile-long piece of this wire has resistance of about

$$\frac{0.02\,\Omega}{1\text{ m}} \cdot 80,000 \text{ m} = 1,600 \text{ }\Omega$$

The round-trip circuit shown in the figure, then, would have resistance of twice this, or 3,200 ohms. By Ohm's law, if a 10-V battery were used at the sender's end, the current that would be created in the receiver's electromagnet would be only

$$current = \frac{voltage}{resistance} = \frac{10 \text{ V}}{3,200 \text{ }\Omega} = 0.003 \text{ A}$$

This is too small a current to drive the electromagnetic printing device at the receiving end. For this reason, a device called a relay was installed in the line at the receiving end. Its purpose was to receive the weak current and convert it into a larger current to be used for driving the printer. Can you think of a way to construct such a relay using only the kinds of parts shown in the figure?

A series of relays could be used for very-long-distance transmission. These were installed at regular intervals along the line. They were automated devices, not requiring human intervention, but each required a battery to operate, because they added energy to the signal.

The first commercial telegraphs were operated in England and the United States around 1845. By 1870 there was a worldwide telegraph network, including undersea cables that could transmit messages between the United States, South America, Europe, Africa, Russia, China, and Japan. Samuel Morse was an American leader in putting such revolutionary systems into practice. The last and final telegraph message was sent on January 26, 2006, when Western Union closed the service after 145 years.

5.10.2   Changing Electric Field Creates Magnetic Field

In the 1860s, James Clerk Maxwell, a Scottish physicist, reconsidered the situation shown earlier in Figure 5.29. He asked what would happen if a single, electrically charged particle passed by a compass needle, as shown in **Figure 5.33**. This is different from the case shown in Figure 5.29, where there is a steady flow of charge in the wire, which remains electrically neutral. Maxwell determined that a magnetic field is caused by a single passing electric charge. This magnetic field, which is temporary, causes a deflection of the compass needle. This was a very important observation! As we shall see later, it is the basis of radio.

Maxwell found a new principle to explain this observation using the idea of fields. Referring again to Figure 5.33, the electric field produced by the charged particle is shown as the dashed line extending from the charge to the compass, with an arrow at its end. The charged particle starts below the plane of the compass and moves upward. Notice that as the charged particle passes by the compass, the electric field vector at the location of the compass changes direction. Its strength also changes. Before the charge gets near the compass, the compass needle points N, as dictated by the Earth's magnetic field. As the charge begins moving upward near the compass (Figure 5.33a), the electric field at the location of the compass is weak and points at an angle in the upward direction. When the charge moves to the center of the plane containing the compass, its electric field at the compass is strong and is pointing horizontally (Figure 5.33b). When the charge has moved above the plane containing the compass (Figure 5.33c), its electric field is again weak and is pointing downward at an angle. The compass needle was deflected by the temporary magnetic field created as the charged particle passed

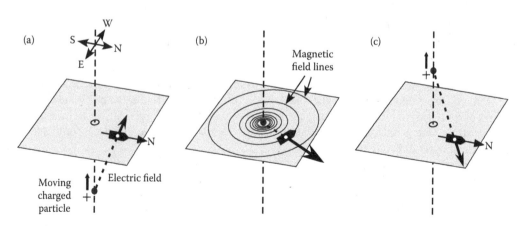

FIGURE 5.33   An electrically charged particle passing by a compass needle creates a changing electric field at the location of the compass. The time-varying electric field creates a magnetic field, which deflects the compass needle.

by. This temporary magnetic field is shown in the figure as the circles surrounding the path of the particle in Figure 5.33b.

It is important to understand that the magnetic force created by the passing charge is not a result of the mere existence of the electric field, but the fact that the charge is passing by the compass. If the charged object were fixed in position at any of the three locations shown in Figure 5.33, no force would be exerted on the compass needle, because a stationary charge does not produce a magnetic field or affect a nearby magnet. It is only when the charge is moving, so that the electric field is changing in its direction and/or strength, that the compass needle feels a force. To describe this, Maxwell devised a new principle:

ELECTROMAGNETISM PRINCIPLE (II)

Maxwell's law: An electric field that is changing in its direction and/or strength creates a magnetic field.

Notice that this principle does not explain why or precisely how a changing electric field creates a magnetic field; it only states that it does.

5.10.3   Changing Magnetic Field Creates Electric Field

In 1831, Michael Faraday, an English physicist, observed that moving a magnet inside a wire coil generates an electric current in the wire. As shown in **Figure 5.34**, no battery was needed. If the magnet was stationary in the coil, no current was generated.

Faraday also observed that an electric current could be generated in a wire coil by using a nearby electromagnet that was not moving. This could be done by rapidly varying the strength of the magnetic field created by the electromagnet. As in **Figure 5.35**, if a switch is suddenly closed, causing a surge of current in the electromagnet, it creates a magnetic field whose strength is changing in time. The "pick-up" coil experiences a time-varying magnetic field from the electromagnet, and an electric current is generated in the light bulb.

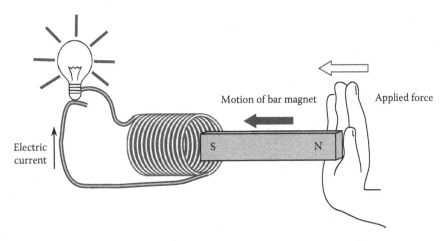

FIGURE 5.34 A magnet being moved by hand in or out of a coiled wire generates electric current in the wire. The bulb lights when current flows.

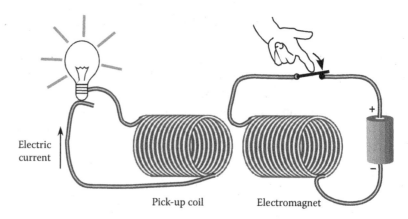

FIGURE 5.35   An electromagnet that is suddenly switched on near a coiled wire generates a temporary electric current in the wire. Repeatedly switching the electromagnet on and off generates a flickering of the light bulb.

Faraday concluded from his experiments the following principle:

ELECTROMAGNETISM PRINCIPLE (III)

Faraday's law: A magnetic field that is changing in time creates an electric field.

The created electric field can induce an electric current in a wire.

Faraday's law, also called the law of magnetic induction, underlies the operation of electrical power generators, electric motors, audio microphones, and many other devices. For example, when a disk jockey at a radio station speaks into a microphone, the air pressure from his voice drives a thin, flexible, plastic diaphragm back and forth at the frequency of the sound wave hitting it. In a so-called *dynamic microphone*, a tiny coil of copper wire is glued to the surface of the diaphragm, and a nearby permanent magnet creates a magnetic field with which the coil can interact by magnetic induction. According to Faraday's law, a magnetic field that is changing creates an electric field. As the copper coil moves into or out of the magnetic field, the electrons inside of the wire experience a changing magnetic field. This creates an electric field inside of the wire, which accelerates the electrons around the coil. The current at the two ends of the coil is carried over a pair of wires to an external electronic amplifier, which boosts the power of the signal to drive a speaker or a radio antenna (see **Figure 5.36**).

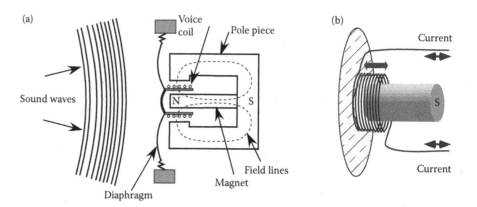

FIGURE 5.36   (a) Microphone design. (b) Detail of diaphragm, voice coil, and fixed permanent magnet, with the outer magnet portion not shown.

THINK AGAIN

If the word *changing* were omitted from the title of this section, the title would be misleading. Why?

IN-DEPTH LOOK 5.4: MAGNETIC MATERIALS AND DATA STORAGE

If you wrap an insulated wire many times around an iron rod, and attach the wire ends to a battery, as in **Figure 5.37**, the rod temporarily becomes a magnet. We say the rod becomes magnetized. To verify this, place a compass near it and watch the compass needle align itself with the rod. This type of electromagnet creates a far stronger magnetic field than does the type with a coil only and no iron rod. Magnetization is the process of making a piece of iron into a magnet by the application of a magnetic field. This type of magnet is called an ***induced magnet***.

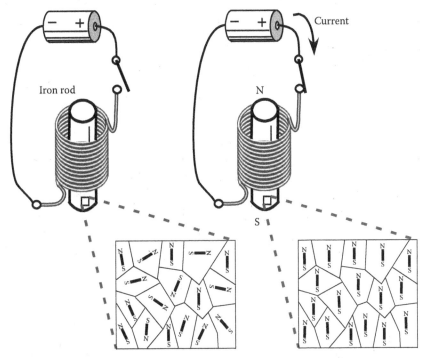

FIGURE 5.37 Electric current in an insulted wire coiled around an iron rod causes the rod to become magnetized (i.e., to become a magnet with N and S poles as shown). Magnetic regions (domains) inside a piece of iron are normally oriented randomly and become aligned in the presence of a magnetic field.

Let us take a closer look at what is happening. Before the current was switched on, the iron rod was not a magnet. Inside iron there are tiny regions, called ***domains***, each of which is a tiny magnet, with north (N) and south (S) poles. These tiny magnets are caused by tiny loops of electric charge flowing around very small regions within the material. Normally—in a nonmagnetized piece of metal—these poles are oriented in a random manner. As a picture, think of a thousand sailboats anchored on a lake, all pointing in random directions. If each boat holds a bar magnet that is glued so its north pole points to the front of the boat, then each magnet would produce a magnetic field in a different direction. In this case, these fields would tend to cancel each other.

The figure on the left in Figure 5.37 shows these tiny magnetic domains in the iron rod, with their north poles pointing in random directions. When the current in the wire coil is switched on, according to Ampère's law, it produces a (weak) magnetic force field

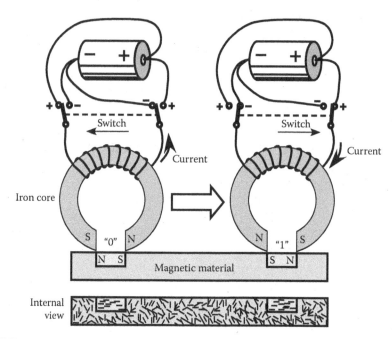

FIGURE 5.38 Electric current in an insulted wire coiled around an iron core creates a magnetic field between the north and south poles. The field magnetizes the recording material below the electromagnet, as shown in the internal view, where short lines indicate magnetic domains.

inside of the iron. Then an amazing thing happens: all of the tiny magnets in the iron are rotated by the field so their north poles are all pointing nearly in the same direction. Now, with all of the tiny magnets oriented in the same way, all of the forces from the tiny magnets act in the same way on any nearby piece of metal. The whole piece of iron becomes a very strong magnet.

Once the domains have been oriented, it takes some energy to undo the orientation. Therefore, after you disconnect the battery, the iron material does not revert completely to a nonmagnetized condition. It tends to stay magnetized.

Magnetic data storage uses Ampère's law and Faraday's law. The magnetic material can be a hard disk or tape having a thin layer of iron or other magnetic particles, such as cobalt, embedded in it. In **Figure 5.38**, a double switch is initially in the left position, connecting the battery to the coil on the electromagnet such that it creates north and south poles at the ends of the core as shown. The created magnetic field magnetizes the material below it as shown. After the recording "head" moves to a different spot above the recording material, the switch can be thrown to the other position, causing a current in the coil in the opposite direction. This causes the magnetization at this spot in the material to orient its north and south poles in the reverse direction. Digital bit values are recorded in the material by choosing a code; for example, a 1 bit value can be represented by a [N S] orientation and a 0 bit value can be represented by a [S N] orientation. By using the proper magnetic materials, the bit values can be stored more or less indefinitely (as shown in **Figure 5. 39**).

Reading of the bit values is performed in a similar manner. The battery is replaced by a capacitor connected to the wire going to the coil. When the head is moved rapidly above a single magnetized region (bit region) on the recording material, a temporary electric current is caused in the wire coil. This is an application of Faraday's law. The current charges up the capacitor plus or minus, depending on which orientation the bit has, [N S] or [S N]. The capacitor can store this charge or bit value for later use by the computer's electronics.

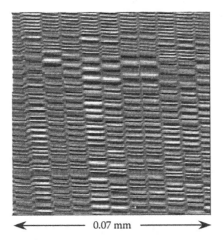

←——————— 0.07 mm ———————→

FIGURE 5.39 An image of a small region on the surface of a magnetic hard disk, showing stored bit values (binary zeros and ones) as dark and light regions. The image was made by placing a tiny magnetic needle tip near the surface and recording the strength of the magnetic force as the tip is moved above the surface. (Image by Tony Alvarez. Courtesy of University of Pennsylvania Nanotech Facility.)

SUMMARY AND LOOK FORWARD

Electric fields are created by electric charges—stationary or moving. Magnetic fields are created only by electric charges that are moving. An electric field occupies the space around an electrically charged object. A magnetic field occupies the space around a bar magnet or an electromagnet. These fields can exist in empty space—there is no need for any physical medium, such as air. This can be verified by repeating the experiments discussed in this chapter inside of a container from which all of the air has been pumped. The fields are real and physical; they contain energy and can transfer energy from one location to another. Electric and magnetic fields are the elements of radio waves, which we will cover in the next chapter.

As we discussed in this chapter, electricity and magnetism are described by a set of principles, called here the principles of electromagnetism to stress that they are a unified set.

The Principles of Electromagnetism

Electricity

(i) Electric charge is a property of particles associated with the fact that electric force acts between particles. Like charges repel, unlike charges attract.

(ii) Electric charge conservation and addition: The total amount of electric charge is conserved. When charges are placed together on an object, the plus and minus values of the electric charges are summed.

(iii) Coulomb's law: The strength of the electric force between two point-like charged objects is proportional to the product of their charge values and inversely proportional to the square of the distance between them.

(iv) Electrical energy is required to move a charged object if it is opposed by an electric force. The amount of required energy is the charge on the object multiplied by the voltage between the object's starting and ending locations.

Magnetism

(i) Every magnet has two poles, called north (N) and south (S). Like magnetic poles repel, and unlike magnetic poles attract.

(ii) A freely rotating compass needle always aligns itself with the direction of the magnetic field at the location of the compass.

Electromagnetism

(i) Ampère's law: Magnetic forces are interactions between electric currents in two objects. The strength of the created magnetic field along a certain field line is proportional to the amount of electric current passing through the loop made by the closed field line.
(ii) Maxwell's law: An electric field that is changing in time creates a magnetic field.
(iii) Faraday's law: A magnetic field that is changing in time creates an electric field.

Prior to 1830 or so, physicists simply collected the facts from many observations of electric and magnetic phenomena. Then they began finding empirical, quantitative rules for calculating these forces in various situations. They found that electric and magnetic forces are closely related. Finally, in 1862, Scottish physicist James Clerk Maxwell discovered a set of mathematical equations that can be used to predict the strengths and directions of the electric fields and magnetic fields.

We will not write Maxwell's mathematical theory here, because to apply it to practical situations requires sophisticated mathematics. It is important to know, however, that by solving Maxwell's equations for different physical situations, engineers can predict and model an enormous number of electronic devices, which form the basis of our technologies. Scientists can use these equations, along with those that describe other basic forces in nature—such as gravitational and nuclear forces—to try to understand what goes on in stars and galaxies, as well as what happened billions of years ago when the universe was young. Thus, the observations described above paved the way for modern science and technology. The behavior of electric and magnetic fields underlies most of this technology, including radio, television, x-rays, magnetic resonance imaging, lasers, fiber optics, semiconductor electronics, and computers.

The next two chapters treat the electromagnetic basis of radio and light used for communication and electronic circuits used for computer operation.

SOCIAL IMPACTS: INNOVATION AND PUBLIC SUPPORT OF SCIENCE

Let us say it is the year 1830 and you are William IV, the King of England. The technology of the time is iron craft, factory machines, and steam engines. Call your greatest scientists together and declare, "I want you to invent television." The scientists, realizing that the word *tele-vision* literally means *distance-seeing*, complain, "We have not the means, your Highness." You—the King—reply, "Then develop the means."

How many years should you give the scientists to develop television (TV)? You could order them to develop TV within, say, 10 years, working in secret with no communication with non-English scientists. Alternatively, you could tell them you do not care how they go about it, as long as it is done within your lifetime or the lifetime of your immediate successor to the Crown.

With hindsight, we can see now that the former edict would have been a failure, because the basic science needed for inventing TV would not be fully discovered until 35 years after your edict. Then, as shown in **Table 5.2**, it took another 41 years for these ideas to be put into practice to operate the first CRT TV in 1906. King William's successor, Queen Victoria, lived until 1901, long enough to have witnessed early rudimentary versions of television. (Although I do not know if she did.) The point of this story is that science and technology go hand in hand—you cannot have one without the other. There is a logical, but unpredictable progression of both together.

TABLE 5.2

Timeline for Physics and the Invention of Television

Date	Physics	Television
1820	Ampère's law of electromagnetism	
1831	Faraday's law of magnetic induction	
1849	Joule's equivalence of mechanical work and thermal energy	
1859		Julius Pluckers' identification of cathode rays
1865	Maxwell's law of electromagnetism	
1884		Paul Nipkow's electromechanical television
1897	Thomson's discovery of the electron in cathode rays	Karl Braun's invention of the CRT oscilloscope
1906		Boris Rosing's electromechanical CRT television
1928		First U.S. commercial television broadcasts by Charles Jenkins

Now let us say it is the year 1945 and you are U.S. President Harry Truman. The Second World War has just ended, and the Secretary of the Navy urges that you support his proposal for government funding of scientific research at universities. You point out that the government has never provided such funding to universities in the past. The secretary argues in favor, as follows:

[Such funding] is of deep significance to the national welfare as well as the national security. ...Research means the search for new knowledge of Nature; development means the application of knowledge. ...Scientific research is one of the highest manifestations of intellectual curiosity. ...All research is beneficial. No quest for truth can be otherwise. Its results may be put to use for good or evil, but to stifle research would be to stifle the main hope of humanity. ...It should be clear that it is a contradiction to speak of directing and controlling research. ...Research to create knowledge, and education to spread it to all

FIGURE 5.40 Professor Charles Townes and James P. Gordon shown with the first maser, which they developed in the early 1950s at Columbia University with support from the Office of Naval Research. The maser was the forerunner of the laser, an invention for which Townes shared the 1964 Nobel Prize in physics. Just as much as TV is, the laser is a direct descendent of Maxwell's 1865 electromagnetic theory. Great, unexpected practical discoveries are often the result of basic research. (With permission of Corbis.)

people, are the basic safeguards of civilization, and the only weapons which will succeed against ignorance, our ultimate enemy. Science now requires more research than private enterprise can support. Universities can no longer depend on the generous endowments of prewar days. [1]

What do you do? Support an unprecedented new use of taxpayers' money to "give" to universities for their own research activities? If so, how do you "sell" the idea to Congress? What are the possible pitfalls? What are the benefits?

In October 1946, 3 months after the Congress and President created the Office of Naval Research (ONR)—the government's first office that funded university research—the Director of the ONR Planning Division stated:

The war brought great advances in the art of computation, and the future is now bright for the mathematicians and those who must use mathematics in their studies. A machine called the ENIAC, built at the University of Pennsylvania for the Army Ordinance Department, can compute the instantaneous position of a projectile and determine its point of fall in a shorter time than the projectile is actually in the air. This machine occupies a room 30 feet by 50 feet and weighs 30 tons. It can multiply two ten-digit numbers in 1/360th of a second. Both the Army and Navy are supporting the research and development of still faster and more flexible machines.

Figure 5.40 shows that ONR funding of faculty at universities resulted in the creation of the first maser, which soon thereafter led to the first lasers being invented.

Questions to Ponder

1. What might have been the consequence if the King had insisted that the TV be developed only within England, with no input from foreigners?
2. What is a typical time lag between a basic scientific discovery and its commercialization? Is this time lag getting shorter or longer as the decades pass?
3. What parts of society, if any, should pay to support scientific research? What parts should pay to support technology development? What case can you make for your answer?
4. Who benefits the most from public support of basic research?
5. What role should corporations play in research and development? Who should pay for their research and development activities—government, stockholders, and/or customers?
6. What role does and should the military play in supporting basic research at universities?

Terms to Research

cathode-ray tube; *Office of Naval Research, history*; *research funding*; *television, history, timeline*

REFERENCES

1. Excerpts from a speech by CAPT Robert Dexter Conrad, Director, Planning Division, Office of Naval Research, University of Illinois, Urbana. October 27, 1946. Taken from *ONR 50th Anniversary*. http://www.onr.navy.mil/about/history.

SOURCES

Bellis, Mary. "The Invention of Television," http://inventors.about.com/library/inventors/bl_television_timeline.htm.

SUGGESTED READING

See the general physics references given at the end of Chapter 1.

KEY TERMS

Ampere (amp, A)
Battery
Capacitor
Coulomb (C)
Domain
Electrical conductivity
Electrical conductor
Electrical insulator
Electrical resistance
Electric charge
Electric circuit
Electric current
Electric field
Electric field line
Electric potential energy
Electromagnet
Electromagnetism
Electron
Field
Induced magnet
Magnetic field
Magnetic field lines
Magnetism
Magnetization
Ohm (Ω)
Proton
RAM
Resistor
Unit
Vector
Volt (V)
Voltage

ANSWERS TO QUICK QUESTIONS

Q5.1 Because A attracts C and repels B, we conclude that C and B are oppositely charged, therefore they would attract. A and B have the same charge, either both plus or both minus, whereas C has the opposite charge.

Q5.2 The piece of dust with an excess of 100 electrons attracts the piece of dust with a deficit of 250 electrons. If these two dust pieces are brought together to make a single, larger dust piece, its net charge is +150 e, that is, positive. A single electron feels an attractive force if it is near this object.

Q5.3 10 amps = $10 \times 6.25 \times 10^{18}$ electrons per second = 6.25×10^{19} electrons per second.

In 2 sec, the number of electrons is $2 \times 6.25 \times 10^{19} = 12.5 \times 10^{19}$.

Q5.4 Work = energy = 5 C × 1.5 V = 7.5 J

Q5.5 In the wire connecting the right plate to the minus end of the battery, the voltage is equal everywhere. There is no net force on the electrons.

EXERCISES AND PROBLEMS

Exercises

E5.1 A metal ball has an excess charge of 1,000 electrons, and a second metal ball has a deficit of 1,000 electrons.

(a) If the two balls are connected by a long, thin piece of copper metal, what will happen to the charge on the negatively charged metal ball?
(b) If the two balls are connected by a long, thin piece of glass, what will happen to the charge on the negatively charged metal ball? Explain.

E5.2 You have three insulated metal balls, A, B, and C, of equal size. You rub a Teflon (plastic) rod with rabbit fur, and then touch the rod to ball A to transfer charge. To charge ball B, you rub a glass rod with silk then touch the rod to ball B. Assume that the quantity of charge transferred in each case equals $-1 \times 10^{16} e$ (recall that $-e$ is the charge of 1 electron). Ball C is uncharged. Consider each case below separately, starting from the same initial conditions described above.

(a) If you connect ball A to ball B through a neutral copper rod, what is the net charge on each ball afterwards?
(b) If you connect ball A to ball B through a neutral glass rod, what is the net charge on each ball afterwards?
(c) If you connect ball A to ball C through a neutral copper rod, what is the net charge on each ball afterwards? (Think of the charges on A repelling each other.)
(d) If you connect ball B to ball C through a neutral copper rod, what is the net charge on each ball afterwards?
(e) If you connect ball B to a copper wire that connects to a copper plumbing pipe that goes into the Earth, what is the net charge on ball B afterwards?

E5.3 Explain why it is scientifically valid (and perhaps amusing) to claim that the Earth's magnetic "north pole" is actually in Antarctica, not in northern Canada.

E5.4 A small copper block has an excess charge of 2,000 electrons. A small gold block has an excess of 1,000 electrons. A small silver block has a deficit of negative charge corresponding to 1,000 electrons.

(a) Describe the forces between each pair of objects if separated by 1 mm. Which force is the strongest?
(b) If the silver block is put into contact with the gold block, describe the force between the copper block and the combined silver/gold object.
(c) If the silver block is put into contact with the copper block, describe the force between the combined silver/copper object and the gold object.

E5.5 Referring to the three blocks in E5.4:

(a) Sketch the electric field arrows (as in Figure 5.7) surrounding each object when isolated. How would a small piece of dust having 100 excess electrons be affected by the field around each object? Which force is strongest (for a fixed distance between charged object and piece of dust)?

(b) As the piece of dust is moved closer and farther from each object, how does the force that is felt by the dust piece change?

(c) Answer (a) in the case that the dust piece has a deficit of 100 electrons.

E5.6 All objects attract each other by the force of gravity, which can be described in terms of gravitation field lines, in analogy to electric field lines.

(a) Make a drawing showing the Earth surrounded by gravitation field lines. Which way do the arrows (forces) point?

(b) There is no repulsion, or repelling force, caused by gravity. If you were to invent the concept of "gravitational charge," how many different types of charges (none, one, or two) would you need to hypothesize to be consistent with the known facts? Explain.

E5.7 A bird can land on a bare metal power line that is held at 10,000 V (relative to the ground) and usually not be harmed. Explain why. Think of a scenario wherein the bird would be harmed when landing on the wire. (*Hint*: There could be a tree branch nearby.)

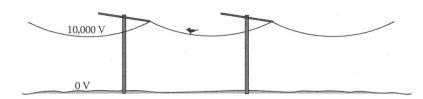

E5.8 Why can it be dangerous to touch electronic components in a circuit, such as a stereo power amplifier circuit, even a significant time after its power has been turned off? *Hint*: Consider the capacitors in the circuit.

E5.9 Two small round stones labeled A and B have the same size and mass and are hanging from thin, nonconducting strings. Initially they are held (perhaps using non-conducting tweezers) in the downward position, as shown. Stone A has a charge of $+1.0 \times 10^{18} e$, and stone B has charge of $+3.0 \times 10^{18} e$. Do the stones feel equal or different strengths of repelling force? (*Hint*: Recall Newton's third law.) After the stones are released from the tweezers, they repel and move away from each other, and are also attracted downward toward the Earth by gravity. Draw a picture similar to the one below, showing the angles of the two strings (after any oscillations stop). (Do not try to calculate the angles; just indicate if one is greater than the other or not.)

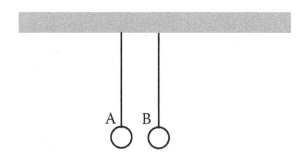

E5.10 Consider three small objects A, B, and C, which are identical, except for having different amounts of electric charge. You carry out separate experiments and find the following results.

(i) At a distance of 10 cm, objects A and B attract strongly.
(ii) At a distance of 10 cm, objects A and C attract weakly.
  (a) Form a hypothesis explaining these results on the basis of amounts of charge (positive or negative) on each object. Give your reasoning.
  (b) Is there only one consistent hypothesis? If not, give another equally good hypothesis.
  (c) How could you determine by experiment which hypothesis is correct? (You may need to introduce new objects, such as Teflon and fur, into the experiment.) For (d) and (e), consider the case that A and B have equal but opposite charges:
  (d) What type and strength of force do you predict between B and C?
  (e) What would be the result if you stuck A and C together to make a single new object and then measured the force between this new object and object B?

E5.11 Two large, round stones have equal amounts of charge, but one is negative and the other is positive, as shown. The stones are fixed in position, and surrounded by nonconducting oil. If a small, light object (test object) is charged and placed at one of the positions labeled (i), (ii), or (iii), the object will feel forces and will move along the electric field line that it is located on. The oil puts frictional drag on the test object, preventing it from accelerating very much, so it moves very slowly.

(a) Make a careful sketch showing the field lines with arrows attached. (*Hint*: See In-Depth Look 5.3.)
(b) On a separate drawing, draw the path taken by the test object for each different starting location, assuming the test object is positively charged.
(c) Do the same, assuming the test object is negatively charged.

E5.12 Nine-volt batteries are connected in two circuits as shown below. Assume the battery maintains its voltage.

(a) If you suddenly close the switch in Figure (i), explain if any electric current will flow immediately after the switch is closed.
(b) If you suddenly close the switch in Figure (i), explain if any electric current will flow a long time after the switch is closed.
(c) After a long time, what will be the relationships between the voltages at points A and B?
(d) Answer (a) and (b) for Figure (ii) and explain.
(e) After a long time, what will be the relationships between the voltages at points C, D, and E?

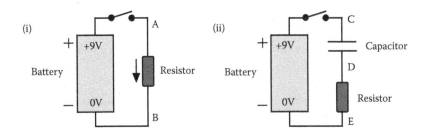

E5.13 If you have an ordinary bar magnet, and you knock it sharply against a hard, dry, nonconducting, concrete floor, you can reduce its magnetic strength. Explain. (Assume the bar does not break.)

E5.14 If you bring a second wire coil, such as that in Figure 5.30, near to the coil shown, both with current flowing, will either loop or both loops feel a force from the other? Explain how the forces depend on the relative location and orientation of the coils.

E5.15 Photocopy or draw the coil and magnetic field lines in the lower half of Figure 5.30. Draw ten small compasses located at various places around and inside of the coil and indicate in which direction the needle of each compass would point. (Assume the compasses do not affect each other.)

E5.16 What electromagnetism principle(s) are behind the operation of:

  (a) microphones (explain)
  (b) loud speakers (You may need to do some research for this.)
  (c) magnetic hard drives (explain)
  (d) magnetic tape recordings (You may need to do some research for this.)

E5.17 Oil can be sprayed into a fine mist made of tiny droplets. Under the influence of gravity, the drops will slowly settle through the air toward the Earth. If the oil mist is exposed to a source of static electricity, a small amount of electric charge can attach itself to each drop. If the mist is located between two metal plates attached to the opposite sides of a battery, the resulting electric force can be adjusted to oppose the force of gravity. Devise and explain a way that this apparatus could be used to prove that electric charge comes in discrete amounts. That is, the charge of an object is always equal to an integer (whole number 0, 1, 2, etc.) multiplied by e, which is the charge on a single electron. Robert Millikan carried out this experiment in 1909.

E5.18 A known joke is a sign on a laboratory door declaring:

$$\boxed{WARNING! — ONE\ MILLION\ OHMS}$$

Why is this amusing to physicists and engineers?

Problems

P5.1 An automobile battery contains about 0.03 kW · hr of energy per kg of battery mass. How much energy (in joules) is contained in a 44 lb (20 kg) battery?

P5.2 A kitchen blender uses 4.5 A of current when operating. How many electrons pass through the motor coil of this device in 1 sec?

P5.3 In an experimental setup, labeled (1) in the figure below, a 9 V battery is connected to two large, fixed metal plates (a capacitor), using metal wires as shown. A test object, with a positive electric charge, is located very near the right plate, as shown. You are holding the test object using long, nonconducting tweezers. You now move the test object from its shown location to a location very near the left positively charged plate.

(a) In moving this object from right to left, do you (using the tweezers) have to do work on the object, or does the object do work on you (i.e., on the tweezers)? Explain.

(b) Is the effect of having moved this test object that its potential energy has increased, decreased, or stayed the same?

(c) In a second experiment, labeled (2) in the figure below, the same experiment is carried out, but in this case the distance between the two plates is larger. Is the change of potential energy in this experiment greater when compared with that in the first experiment? Explain.

(d) Is the magnitude of the electric force felt by the test object in the second experiment greater than that felt by the object in the first experiment? Explain. (*Hint*: It turns out that in this situation the force felt by the object is approximately independent of its location between the plates.)

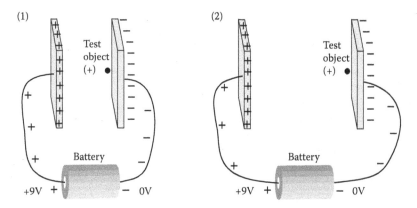

P5.4 A 9 V battery is connected to a resistor as in Figure 5.24.

(a) If the resistor has resistance equal to 50 ohms, what is the current? (*Hint*: Use Ohm's law.)

(b) If the resistor has resistance equal to one Mohm (megaohm), what is the current?

P5.5 Magnetic effects can be used to build computer memory. In the 1960s, computer memory was built by using one small magnet for each bit of information stored. This was called magnetic core memory. Cite two descriptions, in a library or online, explaining how magnetic core memory functions and write a one-page article describing the function. Drawings or other figures should be used and are in addition to the page of text. Identify and cite the source of all information and materials used in your report.

P5.6 Do research online to learn and describe the operating principles of a shakable flashlight, that is, one that requires no batteries and is powered by shaking it back and forth. Explain how the light can stay on for minutes or hours after you stop shaking it. What type of circuit components would enable this operation?

Digital Electronics and Computer Logic

We may in fact lay aside the logical interpretation of the symbols in the given equation; convert them into quantitative symbols, susceptible only of the values 0 and 1; perform upon them as such all the requisite processes of solution; and finally restore them to their logical interpretation.

George Boole
(An Investigation of the Laws of Thought, 1847)

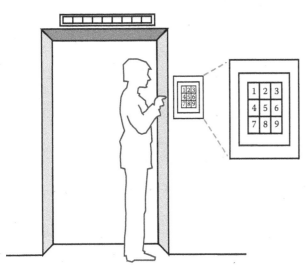

When you push a button calling an elevator, you are setting in motion a complex series of logic operations, which are carried out by automated electronics.

George Boole, the creator of mathematical logic in 1847. He laid the foundations for the theory of computing. (Courtesy of Boole Library, University College, Cork.)

6.1 THE "REASONING" ABILITIES OF COMPUTERS

Computers do not "think." Then how do they manage to perform complex logical tasks like playing chess or identifying an e-mail message as junk? In this chapter, we begin to address the question, "How do computers perform the tasks that we assign to them?" In Chapter 2 we studied binary numbers and the concept of information. To build a computer, we need to devise some physical mechanisms by which computers can use binary numbers, not only to represent information, but also to process it. Processing data means using a set of data to produce a new set of data according to some well-defined instructions. Such a process is called *logic*. It is similar to the human activity of reasoning. A simple example occurs in a calculator when you press the "add" or "+" button. The *input* data are the two numbers you want to add (18, 5). The calculator carries out several logical operations to produce the resulting sum (18 + 5 = 23), which is the *output* data. Before we consider the actual electronic circuits that perform the operations, we will discuss the principles behind logic.

English mathematician George Boole (1815–1864) was one of the founders of the principles of logic. In his honor, the methods used are called *Boolean logic*. In the context of modern computing, Claude Shannon was the most influential scientist who contributed to the theory of logic. In 1938, as a graduate student at Massachusetts Institute of Technology, he submitted his master's thesis that showed how electronic circuits could be used to perform logic. Ten years later Shannon published the paper, "A Mathematical Theory of Communication," which revolutionized scientists' understanding of the concept of information.

6.2   CONCEPTS OF LOGIC

What is logic? In everyday life, it means to take in some information, apply certain rules of reasoning, and produce a decision. For example, you might reason that "if the sky is blue then I will take a walk without an umbrella, but if the sky is cloudy then I will take my umbrella." The color of the sky is the input data and the umbrella decision is the output data. We can make a table showing our logic about the sky and umbrellas:

Input: Sky blue?	Output: Umbrella?
No	Yes
Yes	No

The rules we use to process data or information are called *logic operations*.

A *logic operation* is an elementary rule for arriving at a logical outcome. Three basic operations that can serve as building blocks for all logic operations are NOT, AND, and OR.

In considering complex situations, a diagram is useful to help visualize the logic process. Think of this as a flow chart for making a decision. If you face a complicated decision, such as which college to attend, there might be many factors to consider. Let us say that college B is a better school academically. Your rules for making a decision are:

If college B offers you a scholarship, or if college B's dorms have DSL lines (high-speed Internet connections), then you would attend B. However, if college A offers you a scholarship, but not college B, and if college B's dorms do not have DSL, then you would attend A. If neither A nor B offers a scholarship or DSL, then you would choose neither.

A flow chart for this decision is shown below in **Figure 6.1**. The particular case illustrated is indicated by the circled data.

You should read the logic diagram from left to right. Input information, made up of bits, "flows" along the three lines entering the diagram on the left, and output information flows to the right along the lines labeled "Attend College B?" or "Attend College A?" Each bit travels along a line until it enters a box containing a logic operation, from which an output emerges. A solid black dot on a line shows a point where information

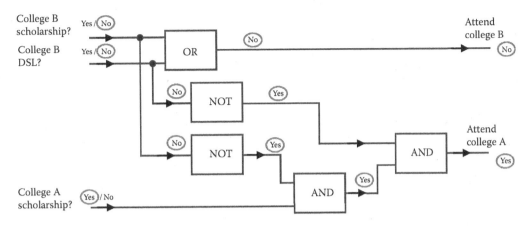

FIGURE 6.1   The big decision. Each line with an arrow represents a YES/NO bit.

is sent from one input source to two logic boxes. A place where two lines cross without a black dot is a bypass (like an automobile overpass); there is no exchange of information at these line crossings. It is common to refer to the lines transmitting the inputs and outputs of the boxes as wires, and to the operations inside the boxes as *logic gates*.

In the case shown in the figure, the upper two inputs equal No and the bottom input equals Yes. Each bit value flows through various gates (boxes), which produce outputs that may be used as inputs for subsequent gates. You should follow the Yes/No bits through the diagram to make sure you understand how this works. Also, check what happens if the uppermost input bit is changed to Yes.

Computers know nothing about human observations, thinking, and feelings, such as the color of the sky or whether a movie is good or bad. Computers can deal only with ones and zeros, that is, binary numbers. So, if they are to do logic for us, we need to translate our problems into their binary language. Therefore, we will consider very simple statements and program our computers to carry out logical "reasoning" about them. Such a statement might be the following: "The value of an input datum, called A, equals zero." Logical reasoning for a computer might be, "If the input A is zero then the output is one, but if the input A is one then the output is zero." Computers can do such reasoning by using simple electronic circuits, which we will discuss later. Computers can also be programmed to do arithmetic using the rules of logic.

To have our computer process the data in the sky/umbrella problem mentioned above, we need to translate it into binary numbers. We can represent the input datum[1] by the symbol A as follows: If the sky is blue we assign the value $A = 1$, whereas if the sky is not blue we assign the value $A = 0$. The output is represented by the symbol Q as follows: The statement "Yes, carry an umbrella" is represented by the value $Q = 1$, whereas the statement "No, do not carry an umbrella" is represented by the value $Q = 0$. Then the computer's logic table is:

A	Q
0	1
1	0

The logic operation illustrated in this table is the NOT operation.

A funny but instructive way to view a logic operation is as a machine, made by placing a trained monkey inside a box, as in **Figure 6.2**. The monkey is trained to examine

---

[1] The word *datum* means a single piece of data.

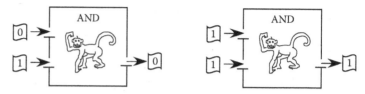

FIGURE 6.2   A trained monkey could perform the AND logic operation.

pieces of paper that are passed into the box through windows, to follow one simple rule concerning what is written on the pieces, to write something on a blank piece according to the rule, and to pass the new piece out through another window. For example, if the monkey is trained to perform the AND operation, he would examine the two pieces of paper, and, only if both pieces had a "1" marked on it would he write a "1" on a blank piece and pass it out. Otherwise, he would write a "0" and pass it out. Then he would get a banana as reward. In theory, an entire computer could be constructed by using trained monkeys inside boxes. The monkeys need not understand the role they are playing in the larger scheme.

6.3   ELECTRONIC LOGIC CIRCUITS

Computers are built using silicon-based *semiconductor* electronic circuits. In later chapters, we will explore the physical basis of such semiconductor circuits. In this section, we discuss simpler methods for implementing logic in order to understand the basic ideas.

To implement logic electronically we use switches. The simplest type of switch is a pushbutton that becomes electrically conducting when pressed by your finger. A finger presses down a small metal bar so that it touches the contact point, allowing electric current to flow from one wire into the other. Because of the presence of a spring that opposes the force applied by the finger, the pushbutton springs back open when released, as shown in **Figure 6.3**. We call this an ordinary switch, or push-ON switch.

We can use a light bulb and a battery to sense whether the switch is pressed or not. When you press the switch in **Figure 6.4**, it closes and electric current is able to flow through the light bulb, making it light up.

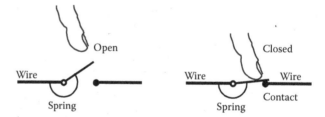

FIGURE 6.3   Push-ON switch, which stays open (nonconducting) when not being pressed and is closed (conducting) while being pressed. It springs open after being released.

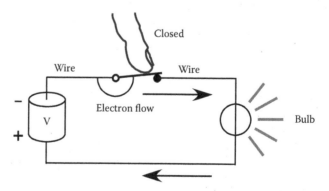

FIGURE 6.4   The bulb lights up only while the switch is pressed.

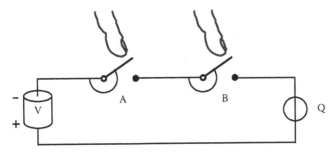

FIGURE 6.5 AND gate, made with two switches in series.

We represent this operation by a logic table. Represent the switch not being pressed by zero (0), and represent the switch being pressed by one (1). Also, represent the bulb OFF by zero (0) and the bulb ON by one (1). Then the table is:

Switch	Bulb
0	0
1	1

This table illustrates a simple one-to-one correspondence, meaning the bulb simply follows, or senses, the position of the switch.

We can use electric circuits to design and make simple logic gates. Consider the case in which two switches are wired in **series**, meaning one after the other sequentially, as in **Figure 6.5**. Because current can flow steadily only in a closed-loop circuit, the bulb in the figure will light only if switch A AND switch B are both pressed. Therefore, this circuit acts as an AND logic *gate*, with 1 representing a pressed switch and 0 representing an unpressed switch. The finger positions act as inputs (0,1), and the bulb acts as the output Q. The logic table for this is:

Logic Table for AND Gate

Switch A	Switch B	Bulb Q
0	0	0
0	1	0
1	0	0
1	1	1

In logic tables, a double vertical line separates inputs from outputs.

To make an OR logic gate, we place two ordinary switches in *parallel* (i.e., side by side), as in **Figure 6.6**. Because current can flow steadily only in a closed-loop

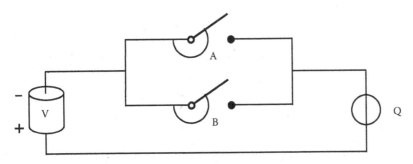

FIGURE 6.6 OR gate, made with two switches in parallel.

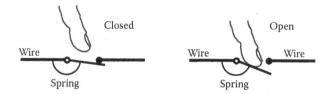

FIGURE 6.7   NOT switch, or push-OFF switch, is CLOSED (conducting) when left unpressed and is OPEN (nonconducting) when pressed, after which it reverts to CLOSED.

circuit, we predict that the bulb in Figure 6.6 will light if either switch *A* OR switch *B* is pressed. This is because only one closed pathway is needed in order for current to flow, and either path can conduct current to the bulb. The logic table is:

Logic Table for OR Gate

Switch *A*	Switch *B*	Bulb *Q*
0	0	0
0	1	1
1	0	1
1	1	1

QUICK QUESTION 6.1

Draw a circuit with two switches that will light up a bulb only if switch *A* is pressed and switch *B* is not pressed.

To construct a NOT gate, we need a new kind of physical switch, which we will call a NOT switch, or a push-OFF switch. In this switch, shown in **Figure 6.7**, a spring is resisting the force of the pressing finger, as in the ordinary switch. Here, in contrast, the rotating metal bar is mounted below the contact point, so that when released by the finger the metal bar touches the contact and current flows. Current is not flowing when the finger is pushing the bar down.

Figure 6.8 shows how to use the NOT switch in a simple NOT gate circuit. The table shows its logical operation, with 1 representing a pressed switch and 0 representing an unpressed switch.

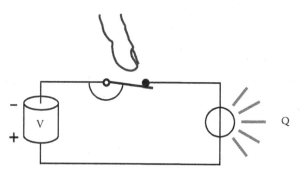

FIGURE 6.8   Push-OFF switch used as a NOT gate.

Logic Table for NOT Gate

Switch *A*	Bulb *Q*
0	1
1	0

6.4   LOGIC OPERATIONS AND DIAGRAMS

Let us leave electronics for the next stage of our discussion and consider logic from a more abstract, mathematical point of view. It is customary to use special symbols to represent the AND, OR, NOT operations. These symbols, along with their logic tables, are shown in **Figure 6.9**. Here A and B represent inputs and Q is an output. The variables A, B, and Q can equal only 0 or 1.

- We can summarize the NOT operation as "If A is *not* 1, then Q is 1."
- We can summarize the AND operation as "Only if A *and* B are 1, then Q is 1."
- We can summarize the OR operation as "If either A *or* B are 1, then Q is 1."
- We can symbolize the NOT operation by the notation $Q = \text{NOT}(A)$.
- We can symbolize the AND operation by the notation $Q = (A \text{ AND } B)$.
- We can symbolize the OR operation by the notation $Q = (A \text{ OR } B)$.

Notice that we could switch the placement of the inputs A and B with no change of the logic operations.

Using the three operations, AND, OR, NOT, any other logic operation can be carried out, for example, the EXCLUSIVE OR (also called XOR). Let us say your parents want you to drive into town with them and with your two little brothers, who are fun individually but are monsters together in the car. You decide that you will go with them if either goes, but not if they both go. This logic is shown in **Figure 6.10**.

The EXCLUSIVE OR operation can be constructed by combining the three elementary operations OR, AND, NOT, as shown in **Figure 6.11**. This circuit is equivalent to the XOR. Notice that we use a small black dot to indicate that we can send an input datum into more than one operation, just as we did in the flow chart in Figure 6.1. We can read this circuit from right to left as saying

$$Q = (A \text{ OR } B) \text{ AND } (\text{NOT } (A \text{ AND } B))$$

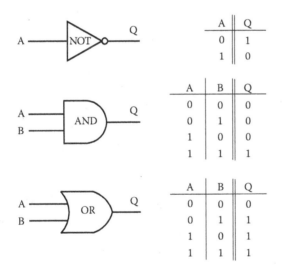

A	Q
0	1
1	0

A	B	Q
0	0	0
0	1	0
1	0	0
1	1	1

A	B	Q
0	0	0
0	1	1
1	0	1
1	1	1

FIGURE 6.9   The symbols and logic tables for the three elementary logic operations.

A	B	Q
0	0	0
0	1	1
1	0	1
1	1	0

FIGURE 6.10   The EXCLUSIVE OR operation, also called XOR.

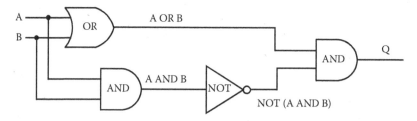

FIGURE 6.11 An equivalent circuit for *A XOR B*.

Another example of combining operations leads to the NAND operation, as in **Figure 6.12**. The NAND operation means "NOT (*A* AND *B*)." It could be used to simplify the diagram in Figure 6.11. The logic table for NAND can be found by the following method: Introduce an extra variable called Q', which is the output of (*A* AND *B*). This variable Q' also acts as the input for the NOT operation. We can make a logic table that has an extra column for the Q' variable. We do not really care about the Q' values—they are just an aid to help us determine the Q values, as follows: $Q' = (A$ AND $B)$. Then $Q = \text{NOT}(Q')$.

FIGURE 6.12 NAND operation.

NAND

A	B	Q′	Q
0	0	0	1
0	1	0	1
1	0	0	1
1	1	1	0

Other simple combinations of operations include the NOR and XNOR, shown in **Figure 6.13**.

The logic tables for NOR and XNOR are:

<table>
<tr><td colspan="3">NOR</td><td></td><td colspan="3">XNOR</td></tr>
<tr><td>A</td><td>B</td><td>Q</td><td></td><td>A</td><td>B</td><td>Q</td></tr>
<tr><td>0</td><td>0</td><td>1</td><td></td><td>0</td><td>0</td><td>1</td></tr>
<tr><td>0</td><td>1</td><td>0</td><td></td><td>0</td><td>1</td><td>0</td></tr>
<tr><td>1</td><td>0</td><td>0</td><td></td><td>1</td><td>0</td><td>0</td></tr>
<tr><td>1</td><td>1</td><td>0</td><td></td><td>1</td><td>1</td><td>1</td></tr>
</table>

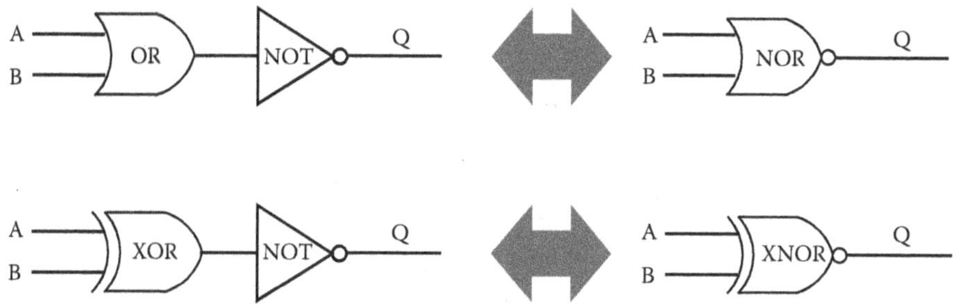

FIGURE 6.13 The NOR operation means "neither *A* NOR *B*." The XNOR operation answers the question "are *A* and *B* the same?"

We see that combining Boolean logic operations is a powerful method for building up complex operations. The type of logic circuits studied in this chapter are called *combinational logic circuits*. There are four rules that define this type of circuit:

Rules for combinational logic circuits:

1. It is forbidden to combine two input wires into one wire.
2. It is allowed to split an input wire into two wires.
3. An output wire from one gate can be used as an input to another gate.
4. The output of a gate, however, cannot be used as an input to the same gate.

The first three rules must always be followed in any logic circuit. The last rule must also be followed if we are talking about combinational logic circuits. Combinational logic circuits cannot be used directly to build memory. This is because the outputs of a combinational logic circuit depend only on the present values of its inputs, not on any previous, earlier values. Therefore, the circuit "forgets" any earlier values. However, if we want to build computer memory, we can remove the last rule, as we will discuss later.

THINK AGAIN

The following logic circuit is not allowed in a combinational logic circuit, because the output of the AND gate is sent back as an input to the same AND gate.

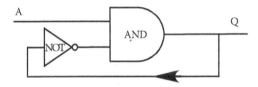

6.4.1 Three-Input Logic Operations

There is no reason a logic circuit could not have three or more inputs. A simple example is a situation in which three brothers agree that they will climb Mount Everest if and only if all three vote Yes. **Figure 6.14** represents this situation.

A more complicated case would be three sisters who agree that if at least two of them vote Yes, then all three will climb Everest. The logic circuit could be drawn as in **Figure 6.15**. Alice (A), Beatrice (B), and Chloe (C) can each put a Yes (1) or No (0) value onto their input line. If A and B say Yes, then Q' will equal 1; otherwise Q' equals 0. If A and C say Yes, then Q'' will equal 1; otherwise Q'' equals 0. If B and C say Yes, then Q''' will equal 1; otherwise Q''' equals 0. If either Q' or Q'' equal 1, then Q'''' equals 1. If either Q''' or Q'''' equals one, then $Q = 1$. The sisters go pack their hiking boots.

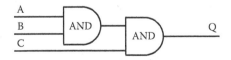

FIGURE 6.14 A logic circuit for a situation in which three brothers agree that they will climb Mount Everest if and only if all three vote *Yes*.

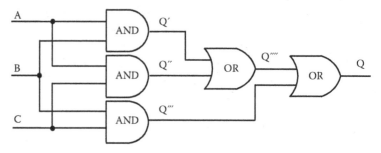

FIGURE 6.15 A logic circuit for a situation in which three sisters agree that if at least two of them vote yes, then all three will climb Mount Everest.

We can make a logic table, shown below, for the sisters' situation. There are three inputs, so there are $2^3 = 8$ possible input combinations. Each sister can vote Yes or No. If Alice votes No (0), there are four possible combinations of Beatrice and Chloe's votes; these are listed in the first four rows. (Rows read across, like rows of seats in a lecture room. Columns read down, like the columns holding up the ceiling in a cathedral.) If Alice votes Yes (1), there are the same four possible combinations of Beatrice's and Chloe's votes.

Inputs			Intermediate Values				Output
A	B	C	Q'	Q''	Q'''	Q''''	Q
0	0	0	0	0	0	0	0
0	0	1	0	0	0	0	0
0	1	0	0	0	0	0	0
0	1	1	0	0	1	0	1
1	0	0	0	0	0	0	0
1	0	1	0	1	0	1	1
1	1	0	1	0	0	1	1
1	1	1	1	1	1	1	1

QUICK QUESTION 6.2

Draw a logic circuit to represent the situation of the three sisters that differs from the one in Figure 6.15 in which they agree that if Chloe and either Alice or Beatrice agree to climb, then all will climb. *Hint*: This case is much simpler than the above case.

■ The column labeled Q' shows the logical values of the output of the uppermost AND gate in the figure. Its output equals 1 only if A and B are each equal to 1.
■ The column labeled Q'' shows the logical values of the output of the middle AND gate in the figure. Its output equals 1 only if A and C are each equal to 1.
■ The column labeled Q''' shows the logical values of the output of the lower AND gate in the figure. Its output equals 1 only if B and C are both equal to 1.
■ The column labeled Q'''' shows the logical values of the output of the upper OR gate in the figure. Its output equals 1 if either (or both) Q' or Q'' are equal to 1.
■ The output column labeled Q shows the logical values of the output of the lower OR gate in the figure. Its output equals 1 if either (or both) Q''' or Q'''' are equal to 1.

By inspecting each row, you can see that the circuit produces a Yes (1) output if two or more of the inputs equal 1.

6.4.2 Building Logic Operations Using the NOR Operation

The NOR logic operation can be used to construct any other logic gate. This proves to be an important fact when constructing practical logic circuits using semiconductors, because a complex logic circuit can be implemented by repeating a single

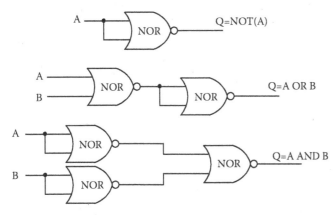

FIGURE 6.16   NOT, OR, AND operations constructed using only NOR gates.

type of device many times in different patterns. Modern computers are built this way. The equivalent circuits are shown in **Figure 6.16**.

The NAND operation can also be used by itself to build up any other logic operation. The proof of this is framed as an exercise at the end of the chapter.

6.5   USING LOGIC TO PERFORM ARITHMETIC

You might wonder what a computer is doing, for example, when it is adding numbers. It is just doing logic! How is this similar to deciding whether to take an umbrella or not, depending on your information about the weather? If you think about what you actually do when you add two numbers in the decimal (base-ten) number system, you will realize that you follow some simple rules: You add two digits (0–9) by using your memory, where you have memorized the answers to many simple operations (e.g., 4 + 8 = 12). Then you carry any value in excess of 9 one column to the left, to the tens column. You probably do this without thinking much about it, but really you are doing a set of logical operations. For example, when you add the base-ten numbers 7,924 and 7,858,

$$
\begin{array}{r}
7924 \\
+7858 \\
\hline
15782
\end{array}
$$

you perform the following steps, illustrated below, from right to left:

(1)	(1)	(0)	(1)	
+0	+7	+9	+2	4
+0	+7	+8	+5	+8
1	(1) 5	(1) 7	(0) 8	(1) 2

- Add 4 + 8 = 12 (all numbers in this example are base-ten). The number 12 means 2 ones and 1 ten. The 2 is called the *sum digit*. The 1 is called a *carry-out*, and it is added into the next column on the left. When it appears at the top of this column, in this case the *tens* column, it is called a *carry-in*.
- Add 1 + 2 + 5 = 8. Here the carry-out equals 0.
- Add 0 + 9 + 8 = 17. Here the carry-out equals 1.
- Add 1 + 7 + 7 = 15. Here the carry-out equals 1. You must create a new place holder one column to the left, so the final result (15782) contains five place holders.

QUICK QUESTION 6.3

You have three coins—a penny, a dime, and a nickel. Make a table showing all of the possible combinations of heads and tails that could occur when you toss all three coins. How many combinations are possible? *Hint*: The table will be similar to the first three columns labeled Inputs in the example above.

QUICK QUESTION 6.4

Construct a diagram for the base-ten addition: 328 + 902.

Computers use binary (base-two) numbers, rather than base-ten numbers, to do arithmetic. Binary numbers are explained in Chapter 2, Section 2.7. To summarize, instead of using as the place values 1, 10, 100, 1,000, etc., as we do in the base-ten system, in the binary, or base-two, system we use as place values 1, 2, 4, 8, 16, etc. For example, the binary number 01101 means there are zero 16s, one 8, one 4, zero 2s, and one 1 (or in base-ten: $8 + 4 + 1 = 13$). Adding numbers in binary requires following slightly different rules than adding in base-ten. Consider using binary numbers to add 0 and 1:

$$0 + 0 = 0, \quad 0 + 1 = 1, \quad 1 + 0 = 1, \quad 1 + 1 = 10$$

We pronounce the binary number 10 as "one, zero." We can also represent these sums in binary as:

0	0	1	1
$+\,0$	$+1$	$+\,0$	$+1$
0	1	1	10

You can see that when adding $1 + 1$ we get 10 in binary (which is 2 in base-ten), and we need to carry the 1-digit into the 2s column. (This is like carrying the 1 into the tens column when adding $4 + 8 = 12$ in base ten.) Next, consider adding binary numbers containing more than one digit. For example, the base-ten calculation $3 + 1 = 4$ is represented in binary as $11 + 01 = 100$, or

$$
\begin{array}{r}
(1) \\
11 \\
+01 \\
\hline
100
\end{array}
$$

When we added the two 1s in the right-most column, we obtained binary 10, with 1 being the carry-out. But just as you do when adding base-ten numbers, you need to add the carry-out to the numbers in the second-most right column. So, move the carry-out to the top of the second column (indicated by the parenthesis) and add it in that column. The number in parenthesis is the carry-in.

For another example, consider adding the binary numbers 01011 and 01011,

$$
\begin{array}{r}
01011 \\
+01011 \\
\hline
10110
\end{array}
$$
or labeling the place holders:

	16s	8s	4s	2s	1s
	0	1	0	1	1
$+$	0	1	0	1	1
	1	0	1	1	0

In this calculation you perform the following steps, from right to left:

(1)	(0)	(1)	(1)	
$+0$	$+1$	$+0$	$+1$	1
$+0$	$+1$	$+0$	$+1$	$+1$
1	(1) 0	(0) 1	(1) 1	(1) 0

- Add $1 + 1 = 10$. This means 1 two and 0 ones. The 1 is the carry-out and is added into the next column on the left, where it is called a carry-in.
- Add $1 + 1 + 1 = 11$. The carry-out equals 1.

- Add $1 + 0 + 0 = 01$. The carry-out equals 0.
- Add $0 + 1 + 1 = 10$. The carry-out equals 1.
- Add $1 + 0 + 0 = 1$. The final result is 10110.

To see how logic operations are used to perform addition, let us denote two binary digits (0 or 1) being added as A and B, and the result being the two symbols S and CO, with S (the sum digit) being the right-most digit of the result and CO being the carry-out. This is shown as:

$$\begin{array}{c} A \\ +B \\ \hline (CO)(S) \end{array} \qquad \text{For example:} \qquad \begin{array}{c} 1 \\ +1 \\ \hline (1)(0) \end{array}$$

The idea is to consider S and CO as two separate numbers that need to be computed for each case. This operation is called a *two-bit adder*.[2] First, let us write the logic table that summarizes all four addition examples:

Two-Bit Adder

A	B	CO	S
0	0	0	0
0	1	0	1
1	0	0	1
1	1	1	0

The above is actually two logic tables in one. For each pair of inputs A and B, we get two outputs, S and CO. You can see that the table corresponds to the four examples:

$$0 + 0 = 00, \quad 0 + 1 = 01, \quad 1 + 0 = 01, \quad 1 + 1 = 10$$

You can also see from the two-bit adder table that CO equals (A AND B), whereas S equals (A XOR B).

How do we construct a logic circuit that can take in two inputs A and B and put out the correct two outputs S and CO? We just split the inputs and send them into two simple gates, an XOR gate and an AND gate, as in **Figure 6.17**. This circuit produces the outputs listed in the two-bit adder table.

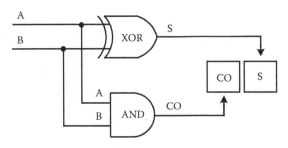

FIGURE 6.17 Two-bit adder, which takes in two single-bit numbers, A and B, and puts out their addition in the form of the sum digit S and the carry-out CO.

The two-bit adder can be used only to add single-bit numbers. In general, we want a circuit that is able to add numbers containing more than one binary digit. For example:

---

[2] Also called a half adder.

QUICK QUESTION 6.5

Construct a diagram similar to the one above for the binary addition $101 + 011$.

$$11$$
$$+01$$
$$\overline{100}$$

This corresponds to the operations:

```
(1)◄┐                              (CI)◄┐
 +1  │    1                         +A   │    1
 +0  │   +1      or in terms of     +B   │   +1
 ──  │  ────        symbols:       ────  │  ────
 10  │  (1) 0                     (CO) S │  (1) 0
     └──┘                                └──┘
```

In the right-most column here, the digits being added are 1s, so the right-most sum digit is 0 and the carry-out is *1*, which then gets moved into the (CI) spot and is added to the second digits $A = 1$, $B = 0$. The sum digit $S = 0$ and the carry-out digit $CO = 1$.

The operation $CI + A + B$ is called a ***three-bit adder***.[3] There are eight possibilities for this scenario. CI can be 0 or 1, and A and B can each be 0 or 1. The following table lists all eight possibilities and the results S and CO. This table can be read as: $CI + A + B = (CO)(S)$.

Three-Bit Adder

CI	A	B	CO	S
0	0	0	0	0
0	0	1	0	1
0	1	0	0	1
0	1	1	1	0
1	0	0	0	1
1	0	1	1	0
1	1	0	1	0
1	1	1	1	1

To implement the three-bit adder operation, we need a circuit that can add three binary digits. This can be accomplished by using two two-bit adders and one OR gate, as shown in **Figure 6.18**. First, A and B are added, producing a carry-out CO_1 and a sum digit S_1, which passes into the second two-bit adder, where it is added to the carry-in CI. This produces the final sum digit S and carry-out CO_2. Note that it is impossible for both CO_1 and CO_2 to equal 1. If either carry-out equals 1, then it produces the final carry-out CO.

To verify case-by-case that the three-bit adder in Figure 6.18 really does work, fill in the blanks in the following logic table.

Three-Bit Adder

CI	A	B	CO_1	S_1	CO_2	CO	S
0	0	0	0	0		0	0
0	0	1	0	1		0	1
0	1	0	0	1		0	1
0	1	1	1	0		1	0
1	0	0				0	1
1	0	1				1	0
1	1	0				1	0
1	1	1				1	1

[3] Also called a full adder.

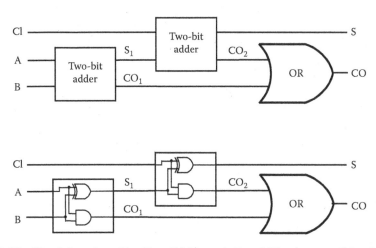

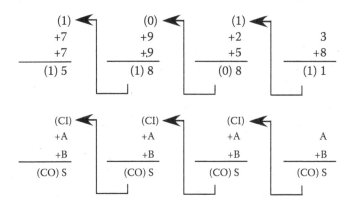

FIGURE 6.18   Circuit for a three-bit adder, with inputs A, B, and CI and outputs CO and S. The upper diagram shows each two-bit adder as a box, whereas the lower diagram reveals the details in each box, taken from Figure 6.17.

How can we use three-bit adders to make a circuit that can add any two binary numbers; for example, $1010 + 0111$? Arrange several three-bit adders side by side, with each passing its carry-out bit to the carry-in input of the adder to its left. This is what you do when adding two base-ten numbers; for example, $7{,}923 + 7{,}958 = 15881$, illustrated here:

(1)	(0)	(1)	
+7	+9	+2	3
+7	+9	+5	+8
(1) 5	(1) 8	(0) 8	(1) 1

(CI)	(CI)	(CI)	
+A	+A	+A	A
+B	+B	+B	+B
(CO) S	(CO) S	(CO) S	(CO) S

To make it easier to draw the proper circuit, let us invent a new way to draw the three-bit adder. Let us indicate the entire three-bit adder circuit in Figure 6.18 by a box labeled 3ba, as shown in **Figure 6.19**. Show the inputs A, B, and CI as arrows going in, and the outputs CO and S as arrows coming out.

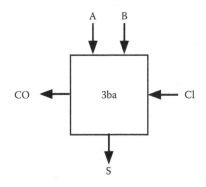

FIGURE 6.19   Symbol for a three-bit adder, with inputs A, B, and CI and outputs CO and S.

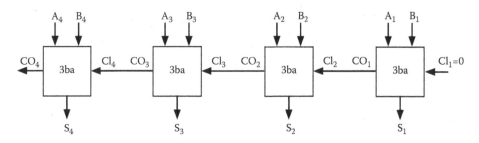

FIGURE 6.20   Cascaded three-bit adders, with inputs A's, B's, CI's and outputs CO's, S's.

For example, we can put four three-bit-adder circuits together in a right-to-left arrangement to add any two four-bit binary numbers. It makes sense to draw it right to left because that is how we do addition by hand. Let us represent the first four-bit number as $A_4\,A_3\,A_2\,A_1$. For example, in a binary number with four bits, 0111, we have $A_4 = 0$, $A_3 = 1$, $A_2 = 1$, $A_1 = 1$. Also choose a second four-bit number, say 1010, with $B_4 = 1$, $B_3 = 0$, $B_2 = 1$, $B_1 = 0$. Then we put the two right-most digits A_1, B_1 into the right-most adder (with a carry-in value $CI_1 = 0$) and obtain the outputs S_1 and CO_1. Then, as shown in **Figure 6.20**, we pass the output CO_1 of the right-most adder to the adder to its left, where it becomes (equals) the input CI_2.

The following example shows the binary addition 1101 + 0101 = 10010. **Figure 6.21** shows the operation of the cascaded adders for this example.

$$
\begin{array}{cccc}
(1)(0)(1) & & (CI_4)(CI_3)(CI_2)(CI_1)(0) \\
1\ 1\ 0\ 1 & & A_4\ \ A_3\ \ A_2\ \ A_1 \\
+0\ 1\ 0\ 1 & & B_4\ \ B_3\ \ B_2\ \ B_1 \\
\hline
1\ 0\ 0\ 1\ 0 & & CO_4\ \ S_4\ \ S_3\ \ S_2\ \ S_1
\end{array}
$$

THINK AGAIN

Notice that the logic flows from right to left in the cascaded adder because that is how we usually visualize this calculation when we perform it ourselves on paper, by hand. This contrasts with the usual way of drawing logic circuits, in which the logic flows from left to right.

6.6   IMPLEMENTING LOGIC WITH ELECTROMAGNETIC SWITCHES

In logic circuits, the output of a gate can be an input for another gate. How can we implement such a situation by using electronic circuits? The lighting of a bulb cannot easily be used by itself as an input to a gate. We need a way to convert the current flowing in

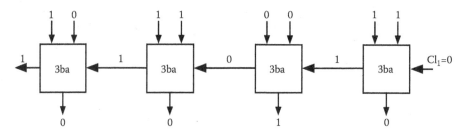

FIGURE 6.21   Cascaded three-bit adders.

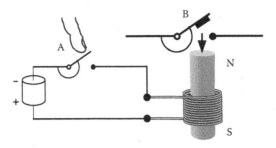

FIGURE 6.22   A *relay*. When the finger, *A*, presses the switch, current flows in the coil circuit, magnetizing the iron rod in the coil, which in turn exerts a magnetic force (the arrow), pulling the "ordinary" switch, *B*, closed.

the light bulb into a force that can push a switch. A physicist will say immediately, "Use an electromagnet!" As in **Figure 6.22**, attach a small piece of iron to the underside of a switch (*B*) that is placed just above an electromagnet. When you press switch *A* with your finger, it activates the electromagnet, which pulls switch *B*, closing it. This setup is called a *relay*, because we can relay the effect of pressing one switch to another.

Relays have various uses. One use is in a keyboard. A switch can be outside of the computer in the keyboard and pushing it activates a switch inside the computer. Another use is for passing the output of one logic gate to the input of another logic gate. In modern computers, switches and relays are not made using electromagnets; in fact, they have no moving parts and are made of semiconductors. We will explore this later.

A relay can also control a NOT switch, as shown in **Figure 6.23**. When the switch *A* is pressed, the electromagnet pulls the NOT switch open. We can express this as $B = \text{NOT}(A)$.

Let us illustrate the method of building complex logic gates by combining elementary gates with relays. A good example is the NAND gate. Recall the equivalent circuit and its logic table:

A	B	Q'	Q
0	0	0	1
0	1	0	1
1	0	0	1
1	1	1	0

In this logic circuit, the intermediate value is $Q' = (A \text{ AND } B)$, which is followed by the operation $Q = \text{NOT}(Q')$. As shown in **Figure 6.24**, we can implement this in an electronic logic circuit by using two ordinary switches in series to make the AND gate, then sending the output of this operation to an electromagnet Q', which can magnetically open the NOT switch. When *A* and *B* switches are both pressed, Q' activates the NOT switch, opening the circuit and turning the bulb OFF ($Q = 0$). If both *A* and *B* are

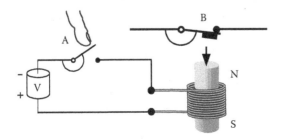

FIGURE 6.23   NOT switch controlled by a relay.

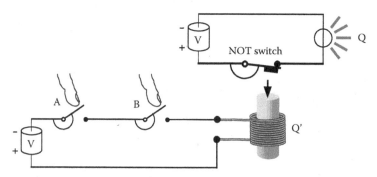

FIGURE 6.24   NAND gate built using a relay to send the output of (*A* AND *B*) to the NOT switch.

not pressed (*A* = *B* = 0), or if only one of them is pressed (*A* = 1, *B* = 0, or *A* = 0, *B* = 1), the bulb remains ON (*Q* = 1). This implements the operation NOT (*A* AND *B*).

Rather than light the bulb, the current from the NOT switch could be sent to another logic gate, as shown in **Figure 6.25**. In addition, *A* and *B* switches could be controlled, not by two fingers, but by two separate gate outputs. The circuit in this figure might only be a small part of a much larger logic circuit.

The devices discussed here are sufficient to illustrate the basic principles of switching and electronic logic and were actually used for automatic control of building elevators until the 1960s. However, we do not really want to build computers using these bulky, power-consuming electromagnetic relays. A computer built this way could fill an entire room but still be less powerful than a present-day $10 pocket calculator. To reach today's level of performance, we need to decrease the size, cost, and power consumption of each logic gate by a factor of at least 1 million! The means for doing this using microelectronics will be discussed in later chapters.

SUMMARY AND LOOK FORWARD

We have gained an understanding of logic operations and how they can be used to perform mathematical and logical tasks. We saw that logical questions can be represented by abstract symbols called inputs and outputs. The symbols (*A*, *B*, *Q*, etc.) can take on values 1, meaning yes (true), and 0, meaning no (false). The logical operations AND,

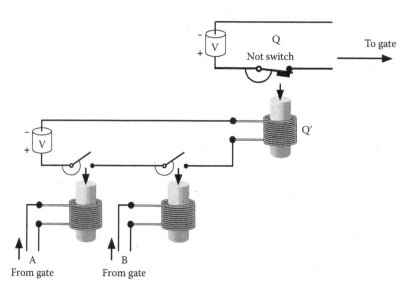

FIGURE 6.25   A general-purpose circuit with inputs *A* and *B* and output *Q* = (*A* NAND *B*), which is equivalent to Q = NOT(*A* AND *B*).

OR, NOT, etc. can be performed on input values to produce output values. These outputs can be interpreted in two ways:

1. As logical answers to questions, or
2. As numbers representing the result of a calculation, such as addition

Combinational logic circuits are a class of circuits that obey a set of rules, which includes the rule that the output of a logic gate cannot be sent back into the input of the same gate. Combinational logic circuits can be analyzed by using a logic table listing all possible combinations of input values along with the corresponding output values. It is useful to be able to list all of the input values in a systematic way in a logic table. All possible logic operations, including NAND, XOR, NOR, etc., can be constructed using AND, OR, and NOT.

Logic operations can be performed with physical devices called switches. Input data are represented by the position of each switch (unpushed = 0, pushed = 1). Outputs are represented by current in a wire (zero current = 0, nonzero current = 1). Electromagnetic relays can be used to transfer the output of a logic gate to the input of another logic gate.

Addition of two binary numbers is performed by a logic circuit that implements the same steps that a person would perform in adding the same numbers. This operation is performed by the circuit called a three-bit adder, which is a cascaded group of two-bit adders. Each two-bit adder does the job of adding the numbers in its column and passing the carry-out value to the carry-in for the next column.

There was a long journey from the theoretical conceptions of George Boole and the mechanical arithmetic devices of Charles Babbage and others, to the building of practical, working computing machines. Electronic computers were not imaginable in Babbage's time (the 1820s) because the physics principles of electricity and magnetism had not yet been fully established. This did not come until the 1860s, as we studied in Chapter 5. Modern computers did not arrive until after the physics of semiconductor materials was developed, as we will discuss in later chapters.

SUPPLEMENTAL SECTION: BOOLEAN SEARCH OF DATABASES

The Internet is a powerful tool for finding information on virtually any subject. For example, the U.S. Patent and Technology Office (USPTO) maintains an online, searchable database of all patents issued by their office at http://www.uspto.gov/patft/help/helpadv.htm. In their instructions on "Nested Quick Expressions," they write:

> You can use the Advanced Search Page to create and execute Quick searches with more than two search terms that use the Quick operators (*OR, AND, ANDNOT*). Along with these operators, you can use parentheses to further clarify your search statement. In the absence of parentheses, all operators associate from left to right.

Rather than using the OR, AND, NOT operators, as we have done in this chapter, the USPTO uses an operation called *ANDNOT*, defined by the combination gate shown in **Figure 6.26**.

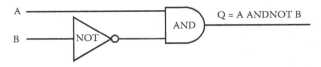

FIGURE 6.26 Construction of ANDNOT gate.

They give several examples:

Example 1

tennis AND (racquet OR racket)
If you enter this query, you will retrieve a list of all patents which contain the terms tennis and either racket or racquet somewhere in the document.

Example 2

television OR (cathode AND tube)
This query would return patents containing either the word television or both the words cathode and tube.

Example 3

needle ANDNOT ((record AND player) OR sewing)
This complex query generates a list of hits that contain the word needle, but does not contain any references to sewing. In addition, none of the hits would contain the combination of record and player.

The two following queries, which are logically the same, return identical results:

- needle ANDNOT ((record AND player) OR sewing) ANDNOT magnetic
- needle ANDNOT ((record AND player) OR sewing OR magnetic)

The search software also allows you to specify that only words in the titles of patents are to be searched. This is done by writing ttl/( ), with a search instruction inside the brackets. For example, ttl/(keyboard AND piano) will return all patents having "keyboard" and "piano" in their title.

The following title searches yielded the number of patents returned, shown in brackets after each:

- keyboard AND ((computer AND ergonomic)) [1,371]
- keyboard AND ((computer AND ergonomic) AND piano) [55]
- keyboard AND (computer AND ergonomic) ANDNOT piano [1,316]

Notice that the numbers in brackets are related by $1,316 + 55 = 1,371$. This must be the case, as you can convince yourself.

SUGGESTED READING

Bierman, Alan W. *Great Ideas in Computer Science—a Gentle Introduction*. Cambridge, MA: MIT Press, 1997.
Epp, Susanna S. *Discrete Mathematics with Applications*. Boston: PWS Publishing, 1995.
Maxfield, Clive. *Bebop to the Boolean Boogie*. Amsterdam: Newnes, 2003.
Murdocca, Miles J., and Vincent P. Heuring. *Principles of Computer Architecture*. Upper Saddle River, NJ: Prentice Hall, 2000. Appendix A: Digital Logic.
Predko, Myke. *Digital Electronics Demystified*. New York: McGraw Hill, 2005.

KEY TERMS

Boolean logic
Combinational logic circuit
Gate

Input/output
Logic
Logic operation (or gate)
Logic circuit
Relay
Three-bit adder
Two-bit adder
Semiconductor

ANSWERS TO QUICK QUESTIONS

Q6.1

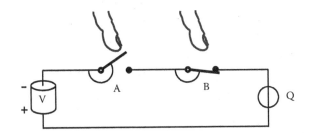

Q6.2 A logic circuit to represent the "three sisters" situation in which they agree that if Chloe and either Alice or Beatrice agree to climb, then all will climb.

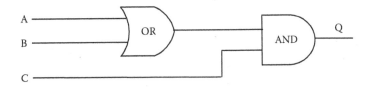

Q6.3 For three coins—a penny, a dime, and a nickel—a table showing all of the possible combinations of heads and tails that could occur when you toss all three coins:

Combinations		
Penny	Dime	Nickel
Heads	Heads	Heads
Heads	Heads	Tails
Heads	Tails	Heads
Heads	Tails	Tails
Tails	Heads	Heads
Tails	Heads	Tails
Tails	Tails	Heads
Tails	Tails	Tails

Q6.4 The diagram for the base-ten addition $328 + 902 = 1,230$ is

$$
\begin{array}{ccc}
(0) & (1) & \\
+3 & +2 & 8 \\
+9 & +0 & +2 \\
\hline
(1) \quad 2 & (0) \quad 3 & (1) \quad 0 \\
\end{array}
$$

Q6.5 The diagram for the binary addition $101 + 011 = 1000$ is

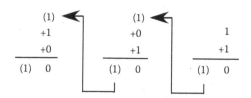

EXERCISES AND PROBLEMS

Exercises

E6.1 Complete the following logic table corresponding to the college admissions decision discussed at the beginning of this chapter.

Inputs			Outputs (Yes/No)	
College B Scholarship?	College B Dorm DSL?	College A Scholarship?	Attend College B?	Attend College A?
No	No	No		
No	No	Yes		
No	Yes	No		
No	Yes	Yes		
Yes	No	No		
Yes	No	Yes		
Yes	Yes	No		
Yes	Yes	Yes		

E6.2 What is wrong with these logic circuits?

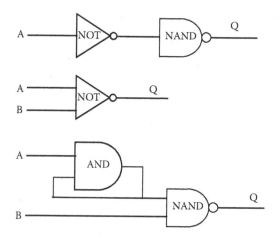

E6.3 Show the logic tables for these logic circuits:

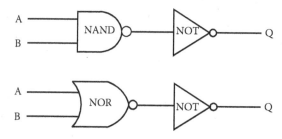

E6.4 Computer memory is based on logic circuits that can store the value of a logic variable or bit for later use. Why can combinational logic circuits alone not be used to construct memory?

E6.5 Others in addition to George Boole contributed to the field of logic, such as Aristotle (384–322 BCE) and Augustus De Morgan (1806–1871). Carry out some research into one or both of these people and write a brief report on your findings.

E6.6 A logic diagram (or circuit) has four binary inputs A, B, C, and D, as shown. How many distinct combinations of input values are there? That is, how many horizontal rows does the logic table need? Explain.

Problems

P6.1 Construct logic tables for the following three electrical circuits.

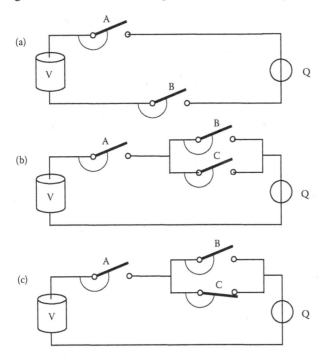

P6.2 Verify that the following logic circuit is equivalent to the XOR gate. This is an alternative circuit to that shown in Figure 6.11. Consider all four possible combinations of inputs (A, B) = (0, 0); (0, 1); (1, 0); (1,1). Make a logic table that also contains the values of Q' and Q'' in each case to explain how you obtained your results.

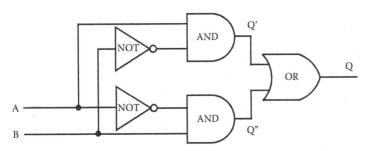

P6.3 Find the logic tables for these two circuits:

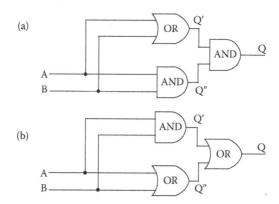

P6.4 For the following circuit, complete the logic table for the eight possible inputs.

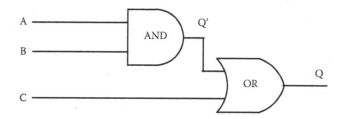

	Inputs		Output
A	B	C	Q
0	0	0	
0	0	1	
0	1	0	
0	1	1	
1	0	0	
1	0	1	
1	1	0	
1	1	1	

P6.5 Design a logic circuit for a padlock (e.g., for locking up a bicycle) that has three inputs A, B, C (wheels or switches), each of which can be set to a value 0 or 1. Design the circuit so that its output equals 1 ("open the lock") only if the three inputs are as follows: $A = 0$, $B = 1$, and $C = 1$.

P6.6 Design and draw a logic circuit (combination of gates) having four inputs, A, B, C, D, and one output Q, that will result in an output $Q = 1$ only if all inputs are 0. For all of the other 15 possible input cases, the output should be $Q = 0$. *Hint*: Refer to the example of the three sisters in this chapter to see how to list all possible input cases when there are three inputs. Extend this reasoning to the case of four inputs.

P6.7 A special type of logic gate is built to have three inputs and one output, as shown. Inside the box is a combinational circuit, which produces the output shown in the table. Find an equivalent logic circuit that would produce this output. Make a complete logic table, showing the inputs, the output, and any intermediate values that occur in your designed circuit.

Inputs			Output
A	B	C	Q
0	0	0	0
0	0	1	1
0	1	0	1
0	1	1	1
1	0	0	1
1	0	1	1
1	1	0	1
1	1	1	1

P6.8 The NAND operation can be used by itself to build up any other logic operation. Show this by the following method: Replace each NOR gate in Figure 6.16 by a NAND gate and draw each resulting circuit. Figure out which of the three resulting circuits implements the NOT, AND, and OR operations.

P6.9 The NAND circuit in Figure 6.24 could also be implemented without using a relay. Find and draw such a circuit using only two push switches and a bulb. Also, draw the corresponding logic diagram using the logic gates AND, OR, NOT.

P6.10 Make a logic table for the following circuit with inputs A and B and output Q. Also, draw the corresponding logic diagram using the logic gates AND, OR, NOT.

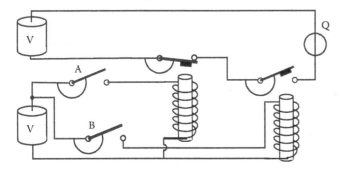

P6.11 Make a logic table for the following circuit with input A and output Q. Also, draw the corresponding logic diagram using the logic gates AND, OR, NOT.

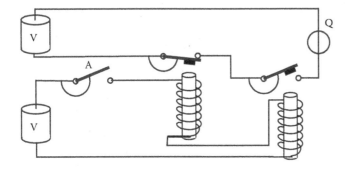

P6.12 Consider a demonstration of using electromagnetic relays to perform logic. First, let us establish some symbols to help us draw such circuits. In the first figure below, a

box with an R indicates a relay, shown in detail on the right. The input arrow labeled
C indicates the wire that controls the relay. The small rectangle with hatching is the
electric ground. (All grounds in such diagrams are connected to each other by wires
that are not shown.) If 12 V is applied to the control wire, the relay is activated, either
pushing closed the switch, or pulling it open, depending on the type of relay. The relay
illustrated here is a push-ON switch.

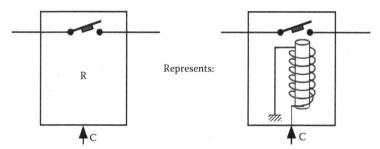

(a) Make a logic table for the following circuit, using *A* and *B* as manual push-
ON switches. The bulb is lit by application of 3 V. If you use logic symbols 0 and
1, make sure you define what these mean (switch closed, bulb ON, etc.) Identify
the proper logic name for this circuit.

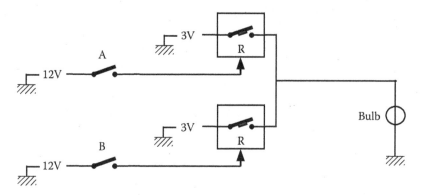

(b) Make a logic table for this circuit, using A and B as manual push-ON
switches. The bulb is lit by application of 3 V. Note that the relay nearest the bulb
is a push-OFF switch. Identify the proper name for this logic circuit.

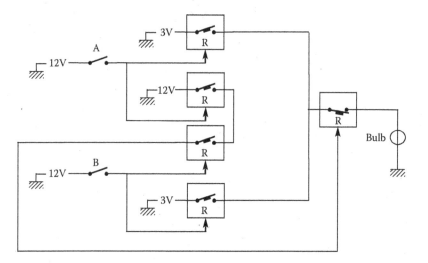

P6.13 Draw an electrical relay circuit that will perform the following logical operation:

$$Q = (A \text{ AND } B) \text{ OR } (\text{NOT } B).$$

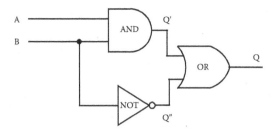

Hints: Recall that the order of the inputs into an AND or an OR gate does not matter. That is, (*A* AND *B*) = (*B* AND *A*) and (*A* OR *B*) = (*B* OR *A*). Split off the *B* input wire and use relays as the inputs for the AND gate and NOT gate. Use relays on the outputs of the AND gate and the NOT gate to couple into the OR gate. Have *Q* be represented by a light bulb.

P6.14 As described in Supplemental Section, the USPTO lets visitors to its website use logical search methods to find information on all U.S. patents. Read the supplemental section, and then log on to the USPTO website (http://www.uspto.gov/patft/help/helpadv.htm#complex) and answer the following:

(a) How many patents contain all of the words "human," "genetic," "engineering," "public," and "reaction?"
(b) How many patents contain all of the words "human," "genetic," "engineering," "public," and "reaction," but do not contain the word "plant?"
(c) How many patents contain in their title all of the words "genetic," "engineering," "public," and "reaction?"
(d) How many patents contain in their title all of the words "human," "genetic," and "engineering?"

P6.15 Determine the logic tables for the three logic circuits shown below. In your tables, include the intermediate values (*D*, *E*, *F*, etc).

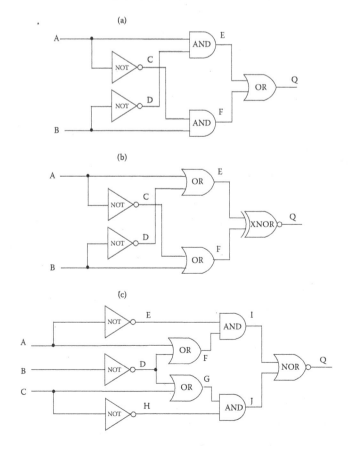

P6.16 The parity of a list of binary numbers (1s and 0s) can be used for error detection and correction. The parity of a list depends on the number of 1s in the list. If there is an odd number of 1s in the list, then the parity of the list is equal to 1. If there is an even number of 1s, then the parity is said to be equal to 0. For example, the four-bit string 0101 has two 1s and thus its parity is equal to 0.

   Below is a logic circuit for determining the parity of a two-bit list. There is also a nearly completed logic circuit for determining the parity of a four-bit list: *A B C D*.

(a) Determine what logic gates should replace the question marks and draw this logic circuit with the correct gates. Also create a logic table for this case.
(b) Generalize (i.e., expand) the four-bit parity circuit to make a circuit that can determine the parity of an eight-bit list: *A B C D E F G H*. Draw the correct circuit diagram.

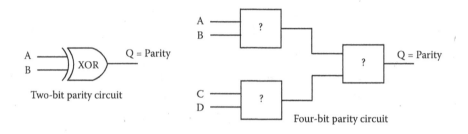

Two-bit parity circuit

Four-bit parity circuit

Two-Bit Parity Logic Table

A	*B*	Parity
0	0	0
0	1	1
1	0	1
1	1	0

P6.17 Consider a logic circuit composed of four full-adders, as in Figure 6.21. We want to use this circuit to add the two binary numbers 1011 + 1010. Make a drawing like Figure 6.21 showing the values of all bits at every stage in the calculation.

(a) What is the answer in binary?
(b) What is the value of this answer in decimal?

P6.18 Verify that the following circuit (copied below eight times) implements the full-adder for two one-bit binary numbers *A* and *B*, and a carry-in digit *CI*. Do this by making a logic table including all eight possible combinations for input values of *A*, *B*, and *CI*; that is, (000), (001), (010), (011), (100), (101), (110), and (111). Also include in the table the outputs *Q* and the *CO* bit. For each circuit, fill in the logic values at every internal location in each copy of the circuit using a different input combination for each circuit. Verify that the logic table you construct for the circuit is the correct one for the full-adder. *Hint*: The internal results are filled in for the first case.

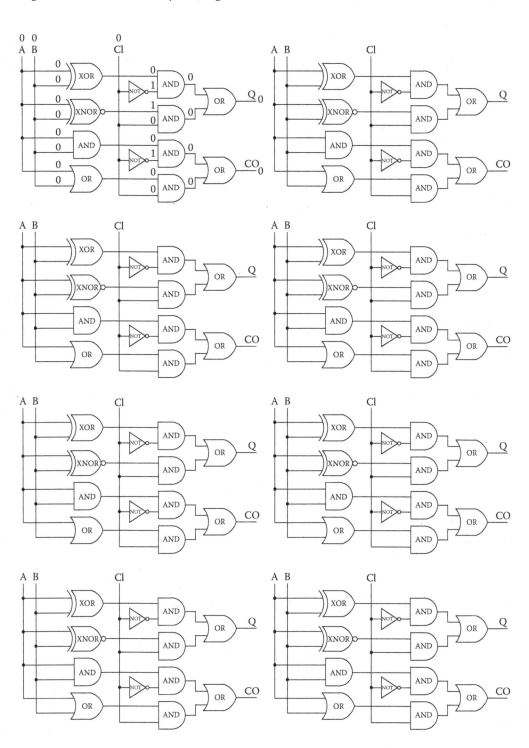

7

Waves: Sound, Radio, and Light

I do not think that the wireless waves I have discovered will have any practical application.

Heinrich Hertz
(ca. 1890)

Heinrich Hertz, the first person to generate wireless EM (radio) waves, in 1888.

Using cell phones on a sunny day. Radio waves, sound, and light all come together.

7.1 COMMUNICATING WITH SOUND, RADIO, AND LIGHT

The Internet consists of millions of computers connected in such a way that they can exchange information (i.e., communicate). They do this using *electromagnetic (EM) waves*, which are carried either through the air, through metal wire, or through thin glass fibers, to the receiver's computer. For example, the digital data that you download from the Internet might be sent from one city to another using *light* waves in an optical fiber. Then it might be converted into *voltages* and sent along a copper wire to a wireless router, where it would be converted into radio waves and broadcast to your computer. The data that you downloaded could represent music. After those data are reconstituted into an analog form, they can drive a stereo speaker, creating sound. The sound waves travel through the air and finally arrive at your ears. We see that the concept of waves plays a key part in these phenomena. We will begin this discussion with the topic of harmonic motion and waves in general, and then move on to sound, radio, and light waves.

7.2   SIMPLE HARMONIC MOTION

A grandfather, or pendulum, clock "counts" time by the number of oscillations the pendulum makes. The simplest pendulum clock is in **Figure 7.1**, showing a metal ball (called a bob) fixed to a rigid rod that is swinging from a low-friction pivot. When at rest, the bob is in the equilibrium position. When set into motion, the bob moves back and forth along an arc. The center picture shows the motion of the bob just after it has been set into motion. The distance the bob travels between its center position and its extreme position is called the ***amplitude*** of the motion.

Remarkably, the time it takes for the bob to move between its two extreme positions is nearly the same regardless of whether the amplitude is small or is large. The story goes that 17-year-old Galileo noticed this behavior while in church, watching the motion of a swinging chandelier. As the service wore on, the pendulum's amplitude grew smaller, yet the time between the plate reaching its extreme position remained the same (he timed the pendulum's motion using his own heartbeat).

If we graph the horizontal position of the pendulum versus time, we see a curve as shown in **Figure 7.2**. By horizontal position, we mean the position of a shadow on the floor below the bob that would be produced by shining a light from above the pendulum. The shadow would move back and forth along a straight line. At any instant of time, the shadow's distance from the equilibrium position would be given by the value of the oscillating curve shown in the graph. Any curve with this specific shape is called a *sine curve*.[1]

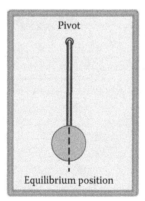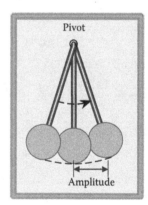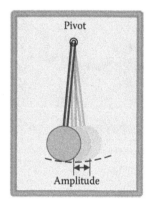

FIGURE 7.1   A pendulum clock "ticks" at the same rate if its pendulum bob swings with larger or smaller amplitude.

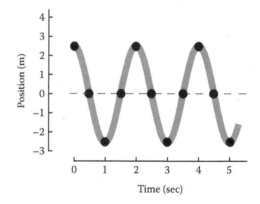

FIGURE 7.2   A pendulum's horizontal position versus time traces a sine curve.

---

[1]   We will not distinguish between sine curves and cosine curves.

The swinging of a pendulum is an example of *simple harmonic motion*, or SHM. A characteristic of SHM is a regular exchange of energy between two forms. In a pendulum clock, an exchange occurs between kinetic and potential energy. When the bob is at its center position, it is moving fast and it carries kinetic energy. As it moves toward either extreme of its motion, it slightly rises against gravity and its potential energy increases. At the same time, the bob slows, and its kinetic energy decreases. When the bob is at an extreme of its motion, it stops moving for an instant, at which time its kinetic energy equals zero—all of its former kinetic energy has been converted into potential energy. Then the bob begins converting its potential energy back into kinetic energy, and the whole cycle repeats.

The mechanism that sustains simple harmonic motion is the continual exchange between kinetic and potential energy.

Other examples of SHM are: (1) the vibrations of the two atoms making up an oxygen molecule, (2) a musical tuning fork, (3) a crystal in a quartz oscillator wristwatch, and (4) an electronic voltage-controlled oscillator in a music synthesizer.

Another example of SHM is a pendulum bob that moves in a circular pattern when viewed from above, as in **Figure 7.3**. Imagine being above the point where the bob is attached and looking down. You would see the bob moving in a circle around a center point on the floor. If a bright lamp were placed near the floor, as shown, the bob would cast a moving shadow on the opposite wall, labeled B.

Say you make a mark on the wall at the location of the shadow, once every half-second. Those marks would be at the locations on the wall shown by the shaded regions. If you use a ruler to measure the positions of the marks and make a graph of those positions, it will look like that in **Figure 7.4**. The center position is labeled 0 (zero), as a convenient reference point. The black dots mark the positions of the shadow at times 0, 0.5, 1, 1.5 seconds, etc. Say that you start making measurements at a time that you call zero (say you start a stopwatch at that time), when the shadow just happens to be at its maximum position ($x = 2.5$ m). For the particular example shown in the figure, the bob moved to its center position ($x = 0$) after 0.5 seconds, then to its maximum negative position ($x = -2.5$ m) at time equals 1 second. It continues in this manner for many oscillations. The curve shows the positions of the bob at all possible times, as if you made a continuous marking on the wall.

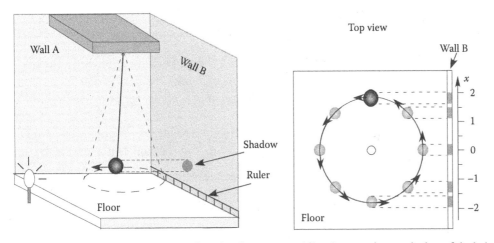

FIGURE 7.3   A pendulum bob moves in a circular pattern, while a lamp projects a shadow of the bob onto Wall B. A ruler is placed along Wall B to indicate the position (x) of the shadow at any time.

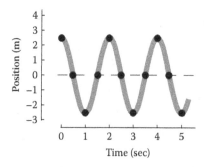 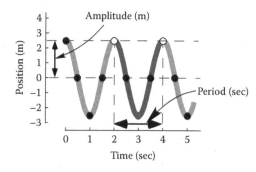

FIGURE 7.4 Graph of the position of the bob's shadow on Wall B at various times. The amplitude equals 2.5 m, and the period equals 2 sec.

THINK AGAIN

If the motion of the swinging pendulum in Figure 7.3 were in the shape of a square or triangle, its shadow would not display SHM, which is only characteristic of circular motion.

The right side of the figure shows the same graph, with the **amplitude** of the motion indicated by a vertical arrow. For this example, the amplitude equals 2.5 meters.

The **amplitude, A,** of a SHM equals the distance from the center position to the maximum position.

The amplitude has units that depend on the particular kind of oscillation you are describing. In our example above, amplitude has units of meters (m).

In the right side of Figure 7.4, the darker shaded portion of the curve between the two open circles is called a **cycle**. A **cycle** of a SHM is the motion between any two successive peaks. Any single portion of the motion having the same duration as a peak-to-peak cycle can also be considered as a cycle (e.g., a portion between two minima, or a portion between two appropriate zero crossings).

The other defining quantity of a SHM is its **period**.

The **period, T,** of a SHM equals the time interval required for one complete oscillation cycle.

The period equals the time between two successive peaks of the motion. In Figure 7.4, a horizontal arrow indicates the period. In this example, the period equals 2 seconds. The period has units of seconds (sec), or equivalently seconds per cycle (sec/cycle). You can use either.

Rather than specifying the value of the period of a SHM, we might prefer to specify its **frequency**. These are two equivalent ways of giving the same information.

The **frequency** of a SHM is the number of cycles that occur per second. The units of frequency are cycles per second, or cycle/sec. The relation between period and frequency is $f = \dfrac{1}{T}$, which implies also $T = \dfrac{1}{f}$.

The frequency unit cycle/sec (or simply, 1/sec) is called **hertz**, abbreviated Hz. This unit is named after German physicist Heinrich Hertz, who discovered wireless EM waves in 1888.

To see the relation between period and frequency, consider an example. If the period of a pendulum is 0.25 sec, or one-fourth of a second, then it should be clear that the pendulum oscillates 4 times per second. That is,

$$f(\text{cycle/sec}) = \frac{1}{T(\text{sec/cycle})} = \frac{1}{0.25 \text{ sec/cycle}} = 4 \text{ cycle/sec} = 4 \text{ Hz}$$

Frequency can be a number other than a whole number. For example, if $T = 5$ sec, then $f = 1/5$ cycles per sec, or 0.2 Hz.

A sports analogy might help clarify the concepts of period and frequency. If a runner takes 3 min to run one lap around a track, then the period T equals 3 min, or one-twentieth of an hour; that is, 0.05 hour. The runner's frequency of lap running equals

$$f = 1/T = 1 \div (0.05 \text{ hour per lap}) = 20 \text{ laps per hour}$$

QUICK QUESTION 7.1

For the SHM shown in the following graph, what are the period, frequency, and amplitude?

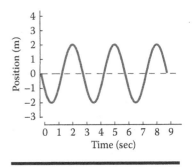

THINK AGAIN

The circular-moving pendulum in Figure 7.3 displays a regular exchange of energy between two forms. This exchange occurs between energy associated with the bob moving in the direction parallel to Wall A and energy associated with it moving in the direction parallel to Wall B. This is different than an exchange between kinetic and potential energies, as is the case in Figure 7.1.

Another example of SHM is a bob attached to the lower end of a metal spring whose upper end is held firmly at some fixed location, as in **Figure 7.5**. A spring can be compressed by pushing its two ends toward each other. When its ends are released, it springs back. If a spring is stretched to be longer than its usual length and then let go, it recoils to reduce its length toward its usual length, which is called the equilibrium length. The bob returns to its equilibrium position and then goes beyond it, etc. These behaviors lead to simple harmonic oscillations of the spring's varying length, illustrated as the dashed curve connecting the positions of the bob shown at different times. The curve shows how the bob's motion would look if the spring were oscillating while hanging from the ceiling of a passing car and you viewed it from a fixed location on the sidewalk. Again, the bob traces out a sine curve.

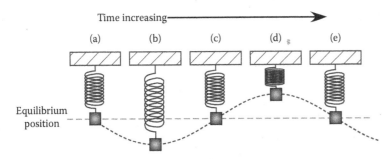

FIGURE 7.5 A spring with an attached mass undergoes SHM, which traces out a sine curve as time advances.

7.3   DAMPED AND COMPLEX HARMONIC MOTION

A SHM will not continue forever, of course. Friction will take its toll, gradually reducing the energy of the pendulum bob's motion. Friction is the agent that is ***damping*** (reducing) the bob's energy. Consider again the pendulum in Figure 7.1. The picture at the center shows the bob's motion after it has started swinging. The picture on the right shows the motion of the pendulum some time later, after friction has reduced the bob's energy, and the amplitude has become less. Eventually, the pendulum will stop moving.

Damped motion of a SHM is graphed in **Figure 7.6**. This shows the amplitude gradually decreasing as time goes on. After a long time, the position settles down to a value equal to zero, its equilibrium position. As the amplitude decreases, the energy of the SHM also decreases. The SHM period and frequency remain constant during damping.

In addition to simple and damped harmonic motion, there is also ***complex harmonic motion***—any pattern of exactly repeating shapes. An example is shown in **Figure 7.7**. This type of motion is described by its period, T, and amplitude, A, but that does not provide a full description of the shape of the oscillation. The period simply tells the time required for the motion to repeat its shape once. Complex shapes like these can carry a lot of information. For example, the tone of a note played on a musical instrument is determined by the shape of the complex harmonic motion it produces.

The word periodic, which simply means repeating, is often used to describe oscillations. Both SHM and complex harmonic motion are examples of ***periodic motion***.

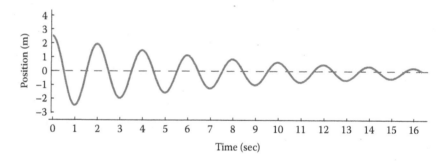

FIGURE 7.6   Graph of the position of the bob's shadow on Wall B at various times. The initial amplitude equals 2.5 m, and the period equals 2 sec.

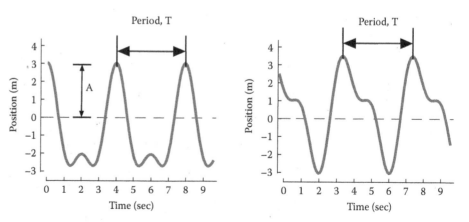

FIGURE 7.7   Graphs of complex harmonic motion.

7.4   DRIVEN HARMONIC MOTION AND RESONANCE

There are two ways to put a pendulum bob or spring bob in motion. If you start it in motion and then allow it to oscillate naturally on its own, its motion is periodic as a result of the restoring force acting on the bob. The period of the bob's natural motion in this situation is called the *resonance frequency*. Another way to cause SHM is to periodically force, or "drive", the bob. An example is pushing a child on a playground swing. Many small pushes can cause her to oscillate with large amplitude, but only if the pushes occur with frequency equal to the frequency at which the child naturally swings when not being pushed. The pusher is providing a force in harmony with the natural oscillation of the child on the swing.

If you subject a pendulum or a spring (an oscillator) to driving by a periodic force, it will oscillate at a frequency equal to the driving frequency. If you drive an oscillator at a driving frequency much higher or much lower than the resonance frequency of the oscillator, it will oscillate with a smaller amplitude. In contrast, if you drive the oscillator at a driving frequency equal to its resonance frequency, it will oscillate with a larger amplitude. This behavior, which applies to any simple harmonic oscillating object, is called *resonance*, and is described by the following principle.

Principle of resonance: The amplitude of an oscillator's SHM is maximized if the frequency of the driving force equals the natural resonance frequency of the oscillator.

Other examples of resonance are:

- The vibration of a violin string when a bow is steadily pulled across it
- The ringing of a crystal goblet when a wet finger is rubbed around its rim
- A radio receiver tuned to select a single station's broadcast frequency among many others
- The crystal in a clock that keeps accurate time, or the electronic oscillator in an all-electronic wristwatch

Why was the chandelier in Galileo's church swinging from side to side? What was supplying the force and energy to make it move? It is unlikely there was someone in the church rafters pushing on its supporting cable. Rather, the chandelier was swinging freely at its natural resonance frequency. How did Galileo's chandelier begin oscillating? The chandelier was put in motion by air currents in the church and vibrations of the rafters holding the cable. The forces from the winds and rafters probably contained many different frequencies, caused by the breezes blowing in through doors being opened and closed, as well as the vibrations of the ground as horse-drawn wagons rumbled by outside. When a pendulum is simultaneously exposed to many different forces, all having different frequencies, it tends to "pick out" and respond only to the force oscillating at its own resonance frequency. This is why the chandelier simply oscillated back and forth at its resonance frequency of 2 Hz.

You can use two experimental methods to determine the resonance frequency of an oscillator. The first is to hit it with a sudden impulse of force. For example, with a bob on a spring resting at its equilibrium position, gently tap it in the upward direction and observe the frequency of the SHM that you create. Another way to measure the resonance frequency of an oscillator is to "drive" it using a periodic force having a variable frequency and observe at what driving frequency the oscillator has maximum amplitude. For example, gently move the top end of the spring up and down in a periodic manner. If you oscillate the top of the spring much faster or much slower than the resonance frequency, the bob will oscillate with smaller amplitude. In contrast, if you oscillate the top of the spring at the resonance frequency, the bob will oscillate with larger amplitude. Using trial and error, you can determine the bob's resonance frequency.

IN-DEPTH LOOK 7.1: RESONANCE FREQUENCIES

A general rule is that materials with greater stiffness have higher resonance frequencies than materials that are less stiff. For example, if you tap a bowl of jelly, the jelly vibrates, perhaps at 5 Hz. If you tap the wall of an empty glass goblet, it vibrates very rapidly, perhaps at 500 Hz. Glass is much stiffer than jelly, so the restoring force felt by the goblet's wall is much greater than the restoring force felt by the different parts of the jelly. A large restoring force creates faster oscillations.

A second general rule is that, all else being equal, an object with small mass will oscillate faster than a similar object with larger mass. Therefore, smaller objects tend to oscillate faster than larger ones. A familiar example is seen in guitars and pianos, where the thicker, heavier, and in some cases longer strings oscillate at smaller frequencies than do the thinner, lighter, and shorter strings.

A third general rule is that oscillators with smaller mass have higher resonance frequencies than oscillators with larger mass. (This is not true for pendulums because the acceleration of an object by gravity does not depend on its mass.) Consider, for example, a bob on a spring as shown earlier in Figure 7.5. If the mass of the bob is increased, the resonance frequency of the oscillation decreases. If a stiffer spring is used, the resonance frequency of the oscillation increases. Quantitatively, the observed results are summarized by the proportionality

$$resonance\ frequency \propto \sqrt{\frac{stiffness}{mass}}$$

The resonance frequency is proportional to the square root of the stiffness divided by the bob's mass.

QUICK QUESTION 7.2

A spring with a 2-kg mass has a resonance frequency equal to 3 Hz. If the mass is replaced by an 8-kg mass, what will its resonance frequency be?

REAL-WORLD EXAMPLE 7.1: CRYSTAL OSCILLATORS AND MICROPROCESSOR CLOCKS

Every computer contains an internal clock for synchronizing its data processing operations. The most common way to build a miniature clock with high precision is to use a quartz crystal as a mechanical oscillator. Quartz is made of silicon dioxide (SiO_2), molecules arranged in a regular pattern—a characteristic of all crystals. In Chapter 4 we discussed a simple model for a crystal, shown in Figure 4.4, in which its atoms are represented as being connected to each other by microscopic springs. This makes it clear that if you tap on a crystal, it will suddenly compress, then spring back and begin vibrating. Because a quartz crystal is very stiff, it has an extremely high resonance frequency. Quartz also has very low internal friction. So, once it begins vibrating, it can continue for a very long time without losing energy too rapidly. Additionally, the mechanical properties of quartz, such as stiffness, are almost unaffected by changes of temperature. These properties make quartz ideal for building tiny clocks for use inside wristwatches and computers and in wireless devices such as cell phones.

The clocks inside computers are based on devices called quartz crystal resonators. The main element in these devices is a small plate of quartz, typically several millimeters in size. Such crystals have resonance frequencies typically in the range of 1 MHz to 200 MHz. Recall that 1 MHz corresponds to 1 million vibrations per second, which is enormously fast for a mechanical object. Compare this to the highest-pitch tone on a grand piano, which is about 4 kHz, 1,000 times slower than the lower range of quartz oscillators.

A thin, rectangular plate of quartz can vibrate in various ways, called **modes**. Four such modes of vibration are shown in **Figure 7.8**. The flexure mode occurs when the plate bends. The flexure mode usually has the lowest frequency, because a rectangular object is usually less stiff against bending than it is against compressing. The extensional mode occurs when the top surface of the crystal moves downward while the bottom surface moves upward, compressing the crystal. (Imagine a rectangular rubber eraser that you lay on a table and push straight down with your hand.) The thickness shear mode occurs when the top surface of the crystal "slides" oppositely to the bottom of the crystal. (Imagine a rectangular rubber eraser that you lay on a table and push down and sideways with your hand.) The face shear mode occurs when opposite corners move either toward or away from each other. Because clocks in computers need to be fast, one of the higher-frequency modes is selected for resonant operation.

Quartz has another remarkable property that makes it perfect for use in computer clocks. When an electric voltage is applied across the crystal, it forces some of the electric charges inside the crystal to move to the crystal's surfaces. This effect is called piezo-electricity. If the plus charges move to the top surface and the minus charges move to the bottom, then these charges attract each other and exert an electric force between the two surfaces, causing the crystal to compress. If the voltage is then suddenly turned off, the crystal will begin to oscillate. This means we can use an electronic circuit, which applies voltages to the crystal, to create vibrations of the crystal.

Piezoelectricity also makes it possible to detect compression or vibrations of a quartz crystal. If a crystal is compressed, charges move within the crystal, creating regions of positive and negative charge. The electric field created by movement of charge can be detected using an electric circuit that is sensitive to changes of voltage.

To control and detect the vibrations of a quartz crystal, the crystal is placed between two electrically conducting contacts, as shown in **Figure 7.9**. A circuit (not shown) is designed to drive the crystal vibrations periodically and then monitor the resulting vibrations. The circuit automatically adjusts its driving frequency to maximize the amplitude of the crystal vibrations. The crystal acts as a natural stabilizing device for the electric circuit, which simply needs to count how many oscillations the crystal makes. If the crystal is manufactured very precisely so that it has a known resonance frequency, then it provides a stable, known oscillation frequency to the electric circuit, which can transfer

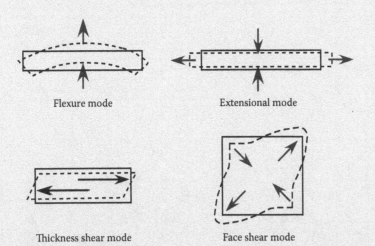

Flexure mode   Extensional mode

Thickness shear mode   Face shear mode

FIGURE 7.8   A thin, rectangular object can vibrae in various modes. (Adapted from http://www.foxonline.com/tech3140.htm.)

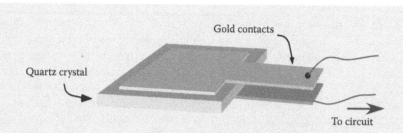

FIGURE 7.9   To make a clock, a quartz crystal plate is mounted between two gold metal surfaces. The gold surfaces are connected to an electric circuit that drives and detects vibrations in the quartz.

this very fast clock "ticking" to the processor in a computer. For example, if the crystal is made to have a resonance frequency equal to 1 MHz, then each time the circuit counts 1 million vibrations of the crystal it updates its clock by 1 sec. Such a clock can also send an electrical pulse to the processor once every millionth of a second, thereby providing extremely precise timing information that the processor can use to synchronize its internal operation.

7.5   WAVES

You know what waves are from experience. You see ocean waves, sometimes occupied by surfers, moving into shore. In **Figure 7.10**, we see waves traveling on the surface of a lake. Less obvious examples of waves include the sound reaching your ears from a music loud speaker, or the vibration of the floor caused by a large truck passing nearby. In all of these examples something is oscillating or vibrating, and that oscillation is traveling from one place to another. The traveling oscillation is a *wave*. Waves have

FIGURE 7.10   Waves created by boats on Jenny Lake in Grand Teton National Park. (Photo by the author.)

several properties that we can describe by a few simple formulas. Understanding these relations is a great help in understanding how sound, radio, and light travel.

A **wave** is a disturbance or an oscillation of an *medium* that travels from the source of the disturbance. A **medium** can be a material, like air, water, or a solid. A medium can also be less tangible, such as an EM field.

As with SHM, the mechanism that sustains a wave is the continual exchange between potential and kinetic energy. By this mechanism, energy in one region of the medium is transferred to adjacent regions of the medium. A wave is created by the action of an outside influence called a source, or driver.

As an example, let us again consider surface water waves. Imagine a long, narrow water canal with a dam at one end, as in **Figure 7.11**. You stand by the dam and hold a flat wooden board above the water, extending across the width of the canal and parallel to the dam. If you push the board up and down on the surface of the water, you will launch a wave that travels away from the dam and down the length of the canal.

Depending on how you push on the board, you can launch either a pulse wave or a simple harmonic wave. If you push the board down just once and release it, you launch a pulse wave, which travels more or less as a lone bump in the water surface, surrounded by one or two little dips in the water level. The board that you push into the water is called the source, or driver. When the wave passes an object floating in the water, the object gets pushed up and down once. In contrast, if you oscillate the board up and down in a smooth, periodic manner, you will launch a series of peaks, all separated by an equal distance from the next, with this whole oscillating pattern traveling at a constant **wave speed** away from the driver board. In the case of periodic driving, the crests and troughs all travel to the right at about the same speed, keeping

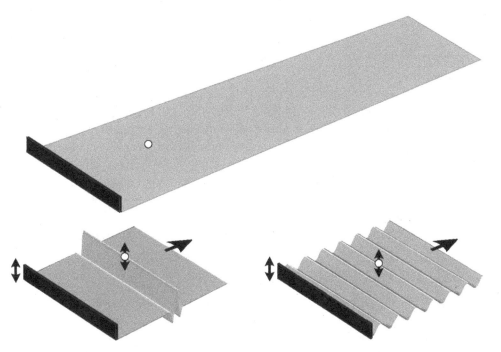

FIGURE 7.11 A wave in water being created by a driver (the black object) being pushed up and down, either once or in a regular repeated manner. As the wave passes by, a small float moves up and down.

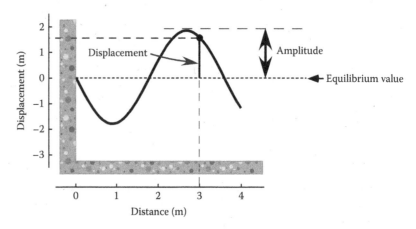

FIGURE 7.12   Side view of a wave in water being created by a driver being pushed up and down.

the shape of the wave unchanged.[2] Each maximum in the water level is called a ***crest***, and each minimum is a ***trough***. Each point on the water (marked by, say, the small floating object) moves up and down and sloshes forward and backward, but does not permanently travel in the direction that the wave is traveling. This means that the wave is not the water, but rather is the pattern of movement in the water.

Waves have the following properties: the ***medium*** for a wave is the substance (e.g., water) or other entity that moves or oscillates as the wave travels. At each position on the wave, we can specify the ***displacement*** of the medium, indicated in **Figure 7.12**. ***Displacement*** is the instantaneous difference of the value of the wave from its value at equilibrium (value at rest). For water waves, the displacement at a certain position is simply the height of the water relative to its height when at rest (the dashed line in the figure). The displacement is represented by a vertical line extending from the horizontal dashed line, indicating the height when at rest, to the height of the wave at that position. The figure shows, for example, that at a horizontal distance of 3 m from the left wall, the wave displacement equals 1.5 m. Note that displacement can be a positive or negative number. The ***amplitude, A,*** of a wave equals the wave's maximum height, measured from the height when at rest; that is, the amplitude equals the maximum value of the displacement.

Waves carry ***energy***. For example, when a boat creates a wave that travels across a lake, the wave can cause a swimmer to be lifted up and dropped down as the wave passes. This energy (measured in units of joules) came originally from the boat's motion and ultimately from the boat's engine. How did the energy get from the boat to the swimmer? It was transferred from one segment of water to a neighboring segment, which passed it on to the next segment, and so on. Although no water flowed from the boat to the swimmer, energy was able to pass along with the motion of the wave in the water. The ***power*** in a wave is the rate at which energy is transferred by the wave from one place to another. Recall that power has units of joules per second, or J/sec.

7.6   SIMPLE HARMONIC WAVES

Figure 7.13 illustrates the formation of a periodic water wave, that is, a wave pattern that repeats itself many times. Figure 7.13a shows a single short pulse traveling from left to right, created by briefly pushing the driver into the water only once. If you

---

[2] If a periodic wave on the surface of water travels a very long distance, the peaks will eventually spread out, but let us ignore this phenomenon.

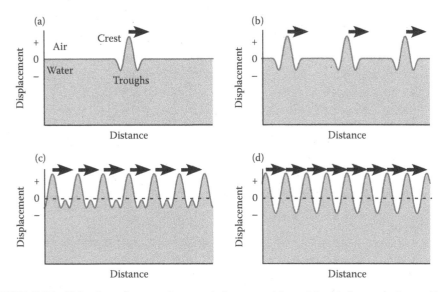

FIGURE 7.13 Side view of a wave in water being created by a driver being pushed up and down. (a) A single push of the driver launches a pulse. (b and c) Repeated pushes launch a series of pulses. (d) Simple harmonic pushing of the driver launches a simple harmonic wave.

repeatedly push the driver into the water, with long pauses between pushes, you could create a sequence of short pulses, as shown in Figure 7.13b. If you repeatedly push the driver into the water, with shorter pauses between pushes, you could create a sequence of closely spaced pulses, as shown in Figure 7.13c. If you repeatedly push the driver into the water, with no pauses between pushes, you could create a sequence of peaks and troughs traveling from left to right, as shown in Figure 7.13d. The wave shown in Figure 7.13d is an example of a ***simple harmonic wave***.

Simple harmonic waves are created by driving a medium with a force (such as pushing the board in water) that oscillates in a SHM. To describe the behavior of the simple harmonic water wave in Figure 7.13, let us represent the displacement of the water level by the symbol d, and the position, or distance, from the dam by the symbol x. Both of

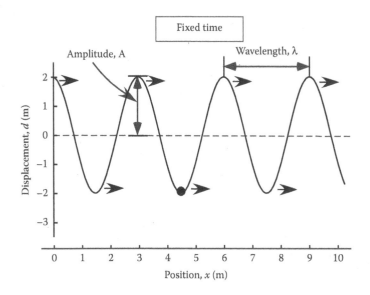

FIGURE 7.14 "Snapshot" graph of wave displacement, d, versus position, x, at a fixed time. The dashed line shows zero displacement—the "height" of the medium when at rest.

the quantities d and x are measured in units of meters. **Figure 7.14** shows a graph of the displacement of the water versus position at some fixed time. This is like a snapshot or a single frame from a movie with time frozen. The amplitude of this wave is 2 m. The crests occur at the points of maximum displacement, d, which equals the amplitude, or 2 m. At the chosen instant of time, the crests are located at positions $x = 0, 3, 6, 9$ m, etc. The troughs have displacement equal to -2 m and, at the chosen time, are located at positions $x = 1.5, 4.5, 7.5$ m, etc.

This wave is periodic—it repeats itself in space every 3 m. The distance between repeating crests (or troughs) is the *wavelength*. As a symbol for the wavelength, it is customary to use the Greek letter λ (pronounced "lambda" and translated roughly as lowercase *L*). (You can find Greek characters in many word processors in the symbols font.) The unit for wavelength is meters.

The *wavelength*, λ, of a simple harmonic wave is the distance, along the direction of travel, from one crest to the next (or from one trough to the next).

There is another way to describe a *periodic wave*—by watching how it changes in time. You could focus your attention on only one position x on the wave; for example, the position of a float indicated by the dark point at x = 4.5 m in Figure 7.14. If you watch the height of the float (ignoring the rest of the wave), you will see the float oscillating up and down in a SHM. In **Figure 7.15**, a graph of the displacement (height) of the float is plotted versus time.

In this example, you can see that a float at the surface starts with displacement equal to -2 m. Then it begins moving up, and takes 3 sec to go from its minimum height to its maximum height and back to the minimum. This time is the *wave period*, or for short the *period*, represented, as before, by the symbol T.

The *wave period, T,* of a harmonic wave equals the minimum time interval required for the wave to exactly reproduce its original form. The *wave frequency, f,* is related to the wave period by $f = 1/T$.

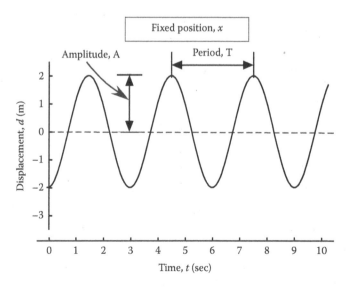

FIGURE 7.15 Graph of displacement for the wave in Figure 7.12, at a fixed position ($x = 4.5$ m), versus time.

It is important to realize that, in the case of a simple harmonic wave, each point on the wave oscillates up and down just as a ball on a spring does. That is:

A *simple harmonic wave* is one whose displacement at each point undergoes SHM.

How far does a crest of a periodic wave travel in a time equal to a single wave period, T? The answer is one wavelength. To visualize this, it helps to draw many images of the wave at different times, as in **Figure 7.16**. Going from one frame of the figure to the next below it is like viewing successive frames in a video. The time advances by one-fourth of a wave period between each frame. A dashed line traces the location of the wave crest (the dark dot). Recall that this dot is not showing the motion of water, which is not traveling from left to right. Rather, it shows the progression of the point on the pattern that is the wave. As the crest moves a distance of one wavelength, a float (the open dot) goes through one complete cycle. We can see that the crest moves a distance equal to one wavelength in the same time that the float goes up and down once.

We can also ask, "How fast is the crest moving?" That is, what is the *speed* of the crest's motion, from left to the right in the figure? (Again, this is not the speed of

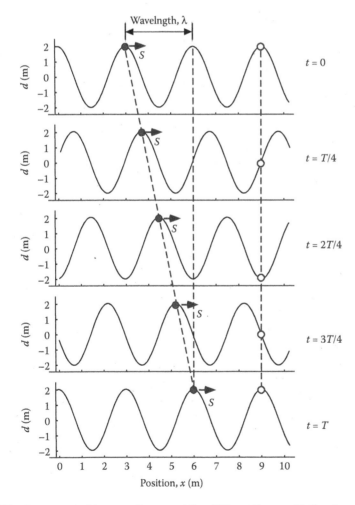

FIGURE 7.16   A sequence of images of a wave at five different times, with time increasing from top to bottom. After one complete cycle ($t = T$), the wave exactly reproduces its original form. In a time equal to one period, the wave crest (black dot) travels one wavelength and the float (open circle) moves up and down one complete cycle. The wave speed is S.

the water, which is not moving to the right.) Speed, or S, has units meters per second. For example, if the wave travels a distance D in a time t, then the speed is $S = D/t$. We know from Figure 7.16 that the wave moves a distance equal to a wavelength, in a time equal to the period, T. Therefore we have the following:

The *wave speed*, S, of a simple harmonic wave with wavelength λ and wave period T is given by $S = \dfrac{\lambda}{T}$.

This can be easily remembered as, "The wave speed is one wavelength per wave period." This equation can also be written using the wave's frequency f, as

$$S = f \cdot \lambda$$

Another nice way to write this relationship is:

$$\lambda = S \cdot T$$

This means that the wavelength λ equals the speed times the period, which is the distance traveled by the wave in a time equal to one period. We can also express the formula as

$$T = \lambda/S, \text{ as } f = S/\lambda, \text{ or as } \lambda = S/f.$$

As an example, if the wavelength is 1.5 m and the frequency is 4 cycle/sec (4 Hz), then the wave speed is

$$S = 4 \text{ cycle/sec} \cdot 1.5 \text{ m/cycle} = 6 \text{ m/sec}$$

If we were to view a water wave looking down from an airplane, we would see a top view, which is useful for understanding the concept of *wave fronts*. Looking from a top view in **Figure 7.17**, we see a wave moving to the right with wavelength and wave speed S. We draw a solid line along the top of each crest and a dashed line along the bottom of each trough. These lines are the wave fronts. As time passes, the whole pattern

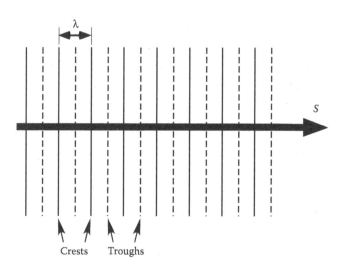

FIGURE 7.17   Top view of a wave, showing the wave fronts as solid lines (crests) or dashed lines (troughs).

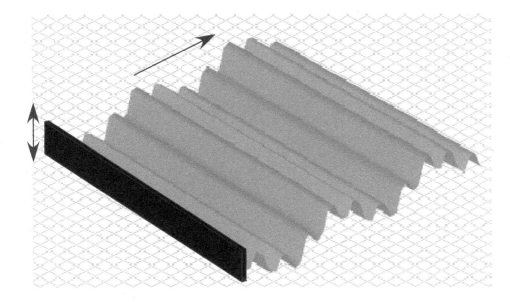

FIGURE 7.18 A complex harmonic wave.

moves to the right, ideally without changing its shape. A wave travels in a direction perpendicular to its wave fronts.

Not all waves are simple harmonic waves. As we saw in Figure 7.1, a wave can be in the form of a single pulse. Waves can also be in the form of complex harmonic waves, an example of which is shown in **Figure 7.18**. Forcing a medium with a periodic but complicated driving creates this wave. A *complex harmonic wave* reproduces itself exactly after the passage of an interval of time equal to the wave period. In this regard, it is similar to a simple harmonic wave, except that its shape is more complex.

7.7 INTERFERENCE OF WAVES

What happens when two waves intersect in some region? At some points, the two waves reinforce each other, leading to larger amplitude. In other places, the waves cancel each other, leading to smaller amplitude. This behavior is called *interference* and leads to regions of larger and smaller displacement than would exist in the absence of interference. This causes many significant effects for sound, radio, and light.

Principle of wave interference: When two waves intersect in some region, their wave displacements add together.

After two waves interfere, they separate again. Then their wave displacements are the same as if there had been no interference.

A nice example of wave interference is seen in the photograph in **Figure 7.19**, which I took from a cliff high above Jenny Lake in Grand Teton National Park. The picture shows waves approaching the shore, which were created by two boats. (Notice the small whitecaps breaking on the beach.) There are two dominant waves traveling across the water's surface. One wave is coming from the upper left of the region in the photo and can be seen most clearly arriving at the shore in the lower right of the photo. The other wave is coming from the upper right region in the photo and can be seen most clearly in the left-center region. In the area where the two waves overlap, a regular interference pattern is created by the waves adding together. Notice that in certain regions the waves

QUICK QUESTION 7.3

In a certain medium, waves travel with speed 10 m/sec. To make a wave having wavelength equal to 2 m in this medium, with what period should you drive it? To what frequency does this correspond?

FIGURE 7.19   Two waves traveling on a lake create an interference pattern, seen most clearly in the lower center region of the photo. (Photo by the author.)

cancel, so the wave displacement there equals zero. This is called *destructive interference*. These regions can be seen as the line-like, flat regions cutting diagonally across the waves near the shore in the lower middle of the photo. These regions are called *nodes,* or *nodal lines*. Such nodal lines occur where the displacement of one wave is positive whereas the other wave's displacement at the same location is negative.

These waves are depicted with computer graphics in **Figure 7.20a** and **b**, which show arrows in the directions each wave is traveling. Figure 7.20c shows addition of the displacements (heights) of the waves. The nodal lines are clearly seen, running parallel to the gray line, which is drawn as a guide for your eye.

The formation of the nodal lines in Figure 7.20 is demonstrated using drawings of wave fronts in **Figure 7.21**. Solid lines indicate the crests of each wave (at a fixed time). In a video of the waves, all of the wave fronts would advance in the direction shown by the bold arrows. Dashed lines indicate the troughs of the waves. The troughs also advance in time. Nodal lines are formed wherever a crest meets a trough (dark circles). At those points, the positive displacement of the crest adds to the negative displacement of the trough, adding to zero. This destructive interference occurs everywhere along the gray line, which is a nodal line. In contrast, at the locations indicated by empty circles, *constructive interference* occurs, creating twice as large of an amplitude at those locations compared with the wave amplitude from one wave alone.

(a)   (b)   (c)

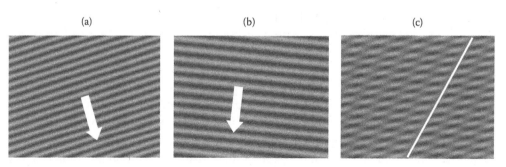

FIGURE 7.20   A computer simulation of the wave interference seen in the photo in Figure 7.19.

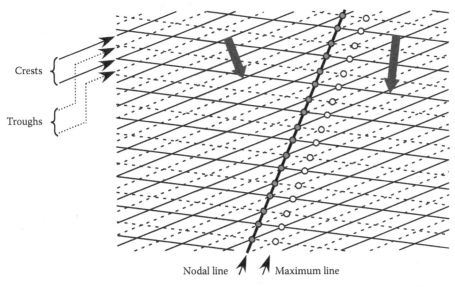

Crests {

Troughs {

Nodal line ↗ ↗ Maximum line

FIGURE 7.21 Diagram of wave fronts demonstrating the wave interference seen in Figure 7.20. Filled circles represent destructive interference; empty circles represent constructive interference.

IN-DEPTH LOOK 7.2: STANDING WAVES

A wave whose crests and troughs do not travel in any direction but simply oscillate between positive and negative values is called a *standing wave*. A standing wave occurs when two waves with the same frequency travel in exactly opposite directions and interfere. An example is illustrated in **Figure 7.22**, which shows six frames from an animation of two traveling waves colliding. Each of these waves has a wavelength equal to 1/3 m. The uppermost-shown wave travels to the right, and the middle-shown wave travels to the left. The arrows attached to each of these show the changing locations of a particular crest on each traveling wave. The wave shown at the bottom of each frame is the result of the two interfering waves adding together. This is the sum wave.

In the first frame (a), the crests of the two traveling waves line up, leading to constructive interference and therefore large amplitude of the sum wave. As time increases, the amplitude of the sum wave begins decreasing, as indicated by the downward-pointing arrow. The decrease is caused by the two waves moving in opposite directions. This is seen in the second frame (b), after the upper wave has traveled one-twelfth of a wavelength to the right and the other wave traveled an equal distance to the left. The third frame (c) is after both waves have traveled two-twelfths of a wavelength. The fourth frame (d) is after both waves have traveled three-twelfths (one-quarter) of a wavelength. Now the two waves add destructively, giving zero amplitude for the sum wave. Frame (e) shows that after each wave travels four-twelfths of a wavelength, the sum wave begins to recover in amplitude. And at five-twelfths (f), the recovery is nearly complete. Complete recovery occurs at six-twelfths (or one-half) of a wavelength. Notice that the crests of the sum wave do not travel to the left or to the right.

Standing waves are important in several technologies: microwave ovens, lasers, and musical instruments, to name a few.

A *resonant standing wave* occurs when a wave medium, such as water, is confined between two rigid walls and is driven at its resonant frequency. For example, if you fill a deep, rectangular baking pan with water, then gently shake it horizontally along the long direction, the sloshing of the water will create a standing wave only if you shake at a resonant frequency. Different frequencies can resonantly drive different standing

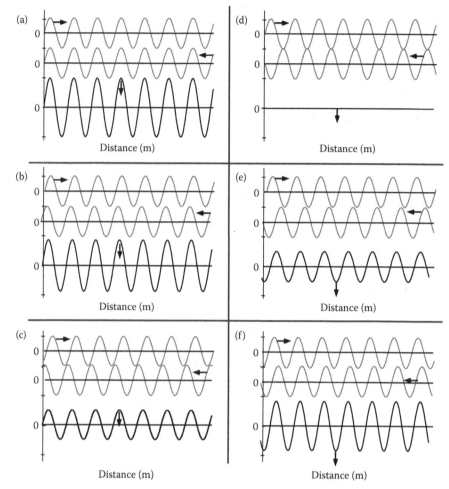

FIGURE 7.22 Creation of a standing wave. The two waves at the top of each frame travel in opposite directions, and their sum is shown at the bottom of each frame. Between each frame, the two waves travel one-twelfth of a wavelength, corresponding in time to one-twelfth of a wave period. The vertical arrows show the direction of motion of a particular point on the sum wave.

waves (unlike for a pendulum, which has only a single resonance frequency). A distinct resonance frequency occurs if some whole number (1, 2, 3, etc.) of half-wavelengths fits exactly between the two walls. In **Figure 7.23a**, there is a single half-wavelength fitting between the walls that are separated by a distance L; that is, $L = 1 \cdot \lambda/2$, or $\lambda = 2L$. In Figure 7.23b, there are six whole wavelengths fitting between the walls; that is, $L = 12 \cdot \lambda/2$, or $\lambda = L/6$.

As usual for a wave, the wavelength λ and frequency f are related by the formula including the wave speed, S, that is $\lambda = S/f$. For example, if the wave speed of water waves in a pan equals 0.6 m/sec, and the pan has a length $L = 0.1$ m, then the lowest-frequency standing wave—that shown in Figure 7.23a—has wavelength $\lambda = 2L = 0.2$ m. Therefore, its frequency is given by

$$f = \frac{S}{\lambda} = \frac{0.6 \text{ m/sec}}{0.2 \text{m}} = 3 \text{ Hz}$$

The standing wave shown in Figure 7.23b has a frequency 12 times higher, or 36 Hz. You can drive some of the possible standing waves in the water pan by shaking the pan at the appropriate resonance frequency, or by pushing the edge of a board in and out of the water's surface.

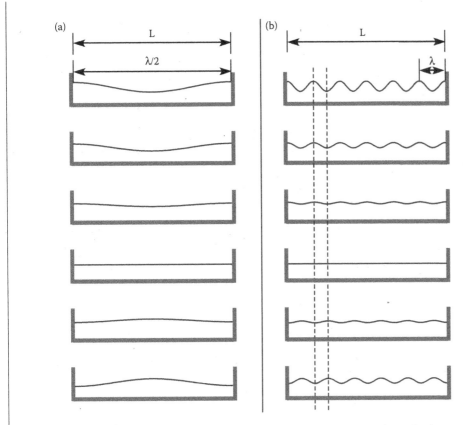

FIGURE 7.23 A resonant standing wave in a pan of water. The crests and troughs do not travel left or right; they move only up and down.

7.8 SOUND WAVES

When you listen to music, it reaches your ears as sound waves traveling through the air. *Sound* is the transfer of energy from a vibrating source through the environment by the vibration of matter. *Sound waves* travel through air in the form of moving oscillations of air pressure. The compression of a gas causes higher pressure, whereas the opposite (rarefaction) causes lower pressure. For sound waves, the wave "displacement" refers to the change of pressure of the air at a particular location. Typically air has a pressure at sea level of one atmosphere (atm), which equals very nearly one bar of pressure, the unit used by the National Weather Service. When a sound wave travels in the air, the changes of pressure are tiny—around one part per million, or roughly 1×10^{-6} atm. The small changes of pressure are positive or negative, relative to the ambient pressure. Sound waves in air travel at a speed approximately equal to

$$speed\ of\ sound\ in\ air = S \approx 345\ m/sec$$

The creation of a sound impulse is illustrated in **Figure 7.24**, which shows a hollow pipe filled with air at room pressure (1 atm), and sealed at the right end by a rubber cork. The length of the pipe is 2.0 m. The seven pictures are frames from an animation showing how the impulse travels. In the first frame, at the top, you hold a cork tightly with your hand near the opening of the pipe and tap on it with a small hammer. This causes the cork to move momentarily toward the pipe and then spring back, without ever touching the pipe. As the cork moves, it slightly compresses the air at the left end

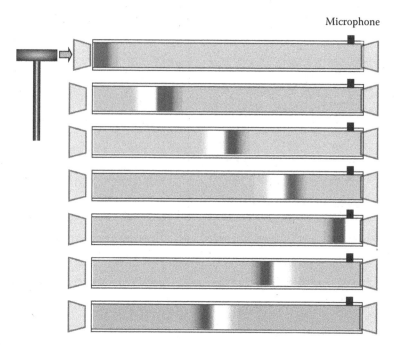

FIGURE 7.24 A hammer taps a cork, which launches a sound impulse moving from left to right in the air inside the pipe. Darker shading indicates higher pressure.

of the pipe. As the cork springs back, it creates a region of low pressure in the air. Air with pressure higher than 1 atm is shown as dark colored, whereas air with pressure less than 1 atm is shown as light colored. The second frame shows the pressure after the impulse has moved a distance away from the cork. The leading edge is high pressure and the trailing edge is low pressure. In the third and fourth frames, the impulse travels down the pipe, maintaining its form. In the fifth frame, the impulse hits the cork at the far end of the pipe, where it reflects. That is, the impulse begins moving toward the left, as seen in the sixth and seventh frames. As the impulse moves to the left, the leading edge (on the left side of the impulse) has high pressure and the trailing edge has low pressure.

The same sound impulse is illustrated in **Figure 7.25**, where each frame shows a graph of the air pressure at each position along the length of the pipe. Where the graph value is above the horizontal line, the air pressure is larger than 1 atm, and where the graph value is below the line, the pressure is less than 1 atm.

The small black rectangle at the right end of the pipe is a microphone—a device that detects air pressure. As described in Chapter 5, a microphone works by electrically sensing the movement of a thin diaphragm that is exposed to pressure changes in the air. **Figure 7.26** shows the microphone's voltage signal (which is proportional to the pressure change) plotted versus time. The pressure reaches a maximum at around 5.5 milliseconds (msec), and goes to zero at around 5.8 msec, after which it goes negative, then back to zero.

The time required for the sound impulse to travel from the left end of the pipe, whose length is 2 m, to the right end, where the microphone picks it up, is measured to be about 5.8 msec, as can be seen in Figure 7.26. This means that the speed of the sound impulse equals:

$$S = \frac{2\text{ m}}{5.8\text{ msec}} = \frac{2\text{ m}}{5.8 \times 10^{-3}\text{ sec}} = 345\text{ m/sec}$$

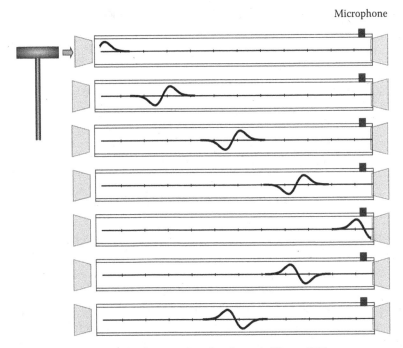

FIGURE 7.25   Pressure graphs for the same impulse shown in Figure 7.24.

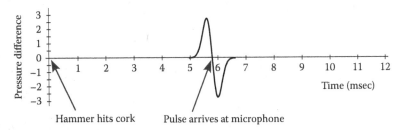

FIGURE 7.26   Microphone signal for the same impulse shown in Figure 7.24. The pressure difference equals zero when the pressure at the microphone equals 1 atm. The unit of pressure difference is 1×10^{-6} atm.

This confirms that the speed of sound, near room temperature and 1 atm of pressure, equals about 345 m/sec. To be more precise, the speed of sound in air is independent of pressure, but increases with increasing temperature by the formula: S[m/sec] = $331 + 0.6 \times$ (Temperature in °C). At 23C, the speed is 345 m/sec.

The two key parameters of a harmonic wave are its amplitude ("loudness" or "volume") and its frequency ("pitch"). The range of frequencies audible to humans is about 20–20,000 Hz. This is the hearing range of the human ear. Let us calculate the wavelengths of typical sound waves. Recall that the wavelength λ equals the speed times the wave period, which is the distance traveled by the wave in a time equal to one wave period. The wave period equals 1 divided by the wave frequency, so for f (frequency) = 20 Hz the period equals

$$T = \frac{1}{f} = \frac{1}{20 \text{ Hz}} = \frac{1}{20 \text{ cycles/sec}} = \frac{\text{sec}}{20 \text{ cycles}} = 0.05 \text{ sec/cycle} = 0.05 \text{ sec}$$

Notice again that the symbol cycles may be omitted from the answer because it is commonly understood. A simple harmonic sound wave with frequency 20 Hz has a wavelength equal to

$$\lambda = S \cdot T = 345 \frac{\text{m}}{\text{sec}} \cdot 0.05 \text{ sec} = 17.25 \text{ m} \approx 17 \text{ m}$$

A sound wave with frequency 20,000 Hz has a period 1,000 times smaller, or

$$T = \frac{1}{f} = \frac{1}{20,000 \text{ Hz}} = 0.00005 \text{ sec} = 5 \times 10^{-5} \text{sec}$$

This can also be stated as $T = 50$ microseconds (μsec), as can be seen from

$$5 \times 10^{-5} \text{ sec} = 50 \times 10^{-6} \text{ sec} = 50 \text{ } \mu\text{sec}$$

Therefore, the wavelength of a 20,000-Hz sound wave equals:

$$\lambda = S \cdot T = 345 \frac{\text{m}}{\text{sec}} \cdot 5 \times 10^{-5} \text{ sec} = 0.01725 \text{ m} \approx 17 \text{ mm}$$

Sound waves are waves of air compression. The air in one region is temporarily compressed and has higher pressure than the adjacent regions. Then this region becomes decompressed and has lower pressure than the adjacent regions. For one region to become compressed, air molecules move into that region from adjacent regions, leaving those regions with lower pressure. The movement of air molecules is an air current and is a form of kinetic energy. Recall from Chapter 4 that high gas pressure means high potential energy, which can be released to do work. (Think of the compressed gas in a can of shaving gel and what happens when you press the button.) The mechanism that sustains a sound wave is the continuous exchange of energy between the potential energy of air compression and the kinetic energy of the moving molecules.

There is a significant difference between surface water waves and sound waves: When a surface water wave travels, the molecules of water move more or less vertically, whereas the wave travels horizontally. This type of wave is called a **transverse wave**. In a **transverse wave**, the medium oscillates in a direction perpendicular to the direction the wave is traveling. In contrast, in a sound wave, the molecules of air move in directions parallel to the direction that the wave is traveling. This is called a **longitudinal wave**. In both cases, energy is transported by the wave from one location to another.

THINK AGAIN

The graphs in Figure 7.25 might appear to represent a transverse wave, because the curve in the graph goes above and below the centerline. But, this is not the case. Recall that in Figure 7.25, the displacement of the curve above the zero line represents the gas pressure at a particular place at a particular time.

IN-DEPTH LOOK 7.3: BEATS

Interference of sound waves causes an interesting effect called **beats**. This is an example of a general phenomenon called frequency synthesis and analysis, which is at the heart of music synthesizers and many key ingredients in Internet technology. We will study the applications of beats in Chapter 8.

If two waves have nearly equal frequencies and they come together in space where they interfere, the phenomenon of beats occurs. Say that two flute players simultaneously try to play the note middle C (261.63 Hz), but one flute is slightly out of tune and generates a wave at frequency 262.63 Hz. What will you hear? As the two wave displacements (air pressure) add in the air at your ear, there are times when the waves add constructively, and times when they add destructively. That is, at times they reinforce, and at other times they cancel. This leads to a pulsating of the loudness of the sound wave. The rate at which the loudness pulsates equals the difference of the two frequencies—in this case the difference equals 261.63 − 262.63 = 1 Hz, or 1 beat/sec. So, the *period of beats*, T_B, is 1 sec.

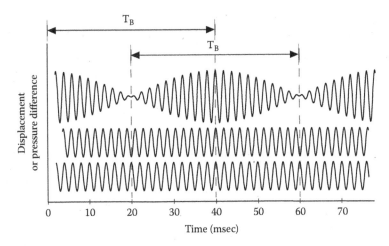

FIGURE 7.27   Illustration of wave beats. The top wave is the sum of the two lower waves. Each wave is graphed on a different vertical axis.

An example illustrates how this beating comes about. In **Figure 7.27**, the sound wave shown in the lower curve oscillates 20 times in 40 msec, so its frequency is $f_1 = 20 ÷ 40$ msec = 500 Hz. The wave shown in the middle curve oscillates 21 times in 40 seconds, so its frequency $f_2 = 21 ÷ 40$ msec = 525 Hz. The difference between the two frequencies is 525 − 500 = 25 Hz. The wave shown at the top is the sum of the two other waves, obtained by adding the displacements of the two at each instant in time. At time $t = 40$ msec, both original waves are at one of their maximums (crests), so when they are added, a large value is obtained. At $t = 20$ msec, one of the original waves is at a maximum whereas the other is at a minimum, so when they are added, a zero value is obtained. They add destructively. At $t = 60$ msec they again add destructively. The period, T_B, of the beats (time between destructive interference times) is seen to equal 40 msec. This equals the inverse of the frequency difference:

$$T_B = \frac{1}{f_2 - f_1},$$

where f_2 is the larger of the two frequencies (so we get a positive result). In our example we find

$$T_B = \frac{1}{525 \text{ Hz} - 500 \text{ Hz}} = \frac{1}{25 \text{ Hz}} = 40 \times 10^{-3} \text{ sec} = 40 \text{ } msec$$

This calculated beat period agrees with the period seen in the graph.

To summarize, the beat period is the time is takes for two waves of unequal frequency to get out of synch (out of phase) and then back in synch (in phase) with each other.

7.9   WIRELESS RADIO WAVES

Radio waves are coordinated oscillations of electric and magnetic fields that can transmit energy from one place to another, even through empty space. When we speak of the electric and magnetic fields together, we refer to them as the electromagnetic (EM) field. Radio waves are a type of *electromagnetic wave*. What is oscillating is not a

FIGURE 7.28 Radio and other EM waves can travel through empty outer space (e.g., to Mars), whereas sound waves cannot travel beyond the Earth's atmosphere, which is the medium for sound. The medium for EM waves is the EM field.

material object or medium but rather the electric and magnetic fields themselves. Radio waves are therefore quite different from sound waves, which require a material medium such as air to travel. For example, a radio wave generated on Earth could cause changes in the actions of the Rover vehicles on Mars, as in **Figure 7.28**. In contrast, a loud sound made in the Earth's atmosphere cannot be "heard" on Mars, because sound waves cannot travel through empty space. (Despite the exciting sound effects we hear in our favorite sci-fi movies.) Another way in which radio waves are different from sound waves is their frequency range. Commercial radio stations operate at very high frequencies—0.5–100 MHz—compared to sound frequencies, which are in the human hearing range 20 Hz to 20 kHz.

Microwaves, television waves, and x-rays are other forms of EM waves. The difference between these types of waves is only their frequencies. EM waves are characterized by wavelength, wave frequency, wave speed, and wave amplitude. For EM waves, the displacement—the analog of height for water waves—is the strength of the EM field (both electric and magnetic fields) at each point in space.

Because EM waves do not require conducting wires to travel from one place to another, their discoverer Heinrich Hertz (1857–1894) called them wireless waves. See his quote at the beginning of this chapter. His prediction about there being no practical uses of wireless waves sounds funny today.

How do we create an EM wave, say a radio wave? We want to create an oscillation in the electric and magnetic fields. Recall from earlier discussions that an *electric field* is associated with *electric charge*, and a magnetic field is associated with moving electric charge. On the basis of the experiments that we discussed in Chapter 5, Maxwell and Faraday predicted that electric charges rapidly moving to and fro could generate EM waves.

We can create an EM wave by grabbing a charged object and rapidly moving it up and down. In **Figure 7.29**, we see an *electron* confined inside a thin, conducting metal wire. Under the influence of a positively charged object located several meters away, the electron can move up or down within the wire, but not side to side. In Figure 7.29a, the electron is attracted toward the charged object by the force *F*. The direction of this force is along the electric field line, as shown. The electron moves as close as possible to the charged object without leaving the confines of the wire. As shown in Figure 7.29b,

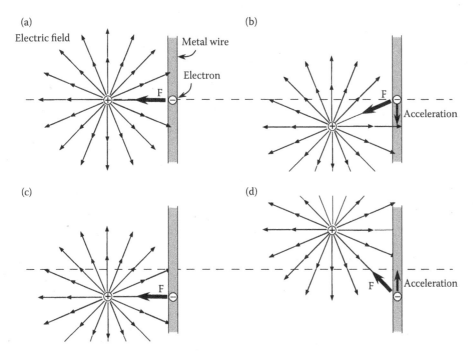

FIGURE 7.29   (a) A distant positively charged object exerts an attractive force on an electron inside a wire. (b through d) When the charged object is moved up or down, the direction of the force on the electron changes, causing the electron to oscillate down and up.

if you move the positive object downward and hold it there, the electron will feel a force F pulling it downward, so it will accelerate downward along the wire and move to the new location shown in Figure 7.29c. Finally, in Figure 7.29b, if you move the positive object upward and hold it there, the electron will feel an upward force and will accelerate upward.

Such changes of electric forces do not act instantaneously. Einstein's theory of relativity (which we will not study in this text) predicts that the influence of an EM field is not instantaneous. In fact, it predicts that nothing can travel with a speed faster than $c = 3 \times 10^8$ m/sec, where c is the **speed of light**.

Consider the situation that initially the positively charged object is at rest above the axis, as in **Figure 7.30a**, so that the direction of the electric force is fixed. Now, suddenly move the charged object to a location below the axis, as in Figure 7.30b. If the electron in the wire was extremely distant, say 1 million meters from the charged object, there is no way that the electron could "know" instantly that the object had moved. That would require an influence traveling faster than the speed of light, c. After some time delay (given by the distance divided by c), the electron would begin to feel the change in the electric field's direction. To illustrate that there is a time delay between moving the positive object and its influence being felt at a distant point, the electric field is drawn in Figure 7.30b as a dashed line with a "kink" traveling at the speed of light toward the electron. This kink indicates a wave traveling on the electric field. As the kink reaches a certain point in space, such as the one labeled P in Figure 7.30c, any electron located there would suddenly begin to feel the electric field's influence.

Now, consider what happens if you oscillate the position of the positively charged object periodically at a frequency f. The moving kink becomes a smoothly oscillating wave of electric and magnetic field lines. The electric and magnetic field **vectors** are organized in space in a wave-like pattern—an EM wave. The oscillating charged

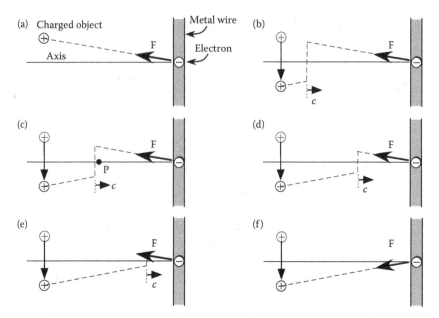

FIGURE 7.30 An electron in a wire influenced by a distant moving charged object. A wave in the electric field travels from the object toward the electron at the speed of light. Time progresses in panels a through f.

object, which acts like a broadcasting antenna, generates this wave. These facts are summarized by:

PRINCIPLE OF EM WAVE GENERATION:

An oscillating electric charge creates traveling waves of oscillating electric and magnetic fields, which can be light, radio, or microwaves.

PRINCIPLE OF THE SPEED OF EM WAVES:

In empty space (vacuum), EM waves travel at nature's speed limit: $c = 3 \times 10^8$ m/sec.

The emitted EM wave has a wavelength $\lambda = c/f$. The speed of light, c, is the absolute speed limit in nature, which applies to every thing or influence.

In air, the speed of EM waves is very close to c, but in mediums, such as water or Earth, the speed of EM waves is significantly slower than c.

THINK AGAIN

The word medium is used in two different ways. For mechanical waves, such as surface water waves, the medium is a material medium. In contrast, the medium for EM waves is simply the substance through which the EM wave travels. We need to keep in mind that this substance is not a true mechanical medium for the EM waves.

You can understand better how EM waves are generated and travel through empty space by considering **Figure 7.31** and recalling the electromagnetic principles in Chapter 5. The figure shows a loop of metal wire, which acts like a broadcasting antenna. If a steady *electric current* passes around the loop, as in Figure 7.31a, then according to Ampère's law, a steady magnetic field is created in the region around the loop. In this case, no electric field is created, because the wire is neutral. On the other hand, if the direction of current in the loop is rapidly alternated between clockwise and counterclockwise, as shown in Figure 7.31b, then the strength and direction of the magnetic field created around the loop changes in time. We know from Faraday's law that a magnetic field changing in time creates an electric field. The electric field line is linked with the magnetic field line created by the current loop, as shown. We also know from Maxwell's law that a changing electric field creates a magnetic field. The magnetic field line is linked with the electric field line. This process goes on and on, in a kind of "leap-frog" manner, creating a wave of electric and magnetic fields that travels readily through air and space.

An EM wave is sustained through the continual exchange between the electric field and the magnetic field, as illustrated in Figure 7.31. That is, the wave is sustained by a continual exchange between electric energy and magnetic energy. This is analogous to

(a) Wire loop with steady current

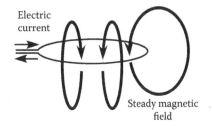

Electric current

Steady magnetic field

(b) Wire loop with oscillating current

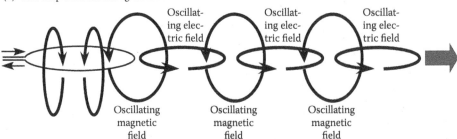

Oscillating electric field

Oscillating electric field

Oscillating electric field

Oscillating magnetic field

Oscillating magnetic field

Oscillating magnetic field

FIGURE 7.31 A steady electric current in a wire loop creates a steady magnetic field. An oscillating current in a wire loop creates a cascade of oscillating electric and magnetic fields, which travel through space as waves.

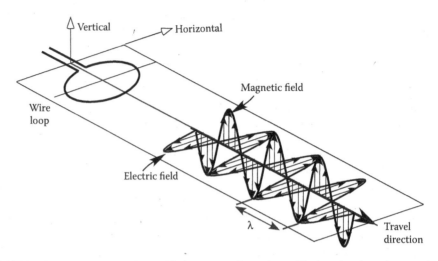

FIGURE 7.32   Far away from the oscillating current loop, the oscillating electric and magnetic fields, which together form a wave, point in directions perpendicular to the direction the wave is traveling, shown by the bold arrow.

the way in which a water wave or a sound wave is sustained by the continual exchange between two forms of energy—kinetic and potential.

For reasons we will not discuss, it turns out that when the EM wave in Figure 7.31 travels very far from the current loop that created it, the electric and magnetic field lines lose their "loopy appearance" and become straight lines, as shown in **Figure 7.32**. At each point along the travel direction there is an electric field vector, whose direction points in the same direction as the electric force and whose length represents the strength of the force on an electron, such as that in the wire in Figure 7.29. There is also a magnetic field vector at each point, and this is perpendicular to the electric vector. The entire pattern moves together with speed $c = 3 \times 10^8$ m/sec.

In EM waves, the wave displacements are the electric and magnetic fields, and they are perpendicular to the direction the wave is traveling; that is, EM waves are transverse waves. This contrasts with sound waves, which are longitudinal waves, in which the motion of air molecules is parallel to the direction of wave travel.

The usual relations between wavelength, wave speed, and wave frequency apply to EM waves, with c replacing the symbol S for speed: $c = \lambda \cdot f$, $\lambda = c \cdot T$, etc. The wave shown in Figure 7.32 has a wavelength indicated by λ. For example, if this wavelength equals 2 m, we can calculate the frequency of the wave. Knowing the speed of light, the frequency of oscillation in this case is

$$f = \frac{c}{\lambda} = \frac{3 \times 10^8 \text{ m/sec}}{2 \text{ m/cyc}} = 1.5 \times 10^8 \text{ cyc/sec} = 1.5 \times 10^8 \text{ Hz}$$

Using the fact that 1 MHz = 10^6/sec, this frequency can also be expressed as 1.5×100 MHz = 150 MHz (megahertz). This is a television station frequency, which is higher than the highest FM radio frequency—about 108 MHz.

REAL-WORLD EXAMPLE 7.2: AM RADIO

Radio waves are used in a wide range of products and technologies: cell phones, cordless phones, music and voice broadcasts, wireless Internet connections, police radar, satellite data transmission, etc. **Table 7.1** lists the frequencies used by the common types of radio transmissions. An interesting trend can be seen in the table: as the demand for higher

TABLE 7.1

Common Radio Frequencies

Radio Type	Frequency Band
AM	520 kHz to 1610 kHz
Shortwave	2.3 to 26.1 MHz
Citizens' Band (CB)	26.9 to 27.4 MHz
FM	88.0 to 108.0 MHz
TV	54.0 to 88.0 MHz and 174 to 220 MHz
Wireless phone	900 MHz
Global Positioning System (GPS)	1.25 to 2.58 GHz
Deep Space Network	2.29 to 2.31 GHz
WiFi (wireless communication IEEE802.11)	2.4 to 2.487 GHz, and 5 GHz

QUICK QUESTION 7.4

What is the wavelength of an EM wave whose frequency is 2×10^7 Hz?

rates of data transfer increases, the frequency of the radio waves used also increases. For example, television requires higher rates of data transfer than does AM radio.

If you look at a radio's dial or tuner, you see that AM radio uses frequencies between 520 kHz and 1,700 kHz, whereas FM radio uses 88 MHz to 108 MHz. The radio waves travel through the air at the speed of light c, so the wavelength is $\lambda = c/f$. Using the frequencies listed in the table, we can calculate that for AM radio the smallest and largest wavelengths are approximately

$$\lambda_{min} = \frac{c}{f} = \frac{3 \times 10^8 \, \text{m/sec}}{1610 \, \text{kHz}} = 186 \, \text{m}, \quad \lambda_{max} = \frac{c}{f} = \frac{3 \times 10^8 \, \text{m/sec}}{520 \, \text{kHz}} = 577 \, \text{m}$$

For FM radio the smallest and largest wavelengths are

$$\lambda_{min} = \frac{c}{f} = \frac{3 \times 10^8 \, \text{m/sec}}{108 \, \text{MHz}} = 2.8 \, \text{m}, \quad \lambda_{max} = \frac{c}{f} = \frac{3 \times 10^8 \, \text{m/sec}}{88 \, \text{MHz}} = 3.4 \, \text{m}$$

How are voice and music transmitted by a radio station and received by a radio? As we discussed in Chapter 5, when a person speaks into a microphone, the air pressure drives a thin, flexible diaphragm back and forth through a magnetic field, creating electric current in a wire coil attached to the diaphragm. The current from the coil is carried by wires to an electronic amplifier, which boosts the power of the signal to drive a broadcast radio antenna. A broadcasting or *transmitting antenna* is a metal wire or rod, the shape of which may be designed for specific circumstances. The operating principle of a broadcasting antenna is that an electric current oscillating in the antenna creates an oscillating EM field, which travels away as a wave. A *receiving antenna* operates in the reverse manner: the incoming radio wave creates electric current in the receiving antenna. These currents are amplified and sent to a loud speaker or headphones. There are many types and shapes of antennas, and each is based on physics principles. Figures 7.31b and 7.32 illustrated one antenna type: a circular loop of wire with an oscillating current.

Another type of antenna that is easy to understand is the quarter-wavelength antenna, shown in **Figure 7.33**. As an example, the antenna rod, with length 75 m, is shown lying horizontally. The broadcasting station feeds an oscillating current with frequency $f = 1,000$ kHz into the rod at its left end, indicated by the black rectangle labeled "driving point." Figures 7.33a through d show graphs of the current and voltage at locations along the rod at four successive times. In Figure 7.33a, the current entering the rod has its maximum positive value. In Figure 7.33b the current has become smaller. In Figure 7.33c the current becomes negative, and in Figure 7.33d it reaches its greatest negative value. Then it returns to zero and again to its positive maximum value, completing one cycle of oscillation. Positive current indicates charge moving in the rightward direction, and negative current indicates charge moving in the leftward direction. As the current runs right and then left along the rod, the voltage varies along the rod. That is, a positively charged particle on the rod would feel its electrical potential energy change from positive to negative as the amount of charge on the rod changes. Both the current and voltage oscillate with frequency f, determined by the driving current from the station. Notice that the voltage varies in time at the right end of the rod, where charge builds up. In contrast, the current is always zero there, because the air around the rod is nonconducting, and current has nowhere to go once it reaches the right end. The current is large at the left end, because that is where the wire from the station connects.

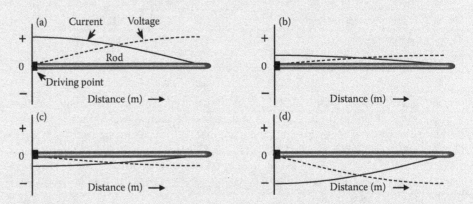

FIGURE 7.33 Current oscillating at the resonance frequency of a quarter-wavelength antenna creates standing waves of voltage and current along the rod.

The voltage on the rod is in the form of a wave, with a wave speed roughly the speed of light c (actually a little slower). The wavelength of this voltage wave is about

$$\lambda = \frac{c}{f} = \frac{3 \times 10^8 \, \text{m/sec}}{1000 \, \text{kHz}} = 300 \, \text{m}$$

The length of the rod making up the antenna is 75 m, or one-quarter of the length of the voltage wave. This ratio of one to four can be seen in the figure, which shows that there is a node (zero) for current at the right end of the rod, and a crest (maximum) at the rod's left end. Under this one-to-four condition, the current sloshes back and forth resonantly in the rod, similarly to water waves sloshing in a shaken pan, as illustrated in Figure 7.23. (The difference is that in the water pan a maximum wave amplitude can occur at both ends, whereas in the antenna a maximum current occurs only at the left end, with a node for current at the right end.)

The resonant behavior of a quarter-wavelength antenna naturally gives rise to large currents and voltages in the antenna rod. This leads to efficient generation of radio waves being emitted, and so is desirable. This is the reason that AM radio broadcasting stations use very long antenna rods or wires, 75 m in this example.

The direction in which radio waves are emitted depends on the orientation of the antenna. The horizontally mounted antenna shown in Figure 7.33 would radiate waves upward and downward, as well as in directions in and out of the plane of the drawing. It would not emit waves directly in the leftward or rightward directions. In contrast, commercial AM radio broadcasting antennas are usually oriented vertically. For example, Figure 7.34 shows the broadcasting antenna of the radio station WSM 650 kHz near Nashville, Tennessee. Built in 1932, this tower has been broadcasting country music, including the longest running radio show—the Grand Ole Opry—since its inception. The antenna is a vertical metal tower, through which current can flow. (The diamond-shaped "Blaw-Knox" design of the steel tower is for mechanical, not electrical, purposes.) Vertical antennas like this radiate waves with equal power in all horizontal directions,

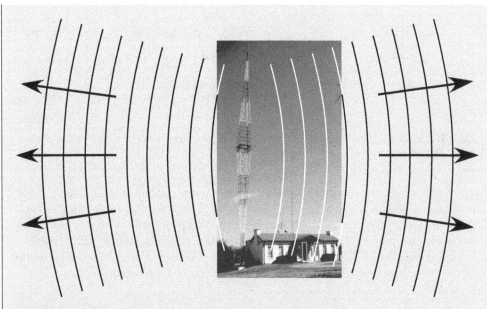

FIGURE 7.34   A vertical antenna emits radio waves mostly to the sides, with little power emitted upward. (Photo of WSM tower by permission of Garrett A. Wollman.)

while emitting very little power in the vertical (upward) direction. This ensures that most of the radio power that is emitted travels toward listeners on the ground rather than toward the upper atmosphere. This behavior is easy to understand: a person looking down on the antenna from directly overhead would not be able to "see" the current flowing up and down the antenna rod. Therefore, a test charge that is directly overhead would not feel any side-to-side electrical forces. To further concentrate the emitted power in the horizontal directions, the height of the WSM antenna (246 m) is made roughly one-half of the radio wavelength (462 m), in contrast to the quarter-wavelength antenna shown in Figure 7.33. We will not consider the detailed mechanisms that determine the directionality of various antenna designs.

In contrast to AM radio, wireless phone transmission typically uses a frequency 900 MHz. This corresponds to a wavelength

$$\lambda = \frac{c}{f} = \frac{3 \times 10^8 \, \text{m/sec}}{900 \, \text{MHz}} = 0.333 \, \text{m}$$

or about 1 foot in length. This small wavelength means phone transmitters can be much smaller than AM radio transmitters and can be mounted in small boxes atop poles or towers.

The operating principle of a receiving antenna is that an oscillating EM field arriving at the antenna creates an oscillating electric current inside of the antenna, which can be amplified and used to drive a speaker. It is usually not practical to use a receiving antenna that is as large as the transmitting antenna. For example, a car radio antenna might be 0.5 m long, far smaller than the 75-m transmitter. Nevertheless, a reasonable amount of power can still be received using a short receiving antenna. A cell phone's antenna is only a few centimeters long, compared to the 20- or 30-cm length of the cell tower antenna. The very weak voltage signal received by the cell phone's antenna must be amplified by an electronic circuit in the phone before it can drive a speaker to make an audible sound. The energy for this amplification process is provided by the phone's battery.

Music and voice information is impressed onto an AM radio wave by a process called *amplitude modulation*, abbreviated AM. The radio station discussed above generates a wave at 650 kHz, called a *carrier wave*. As the antenna is emitting the wave, the wave amplitude is varied in time proportionally to the voltage signal created by the microphone. The microphone creates a voltage with multiple frequencies between 20 Hz and 20 kHz,

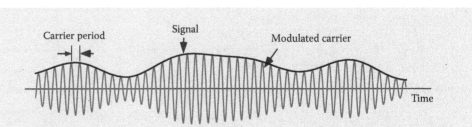

FIGURE 7.35 A slow audio signal multiplies a fast carrier wave to create an amplitude-modulated wave.

which is relatively low compared with the carrier frequency, meaning that the audio signal varies much more slowly than does the carrier wave. This lower frequency range is called the *audio range*. The audio signal and the carrier wave are multiplied together by an electronic circuit to yield a complex wave such as that shown in **Figure 7.35**. The job of a radio receiver is to detect this complex wave and convert it back into a slow audio signal. More details on the workings of radio are discussed in Chapter 8.

7.10   LET THERE BE LIGHT WAVES

Historically, light was "discovered" long before radio. How could it not have been, because people can see light but not radio waves? Galileo and a colleague tried, 400 years ago, to measure the speed of light by having two people stand on different hilltops and using lanterns and stopwatches, but the speed is so great that it cannot be measured in this way. In 1849 the French physicist Armand Fizeau made the first accurate measurement of the speed of light by using a lamp, a rapidly rotating disk with small holes cut into it, and a series of lenses and mirrors on two hilltops separated by about 8 kilometers. He found that light travels with a speed around 186,000 miles per second (about 299,000,000 m/sec). Today's most accurate measurements using lasers gives the value $c = 299{,}792{,}458$ m/sec. In this text, we will use the approximate value 3.0×10^8 m/sec for ease of calculation, because it is very close to the exact value.

James Clerk Maxwell and Michael Faraday were the first to realize that light is an EM wave. During the years 1830 to 1865, they and others developed the electromagnetic principles that we discussed in Chapter 5. Near the end of that time, Maxwell proved mathematically that if these principles were correct, then it should be possible for electric and magnetic fields to create traveling waves by the mechanism that we discussed in the previous section. He used his mathematical theory to calculate the speed that such waves would travel. He found that they should travel with a speed of about 186,000 miles per second. In 1864 Maxwell wrote about this calculated speed, or velocity.

> This velocity is so nearly that of light that it seems we have strong reason to conclude that light itself (including radiant heat and other radiations) is an electromagnetic disturbance in the form of waves propagated through the electromagnetic field according to electromagnetic laws.

It was a huge revelation that electricity and magnetism had anything to do with light. This was one of the most important breakthroughs in the history of physics and quickly led to scientific and technological revolutions such as the discovery of radio waves.

Light is the same as radio waves in nearly all respects, the only difference being in the frequency of oscillation. We are most familiar with visible light—EM waves that

TABLE 7.2

Ranges for Visible Light in Vacuum

Color	Frequency (THz)	Wavelength (nm)
Red	384–482	780–622
Orange	482–503	622–597
Yellow	503–520	597–577
Green	520–610	577–492
Blue	610–659	492–455
Violet	659–769	455–390

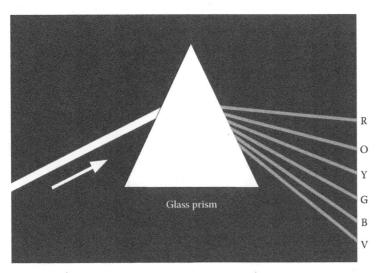

Glass prism

R
O
Y
G
B
V

FIGURE 7.36 A narrow beam of white light is separated into a spectrum of colors: red, orange, yellow, green, blue, and violet.

can be detected by the human eye. The perceived colors of light are determined by their frequencies. For example, red light has a frequency around $f = 4.3 \times 10^{14}$ Hz, or 430 THz (terahertz). This corresponds to a wavelength in air or in a vacuum around

$$\lambda_{red} = \frac{c}{f} = \frac{3 \times 10^8 \, \text{m/sec}}{4.3 \times 10^{14} \, \text{cyc/sec}} = 698 \times 10^{-9} \, \text{m} = 698 \, \text{nm}$$

In comparison, blue light has a frequency around $f = 6.3 \times 10^{14}$ Hz, or 630 THz. This corresponds to a wavelength in air or in vacuum around

$$\lambda_{blue} = \frac{c}{f} = \frac{3 \times 10^8 \, \text{m/sec}}{6.3 \times 10^{14} \, \text{cyc/sec}} = 476 \times 10^{-9} \, \text{m} = 476 \, \text{nm}$$

Table 7.2 gives the ranges of frequencies and wavelengths, in a vacuum, for the colors of visible light. These wavelengths are so small because the frequencies are very large.

Light is separated into its constituent colors when it passes through a glass prism, as in **Figure 7.36**. We will defer discussing the physics behind the operation of a prism until Chapter 13.

The human eye responds to each of these frequencies of light differently. The retina contains three different types of color-sensitive cells, called cones, and each of these responds differently to red, green, and blue light. With only three types of cells, the eye

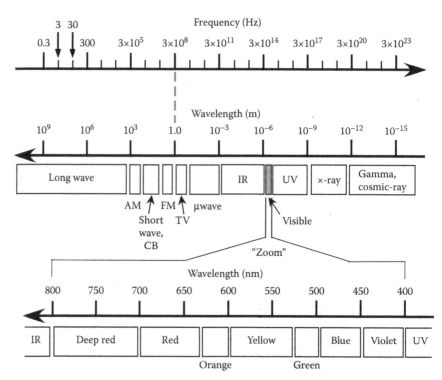

FIGURE 7.37   EM spectrum. Abbreviations: AM, AM radio; CB, citizen's band radio; FM, FM radio; TV, television; μwave, microwave; IR, infrared; UV, ultraviolet.

can distinguish between millions of different colors. This fact provides a clue for how best to design color photography cameras, televisions, and computer monitor screens.

7.10.1   The Spectrum of EM Waves

The ***electromagnetic spectrum*** identifies different types of EM waves by their frequencies (or, equivalently, wavelengths). The spectrum is shown in **Figure 7.37**. The uppermost scale gives the frequency of the EM wave, in units of hertz, or cycles per second. The next scale below gives wavelength. For example, a frequency 3×10^8 Hz corresponds to a wavelength equal to 1.0 m, as indicated by following the dashed line from the frequency scale to the wavelength scale. The visible region of the spectrum occupies a rather narrow range of frequencies or wavelengths. Therefore, this region is blown up in the lowermost scale, showing the colors associated with the different wavelengths.

IN-DEPTH LOOK 7.4: LIGHT POLARIZATION

Light waves have a property called ***polarization***, which refers to the direction in which their electric field oscillates. For example, in the EM wave shown in Figure 7.32, the electric force vectors point in horizontal directions. This is called *horizontal polarization*, and we say that the light is horizontally polarized. In **Figure 7.38a**, the same horizontally polarized wave is again shown, but here for simplicity we do not show the magnetic field, which is always perpendicular to the direction of the electric field. In Figure 7.38b, a *vertically polarized* wave is shown, in which the electric field oscillates in the vertical direction. This type of wave is generated by a wire loop oriented vertically, as shown. In both cases, we are referring to a wave that is traveling along a horizontal direction. To label the two types of polarizations in drawings of this kind, we use the symbols H or V to indicate in which direction the electric field points.

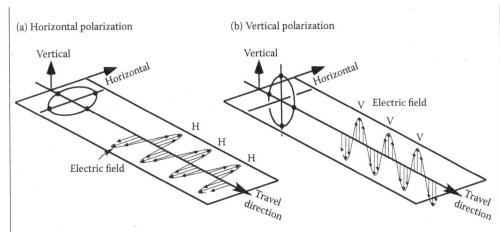

FIGURE 7.38 (a) A horizontally oriented current loop generates a horizontally polarized EM wave. (b) A vertically oriented current loop generates a vertically polarized EM wave.

You can illustrate this for yourself by holding your arm in front of you and pointing your index finger straight out. Imagine that a light wave is traveling in the direction your finger points. Now point your thumb upwards (in the vertical direction). Your thumb points in the direction that the electric force would act on an electron. This is vertical polarization. Now with your finger still pointing outwards, point your thumb sideways (in the horizontal direction). Your thumb points in the direction that the electric force would act. This is horizontal polarization. The property we call polarization refers to the direction of the electric field, not the magnetic field. In an EM wave, the magnetic field always points in the direction perpendicular to the direction of the electric field.

A useful device is called a *polarizer*. A polarizer transmits light of only one type of polarization—horizontal or vertical—and blocks the other, as in **Figure 7.39**. Figure 7.39a shows sunlight, which contains equal amounts of both polarizations, passing through a horizontal polarizer, indicated by the grid of lines. Only the horizontally polarized light transmits through the device, and the other is blocked. Figure 7.39b shows sunlight passing through a vertical polarizer. Only the vertically polarized light transmits through the device, and the other is blocked. Figure 7.39c shows that if horizontally polarized light impinges on a horizontal polarizer, all of it passes through. Figure 7.39d shows that if horizontally polarized light impinges on a vertical polarizer, none of it passes through.

The simplest way to make a polarizer is by stretching many parallel, conducting wires across an open frame. In each of the figures, the grid lines indicate the wires. For example, in Figure 7.39d, the wires are mounted horizontally. Electrons in a horizontal wire can easily move left or right (horizontally), but not up and down (vertically). If the impinging

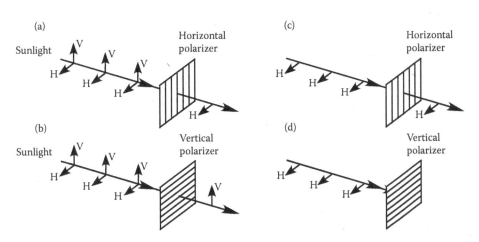

FIGURE 7.39 Light polarizers block one direction of polarization and transmit the other.

light is horizontally polarized, as in this case, the wires absorb energy from the light wave, and no light is transmitted. In contrast, Figure 7.39c shows the wires mounted vertically, making a horizontal polarizer; that is, the horizontally polarized light cannot be absorbed because the electrons are not able to move horizontally in the thin vertical wires. Therefore, the horizontally polarized light passes through the device without being absorbed.

If the polarization direction of the light is neither parallel to nor perpendicular to the direction of the wires, then part of the light is transmitted and part is absorbed. For example, if the angle between the polarization and wires is 45°, then one-half of the light power is transmitted.

A metal wire grid works well as a polarizer for EM waves with long wavelength, such as microwaves, but not for visible light, whose wavelength is much shorter. To make polarizers for visible light, we use long, thin molecules called polymers (i.e., plastic). These molecules have more or less the same shape as a long thin wire, and they preferentially absorb light in which the electric-field polarization is parallel to their long axis. For example, polymer molecules are embedded inside the lenses of polarizing sunglasses.

Polarized sunglasses use the properties of light polarization to block out "glare" (i.e., the bright, blinding reflections of sunlight from shiny horizontal surfaces such as polished car hoods or water in a lake). We say that sunlight is unpolarized because it contains equal amounts of vertically and horizontally polarized light, as shown in **Figure 7.40**. When these two types of light strike a horizontal surface at an oblique angle such as 45°, the horizontally polarized light is more strongly reflected than is the vertically polarized light. This behavior results from the details of the interaction of the light with the atoms at the surface, which is beyond the scope of our discussion. The light that is not reflected is transmitted into the surface and is absorbed there. The reflected light is predominantly polarized in the horizontal direction. A polarizing material can be used as the plastic for the lens of sunglasses. This material blocks (absorbs) all of the horizontally polarized light and transmits the weaker, vertically polarized light. The advantage of using polarized glasses is that the bright glare from shiny horizontal surfaces is blocked, whereas ordinary, unpolarized light is only dimmed partially. You can test whether or not your glasses are polarized by removing them from your face and rotating them by 90°. If they are polarized glasses, then you will see the bright glare being transmitted.

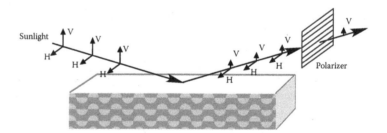

FIGURE 7.40   A horizontal shiny surface reflects horizontally polarized light more strongly than vertically polarized light. The polarizer film in sunglasses blocks the horizontally polarized light.

REAL-WORLD EXAMPLE 7.3: LCD SCREENS

The liquid-crystal display, or LCD screen, in a laptop computer operates using the principles of light polarization, described in In-Depth Look 7.4. The formation of images on a screen using pixels (*pic*ture *el*ements) was discussed in Real-World Example 2.1. Each pixel in the screen contains a tiny switch, like a valve, for light. Behind each of these switches is a steady source of white light, called backlight, commonly produced by a long, thin fluorescent bulb. When a switch is ON (the valve is open), light can pass through it, making a bright spot at that pixel. In the case of a black-and-white or monochrome screen, an ON switch makes a white pixel. When the switch is OFF (the valve is closed), light cannot pass through it, making that spot dark, or black.

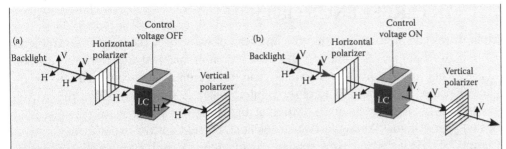

FIGURE 7.41   Liquid crystals can rotate the polarization of light, allowing it to pass through the vertical polarizer.

The light valve inside each pixel is made using a material called a liquid crystal, which is a fancy name for a bunch of rodlike (polymer) molecules suspended in a liquid, the orientations of which can be controlled by applying an electric field to them. The LCD uses a special property of liquid crystals: when they are all aligned with their rod axes pointing along the same direction, these molecules have the effect of rotating the polarization direction of light passing through the liquid crystal. Without going into the details of how the molecules rotate the light's polarization, it is easy to understand how the light valves work.

Figure 7.41 shows the backlight passing through two polarizers with the liquid crystal in between. The polarizers are arranged perpendicular to each other, so ordinarily no light could pass through them both. The first polarizer passes only horizontally polarized light. If the control voltage is OFF, the liquid crystal does nothing to the light, and no light can pass through the second polarizer. When the control voltage is turned ON (e.g., 5 V), the liquid crystals are forced to change their orientation. This causes the polarization of the light to be rotated from horizontal to vertical. Now the light can pass through the second polarizer and light up the pixel.

The brightness of the light passing through the second polarizer can be varied by changing the control voltage. For example, if the voltage is set to a value greater than zero but smaller than 5 V, then the light's polarization is rotated to an in-between angle such as 45°, allowing only a portion of the light to pass through the vertical polarizer.

To make color LCD screens, three subpixels of different colors are grouped together within the area of each individual pixel, as shown in **Figure 7.42**. By controlling the brightness of the light passing through each of these subpixels, millions of distinct colors can be created. Typically, each of the three control voltages is set by using a single 8-bit number, which can represent any of $2^8 = 256$ distinct voltages. This can create 256 different shades of blue, and the same number of shades of the other two colors. Therefore, the total number of different combinations that we can create with the three colors equals $2^8 \times 2^8 \times 2^8 = 2^{24} = 256 \times 256 \times 256 = 16,777,216$, almost 17 million colors.

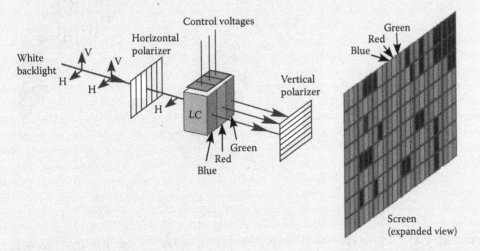

FIGURE 7.42   Three liquid crystals within each pixel control the polarization rotation of three colors of light to make color images on a screen.

7.11   INTERFERENCE OF LIGHT

Light waves can interfere. Just as with any kind of waves (see the lake waves in Figure 7.19), if two light waves come together in a region of space, their displacements add, causing a new oscillating pattern to emerge in this region. It is more difficult to observe interference of light than it is to observe interference of water waves. For this reason, scientists were not certain until 1801 that light actually was a wave of some kind. In that year, Thomas Young, a British scientist, carried out the experiment shown in **Figure 7.43**. He cut two narrow slits into an opaque screen (imagine a sheet of aluminum foil with two slits cut by a razor blade), and sent a light beam from a lamp onto the two slits. The feeble light waves that passed through the slits traveled to a piece of photographic film, where they exposed different parts of the film to create a pattern of bright and dark bars.

Young recognized the pattern of bright and dim regions on the film as being a consequence of wave interference. He explained this by considering that each small slit produces a circular wave emerging from it. He made a drawing like that in **Figure 7.44**, in which he drew one circular wave emerging from each slit, or source, traveling to a piece of film at the far right side. Each crest of a wave is indicated by a solid curve, whereas each trough is indicated by a dashed curve. There is a resemblance between the drawing in Figure 7.44 and the photograph of water wave interference in Figure 7.19. There are bright regions where constructive interference (reinforcement) occurs, and dark regions where destructive interference (cancellation) occurs. Constructive interference occurs at locations where the two crests meet, or where two troughs meet. A few of these regions are indicated by small empty circles, and will appear as the brighter regions on the film. Destructive interference occurs at locations where a crest of one wave meets a trough of another wave. These regions are indicated by small black circles, and will appear as the dark regions on the film.

Although Young's experiment showed conclusively that light was a wave of some kind, it was not until Maxwell's work 60 years later that these waves were proven to be electromagnetic in nature.

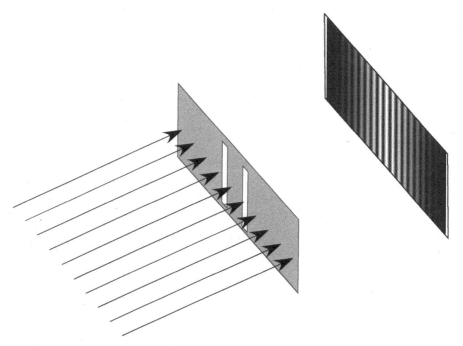

FIGURE 7.43   A light wave hits an opaque screen having two narrow slits in it. The waves that pass through the two slits travel to a piece of photographic film, where they interfere, exposing the film in a pattern of bright and dark bands.

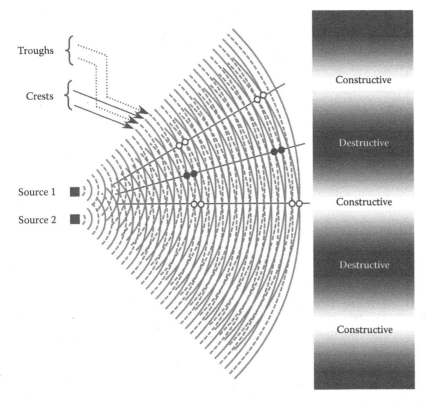

Troughs

Crests

Source 1

Source 2

Constructive

Destructive

Constructive

Destructive

Constructive

FIGURE 7.44 Two circular waves of light emerging from narrow slits labeled 1 and 2 and traveling to a sheet of film, where they merge and interfere to make bright regions and dark regions.

SUMMARY AND LOOK FORWARD

Harmonic waves are simple harmonic motion of an extended medium, described by amplitude A, frequency f, and wavelength λ. The frequency of a periodic wave is the number of cycles that occur each second. The wave period, T, of a harmonic wave is the time needed for the wave displacement at a given position to make one complete oscillation and return to its original value. The wave frequency is related to the wave period by $f = 1/T$, which implies $T = 1/f$. The wave speed, S, of a harmonic wave indicates how fast a wave crest travels. It is given by $S = \lambda/T$. This can also be written as $\lambda = S \cdot T$, which shows that the wavelength λ equals the distance traveled by the wave in a time equal to one wave period. It can also be written as $S = f \cdot \lambda$, with f being equal to the wave's frequency.

When two waves cross in some region, their individual wave displacements add together, a behavior called interference. Regions where the wave amplitude cancels to zero are called nodes. If two waves have nearly equal frequencies and they come together in space where they interfere, the phenomenon of beats occurs.

Sound waves are oscillations of air pressure. In air, sound waves typically travel with a speed $S = 345$ m/sec and have frequencies in the human audible range of 20 Hz to 20 kHz. Light and radio signals are waves of EM fields that travel from one place to another without the need for any physical medium. EM waves are created by rapidly oscillating electrons. They travel in vacuum at the speed $S = c = 3.0 \times 10^8$ m/sec. In air the speed is also very close to this value, but in other mediums, such as water or Earth, the speed of EM waves is significantly slower than c. The EM spectrum identifies different types of waves by their frequencies (or, equivalently, wavelengths). The wavelength of visible light is small—between 400 and 800 nm. Light and radio waves experience interference, just like any wave. As we will see later, interference of waves plays an important role in the concept of bandwidth in communication systems.

It has been recognized for thousands of years that light can be used for communication, but it is only in the past 40 years that we have had the technology to take advantage of its real capability. That technology is the laser. In an optical-fiber communication system, a laser emits light pulses, which represent the binary data (ON = 1; OFF = 0). These pulses are focused onto the end of a thin glass fiber. The light pulses travel in the fiber as far as 100 km before reaching the other end of the fiber, where they are focused onto a light detector that produces a voltage each time a light pulse strikes it. The voltage pulses are sent to a computer memory where they are stored as binary data. In the following chapters, we will discuss the physics behind each of these components of the communication system.

SOCIAL IMPACTS: MUSIC, SCIENCE, AND TECHNOLOGY

When I was an undergraduate student in the early 1970s, I composed a song on the piano. The piano was the "real" kind, with metal strings and ivory keys, because electronic keyboards (synthesizers) were barely available then. There was no reason to try to record the song then. I would have had to lug my parent's 30-lb reel-to-reel tape recorder to the location of a decent piano, borrow a good microphone, and learn to play piano well enough to make the effort worthwhile. Then having it on tape, there would have been nothing to do with it—no sensible or fun way to share it with friends.

The tape recorder of the 1960s and 1970s was, of course, a vast improvement over the earliest recording devices, such as Thomas Edison's phonograph invented in 1877, which recorded by scratching grooves onto a tinfoil sheet wrapped around a rotating cylinder. The slightly later disk phonograph brought recorded music to average people, much to their delight. But the recent explosion of digital music technology is nothing less than breathtaking compared with earlier analog technologies. More than 30 years after composing my little tune, and having played around with laptop computer recording software (such as GarageBand and Pro Tools Free), I was able, in a few hours, to create a great-sounding digital "recording" of the old song (which I had carried in my head). I did not "play" a keyboard; I typed each note into my computer one at a time and then told the computer to play them. I changed the voicing of each computer-generated instrument until I was happy with the sound. I uploaded the song to my website and invited my friends to listen to it. I can load it onto my portable digital music player or a flash memory stick and carry it with me—what fun!

The availability of music technology and its familiarity to so many people—especially younger ones—is undoubtedly affecting our culture and our economy. Recent surveys of high-school students show that the most popular areas for future employment are arts, entertainment, sports, and media. Those same students rated mathematics as the most important subject to study to achieve their career goals. Why math? Perhaps these students know or at least intuit that music and entertainment are forms of information processing, and that means computers and technical know-how.

Where does music technology come from? Mostly not from top musicians (with some exceptions), but from professional or self-taught amateur engineers. Interestingly, Edison resisted using his new technology for recording music, because he thought there would be no market for it. He had in mind voice dictation for legal and business purposes. In the 1920s, the invention of electronic microphones revolutionized the quality of music recording, which in turn strongly affected how the musicians themselves performed and thought about the music. For example, Bing Crosby was the first superstar to reach his audience almost entirely through radio, recordings, and film, rather than through live performance. Starting in the late 1940s, a "Big Bang" of innovation arrived, according to Elektra Records founder Jac Holzman, in the form of three inventions: the tape recorder, the long-playing (LP) phonograph disk, and FM radio. Magnetic tape allowed musicians to edit out bad notes and to separately record portions of a song and combine them into a final song. LPs permitted recording of songs longer than 4 minutes, and in stereo. FM radio improved the fidelity of broadcast music by using a larger frequency range (bandwidth) for transmitting each station's signal than was possible with AM radio. These inventions elevated the sound quality and the availability of recordings to a level that created an enthusiastic community of listeners and musicians. In the 1950s and 1960s, this led to a blossoming of classical, folk, and pop music that we are still witnessing.

Let us back up to 1941, when Les Paul created one of the first solid-body electric guitars at about the same time as Leo Fender's and Adolph Rickenbacher's. Les Paul also introduced recording "effects" when he invented an electronic circuit that could create the sound of delay, which simulates sound waves bouncing from the walls in a large auditorium. Soon, other "effects boxes," such as reverb, chorus, and the flanger, as well as the "extremely loud guitar amplifier," changed the way music was played, recorded, and heard. This also changed the lifestyles of many people.

The new music technology is impacting music in many ways. Consider folk or popular music. Before modern technology, folk songs were meant to be performed by local folk for one's own or one's friends' enjoyment. During the 1940s through 1990s, people increasingly listened to mass-marketed pop radio and recordings. Now it is perhaps reverting, because so many people can now create their own songs using computers. They do this mostly for enjoyment, because only a tiny fraction will make a living at it. Before technology arose, popular musical styles were distinct according to national or regional identities. That is far less true today. In addition, before technology, popular music lagged far behind the more progressive stylistic developments in serious music of early twentieth-century composers such as Stravinsky. That is no longer the case.

The new computer technology raises many questions. For example, what is the proper balance between privately owned intellectual property (song copyrights, etc.) and the public commons (the collection of free information where ideas cross-fertilize)? Music file sharing and song sampling for music composition come to mind. Where is all of this heading?

Questions to Ponder

1. In what ways has the technology of mass communication changed music?
2. How does modern technology empower the academic study of ethnic music, or ethnomusicology?
3. How does modern technology change or expand our understanding of what music is, and what types of sound can or should be considered music? Can a computer be programmed to create or compose "real" music from scratch? Find some examples online.
4. The so-called "youth culture" began during the same period (1940s–1960s) as did the explosion of music recording technology. Is there a connection?
5. What roles did Max Mathews, John Chowning, and Bob Moog play in the story of electronic music?
6. What impacts did technology have on the development of music between 1500 and 1900?

Terms to Research

*Center for Computer Research in Music and Acoustics, electronic music syn-
thesizer, glitch (music), intellectual property rights, peer-to-peer song-com-
posers' websites (such as SoundClick, MySpace, and MacJams), rap music,
sampling, Wendy Carlos*

SOURCES FOR SOCIAL IMPACTS

1. Karr, Rick. *TechnoPop: The Secret History of Technology and Pop Music. Morning Edition*, National
   Public Radio, 2002. http://www.npr.org/programs/morning/features/2002/technopop.
2. Johnston, Ian. *Measured Tones*. London: Taylor & Francis, 2002.
3. Johnstone, Bob. "Wave of the Future." *Wired* 1994, 2.03. http://www.wired.com/wired/archive/
   2.03/waveguides_pr.html.
4. Kleinman, Daniel L. *Science and Technology in Society*. Malden, MA: Blackwell, 2005.

SUGGESTED READING

See the general physics references given at the end of Chapter 1.

M. G. Raymer, and S. Micklavzina. "Demonstration of Boundary Conditions on Sound Impulse Reflections
in Pipes," *The Physics Teacher* 33, 183, 1995.

KEY TERMS

Amplitude
Amplitude modulation (AM)
Audio range
Beats
Carrier wave
Crest
Cycle
Damping
Displacement
Electric charge
Electric current
Electric field
Electromagnetic
Electromagnetic spectrum
Electromagnetic wave
Electron
Energy
Frequency
Hertz
Interference
Light
Longitudinal wave
Medium
Mode
Node
Period
Periodic wave
Polarization
Power

ANSWERS TO QUICK QUESTIONS

Q7.1 Period: $T = 3$ sec. Frequency: $f = 1/(3$ sec$) = 0.33$ cycle/sec $= 0.33$ Hz. Amplitude = 2 m.

Q7.2 The resonance frequency of a mass on a spring decreases as the square root of the proportional increase of mass. If the mass increases by a factor of 4, the frequency decreases by $\sqrt{4}$, or 2. Therefore the new frequency is 1.5 Hz.

Q7.3 The formula $T = \lambda /S$ gives $T = 2$ m$/(10$ m/sec$) = 0.2$ sec. The frequency is $f = 1/T = 1/(0.2$ sec$) = 5$Hz.

Q7.4 The wavelength of an EM wave whose frequency equals 2×10^7 Hz is $\lambda = c/f = (3 \times 10^8$ m/sec$)/2 \times 10^7$ Hz $= 15$ m.

EXERCISES AND PROBLEMS

Exercises

E7.1 Examples of SHM and their typical frequencies are given. For each, determine the period of the SHM.

  (a) An electronic voltage-controlled oscillator in a music synthesizer (988 Hz)
  (b) A quartz crystal in a wristwatch (32,768 Hz)
  (c) The vibration of the carbon and oxygen atoms making up a carbon monoxide molecule $(6.43 \times 10^{13}$ Hz)

E7.2 The top view of an ice rink is shown below. A spring is firmly attached to an immovable wall. A sliding object is firmly attached to the other end of the spring. Assume we can ignore the effects of friction, and that the object moves only in the forward and backward directions, not side to side. When the spring is not stretched or compressed, the object is at rest at "center ice," labeled as zero distance.

(a) You pull the object forward a distance 5 m and hold it still, then at time = 0, you let it go. The spring pulls it toward the fixed wall. You observe that the period of the SHM equals 3 sec. Make a graph of the object's distance from center ice versus time as the object oscillates.

(b) Explain what two forms of energy are continually exchanging as the object oscillates. Elaborate or explain the nature of these forms of energy in this example.

(c) Again the object begins at rest at center ice. You have a very powerful air blower (like a leaf blower) in the location shown. You turn it on, then off, then on, then off, etc. When should you activate the blower and turn the blower off, if you want to cause the largest possible oscillation amplitude for the object using the effect of resonance?

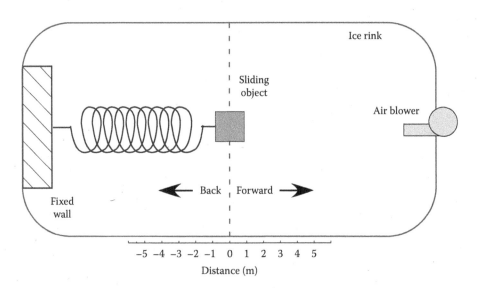

E7.3 Suppose a water wave has a wavelength equal to 3 m and travels with a speed of 7 m/sec. What are its period and frequency?

E7.4 A light wave has a frequency of 5×10^{14} Hz. As all light waves do in vacuum, it travels at the speed of light, $c = 3 \times 10^8$ m/sec. What are the period and wavelength of this light wave?

E7.5 A flute is emitting a tone with frequency 880 Hz. You are at the back of the concert hall, a distance of 40 m from the flute.

(a) How much time does it take the sound to travel from the flute to your ears? (*Hint*: The speed of sound in air is given in the chapter.)

(b) What is the wavelength of the sound wave?

(c) If the air in the hall is warmed up by a few degrees, the speed of sound increases slightly. Assume that the flute player adjusts her playing so the flute still creates the same frequency of 880 Hz. How does this warming affect the wavelength of the sound wave? Explain.

E7.6 A guitar string vibrates and emits a sound wave, which is detected by a microphone whose output voltage is displayed on a screen, as shown.

(a) What is the period of this oscillation in milliseconds?

(b) What is the period of this oscillation in seconds?

(c) What is the frequency of this oscillation in hertz?

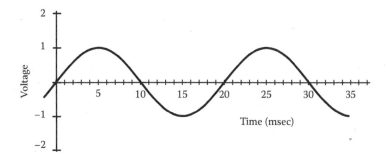

E7.7 Stars are hot balls of gas in outer space that emit light.

(a) It takes light from the star Alpha Centauri, our closest neighboring star, 4.35 years to reach your eyes on Earth. What is the distance from Earth to Alpha Centauri, in meters? in miles?
(b) Discuss the physics of seeing stars. That is, what is going on at the surface of a star that allows you to see it from such a far distance? Explain the microscopic physics behind the star's light emission. *Hint*: A star is hot. What does this mean about the motion of charged particles such as electrons in the star's atmosphere?

E7.8 When a water wave travels on the surface of a lake, the molecules of water move in directions more or less perpendicular to the direction the wave is traveling. This is a transverse wave. Devise a simple example illustrating that although the water molecules are not traveling in the direction the wave is traveling, the wave can transport energy from one location to another.

E7.9 Give several examples of transverse waves, including at least one that is not discussed in the text. Do the same for longitudinal waves.

E7.10 In an EM wave, what is the physical entity that is oscillating?

E7.11 A lightning bolt strikes a hilltop 5 km from your location. How long does it take the light flash to reach your eyes? How long does it take the sound impulse to reach your ears?

E7.12 What is wrong with the following statement? "I turned the light switch on and the room instantaneously became lit up."

E7.13 Say you are in a spaceship 1 million km from Earth, and you want to send an "instant message" or IM to your friend on Earth. By using the fastest form of communication possible according to the known laws of physics, what is the least amount of time it could take to reach your friend?

E7.14 Sir Isaac Newton was the first to experimentally verify that white light is composed of light of different colors. The drawing below is adapted from Newton's own drawing of his experiment. White sunlight passes through a small hole in a screen covering the window of a darkened room. The white light first passes through a lens, to focus the light on an opaque screen after passing through a prism. Newton saw a spectrum of rainbow colors on the screen. He cut a small hole in the screen where the red light struck and allowed the red light to pass through a second prism. He found that the red beam of light was not further broken up into a spectrum, as the white light had been, but remained a single red beam. How does this observation verify his claim about the nature of white light?

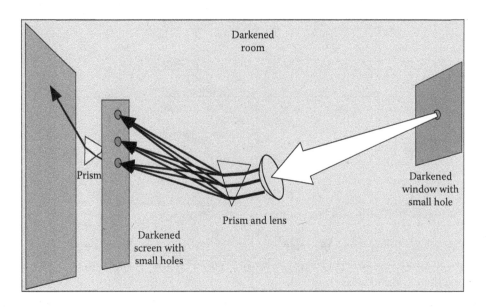

Problems

P7.1 If we know the period or, equivalently, the frequency of a SHM, then we can calculate the number of oscillations, or cycles, that occur in some chosen time interval. Say that we are watching such a motion, and we start a stopwatch at a time t_1, and stop it at time t_2. The number (N_c) of cycles in this time interval ($t_2 - t_1$) is given by the ratio of the time interval to the duration of a single period; that is,

$$N_C = \frac{(t_2 - t_1)}{T} = f \times (t_2 - t_1)$$

(a) If a pendulum oscillates with a period $T = 0.75$ sec, how many cycles would it undergo in a time of 75 sec?
(b) If a different pendulum has frequency of 3.5 Hz, how many cycles would it undergo in a time of 55 sec?

P7.2 In an experiment with a mass (M) on a spring, the oscillation periods (T) for four different masses were measured and the results are:

$$M = 500 \text{ g } (T = 0.745 \text{ sec}); M = 1,000 \text{ g } (T = 1.038 \text{ sec});$$

$$M = 1,500\text{g } (T = 1.257 \text{ sec}); M = 2,000\text{g } (T = 1.450 \text{ sec}).$$

Newton's laws predict (although we did not prove it in the text) that the period of SHM with a linear restoring force is

$$T = \sqrt{M} \times \text{constant}$$

The value of the constant depends on the stiffness of the spring. Verify that this formula is true, to good precision, by calculating the ratio $T / \sqrt{M}$ and showing that it is reasonably constant for all of the masses tested.

P7.3 The speed of sound in air is related to the air temperature, according to

$$\text{Speed (in m/sec)} = 331.5 + 0.6 \times (\text{temperature in Centigrade})$$

This formula is valid in the range −10° C to 100° C. Find the speed of sound in air at these temperatures: −10° C, 0° C, 10° C, 20° C, 30° C, 40° C, 50° C, 60° C, 70° C, and

80 °C. Make a table of these numbers. Make a plot (graph) with speed on the vertical axis and temperature on the horizontal axis.

P7.4 Sound waves are produced by two speakers driven by two separate pure tone generators, one with frequency f_1 and the other with frequency f_2. (Assume the room walls absorb sound, so you can ignore reflections from the walls.)

(a) If $f_1 = 30$ Hz and $f_2 = 31$ Hz, what will you hear?
(b) If $f_1 = 4$ Hz and $f_2 = 5$ Hz, draw graphs showing the oscillations of each frequency (without the other present). Draw a graph showing the oscillations of the combined (added together) oscillations. Label the time axis and indicate the period (use numbers with units) of each oscillation.

P7.5 One way to create the pattern of complex harmonic motion is using a rotating wheel with a smaller rotating wheel attached to it, as shown. The big wheel rotates around its axis and the small wheel rotates around its own axis, which is attached near the big wheel's rim. Part (a) of the figure shows four images of the wheels, separated by equal time intervals, where both wheels rotate at the same frequency. The dashed curve shows the path taken by the black dot on the rim of the small wheel. Part (b) shows a similar sequence of images in the case that the small wheel rotates with a frequency equal to one half of the frequency of the large wheel. For both cases, use the scales at the right to determine the displacement of the black dot at each time, and make graphs of displacement versus time illustrating the motion of the black dot. In each case, extend the time to several times longer than shown below in order to see the overall pattern of motion. It may help to photocopy the figure and make your own modifications for later times.

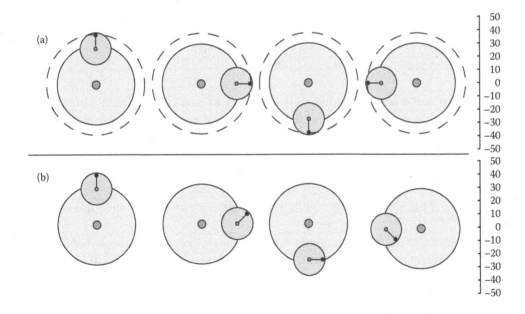

P7.6 As discussed in In-Depth Look 7.1, oscillations of an object on a spring have a natural or resonance frequency, which depends on the stiffness of the spring and the mass of the object. Say you have a 3-kg object attached to a certain spring such that the resonance frequency equals 2 Hz. How does the resonance frequency change if you:

(a) Increase the mass to 6 kg?
(b) Decrease the mass to 1 kg?

(c) Keep the mass the same and change the spring to one with twice the stiffness?

Make graphs of displacement versus time for each case.

P7.7 The speed of sound in a quartz crystal is around 3,300 m/sec.

(a) If the oscillation frequency is 10^5 Hz, what is the wavelength of sound waves in the crystal?
(b) If the crystal is to vibrate in the so-called extensional mode shown in the figure, the crystal must be cut to have a length equal to twice the sound wavelength. What length should the crystal be?

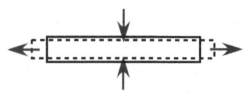

Extensional mode

P7.8 Sound waves travel with speed of about 345 m/sec in air, 5,960 m/sec in steel, and 1,500 m/sec in water. When a sound wave moves from one medium into another, its frequency remains constant but its wavelength changes.

(a) A musical tone having a frequency of 698 Hz is called F_5 on a piano keyboard. If this tone travels through air, what is its wavelength? What is its period?
(b) If this same sound wave encounters a solid steel object, the sound travels into the object, keeping the same frequency. What is the wavelength inside the object? What is its period?

P7.9 The speed of sound in water is 1,500 m/sec.

(a) If you are underwater and a dolphin 80 m away sends out "pings" (short sound clicks) to locate you, how long does it take for a ping to travel to you and back to the dolphin?
(b) If a blue whale sings a pure tone with frequency of 32.7 Hz, corresponding to the musical note C_1, what is the wavelength of the sound wave in the water?

P7.10 Musical tuning forks are used for checking the accuracy of an instrument's tone.

(a) A certain tuning fork is made with two metal prongs so that when it oscillates it creates the tone A_4, which corresponds to a frequency of 440 Hz. What is the period of the oscillation?
(b) Say you have a variable-speed electric toothbrush, whose speed of vibration can be smoothly varied. You fix the toothbrush body in contact with the handle of the tuning fork. You set the toothbrush speed to 410 Hz; then you turn it on. You turn off the toothbrush and place your ear close to the tuning fork and listen for an emitted tone. Then you increase the toothbrush frequency to 415 Hz, and repeat the experiment. You continue to repeat the experiment, each time increasing the toothbrush frequency by 5 Hz until you reach 470 Hz. Describe the results. *Hint*: The toothbrush is designed so that it does not emit much sound when vibrating, but a tuning fork is made to efficiently emit sound.

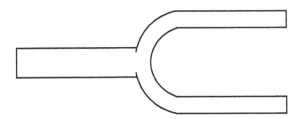

P7.11 The speed of light in a vaccum is exactly $c = 299{,}792{,}458$ m/sec. How long does it take a light pulse to travel the distance from the Moon to the Earth? For a precise distance, use 355,613.97 km, which is the smallest distance that occurred from 1995 to 2005. Give your answers to eight significant digits.

P7.12 Consider two loudspeakers that are driven by the same electronic pure-tone generator. If the frequency of the tone is $f = 690$ Hz, then the wavelength of the sound waves is $\lambda = S/f = (345 \text{ m/sec})/(690 \text{ cyc/sec}) = 0.5$ m. The speakers are at the front of a lecture hall, and are separated by 2 m. You and a friend are sitting a distance d apart in the same row of seats, which is 10 m from the speakers. If only one speaker is on, both of you will hear the same tone and loudness. If both speakers are turned on, one of you could hear loud sound, whereas the other hears quiet sound because of wave interference. Assume the speakers are driven in phase (i.e., synchronously).

(a) Assume your friend is sitting in the seat that lies on a line from the point between the two speakers to the rear of the room. Explain why she will hear loud sound, regardless of the frequency of the tone, assuming both speakers emit the same frequency.

(b) You are sitting a distance d to her side. What distance could you sit from her so you would hear quiet or no sound? To solve this problem, print two copies of the wave shown below on thin, partly transparent paper (or transparency). The drawing represents the sound wave (at some instant in time) emerging from one speaker, with crests of the wave represented by concentric circles separated by one wavelength, 0.5 m. Troughs are not shown in the drawing. Make a ruler along the edge of a separate blank piece of paper, on which you draw tic marks, using the 0.5-m wavelengths on one of your printouts to calibrate your ruler. Place one wave drawing over the other so you can see them both (hold up to a window if necessary.) Adjust the distance between the two sources to be 2 m (4 wavelengths). Use your ruler to measure the distance between the loud and quiet positions. Turn in your wave drawings and your ruler.

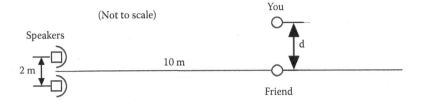

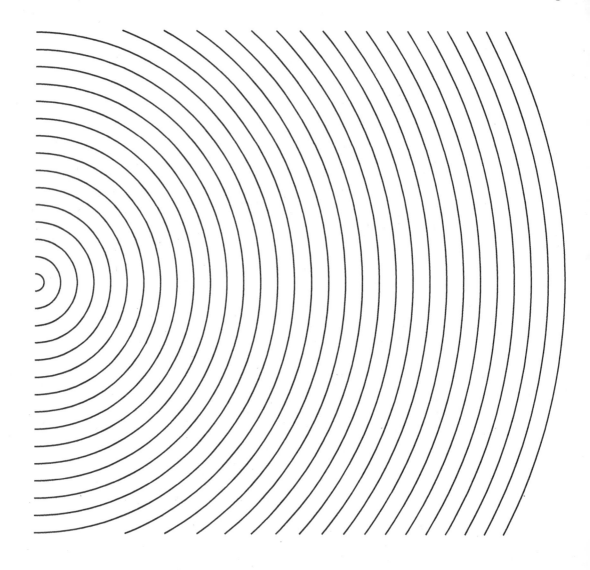

P7.12 Interference can be used to separate colors of light. Consider red light ($\lambda = 750$ nm) and green light ($\lambda = 500$ nm) separately shined onto the two-slit setup in Figure 7.43. Assume that the slits are separated by 2,000 nm. Make two copies on transparent paper of the "red light" circular wave drawn below, place one on top of the other, and shift one so the two sources are separated by a distance corresponding to 2,000 nm. Draw straight lines through the regions where bright light is created by interference. Repeat the process for the green light using the same separation of 2,000 nm between sources. By these drawings, show that red and green light create bright regions in different locations, allowing a person to separate these two colors of light. Note: You can cut the drawings apart for the separate parts of the question.

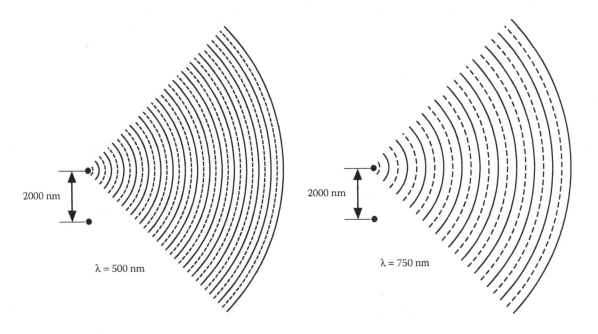

2000 nm

$\lambda = 500$ nm

2000 nm

$\lambda = 750$ nm

8

Analog and Digital Communication

There are no such things as applied sciences, only applications of science.

Louis Pasteur

Twentieth-century t-shirt humor. (Adapted with permission from http://www.thinkgeek.com.)

Jean Baptiste Joseph Fourier (1768–1830), French mathematician who discovered techniques for analyzing and synthesizing signals.

8.1 COMMUNICATION SYSTEMS: ANALOG AND DIGITAL

In communication systems, the goal is to transmit information rapidly with the minimum amount of unwanted noise or error. How can this best be achieved? Familiar forms of communication are sound, telephone, radio, and television (TV). Sound uses the vibrations of air as its physical medium, whereas telephone, radio, and TV use *electromagnetic waves* in metal wire, in air, or in space. Each uses its own particular technology, and each serves different purposes.

In this chapter, we discuss the principles behind the two main types of communication systems—*analog* and *digital*. Recall from Chapter 2 that analog means "smoothly and continuously varying" in a manner analogous to something else, whereas digital means "varying discretely and discontinuously." We will focus on radio—both in its original analog form and in its more recent digital form. Originally, radio was used mostly for voice and music, and these were sent in analog form. The advent of digital radio techniques has allowed all forms of information to be sent via radio; for example, pictures sent to a cell-phone screen.

Some of the principles behind radio and telephones apply broadly and play important roles in physics, mathematics, and engineering. For example, the question "How much information can be transmitted per second through a medium?" could have implications in genetics or cell biology, as well as in the study of radio or Internet technology. The rate of transmitting data through a medium depends on the amount of **bandwidth** the medium has. We will discuss the technical meaning of bandwidth, and why having more bandwidth allows faster data transmission.

8.2   BASICS OF ANALOG RADIO

The meaning of the term *analog* can be understood by examining how sound is transmitted from one location to another in a typical public address system, shown in **Figure 8.1**. Sound is a rapid change of air pressure induced by a vibrating object and travels as waves. When a person speaks into a microphone, a small diaphragm vibrates with frequencies in the **audio range**: 20 hertz (Hz) to 20 kilohertz (kHz). As we discussed in Chapter 5, an **electromagnet** converts the diaphragm's motion into a changing voltage. The microphone's voltage signal is proportional to the air's pressure change. The figure illustrates that the voltage signal varies in time in the same way as the sound wave. The voltage signal is *analogous to* (i.e., a copy of) the pressure signal.

The electrically conducting wire transmits the influence of the voltage from the microphone to an electronic amplifier, which increases the voltage and the power of the electrical signal. The amplified signal then drives an electromagnet attached to the cone in a loudspeaker. As a result, the cone vibrates and causes the air in front of it to move in a manner similar to the original sound wave created at the person's mouth.

An **analog** voltage signal is one that directly follows, or copies, the sound wave signal that it represents.

An analog radio system is similar to a public address system, except that instead of using a wire to transmit the voltage signal to the loudspeaker, the signal is sent between antennas as electromagnetic waves, as shown in **Figure 8.2**. Radio, microwaves, and TV waves are forms of electromagnetic waves, which are characterized by wavelength, frequency, speed, and amplitude. We introduced analog radio in Chapter 7 (Real-World Example 7.2), which you should review now.

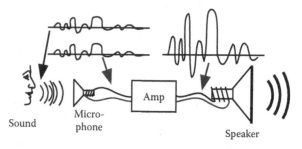

FIGURE 8.1   Analog public address system. Metal wires connect a microphone to an amplifier, which increases the voltage and current and drives a loudspeaker.

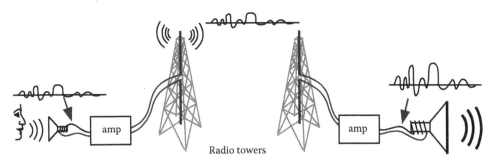

FIGURE 8.2   Analog radio.

AM radio stations do not broadcast their waves at audio frequencies—20 Hz to 20 kHz. Instead, they broadcast at AM radio frequencies—520 kHz to 1610 kHz. If stations were to broadcast directly at audio frequencies, the problem of interference would arise if several nearby radio transmitters (stations) were broadcasting simultaneously. Because waves add together, or interfere, the receiver would simultaneously pick up more than one broadcast. In practical systems, the problem of interference is largely avoided by allocating a separate *channel* to each transmitter. A channel is a portion of a medium that is allocated to one specific stream of information. The division of the medium into channels may be either obvious (separate wires in a bundle of wires) or it may be more subtle (different frequencies on a radio dial). In analog radio systems, this is accomplished by the technique of *frequency multiplexing*. This method uses a separate *carrier wave* with a distinct frequency for each radio channel. A carrier wave is a high-frequency wave with constant amplitude. This allows many separate information channels (broadcasts) to be sent over the same physical medium (EM "airwaves").

Frequency multiplexing is the technique of distinguishing signal channels by using distinct electromagnetic carrier-wave frequencies as identifiers.

A simple example of *multiplexing* is several people talking at the same time. All are using the same air as the transmission medium, and listeners can try to focus on one voice only. Different sound waves can be distinguished by their frequencies. For example, if a woman with a high-frequency voice speaks at the same time as a man with a low-frequency voice, it is easy to focus your listening on just one. In contrast, if two women with the same-frequency voice speak at the same time, it would be harder to distinguish them. If too many people talk at once, it is difficult to select one voice from the others. To overcome this problem, each voice or signal should be carried by a separate channel. For example, in a large crowded room, each pair of talkers could communicate using cell phones. Of course, we want there to be little or no *cross-talk* (interference in one signal from the other) between channels. Engineers realized that the properties of waves allow many signals to be carried on the same transmission medium without necessarily experiencing cross-talk. This follows from the properties of wave interference, discussed in Chapter 7. When two waves come together in a region of space, their wave displacements add. After they separate again, their wave displacements are the same as if there had been no interference.

The idea of radio multiplexing is familiar in "tuning" your radio receiver to receive the signal from just a single station, while avoiding simultaneous reception from nearby stations. What is meant by "nearby"? For two stations that are close to each other on the radio dial, what quantities are actually near each other? The answer is "the carrier frequencies of the separate channels."

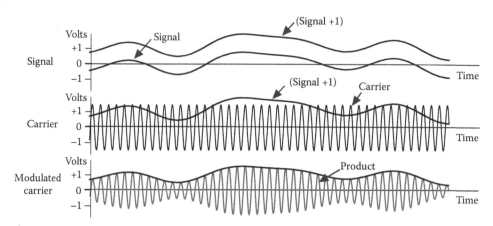

FIGURE 8.3   Amplitude modulation (AM). The top graph shows the audio signal and the signal after being increased by a constant value 1. The middle graph shows the carrier wave and the shifted signal. The bottom graph shows the result of multiplying the carrier–wave values by the shifted signal values.

Figure 8.3 illustrates a carrier wave with ***amplitude modulation*** (***AM***). The top graph shows the low-frequency audio voltage signal. The audio signal has some positive and some negative regions, but never goes below −1 volt (V). To convert this signal to a positive-only signal, it is shifted to higher voltage by adding 1 V to the entire signal. The middle graph shows the carrier wave, which has a high frequency and constant amplitude. Also shown is the audio signal, shifted to higher voltage by 1 V. The bottom graph shows the result of multiplying the carrier wave by the signal wave. This multiplication has the effect of varying, or modulating, the amplitude of the carrier wave by the amplitude of the slower signal wave. At the receiving end, an antenna senses the arriving radio wave, whose signal looks like the curve in the bottom graph. The antenna sends this voltage signal to an electronic recovery circuit, which removes the carrier and leaves the audio signal. (We will not discuss the details of this circuit.) The overall scheme is shown in **Figure 8.4**.

Analog radio uses a high-frequency carrier wave, whose amplitude is modulated to simulate (or emulate) the voltage contour of the signal it is transmitting.

As we discussed in Real-World Example 7.2, each AM radio wave has a carrier frequency somewhere in the range of 520 kHz–1610 kHz. How do we prevent two

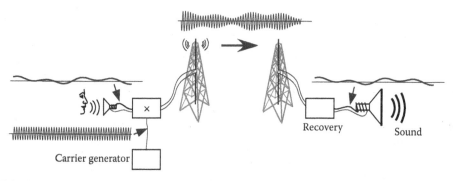

FIGURE 8.4   AM radio system. The circuitry labeled "×" modulates the carrier wave by multiplying it by the audio signal wave. The AM radio wave is broadcast and received, and the audio signal is recovered and sent to a speaker.

broadcast radio signals from interfering? Consider two AM radio broadcasting stations, one transmitting a carrier frequency $f_1 = 600$ kHz, and the other transmitting a carrier frequency $f_2 = 800$ kHz. We know from Chapter 7 that in regions of space where both signals are present, the waves' displacements add or interfere. This causes *beats*—the periodic pulsation of the total wave's amplitude—at the frequency $f_2 - f_1 = 200$ kHz. This does not necessarily mean that a radio receiver will be sensitive to these beats. The recovery circuit in the radio receiver also contains a tunable filter (or tuner), which allows the receiver to select an individual station from among the many signals (having different carrier frequencies) that are in the air at all times.

The question is—how close in frequency can two carrier signals be and still allow the receiver to select an individual station's broadcast without any disturbance or cross-talk from nearby carrier frequencies? The answer to this question will limit the number of stations that can broadcast simultaneously in any geographical region. For example, in Eugene, Oregon, there are twelve AM stations listed as area stations. Their carrier frequencies (in kHz) are: 550, 590, 660, 840, 1050, 1120, 1240, 1280, 1320, 1400, 1450, and 1600. The minimum separation between these frequencies equals 40 kHz; this is to prevent cross-talk. The reason that there are not stations at every 40-kHz interval is that nearby towns have stations, and an effort is made to avoid overlapping frequencies between nearby stations.

To understand why the separation 40 kHz is used, recall that music consists of sound with frequencies up to about 20 kHz. Voice uses a smaller frequency range up to about 4 kHz. AM radio is designed to reproduce audio signals with frequencies up to about 10 kHz. The nature of beats is such that the radio waves will cover a frequency range twice that, or about 20 kHz. This leads to an important point:

Channel bandwidth: Each AM station is allowed to use a channel comprised of a small range (called a band) of radio frequencies covering about 20 kHz (e.g., 590–610 kHz), allowing faithful transmission of music containing frequencies up to approximately 10 kHz. The width of the range of frequencies allocated (in this example 20 kHz) is called the bandwidth of the allocated channel.

Figure 8.5 shows a *frequency spectrum*, which is a graph of the strength or amplitude of radio waves at different frequencies. The spectrum shows two channels, one centered at a frequency of 600 kHz and one centered at 800 kHz. Each channel can use all of the frequencies within 10 kHz on either side of its center frequency, giving a bandwidth for each channel of 20 kHz.

Figure 8.6 shows two carrier waves graphed versus time, one with a frequency of 600 kHz and the other with a frequency of 800 kHz. Each is modulated with audio signals. The idea of frequency multiplexing is that a radio receiver is able to distinguish

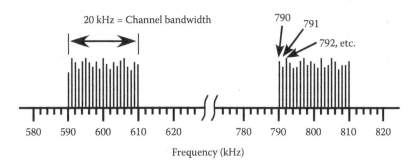

FIGURE 8.5 Frequency spectrum, showing two AM radio channels, each having a bandwidth of 20 kHz.

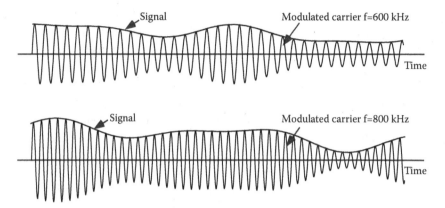

FIGURE 8.6 Two AM waves with different carrier frequencies carrying different modulated audio signals.

the two carrier waves in Figure 8.6 by using an electronic recovery circuit. This will not work properly if the two carriers' frequencies are too close in value, because of beats between them. **Figure 8.7** shows two carriers with frequencies of 760 and 800 kHz. Figure 8.7a in the figure shows the beats in the wave that result from adding the two waves. The beat frequency is 40 kHz; almost low enough to be heard in the audio range. If the carrier frequencies were even closer, say 790 and 800 kHz, then the beats would be in the audio range and would be heard as interference when listening to either channel separately. This illustrates why the channel separation must be greater than approximately 20 kHz.

A useful way to think about modulation of a carrier is the analogy of the way a musician can vary the tone of a note while playing it. Think of a flutist, who plays a steady note, say A440, whose frequency is 440 Hz. She can play it "pure," with no modulation, or she can modulate its amplitude (loudness) up and down, producing tremolo; this is AM. Alternatively, she could modulate its frequency (pitch) back and forth, producing vibrato. This is *frequency modulation* or *FM*.

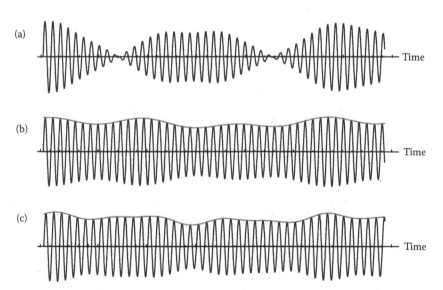

FIGURE 8.7 AM waves (b) and (c) with carrier frequencies, 760 kHz and 800 kHz, that are close enough that beating between them occurs just beyond the audio frequency range. Part (a) shows the beats occurring when the two waves are added together.

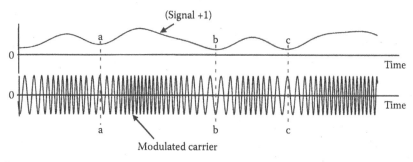

FIGURE 8.8 Frequency modulation (FM). The upper graph is an audio signal, after being shifted by a constant value 1. The lower graph shows a carrier wave whose frequency is varied in accordance with the signal wave.

Instead of broadcasting using AM techniques, radio can be broadcast using FM, as illustrated in **Figure 8.8**. The instantaneous frequency of the carrier is varied, or modulated, according to the value of the signal at each instant. Times labeled a, b, and c are times when the carrier frequency is smallest, making the carrier oscillations slower at those times. For FM radio, the carrier frequency is much higher (88–108 MHz) than for AM radio (520–1610 kHz). Broadcasting and receiving FM transmissions takes place as shown in Figure 8.4, although a different kind of recovery circuit must be used. A major difference is that AM is restricted by the Federal Communications Commission[1] to a frequency range with total bandwidth equal to 1090 kHz (i.e., 520 to 1610 kHz), whereas FM has a much wider range, 88 to 108 MHz, with a much larger total bandwidth, 20 MHz. We will see below that this larger bandwidth gives FM much better fidelity for sound reproduction.

8.3 BASICS OF DIGITAL RADIO

The modern alternative to analog radio is digital radio, in which computers pass digital signals between one another. Computers "speak" to each other using a simple set of symbols, or alphabet. This alphabet, called binary, consists of only two characters, as we discussed in Chapter 2. We could call the two used characters A and B, or we could call them ☺ and ☹. It really does not matter. Computers use as their two characters *one* (1) and *zero* (0).

A nice illustration of the difference between analog and digital representations of music is shown in **Figure 8.9**. The magnified photograph of the surface of a typical vinyl LP (long-playing) recording disc shows continuously varying tracks in which the playback needle moves, creating an analogy of the original sound wave. The magnified photograph of the surface of a typical compact disc, or CD, shows a binary representation of the strength of the sound wave.

How does digital radio work? As shown in **Figure 8.10**, when a person, say Alice, speaks into a microphone, an analog voltage is produced in the wire. This voltage goes to Alice's computer, where an electronic circuit converts it into a list of *ones* and *zeros*; that is, binary data. A *one* is represented by a higher voltage (e.g., 9 V), whereas a *zero* is represented by a lower voltage (e.g., 0 V). These data are transmitted across a network of computers and are received by Bob's computer, where the *ones* and *zeros* are converted back to a continuous analog voltage, which drives Bob's loudspeaker.

The process of converting a continuously varying voltage or signal into discrete values is called signal sampling. Say the voltage from the microphone varies in time

---

[1] The Federal Communications Commission (FCC) is a United States government agency. The FCC is charged with regulating interstate and international communications by radio, television, wire, satellite, and cable.

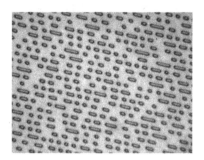

FIGURE 8.9   Music on a vinyl record (left) is represented as an analog signal, in which the shape of the grooves is analogous to the displacement of the recorded sound wave. Music on a CD (right) is represented as a binary signal, in which each bit value is represented by a flat spot or a pit on the CD surface. (Vinyl photo courtesy of Carl Haber, Lawrence Berkeley National Laboratories. CD photo courtesy of Jerry Gleason, University of Oregon.)

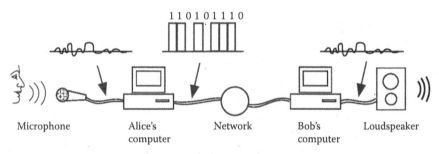

FIGURE 8.10   Internet radio. Each connection shown here is a pair of wires.

as shown in **Figure 8.11**. Such a variation might come from a person whistling a pure musical tone into the microphone. The air near the person's mouth vibrates smoothly in time, causing the air pressure to vary similarly. This causes the voltage to vary in much the same way. An electronic circuit inside Alice's computer samples the value of the voltage (the number on the vertical axis in the graph) at many successive points in time, as shown in the lower part of the figure. *Sampling* means measuring something at a series of discrete times and recording the values as a list of numbers. Using sampling, we can represent the signal in Figure 8.11 by a list of numbers; for example,

10, 9, 8, 5, 3, 1, 0, 1, 2, 5, 7, 9, 10, 9, 8, 5, 3, 1, 0, 0, 2, 4, 7, 9, 10, 10, 8, 6, 3, 1.

In this example, the sampled values were rounded off to the nearest whole number. The more precise that the measuring system is, the more digits the numbers will have. For example, 10.0, 9.3, and 7.6, have higher precision than do 10, 9, and 8. The idea of precision was discussed in Chapter 2. Using these rounded sampled values, the signal is represented by the series of points shown in **Figure 8.12**. In general, some information is lost in the sampling process, leading to a distorted version of the signal, seen in the fact that the points do not match up precisely with the original curve.

We can represent the above list of voltage samples (the decimal numbers 10, 9, 8, 5, 3, 1, 0, 1, 2, 5, 7, 9, 10, etc.) using binary representation. Let us agree to use four bits for each number, so the list becomes:

1010, 1001, 1000, 0101, 0011, 0001, 0000, 0001, 0010, 0101, 0111, 1001, 1010, etc.

The computer needs to "know" that it is dealing with decimal whole numbers (no digits to the right of a decimal point) and that these are being represented by 4-bit numbers.

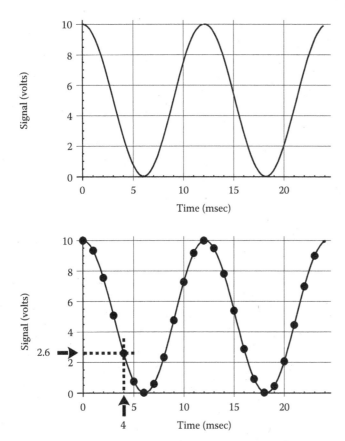

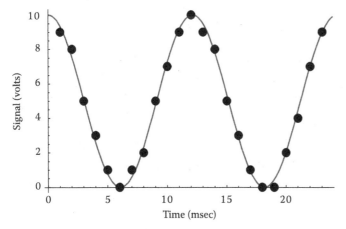

FIGURE 8.11   A voltage varying in time, and the same signal showing the sampled values at time intervals of 1 msec. For example, when the time equals 4 msec, the voltage equals 2.6 V, as indicated by the arrows.

FIGURE 8.12   A graph showing the samples (indicated by dots) that the computer stores and uses to represent the original smoothly varying voltage.

The computer also needs to know that the list starts at a certain time. This set of rules is an example of a **protocol**.

A **protocol** is a set of rules for interpreting a list of binary numbers. The protocol specifies the starting digit, the number of bits per number, whether the number is a whole or fractional number, the voltage values to be used to represent each bit value, error correction scheme, etc.

Given that the sender and receiver agree on this protocol, we can omit the commas and spaces and write the binary data as:

1010100110000101001100010000000100100101011110011010, etc.

Now we have constructed a list of solely ones and zeros that represents the original voltage signal in **Figure 8.11**.

THINK AGAIN

When a computer records a set of sample values, such as those plotted as points in Figure 8.12, it does not know anything else about the signal. It does not know if the actual signal is a smoothly varying signal like the gray line in Figure 8.12, or a kinky-looking signal, such as would be obtained by connecting the points in the figure by straight lines.

How does Alice's computer send these *ones* and *zeros* to Bob's computer? This is part of a more general, very important, question: How can we represent abstract, mathematical concepts like *zero* or *one* by something in the physical world? In Chapter 2, we discussed using our fingers; those are certainly physical things. They are convenient for representing digital information. Computers use discrete voltage levels as their "fingers" for counting. Recall that the word *digit* means *finger*. To represent binary numbers, computers need to use only two different voltages. Any two values would do. Let us adopt the convention that a *zero* is represented by zero volts (0 V), and a *one* is represented by nine volts (9 V). More generally, we could call these two voltage values "low" and "high." Furthermore, the actual voltage values do not need to be extremely accurate—any voltage close in value to 9 V (e.g., 9.1 V) will be recognized as a *one*, and any voltage near 0 V (e.g., 0.2 V) will be recognized as a *zero*. This insensitivity to precise voltage values leads to the robustness of digital systems, which makes them less prone to making data errors. The designation of voltage ranges for the two bit values must also be part of the protocol.

An illustration of a system for sending digital information between computers is shown in **Figure 8.13**, which shows a telegraph-like technique. A 9-V battery is connected to a switch (a momentary push-ON switch, which snaps back open when not being pressed), and a twisted pair of plastic-insulated wires that travel to the receiving station. Twisted pairs of wires are used to avoid the wires acting as an antenna. Recall from Chapter 5 that an open current loop acts as an antenna when magnetic field lines penetrate the loop. In contrast, a tightly twisted pair has no open loop for magnetic fields to penetrate. Each time the switch is pressed closed, the wire pair has 9 V applied across it. This voltage is relayed to the distant electronic circuit labeled "memory."

The sequence of low and high voltages used to send the first numbers making up the original signal in the above example is shown in **Figure 8.14**. Bob's computer

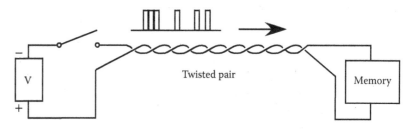

FIGURE 8.13 A telegraph-like system for sending digital information.

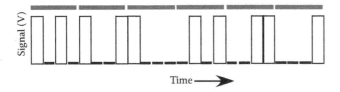

FIGURE 8.14 Twenty-four bits, or *ones* and *zeros*, represented by high (9 V) and low (0.1 V), corresponding to the first six numbers in the sampled signal. Each group of four bits (indicated by horizontal gray bars) corresponds to a number in the range 0 to 9. The horizontal axis is time.

QUICK QUESTION 8.1

Using the same 4-bit protocol as in the above example, draw the voltage sequence representing the four base-ten numbers 2, 4, 0, 9, following the format in Figure 8.14.

will receive the *ones* and *zeros* by measuring the voltage at each pre-assigned time in the series, and will then use the agreed-upon protocol to decode the information. You should check for yourself that the sequence of bits shown in the figure correctly represents the first six numbers in the original sequence in our above example.

Next, Alice's computer transmits the list of bits representing the sampled numbers to Bob's computer. Bob's computer reconstructs the original analog voltage signal and sends this voltage signal to a loudspeaker, which generates a replica of the original sound waves.

The telegraph-like technique shown in Figure 8.13 does not use any carrier wave, and therefore this signal cannot be transmitted as a radio wave through air. Recall that to generate radio waves, you need to oscillate electric charges in an antenna at a radio frequency, for example, 520 kHz. Often we want to send digital data in wireless form; for example, to operate cell phones or wireless Internet connections. We can do this, as with analog radio, by using AM. Text or voice messages being sent to different cell phones are distinguished by their carrier frequencies. To prevent these digital transmissions from interfering with music radio broadcasts, carrier frequencies outside of the AM and FM bands are used for cell phones and wireless Internet. Carriers around 900 MHz are commonly used for cell phones, and carriers around 2,500 and 5,000 MHz are used for wireless Internet (WiFi).

In **Figure 8.15**, a radio-frequency carrier wave is modulated by a sequence of binary data, labeled as the signal voltage. The duration of each data pulse is called the **data period**, which we denote by T_D. A smaller data period allows us to send more bits per second. That is, the rate of transmitting data, or **data rate**, becomes larger. In general, the data rate equals the inverse of the data period, that is:

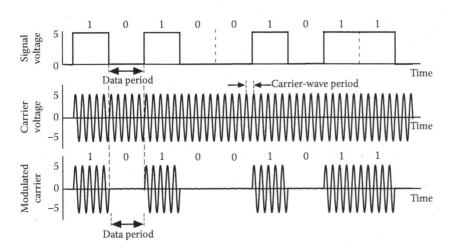

FIGURE 8.15 A binary data list (top) is impressed onto a radio-frequency carrier wave (middle), to create the digital-modulated carrier wave (bottom). In this example, the data period is 5 times larger than the carrier-wave period, meaning that five carrier oscillations occur within each data pulse.

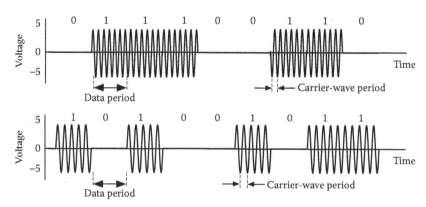

FIGURE 8.16 Two binary data streams, impressed on two carrier waves with distinct frequencies, can be distinguished by a receiver.

$$data\ rate = \frac{1}{data\ period} = \frac{1}{T_D}$$

The carrier wave is shown in the middle graph and has a period equal to $T = 1/f$, where f is the carrier-wave frequency. Notice that the carrier period is smaller than the data period. When the data pulse sequence is multiplied by the carrier wave, the modulated carrier wave shown in the bottom graph is created. This is the radio signal that is broadcast by the antenna. It contains two types of information: the data sequence and the identity of the carrier, which is determined by its frequency.

As an example, if the duration of the data pulses is 1/1000 of a second (sec), or 0.001 sec, then in a time of 1 sec you can send 1,000 pulses. The data rate is then 1,000 pulses/sec, which we could also state as 1,000 Hz or 1,000 bits per second (bps), or 1,000 bps. It is common to use the term **baud** to mean 1 bps, after French engineer J. B. F. Baudot. For example, if a phone modem has a data rate of 52,000 bps, we say this rate equals 52 kbaud.

To emphasize the idea of frequency multiplexing for digital channels, **Figure 8.16** shows two modulated carriers, each with a distinct carrier frequency and different binary data being transmitted. Both waves can be present in the same medium, as long as the receiver has the ability to select one frequency channel for reception.

We summarize as follows:

Digital broadcast radio uses a high-frequency carrier wave, the amplitude of which is modulated to represent the binary-number sequence corresponding to the sampled voltages in the signal it is transmitting.

8.4   THE PRINCIPLE OF CARRIER MODULATION

An important question is: How rapidly can we transmit data through a medium? That is, how fast can we modulate a carrier wave with a stream of data, whether it is analog or digital? The answer depends on an important principle:

Principle of carrier modulation: We cannot modulate a carrier wave at a frequency higher than the frequency of the carrier, without destroying the identity of that carrier.

In the example shown in Figure 8.15, there are five carrier oscillations within each data pulse. This means that the rate of modulation is about 5 times smaller than the carrier frequency. For example, if the carrier frequency equals 5,000 Hz, then the modulation rate is seen from the figure to be about 1,000 Hz. This allows the wave to be identified by its carrier frequency, while still allowing us to send data using this carrier. The integrity of the carrier is needed if we are to use multiplexing; that is, the technique of distinguishing different broadcast signals by having unique carrier frequencies.

Principle of carrier modulation, version 2: Each data pulse must contain at least one complete cycle of the carrier wave, so that the wave can be identified using its carrier frequency.

The limiting situation is illustrated in **Figure 8.17**, where each data pulse has duration equal to one period of the carrier wave. In the lower example, the carrier frequency is 5 times greater than in the upper example. This allows it to carry 5 times more data pulses in any given time interval than is possible using the slower carrier frequency shown in the upper example.

If we tried to modulate a carrier wave using a series of data pulses, each of which is shorter in time than the period of the carrier oscillation, the receiver would no longer be able to determine the frequency of the carrier wave. This information would be lost. The safe rule of thumb is: Do not modulate a carrier wave at a frequency higher than the carrier frequency. This rule tells us the theoretical maximum data rate that could conceivably be sent through a particular channel.

The upper limit for the rate of transmitting data across a particular channel, the carrier frequency of which equals f, cannot exceed approximately the value f.

This means that we can send data at higher rates when using carrier waves with higher frequencies.

In practice, AM radio stations modulate their carrier waves at a frequency about one-hundreth of the carrier frequency. This allows many distinct carrier frequencies

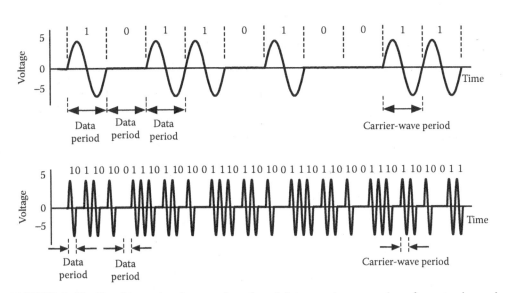

FIGURE 8.17 Top: Binary data (zeros and ones) modulate a carrier wave, whose frequency is equal to the data rate. Bottom: Increasing both the carrier frequency and the data rate.

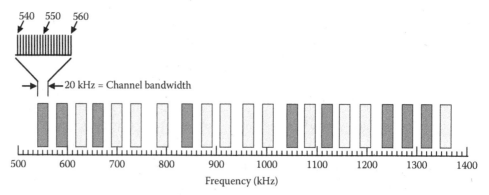

FIGURE 8.18 The AM radio stations broadcasting in a certain city are shown as dark gray rectangles, whereas stations broadcasting in nearby towns are shown in light gray. The position of each rectangle indicates the station's carrier frequency, whereas the width of a rectangle indicates the bandwidth that is allocated to that station. An expanded view of a single station's frequency band is also shown.

to be used, one for each station. For example, an AM station broadcasting at a carrier frequency 1000 kHz is allowed to modulate the amplitude of its carrier with frequency 10 kHz, but no faster. For example, this ensures that this station will not interfere with another station that is broadcasting at a nearby carrier frequency of 1040 kHz. This situation is illustrated by the frequency spectrum shown in **Figure 8.18**. Each AM station that is broadcasting in a certain region (near Eugene, Oregon) is represented by a gray rectangle centered at its carrier frequency (550 kHz, 590 kHz, etc.) The width of each rectangle indicates each station's channel bandwidth, which equals 20 kHz. For example, the station centered at 550 kHz can use every frequency between 540 and 560 kHz.

FM radio broadcasts are broadcast using much higher carrier frequencies, around 100 MHz. If you operate an FM radio station, you are allowed to modulate the wave at frequencies up to about one-fiftieth of the carrier frequency, or about 2 MHz. You can see that FM radio can have a higher data rate than AM radio.

Above, we stated that the width of the range of frequencies allocated to a particular channel is called the bandwidth of the allocated channel, or the channel bandwidth. In the AM radio example, the bandwidth of each channel equals 20 kHz. The channel bandwidth is roughly equal to the maximum allowed modulation rate for that channel.

To summarize, a radio station is allowed to use a fixed band (or range) of frequencies, which includes the carrier frequency. The width of the band allocated is called the channel bandwidth of that channel. The wider the band (larger the bandwidth) is, the greater the data rate can be. The largest value that a channel's bandwidth can possibly be equals the carrier frequency itself. In the following sections, we explore these concepts more fully.

8.5   SIGNAL SYNTHESIS, ANALYSIS, AND BANDWIDTH

The maximum rate of transmitting data on a channel depends on the bandwidth of the channel. In typical communication systems such as radio and TV, the bandwidth allocated to each channel is set by mutual agreement of those using the system. That is, the bandwidth allocation is part of the protocol. This section explores the physics and mathematics behind the concepts of bandwidth and rate of data transmission. In discussing this, we will learn about a very general concept—spectral analysis of signals.

The idea of spectral analysis of signals can be illustrated using the familiar idea of musical notes written on sheet music. A chord is comprised of three frequencies played simultaneously. **Figure 8.19** shows the notation for several chords, with each oval

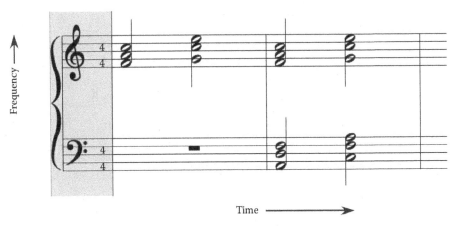

FIGURE 8.19 Chords shown in musical notation, which we can interpret as a graph of frequency versus time. To play all of the notes shown in the last two chords, a larger bandwidth is needed than is needed to play the earlier two chords.

indicating a different note on a piano keyboard. Higher-frequency notes are indicated by ovals closer to the top of the figure. You can think of this diagram as being a graph of frequency versus time, where the vertical axis is frequency and the horizontal axis is time, as indicated. First, two chords are played in sequence. Each of these chords spans a small frequency range; that is, a small bandwidth. Later, the same chords are played again, simultaneously with two other chords at lower frequency. To play these four chords requires a larger bandwidth than to play only the two chords originally shown. You can visualize bandwidth by considering two piano keyboards, shown in **Figure 8.20**. The upper keyboard has fewer keys than the lower keyboard, and so spans a smaller bandwidth than does the lower keyboard.

An important, general principle of signal synthesis and analysis is named after Jean Baptiste Joseph Fourier, the French mathematician pictured at the start of the chapter, who pioneered the mathematical techniques for analyzing and synthesizing signals.

Fourier's principle: Any wave motion or other vibration can be made up, or synthesized, by adding together many waves of different frequencies. The wave's frequency spectrum is like a *spectral recipe* for making that wave and tells what frequency ingredients are present, and how strong (loud) each frequency component is. Every wave motion or other vibration can also be broken down, or analyzed, in terms of its frequency spectrum.

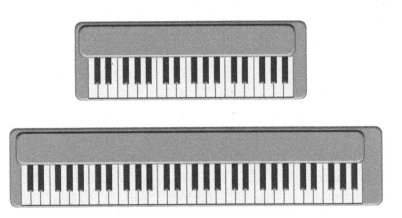

FIGURE 8.20 The number of keys on a piano keyboard indicates the bandwidth of that keyboard.

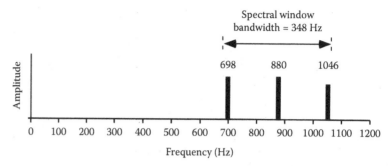

FIGURE 8.21   The frequency spectrum of the F chord, comprised of notes F-A-C.

In the case of sheet music, the written notes define the spectral contents, or recipe, for each chord played. For example, the chord F—the first chord written in Figure 8.19 is comprised of three frequencies: F (698 Hz), A (880 Hz), and C (1,046 Hz). The frequency spectrum for this chord is shown in **Figure 8.21**. A vertical bar indicates the frequency and amplitude (strength or loudness) for each frequency component of the chord. In the example shown, the highest frequency, 1,046 Hz, has smaller amplitude than the other two. The bandwidth is the width of the spectral window occupied by the chord's frequencies. In this case the bandwidth equals 1,046 – 698 = 348 Hz.

Let us further explore how the concept of bandwidth arises in the context of digital transmission. Say we want to send a series of data pulses (*ones* and *zeros*) by amplitude modulation of a carrier wave. As a simple example, let us consider a carrier wave with frequency 20 kHz, rather than at a radio frequency. The lower frequency makes it easier to visualize examples by graphing the signals. The carrier wave, before modulation, carries no information (except for its identity, as expressed by its frequency, 20 kHz). This is shown in **Figure 8.22**. Notice that here the units of frequency are kHz.

If we want to send a list of binary data, say (1, 0, 1, 0, 0, 1), we can do this by modulating the carrier with these data, as illustrated earlier in Figure 8.15. Such a modulated-carrier wave can be synthesized by adding together several waves having frequencies that are near, but not equal to, the carrier frequency. This is a generalization of the idea of beats involving only two frequencies, which we discussed in Chapter 7, In-Depth Look 7.3, which you should now review.

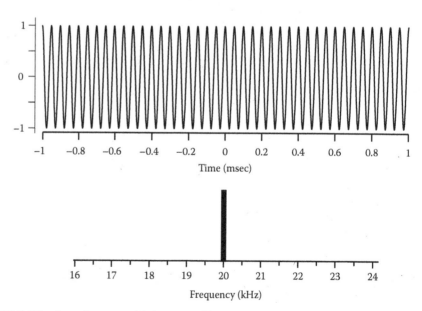

FIGURE 8.22   A carrier wave with frequency 20 kHz, and its frequency spectrum.

Signal bandwith is the width of the spectral range (or window) needed to synthesize the signal.

How large a signal bandwidth is needed to synthesize and transmit a list of binary data? First, consider how to synthesize individual signal pulses, for representing bit values, by adding together waves of different frequencies. As we discussed in In-Depth Look 7.3, if two waves come together in some region, the wave amplitudes add or interfere. If the two waves have slightly different frequencies, then at one instant in time, the waves add constructively and the net wave has a greater amplitude, whereas later the waves add destructively, causing a lesser amplitude of the net wave. By adding more than two waves with unequal frequencies, we can create or synthesize short wave pulses, as we now demonstrate.

Figures 8.23 and **8.24** show two efforts to synthesize a signal pulse. The frequency and strength of each wave to be added is represented as a vertical bar. The central peak, at 20 kHz, is the carrier. In Figure 8.23, three waves are added—the carrier and two equally spaced neighboring frequencies, which we call side bands. This results in signal pulses whose duration equals approximately 1 millisecond (msec). The bandwidth B of this signal is 1 kHz.

If we wish to make a pulse that is shorter in duration than that in the previous example, we can do so by adding in more frequencies over a wider range. In Figure 8.24, 11 waves are added—the carrier and five pairs of side bands. The spectrum for this signal is given in **Table 8.1**, which lists the frequency and amplitude of each wave that is added to make the signal. You can think of this as a recipe: Take a carrier at 20 kHz with amplitude 1.0; add side bands at frequencies 19.6 and 20.4 kHz with amplitudes of 0.97; add side bands at frequencies 19.2 and 20.8 kHz with amplitudes of 0.89, etc. Mix together and you will make a beautiful pulse whose duration equals about 0.25 msec. The bandwidth B of this signal is $22 - 18 = 4$ kHz. The particular recipe used in this example is not unique, but is used here because it produces nice looking pulses for the figure.

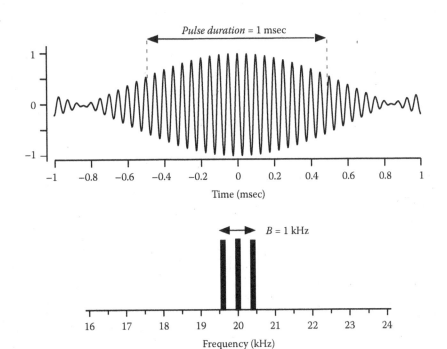

FIGURE 8.23 Top: A data pulse synthesized from the spectral recipe shown in the lower plot. The bandwidth is $B = 1$ kHz. The pulse duration is 1 msec.

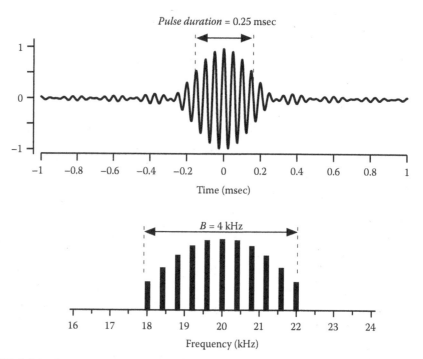

FIGURE 8.24 Top: A data pulse synthesized from the spectral recipe shown in the bottom plot. The bandwidth is $B = 4$ kHz. The pulse duration is 0.25 msec.

TABLE 8.1

Spectral Recipe for the Signal in Figure 8.24

Amplitude	0.38	0.59	0.76	0.89	0.97	1.0	0.97	0.89	0.76	0.59	0.38
Frequency (kHz)	18.0	18.4	18.8	19.2	19.6	20.0	20.4	20.8	21.2	21.6	22.0

The examples just discussed illustrate a general result of Fourier's theory:

The duration of a synthesized pulse is given roughly by the inverse of its spectral bandwidth, that is,

$$pulse\ duration = \frac{1}{bandwidth} = \frac{1}{B}$$

For example, for a bandwidth equal to 4 kHz, the pulse duration equals $1 \div (4\text{ kHz})$, or 0.25 msec, as in Figure 8.24.

Although the signal wave synthesized in Figure 8.24 is not square-shaped, as is each data pulse in Figure 8.15, this is not a problem. The wave already looks enough like a data pulse to serve just fine in a digital communication system. How many data pulses of this shape could we send per second, without the pulses overlapping in time so badly that they could not distinguished? Roughly speaking, the pulses could be placed with their centers separated in time by at least one pulse duration and still be separable in a measurement. Nine such pulses (some with zero amplitude) are shown in **Figure 8.25**. It shows a series of data pulses arriving one after the other, representing the binary data (1, 0, 0, 1, 1, 1, 0, 1, 0). The separation between the centers of adjacent data pulses is the data period. In this example, the data period is 0.25 msec. In this case, the number of pulses per second equals:

$$one\ pulse\ per\ 0.25\,\text{msec} = \frac{one\ pulse}{0.25\,\text{msec}} = 4{,}000\ \text{pulse/sec}$$

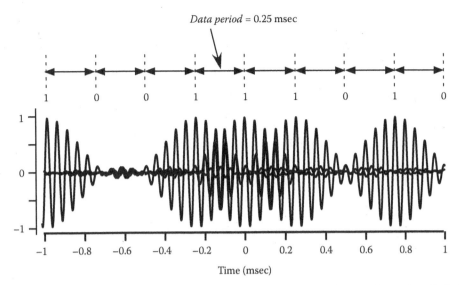

FIGURE 8.25 A series of partially overlapping data pulses representing the binary data (1, 0, 0, 1, 1, 1, 0, 1, 0).

The number of nonoverlapping pulses that can be sent per second equals the bandwidth $B = 4$ kHz. We see that in this example the data rate (number of bits per second, or baud) equals the bandwidth. For example, if we were to increase the bandwidth by, say, a factor of 2, the data period would be cut in half, and the data rate would be doubled. This insight tells us that the wider the range of frequencies occupied by the spectral recipe (Figure 8.24), the higher the data rate can be.

8.6 MAXIMUM DATA RATE

The data rate of a channel is the amount of information (bits) sent per second. The *maximum data rate* is the maximum amount of information (bits) that can be sent per second. The maximum data rate depends only on the bandwidth of the channel, not on the frequency of its carrier. In the above example, the data rate equals the bandwidth. It turns out that by using clever techniques based on Shannon's theory of information, engineers can achieve data rates twice higher than in our example. But, there is an absolute upper limit to the data rate that can be achieved in any communication channel with a fixed bandwidth.

Principle of maximum data rate: For a communication channel with a fixed bandwidth B, the data rate (number of bits per second) that can be transmitted on this channel can equal but not exceed $2B$.

This is called the Shannon-Nyquist Channel-Capacity Theorem, after Claude Shannon and the Swedish-American scientist Harry Nyquist.

As an example, say an AM radio station is allocated a channel with a bandwidth of 10 kHz. If this channel were to broadcast digital information, the maximum rate at which it could do so is 20,000 bps. So, the station's data rate is 20 kbaud. It is interesting to note that if all of the AM-radio bandwidth ($1610 - 520 = 1090$ kHz) were to be allocated to a single AM station, it could transmit at a data rate of $2B = 2 \times 1090$ kHz = 2180 kHz, or 2.18 million baud.

There is a simple way to appreciate this result for a digital system, by analogy with the piano keyboards we discussed earlier. Consider a keyboard with only eight keys.

How many different distinct chords could you play on this keyboard? Because each key can either be pressed or not pressed, it represents a binary number, *one* or *zero*. We know that with 8 bits, we can express $2^8 = 256$ distinct numbers (decimal 0, 1, 2, ... 255). There are 256 distinct chords you could play. If you played one chord per second, you would transmit information at a rate of 8 bps or 8 baud. Next, consider a keyboard having 16 keys. There are $2^{16} = 65,536$ distinct chords you could play. If you played one chord per second, you would transmit information at a rate of 16 bps or 16 baud. That is, doubling the bandwidth (number of keys) also doubles the data rate.

To summarize, signal bandwidth is the width of the spectral window occupied by the spectral recipe that is needed to represent a certain signal. The larger the signal bandwidth is, the smaller the data period becomes, and the larger the data rate becomes. A signal can be sent through a channel only if that channel has a *channel bandwidth* at least as large as the signal bandwidth.

8.7   FREQUENCY MULTIPLEXING AND BANDWIDTH

Now that we have gained a better understanding of bandwidth, let us again discuss *frequency multiplexing* for radio systems. For a practical communication system, an important question is: What conditions are needed for two stations to be able to transmit data on the same medium without interference, or *cross-talk*? To prevent cross-talk, the spectral windows occupied by these two signals must not overlap, as they do in **Figure 8.26a**, where the carrier frequencies are 595 and 605 kHz. This overlap would be a serious problem, because a radio receiver could not distinguish to which channel the signal data belonged. On the other hand, if the carrier frequencies are 595 and 615 kHz, as shown in Figure 8.26b, then the receiver can cleanly separate the signals with no interference. Because there is no spectral overlap, the two signals can coexist on the same medium without any cross-talk.

The conclusion is: For a fixed signal bandwidth, to avoid cross-talk we need to choose the carriers representing different channels far enough apart in frequency so that their spectral windows do not overlap. This limits the number of separate channels that can fit into a given total bandwidth. As we discussed earlier, this is called

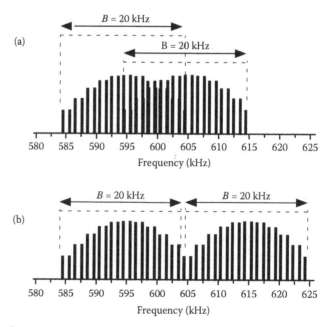

FIGURE 8.26   Spectral windows of two signals (a) showing overlap and (b) with no overlap.

frequency multiplexing. We can now estimate how many such distinct channels can be put onto a single medium.

Let us say that the physical properties of the medium give it a fixed total bandwidth that can be used. This total bandwidth is divided into several allocated channels. If each channel occupies a bandwidth B, then the number of channels that can be accommodated is:

$$number\ of\ channels = \frac{total\ bandwidth\ of\ medium}{B}$$

As an example, in the United States, AM radio has a total spectrum allocation of 520 to 1610 kHz, or a total bandwidth equal to $1610 - 520 = 1090$ kHz. So the maximum number of channels having 10-kHz bandwidth is $1090 \div 10 = 109$ channels. In practice, a smaller number of channels are allocated in any geographic region to avoid overlaps with nearby communities and to make it easier to use low-cost receivers to separate the channels.

IN-DEPTH LOOK 8.1: SIGNAL RECONSTRUCTION

Recall that when we discussed Internet radio, as in Figure 8.10, we said that Bob's computer, after receiving the binary data sent by Alice, could reconstruct the analog music signal from the received digital data. An interesting question is: how *well* can Bob's computer do this reconstruction? It turns out that the concept of maximum data rate applies here as well. Claude Shannon (the same Shannon who introduced the concepts of information and electronic logic circuits) proved a mathematical statement about sampling.

Sampling theorem: If a signal contains frequencies only within a spectral window with bandwidth equal to B, and if the signal is sampled at a rate of 2B and with perfect precision, then the discrete sample values can be used to exactly reconstruct the original signal.

For example, say that the signal shown as the solid curve in **Figure 8.27** represents a music signal that was recorded using an electronic circuit that limits the bandwidth of the recording to $B = 20$ kHz. That is, the recorded frequency range is 20 Hz to 20 kHz, which is sufficient for high-fidelity music. The dots show the sampled values, with one sample being taken every 25 µsec. This corresponds to a sampling rate of 40 kHz; that is, 40,000 samples per second.

$$sampling\ rate = \frac{1}{25\mu\sec} = \frac{1}{25 \times 10^{-6}\sec} = 0.04 \times 10^{6}\frac{1}{\sec} = 40,000\ \text{Hz}$$

Alice's computer transmits only the sampled values to Bob's computer. Nevertheless, Bob's computer can reconstruct the original music signal exactly. This seems remarkable, but it is true, although we will not explain the proof or the method actually used for reconstruction.

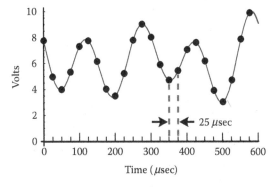

FIGURE 8.27   Analog voltage signal and digital sampled values.

SUMMARY AND LOOK FORWARD

In this chapter, we studied several main ideas about signals and information. We can represent a time-varying signal as either an *analog* (continuous) variation of voltage, or as a *digital* (discrete) variation of voltage. In the analog representation, the voltage directly follows the signal voltage. In the digital representation, the signal is *sampled*, or measured, at selected times and a number represents the value of each sample. In modern communication systems, the sampled values are represented using *binary* numbers. When data are transmitted across a communication channel, the sender and receiver must agree on a certain *protocol*—a set of rules by which the numbers will be formatted, sent, and interpreted.

We discussed analog and digital radio, both of which use the technique of frequency multiplexing to send separate streams of information using different channels. Broadcasting stations operate on different carrier frequencies. Each radio station is allowed a fixed band (or range) of frequencies. The width of the range of frequencies allocated is called the bandwidth of the allocated channel. The larger (wider) the bandwidth is, the larger the data rate can be. Signal bandwidth is the width of the spectral window occupied by the spectral recipe that is needed to represent a certain signal. A signal can be sent through a channel only if that channel has a bandwidth at least as large as the signal bandwidth.

Analog radio and digital radio both have certain advantages and disadvantages.

- ■ **Advantages of analog radio broadcasting:**
  - ○ Analog technology is simple.
  - ○ It is usually not subject to interruptions.

- ■ **Disadvantages of analog radio broadcasting:**
  - ○ AM and FM radio systems can typically broadcast only over a small geographical region.
  - ○ Radio broadcasts cannot easily include other content in addition to audio.
  - ○ Radio spectrum (bandwidth) is a fixed and precious commodity, which limits the amount of information that can be carried.

- ■ **Advantages of digital radio broadcasting:**
  - ○ Can broadcast from anywhere in the world.
  - ○ Can reach many areas that have poor analog radio reception.
  - ○ Can include other content in addition to audio in the broadcast.
  - ○ Can obtain better sound quality using a "fast" enough computer system.

- ■ **Disadvantages of digital radio broadcasting:**
  - ○ Subject to interruptions of Internet connections.
  - ○ Sound quality may not be as good as with analog radio, because it depends on data rate ("speed") of Internet connection.
  - ○ Data rate is not consistent across the Internet.

There is a physical limit to the total bandwidth that can be achieved in any channel medium. The limit is set by the highest carrier frequency in the band. For example, if we adopt a technology that uses carriers with frequencies up to 108 MHz (the top of the range for FM), then we could not have a bandwidth greater than 108 MHz. Indeed, this would include all frequencies from zero up to 108 MHz, and there are not any more megahertz to be had in this range. The only way to increase the data rate is to use higher and higher carrier frequencies. This is the motivation behind using visible light as a carrier for communication systems. Recall that the frequencies of visible light are approximately 10^{14} Hz; that is, 100 million MHz. This is the reason that optical fibers can carry such large amounts of data per second, as we will discuss in later chapters.

SUGGESTED READING

A popular account of radio and the Internet:
Naughton, John. *A Brief History of the Future.* Woodstock, NY: Overlook Press, 2000.

An excellent text on telecommunications, with discussions of bandwidth and signals, at the same level as the present text:
Rogers, Alan. *Understanding Optical Fiber Communications.* Boston: Artech House, 2001.

Goldsmith, Andrea. *Wireless Communications.* New York: Cambridge University Press, 2005.

History of radio, television, and the Internet, by the Federal Communications Commission (FCC) at http://www.fcc.gov/omd/history.

KEY TERMS

Amplitude modulation (AM)
Analog
Audio range
Baud
Bandwidth
Beats
Carrier wave
Channel
Channel bandwidth
Cross-talk
Data period
Data rate
Digital
Electromagnet
Electromagnetic wave
Frequency modulation (FM)
Frequency multiplexing
Frequency spectrum
Multiplexing
Protocol
Sampling
Sampling theorem
Signal bandwidth
Spectral recipe

ANSWERS TO QUICK QUESTIONS

Q8.1 The voltage sequence representing the four decimal numbers 2, 4, 0, 9 is 2 = (0010), 4 = (0100), 0 = (0000), 9 = (1001), or graphically:

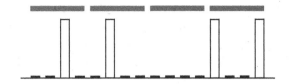

EXERCISES AND PROBLEMS

Exercises

E8.1 Find and list the carrier frequencies of all AM and FM radio stations in the city or town where your school is located. Try the websites Radio Locator or On the Radio: http://www.radio-locator.com or http://www.ontheradio.net/Stations.aspx.

E8.2 Imagine that you are an entrepreneur whose task is to fund and set up a new radio station. Give all of the arguments in favor of using the Internet to broadcast the signal, rather than using traditional radio-wave broadcasting. Next, take the viewpoint of a critic who argues the reverse case. Using the Internet, do some further research on this question and write up the arguments in the form of an imagined debate between the two opposing proponents.

E8.3 What is a protocol, and how is it used in digital communication? Think of an example of a protocol that is not discussed in this chapter; it does not need be one used in Internet communication.

E8.4 The graph of the analog signal below represents a musical tone with frequency 70 Hz. Your task is to digitally sample this signal at two different sampling rates.

(a) Make a table listing the sampled voltage values corresponding to a sampling rate of 1,000 samples/sec. (i.e., one sample every 0.001 sec, starting at time = 0). Photocopy the empty graph below and plot the points from your table.

(b) Make a table listing the sampled voltage values corresponding to a sampling rate of 250 samples/sec. On a second copy of the empty graph, plot the points from this second table. On each of your graphs, connect the dots with straight lines to give the representation of the digitized signal. Which of the sampling rates best reproduces the original signal?

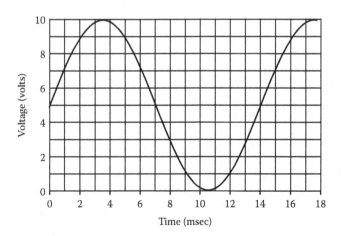

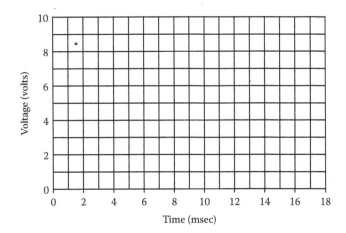

E8.5 Why do compact disks (CDs) use a sampling rate of 44 kHz? If the sampling rate were reduced to 22 kHz, would the CD be able to store high-quality recordings of people speaking? Would the CD be able to store high-quality music recordings? Explain.

E8.6 Follow Figure 8.3 as an example to fill in the missing curve in the graph below. The missing curve equals the product of the carrier wave times (Signal + 1).

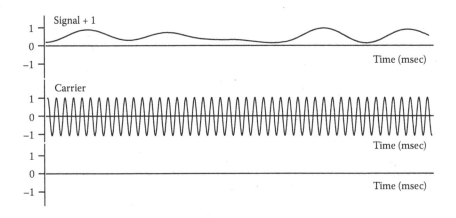

E8.7 Answer E8.6 for the case of frequency modulation.

PROBLEMS

P8.1 A digital signal is: 1, 0, 1, 1, 0. We want to send this sequence of binary numbers using amplitude modulation of a carrier, the frequency of which is 3 kHz. The protocol is that a *zero* bit is represented by 1 V and a *one* bit is represented by 5 V, and each bit value will be held for a duration of 1 msec. Draw the signal wave as a function of time, using Figure 8.14 as an example, being careful about how many carrier cycles occur within the duration of each data period. Label all axes and give units.

P8.2 You are sending binary data pulses using a wireless system. The duration of each bit-representing pulse equals 200 nsec (1 nsec = 10^{-9} sec). What is the data period, given in seconds? What is the data rate, given in bits per second? Given in baud? Given in kilobaud? Given in megabaud?

P8.3 A microphone picks up a musical signal and creates a voltage that varies in time as shown in the graph. Your job is to convert this signal to decimal numbers.

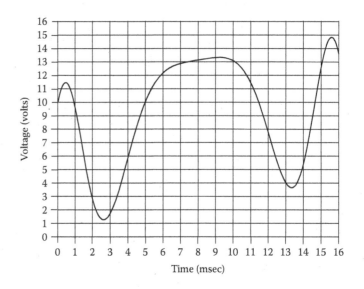

(a) At each millisecond time point (0, 1, 2, etc.) determine the value of the voltage, rounded off to the nearest whole number, 1, 2, 3, etc.

(b) The sampling in part (a) has a sampling rate of 1,000 samples/sec. Repeat part (a), now using a sampling rate of 500 samples/sec. (Again do the first sample at time equal to 0.)

P8.4 If a signal is sampled at too low of a rate, then artificial low-frequency signals can appear in the data. An analog voltage signal is shown below. Make two photocopies of this figure and follow the instructions below:

(a) Draw points on the graph corresponding to sampling this signal with a sampling rate of 2,000 samples/sec. Make the first sample at time equals zero. *Hint:* Recall the time between samples equals 1 divided by the sampling rate.

(b) Draw points on the graph corresponding to sampling this signal with a sampling rate of 666 samples/sec. Make the first sample at time equals zero. Connect the dots indicating the sampled points to make a smooth curve through them. What is the apparent frequency of the smooth curve that you drew? This is the false signal, called an alias (i.e., a false identity).

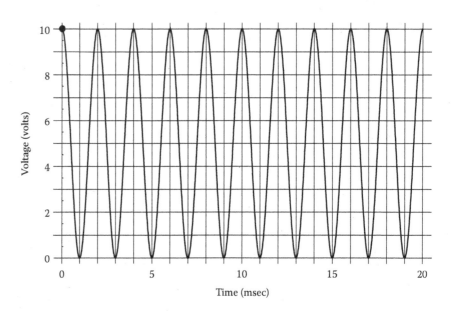

(c) What is the lowest sampling frequency you could use in this case to represent the signal with the correct frequency and avoid aliasing? How does this compare to the actual frequency of the signal? Comment on this result in the context of the sampling theorem.

P8.5 A digital voltage signal is shown below. Notice that some voltages are "high" (i.e., above 5 V) and some are "low" (below 5 V). Let us say that the computer interprets any signal above 5 V (called the threshold) as a binary "1" and any signal below 5 V is considered a binary "0". Assume the sampling rate is 1 bit per millisecond. Take the voltage reading every millisecond starting at 0.5 msec. Determine the binary string; that is, the list of binary digits, associated with this signal.

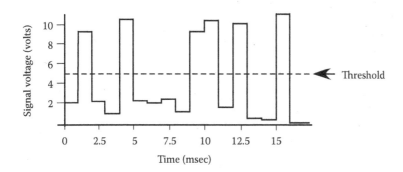

P8.6 You are sending data pulses on a radio channel that has a bandwidth equal to 6 kHz. What is the shortest duration of pulses you could send using this channel? What is the largest number of pulses you could send in a time interval of 10 sec?

P8.7 You are sending binary data pulses on a channel that has a bandwidth equal to 60 MHz. What is the largest data rate that could be achieved with this channel? How many bits could you send in 1 minute using this maximum rate?

P8.8 You are allocated a radio spectrum band between the frequencies 2.0×10^6 Hz (2.0 MHz) and 2.8×10^6 Hz (2.8 MHz) for your use. You place as many radio channels in this interval as you can. Each channel has bandwidth equal to 50 kHz, and the center carrier frequencies are 50 kHz apart, so these channels barely touch adjacent channels in the spectrum. How many channels can you use? What is the maximum data rate of each channel? And what is the maximum total data rate of all channels combined?

P8.9 You are receiving binary data pulses at a rate of 50 kbaud. Is this fast enough to exactly reconstruct the signal in Figure (a) below? In Figure (b) below? Explain carefully using numbers to verify your conclusion.

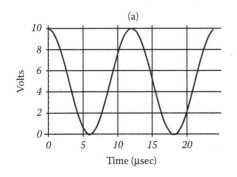
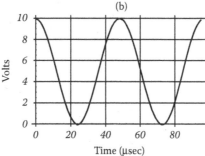

9

Quantum Physics of Atoms and Materials

The first postulate enunciates the existence of stationary states of an atomic system. The second postulate states that the transition of the system from one stationary state to another is ... accompanied by the emission of one quantum of ... radiation.

Niels Bohr
(1913)

Niels Bohr, Danish physicist who in 1913 discovered the quantum model of the atom and the relation of an atom's change in energy to the light emitted or absorbed by it.

Physicists Dawn Meekhof and Steve Jefferts with their atomic clock, which would neither gain nor lose 1 sec in 60 million years! Their clock uses the quantum properties of cesium atoms to provide its extreme stability. (Courtesy of the National Institute of Standards and Technology. Copyright Geoffrey Wheeler, 1999.)

9.1   ATOMS, CRYSTALS, AND COMPUTERS

Modern computers are made with semiconductor-based electronic circuits, which can act as switches, enabling binary data to be stored and logic operations to be performed. Semiconductor-based circuits are made of silicon *crystals* with small amounts of other elements added to control their electrical properties. Engineers invented modern computers using an understanding of how electrons flow in crystals. This required an understanding of the basic properties of atoms and how they combine to form crystals. Gaining a proper understanding of atoms and crystals requires us to learn more about the properties of electrons.

In Chapter 5, we discussed the origins of magnetic forces, which, according to Ampère's law, arise solely from moving electric charges. We learned that magnetic data-recording materials such as iron contain many tiny magnetic regions, called *domains*. If these domains are all aligned and kept in a common direction, the iron becomes a magnet, allowing us to store a data bit value. A question remained, however: Why is each microscopic domain a permanent magnet itself? The answer lies in atomic physics, namely in the motion of electrons within atoms.

Between 1900 and 1930, there was a rapid and incredibly important advance in scientists' understanding of the properties and behavior of electrons. They discovered, through careful analysis of experiments, a set of quantum physics principles describing the behavior of microscopic objects—in particular, electrons in atoms. They found that the physics rules for the behavior of microscopic objects are in some ways radically different from those expounded in the nineteenth century by Newton to explain the behavior of large objects such as baseballs or the Moon. They found that microscopic objects obey different laws of motion than do macroscopic ones. This discovery rather shocked them, so much so that some of the founding fathers of the then-new principles—including Einstein—never completely accepted them as correct. Nevertheless, physicists persisted and confirmed that the then-new quantum principles are indeed correct. From the results of many experiments, combined with clever mathematics, physicists developed a theory that allows us to understand and predict electrons' properties very accurately. Without using mathematics, we can state the main principles in words and illustrate them using pictures. This will allow us to build up a set of rules and guidelines that provide a mental picture of how electrons behave in atoms.

We will learn how to read the Periodic Table of the Elements, which summarizes the structure and properties of atoms—the building blocks of matter. We will see how atoms combine to form crystals, and how the atomic structure of each type of crystal determines its electrical properties, that is, whether it is a good electric-current conductor, insulator, or in between. In the following chapters, we will develop models for the operation of semiconductor devices and see how semiconductor computer logic works.

To start at the beginning, let us consider the surprising properties of electrons, and how their behavior leads to understanding the structure of atoms.

9.2   THE QUANTUM NATURE OF ELECTRONS AND ATOMS

Before we discuss the structure and behavior of atoms, we need to review a few descriptive facts.

- A *proton* is a tiny object having a small mass and positive (+) electric charge.
- A *neutron* is a tiny neutral object (zero electric charge) having approximately the same mass as a proton.

- An *electron* is an even tinier object having negative (–) charge and mass about 1/(2000) that of a proton or neutron.
- A *nucleus* is made of protons and neutrons bound tightly together by so-called nuclear forces, which we will not discuss in this text.
- An *atom* consists of a nucleus and one or more electrons moving around it.
- An *element* is a substance made of a single kind of atom.

As we discussed in Chapter 5, the net charge of an object equals the sum of the charges of all particles making up the object. This means, for example, that the net charge on a nucleus equals the number of protons in that nucleus, because the other particles in the nucleus—the neutrons—have zero charge. Normally, atoms are neutral: they have zero net charge. This means that the number of electrons surrounding the nucleus equals the number of protons in the nucleus.

Many of us have a mental picture of an atom—a small, hard nucleus at the center, with electrons orbiting around the nucleus like wee planets orbiting around a tiny sun. We can think crudely of the electrons as particles moving in an *orbit* around the nucleus, as shown in **Figure 9.1** for the case of a helium atom. This picture of the atom is somewhat naive and is not entirely correct. In fact, no one has a truly satisfactory mental picture of exactly how electrons behave, although we do have a good mathematical theory describing their behavior.

By studying how atoms absorb and emit light of different colors, and how electrons travel through electric and magnetic fields, scientists discovered after 1900 that atoms and electrons do not obey the classical principles of mechanics that were put forth by Newton and described in Chapter 3. Scientists of the time could not understand how the basic "laws of motion," which were so successful in describing the motions of typical large objects, could fail when applied to atoms.

Perhaps, with hindsight, it is not so surprising that electrons don't follow Newton's laws. The objects that we can see directly—those at the human scale—do obey Newton's laws. By *human scale* we mean the scale of baseballs, racing cars, and space shuttles. Newton's theory is extremely accurate for large, slow-moving objects, but fails when the object is on the scale of electrons and atoms; that is, roughly a billion–billion times less massive than a baseball. Newton did not take into account the behavior of such tiny objects when he formulated his laws, because at that time nothing was known about such objects. Any successful theory of atoms must take such behaviors into account.

One of the limitations of the naive Newtonian view of the atom was the faulty assumption that an electron is actually a *particle*, as illustrated by the dots in Figure 9.1. What is meant here by "particle"? A particle is an entity or thing with mass, a definite location in space, and a definite speed. Surprisingly, this description does not apply to electrons. It is not simply that we lack information about where an electron is at a particular moment. Rather, the very concepts of location and speed are not strictly appropriate to electrons. It is as if the electron is spread or smeared throughout some region in space, rather than being at a specific place. This is one of the mysterious properties referred

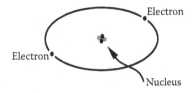

FIGURE 9.1   Naïve picture of a helium atom, showing two electrons orbiting around a nucleus, comprised of two protons (black) and two neutrons (gray). The drawing of the atom is not drawn to scale.

to as the "quantum nature of electrons," which distinguishes them from the classical concept of particles used by Newton and his followers. We can crudely represent the spread-out nature of electrons by drawing a fuzzy region as in **Figure 9.2**.

Although we need to keep this spread-out picture in mind, it is cumbersome to draw it in this way, especially when there are many fuzzy orbits that need to be drawn. So we will use the simpler style of drawing shown in Figure 9.2a to symbolically represent the more accurate picture in Figure 9.2b.

We should wonder what the spread-out picture of an electron really represents. The mathematics of *quantum theory*, which we will not study here, shows that a spread-out electron behaves in some ways like a *wave*. A wave—such as waves in the ocean—is not located at a particular position. A water wave is made of many separate water (H_2O) molecules, moving in an organized pattern. In contrast, the electron wave is associated with only one electron. We believe in the validity of this wavelike description because the mathematics that goes along with it is in excellent agreement with all of the experimental observations on electrons.

This description of an electron might seem strange, but physicists have an interpretation of the meaning of the electron's wave. The wave's amplitude in a region of space tells us the likelihood that the electron will be found in that region. In Figure 9.2b, the darker shaded regions are the places with higher likelihood for the electron to be located. Before we make the measurement, the electron is not at a definite location, but the very act of measuring causes the electron to appear at a definite location.

To further develop the water-wave analogy for the electron in an atom, consider the surface of water in a drinking cup. The wave is confined within the cup. The pattern illustrated in **Figure 9.3** is a circular wave, rotating counterclockwise as time goes on. In the example shown, there are eight wave peaks around the edge of the circular

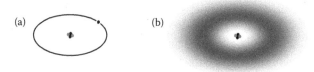

FIGURE 9.2 (a) Naïve classical picture of an electron orbit as a localized particle traveling around a localized path. (b) Quantum picture of an electron orbit as a spread out region in space. The darker the shading, the more likely it is to find the electron at that location.

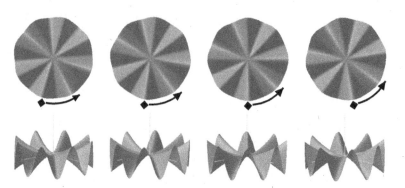

FIGURE 9.3 Frames (left to right) showing a rotating circular wave, in top view and side view. The diamond labels a particular spot on the wave, showing how it rotates in time. In this simple model of an electron's wave, the likelihood that the electron is in some region is highest at the edges of the circular region, where the amplitude is greatest.

pattern. This means that the wavelength along the edge equals one-eighth of the circumference of the circular edge of the pattern.

THINK AGAIN

When you think of a wave, such as a water wave, you usually think of many particles (H_2O molecules) moving in an organized pattern. However, the wave describing an electron corresponds to only a single electron. This is very different from the idea of a wave in classical physics.

An analogy that is simpler to visualize is that of a water wave traveling around a circular canal, as in **Figure 9.4**. In this example, the wave travels around the canal in the clockwise direction and has 16 wavelengths fitting precisely around the circular length of the canal. This leads to constructive interference of the wave when it goes around once and meets up with its "tail." This reinforces and makes a stable wave.

The condition for stability of an electron wave is shown in **Figure 9.5**. For a wave moving in a circular path to be stable, there must be an integer number of wavelengths exactly fitting around the edge circumference. If instead the wavelength equaled, for example, 1/(8.5) of the edge circumference, as shown in the middle of Figure 9.5, the wave would not constructively reinforce itself; rather it would tend to cancel, leading to an unstable wave. This means that only certain discrete wavelengths are allowed for stable circular waves of a given circumference. (*Discrete* means distinct or unconnected.) The figure also shows a wave with 20 wavelengths fitting around a somewhat larger circumference; this is also a stable wave.

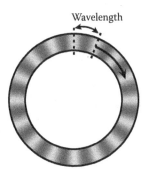

FIGURE 9.4   A water wave traveling around a circular canal. The circumference of the canal must equal an integer number of wavelengths (in this example, 16), otherwise the wave cannot be continuous and stable.

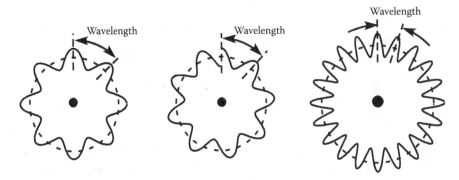

FIGURE 9.5   Constructive interference of electron waves. The circumference of the edge of an orbit must be an integer number of electron wavelengths, otherwise the wave cannot be continuous and stable.

Niels Bohr, in the chapter-opening quote, called these stable conditions stationary states. We refer to the discreteness of wavelengths by saying that the values of the wavelengths are "quantized." This is the origin of the term quantum physics. The quantized nature of a wave's wavelength is a result of its being confined to a small region—in the water case, the region of the cup. In the case of atoms, the electron is confined to the small region around the nucleus.

The electron's wave has a frequency as well as a wavelength. For the simple model in Figure 9.5, the wave's frequency could be observed by sitting at a fixed point on the outer edge and counting the number of oscillations of the passing wave's displacement during a certain time interval. As for any wave, a decreased wavelength means an increased frequency, although the precise relation depends on the type of wave and the shape of the small volume to which it is confined. Because the electron's wavelength in an atom is quantized, its possible frequency values are also quantized.

According to quantum theory, an electron's wavelike motion determines its energy. When an electron is confined to the volume of an atom, this motion is quantized, and therefore the electron's energy is quantized. This means that when an electron is confined to the volume of an atom, its energy can take on only certain discrete values. This behavior is quite unlike a moon orbiting around a planet. Such a moon can have any energy as it flies, depending on how fast it moves; that is, the energy of an orbiting moon is not quantized. The mathematics of quantum theory makes it possible to create accurate pictures of the electron's wave within a hydrogen atom. A few examples of electron waves of different energies are shown in **Figure 9.6**.

The *quantization* of an electron's energy, first proposed by Niels Bohr in 1913, is a remarkable property, totally outside the realm of classical, Newtonian physics. Its discovery led to a revolution in our understanding of the nature of atoms, molecules, and crystals and paved the way for developing computer technology. It was arrived at—not by purely intellectual reasoning—but by thinking hard about how to understand the results of experiments carried out around 1900. Next we review some of those experiments.

THINK AGAIN

When we say that the electron behaves in a discrete manner, we do not mean that its position is discrete (that would be more like a particle than a wave). We mean that the electron's energy is discrete, or quantized.

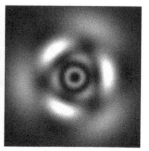

FIGURE 9.6   Realistic computer-generated images of electron waves in a hydrogen atom. (Created using *Atom in a Box,* http://daugerresearch.com/orbitals. With permission of Dauger Research, Inc.)

9.3 THE EXPERIMENTS BEHIND QUANTUM THEORY

How do we know that these claims about electron behavior are true? Three crucial experiments that were carried out around 1900 paved the way for the discovery of the quantum nature of electrons and other particles.

9.3.1 The Spectrum of Light Emitted by a Hot Object

When a piece of any material, such as a **metal** light-bulb filament, is heated to a high temperature, it glows and emits light. When it is not very hot, it emits mostly red and infrared light. The hotter it is, the more yellow and blue is the light. When heated to very high temperatures, it emits light of all colors, so it looks white. Light can be analyzed for the different colors it contains, using a prism or other device to spread out the colors into a **spectrum**. As shown in **Figure 9.7**, light from an incandescent light bulb is passed through a narrow slit in an opaque screen to make a narrow beam of light, and then is spread out by a prism. A smooth, continuous spectrum is observed.

In 1900 Max Planck, a German scientist, found that the prediction for this spectrum based on the theories of Newton and Maxwell did not agree with experiments. It did not correctly predict the relative amounts of light at different colors in the spectrum illustrated in Figure 9.7. Those theories predicted blue light that was far too intense relative to the intensity or brightness of the red light. Resolving this discrepancy led to a revolution in our understanding of the physics of the universe. Max Planck found that he could alter the theory of Newton and Maxwell by making a radical assumption about how light behaves. He hypothesized that light cannot exist in continuous amounts of energy, but rather comes only in indivisible, discrete bundles of **electromagnetic** (EM) energy. We call these energy bundles **photons**, as will be described in more detail in Chapter 12. Planck found that by making this alteration to the old ("classical") theories, he could derive a formula that accurately predicts the relative intensity of each color in the spectrum of light emitted by a hot object.

Planck was on to something, but he did not know precisely what. In fact, he spent the next couple of decades trying to wriggle out of his 1900 hypothesis, thinking that it was simply an accident of the mathematics. That is, he looked hard for an alternative, less radical explanation for the spectrum of colors that did not use the idea of light energy bundles. His hard work failed. Since then, thousands of physicists have looked for convincing alternative theories, but they have also failed. Their failure greatly strengthens

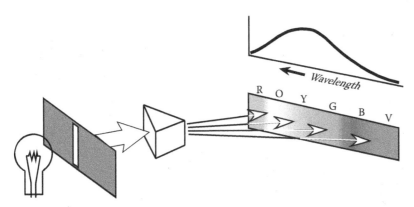

FIGURE 9.7   The spectrum of light from a hot metal filament in an incandescent light bulb is made up of smooth, continuous bands of colors: red, orange, yellow, green, blue, and violet.

physicists' confidence that no other satisfactory explanation exists. It seems that we are stuck with the idea of photons, that is, the idea that the energy in light comes in little indivisible amounts, lumps, or bundles. This means that light cannot simply be described as a wave of EM fields, as we implied in earlier chapters. The classical wave picture of light is not completely wrong, but it did need to be refined with the quantum theory.

9.3.2   Sharp-Line Atomic Lamp Spectra

In contrast to the case of an incandescent light bulb, if we analyze light emitted by an atomic-vapor lamp, such as those in a neon display light or a yellow sodium streetlamp, we observe sharp, discrete lines of color. We discussed atomic-vapor lamps in In-Depth Look 5.2. The experiment showing the discrete lines of color is shown in **Figure 9.8**. Linelike spectra of this type were observed as early as 1885 by Jakob Balmer, a science teacher in a Swiss girls' school. He found a mathematical formula that matched the pattern of the line positions in the spectrum, but he had no explanation for why this pattern arose. This effect was a mystery to scientists at the time. The theories of the time, based on Newton's theory and Maxwell's theory, predicted a smooth, continuous spectrum, such as that seen in Figure 9.7.

In 1913 Niels Bohr proposed that these sharp lines of color were associated with light given off by an atom when it suddenly loses energy, accompanied by a "jump" of the electron from a high-energy orbit to a low-energy one. His ideas are summarized in the quote at the beginning of this chapter. There he talked about the stationary states, or orbits. These are analogous to the stable circular waves illustrated in Figure 9.5. Bohr hypothesized that, if the possible states have discrete energies, then the light given off when an electron jumps from one state to another would have discrete colors. The colored light is given off in bundles called photons, as postulated earlier by Planck. Bohr reasoned that one could learn about the nature of the electron orbits by analyzing the pattern of the colored lines that Balmer had studied earlier.

The principal difference between an atomic-vapor lamp and an incandescent lamp, discussed in the previous section, is that in the vapor lamp all atoms are moving freely as individual particles in the vapor. Their light emission is therefore characteristic of isolated atomic properties. In contrast, in an incandescent lamp, the atoms make up a solid metal, with properties quite different from those of isolated atoms.

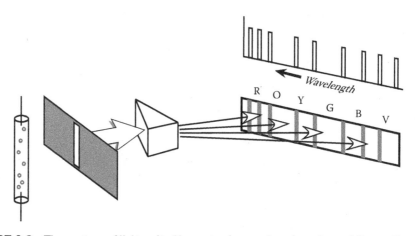

FIGURE 9.8   The spectrum of light emitted by an atomic-vapor lamp is made up of discrete lines of color.

IN-DEPTH LOOK 9.1: SPECTRUM OF HYDROGEN ATOMS

An important part of the scientific process is using mathematical formulas to represent the results of physical observations. By using such formulas, scientists can better recognize regular patterns in a seemingly complex collection of observed data. As a consequence, a mathematical theory can often be devised to summarize the results. In the case of the spectrum of the hydrogen atom, this procedure led to a whole new type of science—quantum physics—along with its remarkable implications for science and technology.

When a glass tube containing hydrogen atoms is used as an atomic vapor lamp, as in Figure 9.8, sharply defined colors are emitted, rather than a continuous spectrum as with a hot-filament lamp. Many of these sharp color lines are illustrated in **Figure 9.9**, along with the value of the wavelength associated with each "color" of EM radiation. Recall that the visible region of the spectrum is roughly 400–800 nm, and only about four of the hydrogen lines fall in this range. The rest are in the infrared or ultraviolet regions.

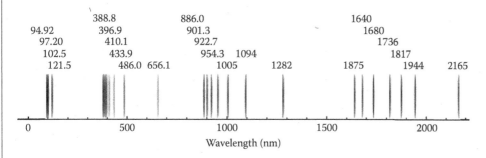

FIGURE 9.9   Sharp lines of color appearing in the spectrum of light emitted by hydrogen.

In 1890, a Swedish physicist, Johannes Rydberg, found that he could represent all of the wavelengths shown in the figure by a single, simple formula:

$$\lambda = \frac{91.12671 \text{ nm}}{\left(\dfrac{1}{m^2} - \dfrac{1}{n^2}\right)}$$

In this formula, 91.12671 nm is called the Rydberg constant. The variables m and n take on integer values (i.e., $m = 1,2,3\ldots$, and $n = 1,2,3\ldots$), but with the restriction that n is always greater than m. The first four wavelengths are calculated as:

$$\lambda = \frac{91.12671 \text{ nm}}{\left(\dfrac{1}{1^2} - \dfrac{1}{5^2}\right)} = 94.92 \text{ nm}, \quad \lambda = \frac{91.12671 \text{ nm}}{\left(\dfrac{1}{1^2} - \dfrac{1}{4^2}\right)} = 97.20 \text{ nm}$$

$$\lambda = \frac{91.12671 \text{ nm}}{\left(\dfrac{1}{1^2} - \dfrac{1}{3^2}\right)} = 102.5 \text{ nm}, \quad \lambda = \frac{91.12671 \text{ nm}}{\left(\dfrac{1}{1^2} - \dfrac{1}{2^2}\right)} = 121.5 \text{ nm}$$

It is remarkable that such a simple formula can be used to calculate precisely the spectrum lines of hydrogen, and this fact provided a powerful clue to the physics "sleuths" of the time about the nature of atoms. Niels Bohr used this clue to create the quantum theory of electron motion in atoms. He and others formulated the quantum principles, described in the following sections, specifically to create a theoretical model that could explain the reasons behind Rydberg's simple formula.

9.3.3 Electron Scattering from Crystals

A third piece of evidence for the wavelike properties of electrons was found in 1927, when two American physicists working at AT&T Bell Telephone Laboratories, Clinton Davisson and Lester Germer, observed an astounding effect in their laboratory. They fired a beam of electrons (similar to the electron beam that we still sometimes use to create images on a television cathode-ray tube) at a crystal of nickel. A crystal is a regular array of atoms, analogous to a regular array of slits. (Recall the discussion about wave interference in Section 7.11.) They found, as expected, that some of the electrons bounced off (scattered) from the nickel atoms. They measured the pattern made by the scattered electrons. If electrons were tiny particles, one would anticipate that the observed pattern would be easily explained by Newton's theory of forces and acceleration. Instead, they found a pattern of scattered electrons that looked completely different from this expectation. The pattern they saw looked like a wave-interference pattern.

In the same year, George P. Thomson in England sent a beam of electrons into a very thin foil of aluminum (Al), and observed interference patterns where the transmitted electrons hit a screen. The circular-shaped interference patterns that Thomson recorded are shown in **Figure 9.10**, which also shows interference patterns produced when x-rays were passed through a similar Al foil. It was well established by then that x-rays

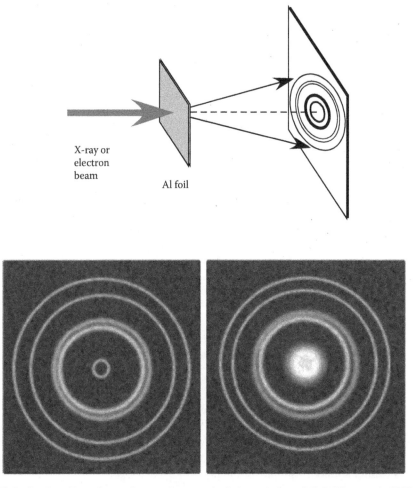

X-ray or
electron
beam

Al foil

FIGURE 9.10 Top: Experimental setup. Bottom: Artist's rendering of G. P. Thomson's 1927 recordings of interference, or diffraction, patterns observed when x-rays (left) or electrons (right) pass through thin aluminum foil.

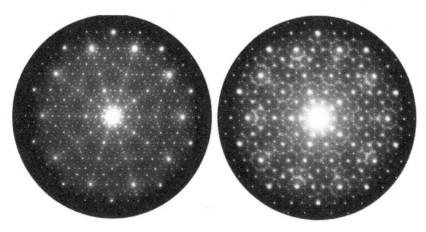

FIGURE 9.11 Electron interference patterns observed when electrons pass through two types of AlCoNi crystals. In both cases 5-fold symmetry of the intensity distribution may be observed. (Courtesy of Conradin Beeli, Swiss Federal Institute of Technology. With permission.)

are EM waves, so it was not surprising to see interference with them. In contrast, it was astonishing to see interference with the electron beam. It seemed that the electrons interacted with the crystal as if they were waves, or as if they were being guided by a wave. The phenomenon of interference that occurs when a wave passes through a material is called diffraction.

The experimenters learned that 3 years earlier, in 1924, a young French physicist, Louis de Broglie, had precisely predicted the behavior of electrons that the Americans and British had observed. De Broglie had hypothesized that electrons simultaneously have properties of both particles (called corpuscles) and waves. As de Broglie said some years later, in his Nobel Prize address, "the existence of corpuscles accompanied by waves has to be assumed in all cases." This hypothesis, which has since been proven true, opened the gates to a mathematical formulation of quantum theory. In 1926, Austrian physicist Erwin Schrödinger developed a mathematical theory that updated the venerable theory of Isaac Newton. With Schrödinger's theory and with Davisson's and Thomson's experiments, the quantum revolution had arrived.

Since it was discovered in 1927, electron diffraction has been an important tool for studying the structure of crystals. The internal structure of the crystal lattice is revealed by diffraction patterns, sometimes strikingly beautiful, as those shown in **Figure 9.11**.

9.4   THE SPINNING OF ELECTRONS

Let us return to our solar-system model of a helium atom with two electrons orbiting around the nucleus-like planets orbiting around a sun, shown in more detail in **Figure 9.12**. As the Earth orbits around the Sun, it also spins around its north-south axis. In analogous fashion, an electron spins around its axis as it travels around the atom's nucleus. Each electron can spin either clockwise or counterclockwise. We cannot literally describe electrons as particles that spin around their axis. Such a picture is the best we can do in describing electrons without using the full mathematical methods.

The spin of electrons plays an important role in determining the structure of the various kinds of atoms. We will discuss how this works after introducing the principles of quantum physics in the next section.

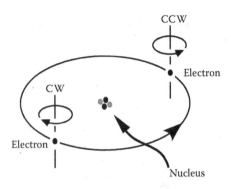

FIGURE 9.12 Naïve model of a helium atom, with two electrons orbiting around a nucleus. The vertical lines indicate the axis of internal rotation or spin for the electrons, which are shown spinning in opposite directions—one clockwise (cw) and the other counterclockwise (ccw). The spin of an electron is roughly analogous to the spin of a basketball.

9.5   THE PRINCIPLES OF QUANTUM PHYSICS

To summarize the above discussions, experiments confirmed that electrons, which were classically thought of as particles, actually have properties of particles *and* waves. In addition, light cannot be thought of simply as a wave (as it was in earlier, classical times), but is comprised of bundles of energy. Clearly, these entities cannot properly be called particles or waves. Some people jokingly call them *wavicles*. For our discussions, we will continue to use the terms particles and waves, keeping in mind that these terms are being used loosely.

To explain the observed behaviors of electrons and photons, scientists found a set of rules that describe objects at the atomic scale. These rules are the principles of quantum physics. They form the basis of quantum theory, which is an extremely successful physical theory. By using the theory, scientists can predict with incredible accuracy the outcome of virtually any experiment that can be done with electrons and atoms. This is despite the fact that the theory is so counterintuitive. The mathematics of quantum theory was developed in the 1920s and 1930s by Niels Bohr in Denmark (opening photo), Erwin Schrödinger in Austria, Wolfgang Pauli and Werner Heisenberg in Germany, Paul Dirac in England, and others. It later underwent refinements by Richard Feynman and Julian Schwinger in the United States, Shin-Ichiro Tomonaga in Japan, and others. All of these received the Nobel Prize in physics for their contributions to quantum theory. Here we will not use the mathematical quantum theory to get a basic understanding of electrons and atoms. We will instead work with a set of principles that can be stated in words. Please realize, however, that it is only through the detailed study of the mathematical predictions and corresponding experiments that we know these principles to be correct.

In all, we will introduce six quantum principles. The first three, stated here, deal with the properties of electrons in atoms. We have already discussed the concepts behind the first two principles.

QUANTUM PHYSICS PRINCIPLES (I–III)

(i)   **The wave nature of electrons:** An electron is a small entity or "thing" that cannot be thought of as a tiny particle, because it cannot be said to have a definite location and a definite speed. Although an electron has mass—a particle-like property— it also has wavelike properties, with the amplitude of the electron wave determining the likelihood of finding the electron in a certain region.

(ii)   **Quantization of energy:** An electron in an atom can have only certain discrete energies; that is, its energy is quantized. The wave patterns associated with the electron's stable motion in the atom are called stationary or stable orbits.

(iii)   **Exclusion Principle:** At most two electrons can occupy the same orbit. Furthermore, for two electrons to occupy the same orbit, they must be spinning in opposite directions around their axes.

The meaning of the third principle—the Exclusion Principle—can be understood by referring to Figure 9.12, which shows each electron spinning around its own vertical axis, in opposite directions. Two electrons in the same orbit simply cannot be spinning in the same direction. That is "forbidden" by the laws of physics, meaning simply that we do not ever observe it. This is different from two moons orbiting a planet, which could be in the same orbit and be spinning in the same direction. An electron's spinning behavior is not easy to grasp intuitively, because an electron is not really a particle, like a tiny basketball. Nevertheless, we draw the spinning electron as if it were a tiny particle to give us some picture to visualize. The Exclusion Principle is known to be correct from examining the structure of the atoms or elements in the Periodic Table, as we shall do in the following section. In fact, it was hypothesized by its discoverer, Wolfgang Pauli, for this very purpose—to explain the nature of the elements using quantum theory.

An important concept is the *state* of an electron. The word *state* means *condition*, or "how something is." The state of an electron is determined by which orbit it is in and in which direction it is spinning. For example, in the lowest-energy orbit, there are two possible electron states—each corresponding to the same orbit but distinct spin directions. Using the language of electron states, we can rephrase the Exclusion Principle as: *No two electrons can have (be in) the same state.*

We can draw pictures to illustrate and explore the consequences of the three principles. Let us use the naïve, easy way of drawing electron orbits, as shown in Figure 9.1. Here we also draw the nucleus as a single dot for simplicity. **Figure 9.13a** shows two electrons moving around a nucleus in the same orbit, with opposite spin directions—clockwise (cw) or counterclockwise (ccw) with respect to a vertical axis of rotation. These two electrons have the same energy—the lowest energy allowed for an electron in an atom. This lowest-energy orbit makes up what is called the first *energy shell*.

Figure 9.13b shows four electrons within an atom. Two are in the lowest-energy shell, and two are in the next-to-lowest shell, with a larger orbit diameter, which we call the second energy shell. Only one of the four orbits in the second shell contains

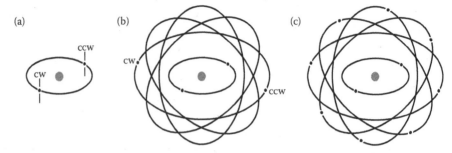

FIGURE 9.13 Electron orbits in an atom. (a) Two electrons in the lowest-energy (first) shell. (b) Four electrons, filling the first shell and partially filling the second shell. (c) Ten electrons, completely filling the first and second shells.

electrons—the other orbits are empty. Although the fourth electron is shown as being in the same orbit occupied by the third electron, this need not be the case. It can go into one of the other unoccupied orbits in the second shell, all of which have about the same energy. This is not an important distinction for our purposes. As always, any two electrons in the same orbit necessarily have opposite spin directions. Figure 9.13c shows ten electrons within an atom. Again, two (the maximum number) are in the lowest-energy shell, and eight are in the second energy shell. The second shell has four distinct orbits, which have equal energies but are distinguished by the details of their orbit shapes, which are represented here simply by different orientations.

Remember that the actual orbit shapes are complicated three-dimensional patterns. Although the drawings are shown as flat, two-dimensional pictures, the actual orbits are three-dimensional. They can be visualized as filling a sphere, as in **Figure 9.14**.

The first three quantum principles, along with some advanced mathematics that we will not discuss, correctly predict how many orbits are within each shell.

- The lowest-energy shell has one orbit (two states). So, according to the Exclusion Principle, this shell can contain at most two electrons (one per state).
- The second-lowest-energy shell has four orbits (eight states). So this shell can contain at most eight electrons.

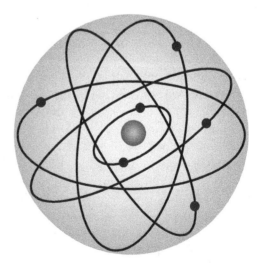

FIGURE 9.14 Three-dimensional drawing of electron orbits in an atom.

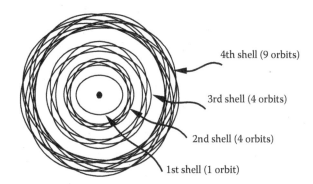

FIGURE 9.15 Naïve representation of all possible electron orbits in the first four energy shells.

- The third-lowest-energy shell also has four orbits (eight states). So this shell can contain at most eight electrons.[1]
- The fourth- and fifth-lowest-energy shells each have nine orbits (eighteen states). So these shells can contain at most eighteen electrons each.

The first four shells are illustrated in **Figure 9.15**. Orbits within the same shell have equal, or nearly equal, energies. Shells with lower energies and smaller orbits are called *inner shells*, and those with higher energy and larger orbits are called *outer shells*.

9.6   BUILDING UP THE ATOMS

One of the early triumphs of quantum theory was its ability to explain the structures of the known atoms and to explain why each element appears where it does in the periodic table of the elements, which is shown in **Table 9.1**. Chemists had previously constructed this table on the basis of how different elements combine in chemical reactions. Each element in the Periodic Table is identified as a different atom (hydrogen, carbon, silicon, gold, etc.). Now we know that, for neutral atoms, the number of electrons and the number of protons is equal. This number is called the ***atomic number***, and is denoted by N. In the Periodic Table, N is given directly above each element's name. The atoms are listed in order of increasing mass per atom, with hydrogen (H) being the lightest. The number below each element's name gives the atom's mass in units called atomic mass units. One atomic mass unit is approximately equal to the mass of one proton.[2] The H atom has one electron and one proton and no neutrons, so its mass (1.008) very nearly equals that of one proton. The carbon atom has six electrons, six protons, and six neutrons, and its mass (12.01) approximately equals that of twelve protons.

The key to determining the structure of atoms is the following two rules:

1.   Electrons will occupy one of the lowest-energy orbits available (i.e., not fully occupied).
2.   If an orbit is already occupied by two electrons (with opposite spin directions), then another cannot join them in that orbit.

The first rule is consistent with common experience that normally objects will move to the lowest energy they can reach. This is a rather general property of nature. An analogy is a ball rolling down a grassy hill to the lowest level. The second rule is again the Exclusion Principle.

QUICK QUESTION 9.1

An atom has 37 electrons. How many protons does it have, assuming the atom is overall neutral?

---

[1] Standard texts use the term *shell* in a different manner than used here, but for our purposes our usage is fine.
[2] More precisely, one proton has mass equal to 1.007 atomic mass units.

TABLE 9.1
Periodic Table of the Elements

Group / Period	1	2	3	4	5	6	7	8	9	10	11	12	13	14	15	16	17	18
1	1 H 1.008																	2 He 4.003
2	3 Li 6.941	4 Be 9.012											5 B 10.81	6 C 12.01	7 N 14.01	8 O 16.00	9 F 19.00	10 Ne 20.18
3	11 Na 22.99	12 Mg 24.31											13 Al 26.98	14 Si 28.09	15 P 30.97	16 S 32.07	17 Cl 35.45	18 Ar 39.95
4	19 K 39.10	20 Ca 40.08	21 Sc 44.96	22 Ti 47.88	23 V 50.94	24 Cr 52.00	25 Mn 54.94	26 Fe 55.85	27 Co 58.47	28 Ni 58.69	29 Cu 63.55	30 Zn 65.39	31 Ga 69.72	32 Ge 72.59	33 As 74.92	34 Se 78.96	35 Br 79.90	36 Kr 83.80
5	37 Rb 85.47	38 Sr 87.62	39 Y 88.91	40 Zr 91.22	41 Nb 92.91	42 Mo 95.94	43 Tc (98)	44 Ru 101.1	45 Rh 102.9	46 Pd 106.4	47 Ag 107.9	48 Cd 112.4	49 In 114.8	50 Sn 118.7	51 Sb 121.8	52 Te 127.6	53 I 126.9	54 Xe 131.3
6	55 Cs 132.9	56 Ba 137.3	57 La 138.9	72 Hf 178.5	73 Ta 180.9	74 W 183.9	75 Re 186.2	76 Os 190.2	77 Ir 192.2	78 Pt 195.1	79 Au 197.0	80 Hg 200.5	81 Tl 204.4	82 Pb 207.2	83 Bi 209.0	84 Po (210)	85 At (210)	86 Rn (222)
7	87 Fr (223)	88 Ra (226)	89 Ac (227)	104 Rf (257)	105 Db (260)	106 Sg (263)	107 Bh (262)	108 Hs (265)	109 Mt (266)	110 — 0	111 — 0	112 — 0		114 — 0		116 — 0		118 — 0

Lanthanide series	58 Ce 140.1	59 Pr 140.9	60 Nd 144.2	61 Pm (147)	62 Sm 150.4	63 Eu 152.0	64 Gd 157.3	65 Tb 158.9	66 Dy 162.5	67 Ho 164.9	68 Er 167.3	69 Tm 168.9	70 Yb 173.0	71 Lu 175.0
Actinide series	90 Th 232.0	91 Pa (231)	92 U (238)	93 Np (237)	94 Pu (242)	95 Am (243)	96 Cm (247)	97 Bk (247)	98 Cf (249)	99 Es (254)	100 Fm (253)	101 Md (256)	102 No (254)	103 Lr (257)

Source: Adapted from Los Alamos National Laboratory's Chemistry Division. In the online version, each element symbol links to a web page with more information. http://pearl1.lanl.gov/periodic/default.htm.

TABLE 9.2
Some of the Low-Atomic-Number Elements

Atomic Number N	Symbol	Name	Number of Electrons in Shells					Common Pure Form
			1	2	3	4	5	
1	H	Hydrogen	1					Gas (molecular)
2	He	Helium	2					Gas (atomic)
3	Li	Lithium	2	1				Metal
4	Be	Beryllium	2	2				Metal
5	B	Boron	2	3				Metal
6	C	Carbon	2	4				Various
7	N	Nitrogen	2	5				Gas (molecular)
8	O	Oxygen	2	6				Gas (molecular)
9	F	Fluorine	2	7				Gas (molecular)
10	Ne	Neon	2	8				Gas (atomic)
11	Na	Sodium	2	8	1			Metal
12	Mg	Magnesium	2	8	2			Metal
13	Al	Aluminum	2	8	3			Metal
14	Si	Silicon	2	8	4			Semiconductor
15	P	Phosphorus	2	8	5			Various
16	S	Sulfur	2	8	6			Crystal
17	Cl	Chlorine	2	8	7			Gas (molecular)
18	Ar	Argon	2	8	8			Gas (atomic)
19	K	Potassium	2	8	8	1		Metal
20	Ca	Calcium	2	8	8	2		Metal
21	Sc	Scandium	2	8	8	3		Metal
22	Ti	Titanium	2	8	8	4		Metal
31	Ga	Gallium	2	8	8	13		Metal
32	Ge	Germanium	2	8	8	14		Semiconductor
33	As	Arsenic	2	8	8	15		Various
34	Se	Selenium	2	8	8	16		Various
35	Br	Bromine	2	8	8	17		Various
36	Kr	Krypton	2	8	8	18		Gas (atomic)

The numbers of states in each of the shells—described in the previous section—give the order of the elements in the Periodic Table. A list of examples is given in **Table 9.2**. The first (lowest-energy) shell corresponds to the hydrogen (H) atom, which has one electron, and helium (He), which has two electrons. These are illustrated in **Figure 9.16**.

Hydrogen and helium appear in the top row of the Periodic Table. The second row contains atoms having the second-lowest shell partially or completely occupied. This row therefore has eight different atoms (elements): lithium (Li), with three electrons, through neon (Ne) with ten electrons. The neon atom has two electrons in the first shell and eight electrons in the second shell, making these shells filled. For the next element—sodium (Na)—the eleventh electron must be in the third shell. By the time we reach argon (Ar) the third shell is filled for a total of $2 + 8 + 8 = 18$ electrons. Notice that the atomic number, N, of argon is 18 in the table. This number tells us how many electrons (and protons) each atom has. The number of neutrons is less important for our purposes—in the lighter elements there are roughly equal numbers of neutrons and protons, whereas in the heavier elements there are more neutrons than protons.

QUICK QUESTION 9.2

An atom has five electrons. Give the name of this atom and draw its orbit picture analogous to those in Figure 9.16.

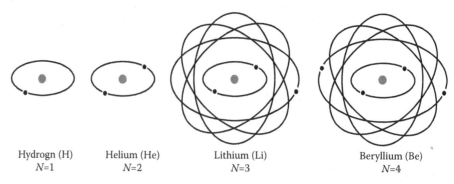

Hydrogn (H)   Helium (He)   Lithium (Li)   Beryllium (Be)
$N=1$ $N=2$ $N=3$ $N=4$

FIGURE 9.16 The atoms hydrogen, helium, lithium, and beryllium have one, two, three, and four electrons, respectively. Each electron added goes into the lowest-energy shell that is not fully occupied.

A type of diagram that we will use extensively in our study of semiconductors is the atomic energy level diagram. This is made by drawing a scale (like on a ruler) in the vertical direction indicating the total (kinetic plus potential) energy of the electron states. The horizontal axis or direction has no meaning, just as it has no meaning on a ruler. As shown in **Figure 9.17**, the lowest-energy shell has two states with nearly equal energies, then there is an *energy gap*, a range of energies in which electrons are forbidden. At higher energy, above the gap, is the second shell with eight states having nearly equal energies, followed by another gap. At still higher energies is the third shell with eight more states having nearly equal energies.

A good way to think of this diagram is that it portrays a staircase, as in **Figure 9.18**. To increase its energy, an electron must "jump" up a full step. An electron is not allowed to be between steps. The concept of energy and the fact of its conservation are key ingredients in the quantum theory. These concepts are among the few that carried over from the classical, Newtonian theory to the quantum setting.

We can use the energy-level diagram to represent the structure of different atoms, as in **Figure 9.19**. Hydrogen has atomic number $N = 1$, meaning it has one electron, shown as the dark dot in the lowest-energy state. Helium has $N = 2$, meaning it has two electrons, shown as two dots, one in each of the two lowest-energy states. Lithium has $N = 3$, and its third electron must have an energy much higher than the lowest two, because there is a large gap in the energy level diagram above the second lowest state. Boron has $N = 5$, and its fourth and fifth electrons have energies only slightly higher

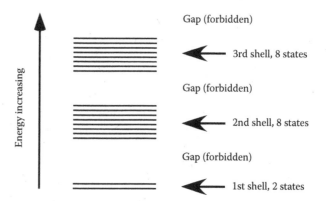

FIGURE 9.17 Energy-level diagram for the total (kinetic plus potential) energies of electrons within atoms. Energy increases in the vertical direction.

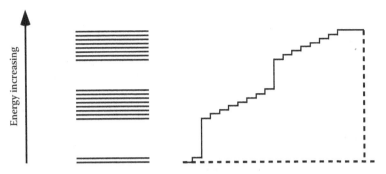

FIGURE 9.18   Energy levels of an atom viewed as a staircase of increasing total energy.

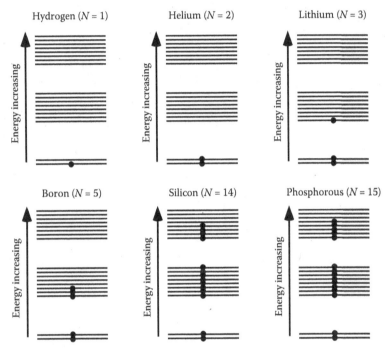

FIGURE 9.19   Energy-level diagrams for some atoms. Although each diagram is drawn with the same spacing between energy levels, there are actually different spacings for different atoms.

than its third electron. Silicon ($N = 14$) and phosphorous ($N = 15$) are important atoms for semiconductors, and their highest-energy electrons are in the third energy shell. A way to think of each diagram in Figure 9.19 is as a high-rise building: each level is a floor of the building, which may be occupied or not, as indicated by a room light viewed from outside.

What are typical values of the energies of the atomic levels? This will be useful later when we discuss lasers and light detectors in the context of optical communication. As an example, in hydrogen, the difference of energies between the lowest shell and the second-lowest shell is about 1.6×10^{-18} joules (J), or 1.6 attojoules (aJ). According to the metric system, 1 aJ equals 10^{-18} J. This is a very small energy, reflecting the fact that atoms are microscopic objects. It seems astonishing that even differences between such small energies can be easily observed (e.g., in the colors of light in the spectrum of a hydrogen lamp).

REAL-WORLD EXAMPLE 9.1: FLUORESCENT LAMPS

The liquid-crystal display (LCD) screen in a laptop computer is illuminated from within the computer. What type of light source is inside, and why is this particular source used? Laptop designers want a light source that is efficient, consumes (or converts) little electrical energy from the battery, and generates as little heat as possible. Two types of light sources could satisfy these requirements—light-emitting diodes (LEDs) and fluorescent lamps. At present, most laptops use fluorescent lamps, because these are cheaper than LEDs to produce. This is a much better choice than incandescent bulbs, which heat up and waste a lot of energy, generating invisible infrared radiation rather than visible light.

A fluorescent lamp is closely related to the atomic-vapor lamps that we discussed earlier in In-Depth Look 5.2. **Figure 9.20** shows an oscillating voltage source, which causes electric current to flow through a gas mixture containing several types of atoms, typically mercury and argon. The electrons making up the current collide with the mercury atoms, exciting the electrons inside these atoms to higher-energy shells. After a brief time, those electrons drop down to a lower energy level, giving off light of various sharply defined colors, primarily ultraviolet (UV). To alter the atomic lamp so that it gives off white light, the manufacturer coats the inner walls of the glass tube with a special substance called a phosphor. This substance is the same as glow-in-the-dark, or fluorescent, paint. The phosphor absorbs the UV light, and re-emits this energy as white light, that is, light containing all visible colors. Fluorescent lamps are typically about 5 times more efficient than incandescent bulbs in converting electrical energy into visible light energy.

How does electric current flow through a gas? Normally the atoms in a gas, such as air, are electrically neutral; that is, each atom has zero net charge and there are no electrons roaming freely around in the gas. In such a situation, a gas is a very good insulator—if a moderate voltage is applied across a region of gas, no current will flow. To make a gas into a conductor, it must be ionized. This means that many of the atoms must be converted into ions by removing one or more electrons from each atom, leaving it positively charged. Ionized gas is called a plasma. The electrons that were removed can flow freely through the plasma if a voltage is applied across it. The positively charged atomic ions can also move—in the direction opposite to that of the electrons—and thereby can also carry current. Finally, any electrons that are newly introduced into the plasma from a negatively charged metal electrode can flow through the gas to another positively charged electrode.

Before a fluorescent lamp is turned on, the gas is neutral, so even a medium-sized voltage would not create the current needed to light the bulb. Therefore, when the lamp is first turned on, a control circuit applies a very high voltage (several hundred volts) for a

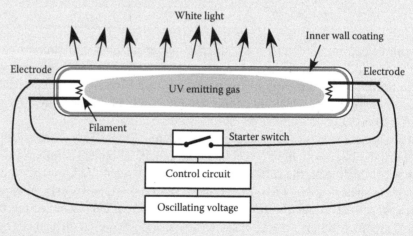

FIGURE 9.20   Fluorescent lamp.

brief instant to the electrodes, ionizing the gas and making a plasma, which then conducts current. The control circuit also momentarily closes a starter switch, which causes current to flow through thin wire filaments, which form the two electrodes. The filaments heat to very high temperature, aiding the emission of electrons from the filaments into the plasma. Once the lamp has been started, the starter switch is opened, because the plasma can now sustain itself without external heating of the filaments. This type of operation is used for room lighting, where fairly high powers are needed.

Most laptop computers use a thin, straight fluorescent lamp to illuminate the LCD screen from behind, in an arrangement called backlighting. The operation of LCDs was discussed in Real-World Example 7.3. To avoid generating too much heat in the laptop, a so-called cold cathode lamp is used. That is, without heating the filament to a high temperature, small lamps can be operated at sufficient power to illuminate an LCD screen.

9.7   ELECTRICAL PROPERTIES OF MATERIALS

The physical properties of atoms are what make the elements useful for different purposes. Iron is a strong metal that rusts easily, gold is a weak metal that resists rust, copper is an excellent electrical conductor, and magnesium is a soft metal that burns easily. The number and arrangement of the outer-shell electrons determine the physical properties of the elements. The inner-shell electrons are tightly bound to the nucleus and can do little.

For our purposes in discussing the basis of computers, the most important property is *electrical conductivity*. Recall from Chapter 5 that electrical conductivity is the ability for electrons to flow through a certain material when pushed by an electric force associated with a voltage. Materials are classified as one of the following:

- *Conductors* (metals)—materials with high conductivity
- *Insulators* (diamond, quartz, glass, some ceramics)—materials with nearly zero conductivity
- *Semiconductors* (silicon)—materials with small conductivity, which is controllable

To understand why high electrical conductivity arises in some materials but not others, we need to understand the nature of the electron energy states in crystals and other solids. The behavior of electrons in crystals is quite different than in isolated atoms, in that each electron moves near several nuclei, rather than near just one nucleus.

A crystal is composed of a regular three-dimensional array of atoms. **Figure 9.21** shows a small region of a lithium ($N = 3$) crystal, with each small circle representing a

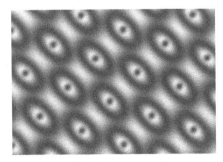 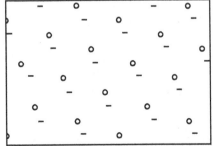

FIGURE 9.21   Crystal structure with nuclei plus inner electrons (circles), and electrons (dashes).

nucleus along with its two inner electrons. Each oval-shaped cloud represents an outer-shell electron in a smeared-out orbit. We are only concerned with the electrons in the outermost shell, because they determine the electrical properties. In the right side of the figure, for simplicity, we represent each nucleus with its two inner electrons by a small circle and each outer-shell electron by a dash, or minus sign.

In Chapter 5, it was stated that electrons move easily between objects, whereas protons do not. The reason for this is that the protons are within the atomic nuclei, and these are bound tightly by the chemical bonding that holds the crystal together. In contrast, the outer-shell electrons are spread out in the crystal, with overlapping orbits, and in some materials can move through the crystal.

When a large number (N_A) of atoms assemble into a crystal, the number of electron states, (i.e., energy levels) is N_A times larger than for a single atom. In a typical crystal 1 centimeter on a side, the number of atoms is about 10^{22}. This is ten-thousand billion-billion atoms. In this crystal, each atom provides one set of states of the type shown above in Figure 9.18. Therefore, the crystal as a whole has 2×10^{22} states that correspond to the first shell. Likewise, the crystal has 8×10^{22} states that correspond to the second shell, etc. For each shell, the number of states is 10^{22} times larger than for a single atom. Imagine trying to draw an energy-level diagram with this many states. The states are so close together, that we can cannot draw a separate line for each; we can only draw continuous "bands" of energy. An *energy band* is a region of very closely spaced energy levels that are allowable for an electron in a crystal. For energies outside of the bands, electrons are forbidden; these regions are the *energy gaps*.

Figure 9.22 shows the energy-level diagram for crystalline lithium. Energy bands are shown as boxes. Because there are far more electrons than we can draw, we use a schematic drawing using connected dots to represent electrons. In this way we can indicate if an energy band is filled or only partially filled. Each dot in this example represents 10^{22} electrons. The figure also shows on the right a cartoon picture of the energy bands, indicating that the energy within a band increases smoothly and continuously, but in the gaps no energy levels exist. An electron simply cannot have energy in one of the gaps. As before, there is no meaning to the horizontal axis.

THINK AGAIN

In Figure 9.22, which of the Quantum Principles explains that the electron in the second shell cannot lose energy and fall into the first shell?

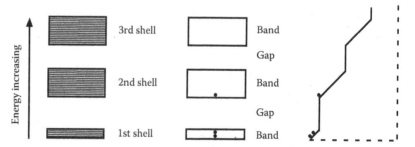

FIGURE 9.22 Energy-level diagram of crystalline lithium ($N = 3$). Each dot represents a large number (for example, 10^{22}) of electrons. The lowest-energy band is filled (completely occupied) with electrons, and the second band is partially filled, making solid lithium a metal. The right side illustrates that the energy of the electron states is continuously increasing, like a ramp as opposed to a staircase.

9.7.1 Conductors

Crystalline lithium is a metal. Recall that this means it has high electrical conductivity, which means that electrons can easily flow through it when pushed by a voltage. It is a conductor. Why is lithium, with three electrons per atom, a conductor? It is because the second lowest band is partially empty. Crystals that have partially empty energy bands are metals, and have high electrical conductivity, as we now explain.

Consider attaching two wires to either end of a solid piece of lithium and attaching the wires to a battery, as shown in **Figure 9.23**. Electrons will flow from one side of the crystal to the other because lithium is highly conducting. Under the influence of the battery's voltage, electrons feel a force, and they accelerate; their speed increases and they move through the crystal to the other side. To do this, each electron must gain some energy when it accelerates. Recall that an electron at rest has zero kinetic energy, whereas a moving electron has nonzero kinetic energy. Notice that the crystal remains neutral at all times, because for each electron that enters it at the negative side, one departs from the positive side. There is no buildup of extra electrons.

In terms of the energy-level diagram, the acceleration is visualized as follows. Each electron in the partially empty band can gradually gain some energy, upon which it gradually goes "up" the vertical energy axis in the diagram, as indicated by the arrow in **Figure 9.24**. In contrast, the voltage cannot accelerate electrons that are in the lower, filled band. This is because the small voltages used (up to about 100 V) cannot impart enough energy for them to "jump the gap" and go into the second shell. The point is

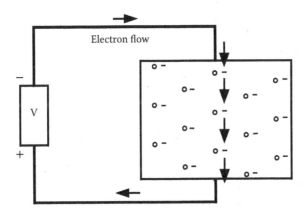

FIGURE 9.23   Electrons accelerating and moving through a lithium crystal.

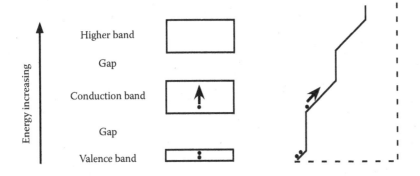

FIGURE 9.24   Outer-shell electrons in a metal accelerating under the influence of a small voltage. Inner-shell electrons (those in the lower, filled band) cannot accelerate.

that electrons cannot accelerate gradually through the forbidden gap energies. They cannot gradually gain energy and are therefore fixed with the energy they originally had. Also, electrons cannot accelerate within the lowest band. The Exclusion Principle prevents this, because if acceleration were to occur, there would be more than one electron in some of the states in the lower band. Electrons also repel one another by the electric force, but this is not the fundamental reason that no two electrons can be in the same state. Rather, it is a fundamental property of electrons. The Exclusion Principle "blocks" electrons from accelerating in a completely filled band. Physicists use the term *valence band* for the highest-energy band that is completely filled, and *conduction band* for the band just above the valence band. In the case of a metal, the conduction band is partially filled. Electrons in the conduction band are free to move, whereas electrons in the valence band are not.

THINK AGAIN

When explaining why a metal is a conductor, we should not say it is because there is a lot of unoccupied space for the electrons to flow; it is because there are many unoccupied energy levels into which the electrons can be accelerated.

THINK AGAIN

When current flows in a conductor, there are no extra electrons in the metal. If there were, the metal would have negative net charge, which it does not. Rather, electrons that are already in the metal before the voltage is connected are made to move through and out of the crystal by the voltage while being replenished by new electrons from the battery.

It is important to realize that we are looking at two different types of diagrams. Figure 9.21 is a drawing of the actual crystal in space, like a photograph taken through a very powerful microscope. An example of such an image of a real silicon crystal is shown in **Figure 9.25**. On the other hand, the energy-band diagrams, such as Figure 9.24, are not pictures in space. They are graphs showing the energies of electrons. Recall the comparison to a high-rise building, shown here in Figure 9.25. In these diagrams, the vertical location of each dot tells us the energy of each electron in the atom, just as the vertical position of a lighted window in the building could tell us what floor level a person is on.

9.7.2 Insulators

Quartz and glass are insulators, meaning they have very low electrical conductivity. This arises from the way the energy bands are filled, or not filled, in these materials. Insulators have energy-level diagrams as shown in **Figure 9.26**. The two lower-energy bands are completely occupied with electrons. There is a large gap between the highest filled band and the next empty band. By large, we mean approximately 0.5 aJ or larger. When a voltage (not too great) is applied across an insulator, electrons cannot accelerate, because there are no empty energy levels immediately above the highest occupied energy level. Therefore no electrons can move or flow. Again, the Exclusion Principle "blocks" the electrons from accelerating in a completely filled band. Note that if too

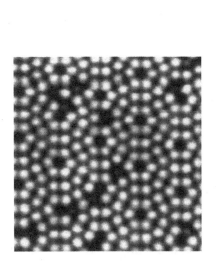

FIGURE 9.25   A laboratory image of the surface of a silicon crystal, revealing atoms forming a regular pattern, along with some "defects" where atoms are missing. This image is taken, not with a regular microscope, but by slowly dragging the tip of a tiny needle across the crystal's surface and recording the force felt at each location. Lights in the windows of a high-rise building indicate which floors are occupied. (Silicon image by Juergen Koeble. Courtesy of Omicron NanoTechnology, Germany.)

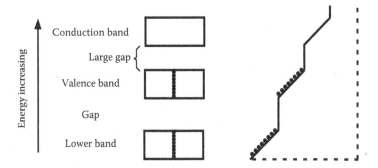

FIGURE 9.26   Every energy band of an electrical insulator is either all filled or all empty. Electrons cannot accelerate when voltage is applied.

large a voltage (e.g., greater than 10^6 V) is put across an insulator, the crystal structure can be damaged, allowing current to flow. This "breakdown" behavior is normally not useful for operating electronics.

Diamond and quartz are crystals. Although glass is a noncrystalline solid, its energy diagram looks similar to that shown in the figure, explaining why it also is an insulator.

9.7.3   Semiconductors

The most useful semiconductor is the silicon crystal. **Figure 9.27** shows the structure of a silicon (Si) crystal. Each circle represents a Si atom nucleus plus all inner-shell electrons. As we can see by referring to Figure 9.19, the Si atom has four outer-shell electrons. In Figure 9.27, these electrons are shown as four minus signs, pointing away from each Si nucleus. Because the atoms are close together, the orbits on different atoms overlap, and there are eight electrons close to each atom.

A Si crystal is a semiconductor, meaning it has low, but not zero, electrical conductivity. A semiconductor has small conductivity because it has a small energy gap just

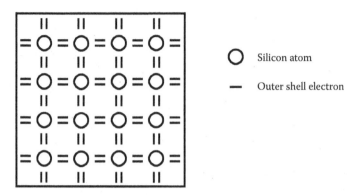

FIGURE 9.27 A silicon crystal is made of a regular array of Si atoms, each with four outer-shell electrons. Each outer-shell electron is indicated by a dash (minus sign).

above the highest filled band, as shown in **Figure 9.28**. The values of semiconductor band gaps fall in the range of 0.01–0.5 aJ. By looking at the energy-level diagram for a single Si atom in Figure 9.19, one would not expect that such a gap would be present in the Si crystal, but experiments show that there is a small gap, nevertheless. Although the gap is correctly predicted by quantum theory, for our purposes the existence of the small gap is more important than the cause of the gap. In-Depth Look 9.2 explains the physical origin of the gap.

IN-DEPTH LOOK 9.2: ORIGIN OF THE ENERGY GAP IN SILICON CRYSTALS

The small energy gap in Si crystals is explained as follows. As shown in Figure 9.27, in a crystal the atoms are packed closely together, so the orbits on different atoms overlap. This means that there are eight electrons close to each atom. Therefore, from the point of view of each Si atom, it "feels" as though its third energy shell is filled with the maximum eight electrons. If any extra electrons are put into the crystal, they have a hard time finding a "place" to go, because it seems as though all orbits are filled. For this reason, it takes a somewhat higher energy to place an additional electron into the Si crystal than would be required to place the same electron into an isolated Si atom. Therefore, if a small voltage is applied across the crystal, electrons are not able to move freely and there is no conduction (at low temperature). This fact is reflected by the existence of a small energy gap, as shown in Figure 9.28. The overlapping of electron orbits, which gives rise to the small energy gap, also causes chemical bonding, which holds the Si crystal tightly together.

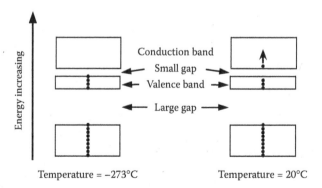

FIGURE 9.28 At absolute zero temperature (–273°C), the energy bands of a semiconductor are either completely filled or completely empty. Electrons cannot accelerate when a voltage is applied. At room temperature (20°C), the thermal energy in the crystal raises some electrons to the band above the small gap, making the crystal a conductor.

When a voltage is put across a semiconductor at room temperature, a small current flows. Why? It might appear that no electrons could be accelerated, because of the presence of the gap. This would be true at extremely low temperatures near absolute zero (–273°C, or 0 K). No electron would have enough energy to jump the gap. But, because the gap is small, if the crystal is at room temperature there are a few electrons that are not in the lowest possible energy levels. This is because the thermal-energy in the crystal gives some of the electrons enough energy to jump the gap and go into the previously unoccupied upper band. It is interesting that thermal energy can cause an electron to jump the gap and go to a higher energy band, but voltage from an applied battery cannot do this. We will clarify this in Chapter 12. (It turns out that thermal energy can also produce infrared light, and it is this that actually causes the electrons to jump the gap.)

With this insight, we can update the rule given earlier in the chapter as:

State occupation rule: At very low temperature, an electron goes into the lowest unoccupied state available. At higher temperatures, such as room temperature, a relatively small number of electrons can gain enough thermal energy to go into a higher energy state, or band.

The important difference between an insulator and a semiconductor is the size of the gap separating the highest-energy filled band (valence band) and the next higher band (conduction band). Because this gap is quite small in a semiconductor, there is a small temperature-induced current. The smallness of the gap also gives us the opportunity to modify the properties of semiconductors by adding small amounts of other elements into the crystal. This is a crucial technique for making electronic circuits with semiconductors, and is the topic of the next chapter.

THINK AGAIN

When explaining why a warm semiconductor can conduct electrical current, we should not say it is because there is a lot of unoccupied "space" for the electrons to flow; it is because there are many unoccupied *energy levels*, into which the electrons can be accelerated.

THINK AGAIN

Keep in mind that some crystals are conductors (e.g., gold and copper), whereas other crystals are insulators (diamond, quartz). In addition, some noncrystalline solids are insulators (glass), whereas some other noncrystalline solids are conductors (some specialized plastics).

IN-DEPTH LOOK 9.3: ATOMIC NATURE OF MAGNETIC DOMAINS

In Chapter 5 we discussed magnetic materials. In a magnetized piece of iron are small magnetic domains, which act like tiny permanent magnets. If you take any piece of non-magnetized iron and grind it into small microscopic grains, you will find that each grain is a tiny magnet. They will be attracted to any nonmagnetic piece of iron by the induced-magnetization effect described in In-Depth Look 5.4.

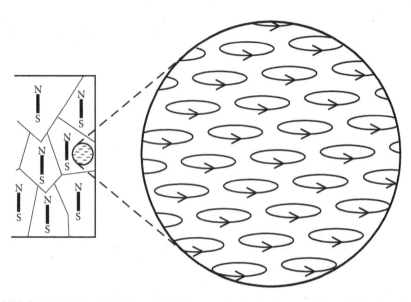

FIGURE 9.29   Magnified view of the interior of a single domain of iron. Each atom is represented as a tiny current loop.

Consider a domain made up of a million atoms. You might wonder, "Why is this tiny piece of metal a magnet?" If you keep grinding this tiny speck of metal smaller and smaller, so you end up with individual atoms, you will find that each iron atom is by itself a magnet! This is not true of, say, aluminum atoms; there is something special about iron. Each iron atom can be viewed as containing an atom-sized current loop. **Figure 9.29** shows an imagined blow-up of the interior of a domain of iron. Each atom is represented as a tiny current loop. This is a reasonable picture, because atoms are made of electrons orbiting around a central point. The electrons carry an electric current around a loop-shaped path. According to Ampère's law, electric current moving around a loop creates a magnetic field. Because all of the atomic current loops are aligned in same direction, their summation creates a stronger magnetic field than would otherwise be the case.

SUMMARY AND LOOK FORWARD

Atoms are nature's building blocks, and their internal structure determines the properties of conductors, insulators, and semiconductors—the materials in computer circuits. Surprisingly, when physicists discovered the physics of atoms and electrons in the early twentieth century, they found that atom-sized objects obey different physical laws than do large objects. They found that electrons cannot correctly be thought of as tiny particles, because they have wave-like as well as particle-like properties. An electron has no definite location. It is more meaningful to talk about the likelihood of finding the electron in various selected regions than to talk about its actual position. As a consequence of its wave-like properties, an electron in an atom must have certain discrete energies, described by groups of orbits organized into energy shells. The energy is quantized.

Each element has a distinct number of electrons—its atomic number, N. The Exclusion Principle says that at most two electrons can occupy any orbit, and if two electrons share an orbit, they must have opposite directions of spin. Each electron energy shell has a maximum number of electrons that it can house (2, 8, 8, 18, 18, etc.).

The electrons in the outer shells of each atom in crystals and other solids are close to neighboring atoms. The number of electrons in the highest-energy occupied shell determines the physical properties of solids. Electrical conductivity is the ability for electrons to flow through a material when pushed by an electric force associated with a voltage. Materials are classified as one of the following: insulators, semiconductors, or conductors, depending on how well they conduct. Using the quantum principles, we can predict which crystalline elements are conducting or insulating.

- Metals are crystals that have their highest-energy occupied band partially empty. This makes them electrical conductors, since the electrons can be accelerated by an externally applied voltage.
- Insulators are crystals that have all of their occupied shells completely filled, with a large energy gap between the highest filled shell and the next lowest shell. They do not conduct electrical current, because the Exclusion Principle prevents electrons from accelerating into already-occupied energy levels.
- Semiconductors are crystals that have all of the occupied shells completely filled, with a small energy gap between the highest filled shell and the next higher shell. At low temperature semiconductors do not conduct electrical current, because the Exclusion Principle prevents electrons from accelerating into already-occupied energy levels. At room temperature they can conduct, because some electrons are boosted into higher, non-filled shells by thermal energy in the crystal.

Where is this discussion leading in our quest to understand computers and the Internet? We know vastly more about electrons and atoms than was understood only four generations ago. In parallel with this intense scientific effort, engineers have made astounding progress in applying the discovered basic knowledge to building electronic circuitry that is small, fast, and energy efficient. Our next step will be to study how the atomic nature of crystals is used to build electronic switches and other devices that make up computer logic circuits and electronic memory.

SOCIAL IMPACTS: SCIENCE, MYSTICISM, AND PSEUDO-SCIENCE

We are to admit no more causes of natural things than such as are both true and sufficient to explain their appearances.

Isaac Newton
(1687)

Man seeks to form for himself, in whatever manner is suitable for him, a simplified and lucid image of the world, and so to overcome the world of experience.

Albert Einstein
(1937)

What evidence would it take to prove your beliefs wrong?

Steven Dutch
(2007)

Some years ago, a man came to my office at the university and announced that he had invented a machine that produced energy using no fuel or other external source of energy. He said it involved "chaos" and extracted energy from the rotations of magnets in some mechanical contraption, and he wanted to discuss the physics behind it. I politely said that if it were real, then I would like to see it and try to confirm its operation. I was not much worried that he would take up my offer; he replied that it was in his barn in another town, and in any case, he needed to keep the details secret. I said in that case, "I can't help you." He left. I believe that this man was practicing pseudo-science. He used scientific terms such as energy, magnetism, and chaos and said he had done experiments that supported his claims. Maybe he had, maybe he had not. Without independent testing, there is no way to know.

Pseudo-science means *fake science*. It is defined as any kind of intellectual or experimental exploration that is claimed to be scientific but actually does not follow the widely agreed upon methods of science. In Chapter 1, Section 1.4 we discussed the scientific method, which is the process by which repeated observations of physical events allow people to discover regular rules, laws, or principles about such events. The scientific method necessarily involves making quantitative measurements to represent the results of the observations being reported. A "good" scientific theory is a theory about physical events that has been well tested and is seen not to make false predictions.

To be believable, the results of observations must be repeatable. Any person following the procedure described should be able to observe the same or similar results. Therefore, if a person refuses to tell anybody else what procedure he or she used, then the results probably should not be called scientific. This applies to my office visitor.

An additional requirement is that for any theory to be scientific it must be testable. As well put by geologist Steven Dutch, "Refutability is one of the classic determinants of whether a theory can be called scientific." [1] That is, if there is no possible test or evidence that would convince a believer that his or her favorite hypothesis is wrong, then that idea can not be called scientific.

It is difficult to mark precisely the boundary between science and pseudo-science. This is called the demarcation problem in the philosophy of science. For example, it seems possible to apply the scientific method to the question of unidentified "flying" objects, or UFOs. There are certainly some researchers (mostly amateur) who do apply the scientific method to this question. But, it is apparent that no definitive conclusions that could be called laws or principles have emerged to help us understand why so many people report unidentified objects and to what these observations correspond. Pseudo-science enters at this point—in the absence of any confirmed scientific results—and exploits the gullibility of some people who cannot distinguish, or do not want to distinguish, between real science and pseudo-science.

Peoples' gullibility perhaps comes from their need, which all people have, as Einstein says above, to "form for himself, in whatever manner is suitable for him, a simplified and lucid image of the world, and so to overcome the world of experience." That is, to be human is to be curious and to try to understand. The need to understand can lead to self-delusion. A goal of science is to help people avoid gullibility and self-delusion. (This discussion has nothing to do with whether one has religious or spiritual beliefs, which are independent of science, according to many careful thinkers. If, on the other hand, people try to invoke scientific methods to address fundamental religious questions, then most likely it turns into pseudo-science.)

Although there may be no sharp line between science and pseudo-science, various thinkers have drawn up lists of "warning signs," which indicate a high likelihood that pseudo-science is being presented. Most pseudo-science claims do not suffer from all or even most of these, but when you see more than one cropping up, beware.

Ten Warning Signs for Detecting Pseudo-Science

1. Pseudo-science usually begins with an idea based on wishful thinking, and then claims to have found evidence supporting it. (Would it not be great if there were a source of unlimited, free energy?)
2. Pseudo-scientists usually ignore all evidence that might refute their claims.
3. Pseudo-scientists do not believe they need to give examples of testable predictions that logically follow from their claims.
4. Pseudo-scientists do not carry out careful, repeatable experiments and share the results in a way that would enable others to repeat them as a further test.
5. Pseudo-science is often based on a claim that certain tenets of established science are wrong, not realizing that if the claim being made were true, it would undermine hundreds of other well established laws and tested theories. Established science is an interconnected web of knowledge, and changing major pillars of it is not so easy.
6. Pseudo-scientists often work alone or in isolation, with no one to check their work.
7. Pseudo-scientists often claim that some powerful organization, such as the government or the auto industry, is trying to suppress or unfairly criticize their work.
8. Pseudo-scientists often use technical-sounding words like energy or vibration, without clearly defining what they mean.
9. Pseudo-scientists often cite alleged "experts," who turn out either to be pseudo-scientists themselves or legitimate scientists with no expertise in the field that is most relevant.
10. Pseudo-scientists often use an argument that appeals to the "unknown," rather than to known scientific laws and principles. They might argue that because no one has explained some phenomenon (such as the origin of life on Earth), their harebrained ideas may be valid.

Of course, scientists can be wrong, too. In fact, admitting that one is wrong when evidence points in that direction is a hallmark of a good scientist—but not of a pseudo-scientist.

Unfortunately, the media often report stories about pseudo-science as if it were real science. This should be intolerable, but, "Hey! —It sells advertising space." Rather than laughing it off, educated readers would do us a favor by writing rebuttals to nonsense when it appears in legitimate media.

An example of a broad area of pseudo-science is the purported connection between quantum physics and spirituality or mysticism. As a subject of "pop-philosophical" speculation, "quantum spirituality" can be a lot of fun, but at its worst, it exploits an appeal to the unknown, rather than to known scientific laws and principles (see warning 10 above). Claims made that spirituality and quantum physics are somehow connected miseducates the public about the fascinating subject of quantum physics. The following quote sets the stage for some of the present confusion:

The discovery of quantum mechanics is one of the greatest achievements of the human race, but it is also one of the most difficult for the human mind to grasp … It violates our intuition—or rather, our intuition has been built up in a way that ignores quantum-mechanical behavior.

Murray Gell-Mann
(1994)

Gell-Mann is a Nobel-winning physicist and yet even he finds some of quantum physics hard to grasp and in a sense mysterious. Pseudo-scientists will point out that consciousness and spirituality are also hard to grasp and mysterious. Therefore, they will argue, "Ah-hah! These topics must be talking about the same thing!" This, of course, is a fallacy. Two mysteries need not be related. In particular, no repeatable, verifiable experiments have been conducted that convince the scientific mainstream that there is any connection in this case. Although I feel that it would be nice if there were a connection, I must recognize that science is plodding, and it slowly arrives at deep and true insights. One cannot take shortcuts via wishful thinking. This is well explained in an article in *What is Enlightenment* magazine [2].

It seems that the majority of quantum physicists see no need for the injection of human consciousness into the mathematical formalisms that form the basis of their science. As Ken Wilber pointed out 20 years ago, even the founding fathers of quantum physics/mechanics—Max Planck, Niels Bohr, Werner Heisenberg, Erwin Schrodinger, Sir Arthur Eddington, etc. (who were all self-proclaimed mystics) strongly rejected the notion that mysticism and physics were describing the same realm. The attempt to unify them is, in the words of Planck, "founded on a misunderstanding, or, more precisely, on a confusion of the images of religion with scientific statements."

By the way, the man who came to my university office 18 years ago never returned, and the world never did hear of his self-proclaimed revolutionary energy-creating "chaos" device.

REFERENCES

1. S. Dutch, at the University of Wisconsin-Green Bay, http://www.uwgb.edu/DutchS/index.html. He also writes, "There is no single Scientific Method. Rather, I believe we must think of a battery of methods that have proven useful. Testing of scientific ideas can include the classical experimental method, replication, attempted refutation, prediction, modeling, inference, deduction, induction and logical analysis."
2. Huston, Tom. "Taking the Quantum Leap... Too Far? Not Just a Movie Review of What the Bleep Do We Know!?" *What is Enlightenment*, October–December, 2004.

SOURCES

The following sources discuss how to distinguish pseudo-science from science:

Coker, Rory. "Distinguishing Science and Pseudoscience," *Quackwatch* (2001). http://www.quackwatch.org/01QuackeryRelatedTopics/pseudo.html.

Gardner, Martin. *Did Adam and Eve Have Navels?: Discourses on Reflexology, Numerology, Urine Therapy, and Other Dubious Subjects.* New York: W.W. Norton & Company, 2000.

Hines, Terence. *Pseudoscience and the Paranormal.* Amherst, MA: Prometheus, 1988.

Park, Robert L. *The Seven Warning Signs of Bogus Science* (January 31, 2003). *The Chronicle Review*, 49, no. 21, B20. http://chronicle.com/free/v49/i21/21b02001.htm.

Sagan, Carl. *The Demon-Haunted World—Science as a Candle in the Dark.* New York: Random House, 1995.

Schick, Theodore, Jr., Lewis Vaughn. *How to Think about Weird Things: Critical Thinking for a New Age.* Mountain View, CA: Mayfield, 1995.

Shermer, Michael. *The Borderlands of Science: Where Sense Meets Nonsense.* Oxford, UK: Oxford University Press, 2001.

Simanek, Donald E. "What is Science? What is Pseudoscience?" (2005). http://www.lhup.edu/~dsimanek/pseudo/scipseud.htm.

Wilber, Ken. *Quantum Questions.* Boston: Shambhala, 2001.

The image in Figure 9.11 appeared in Ritsch, S., et al., *Philosophical Magazine Letters,* 78, 67–75, 1998. With permission of Taylor & Francis.

SUGGESTED READING

Popular-level discussions of the experiments carried out during 1900–1927 that led to our present understanding of quantum theory are given in:

Asimov, Isaac. *The History of Physics*. New York: Walker, 1984.
Gribbon, John. *In Search of Schrödinger's Cat*. New York: Bantam, 1984.
Motz, L., and J. H. Weaver. *The Story of Physics*. New York: Avon Books, 1989.

KEY TERMS

Atom
Atomic number
Conduction band
Conductor
Crystal
Domain
Electrical conductivity
Electromagnetic
Electron
Element
Energy band
Energy gap
Energy shell
Exclusion Principle
Glass
Insulator
Metal
Neutron
Nucleus
Orbit
Particle
Photon
Proton
Quantization
Quantum theory
Semiconductor
Spectrum
State
Valence band
Wave

ANSWERS TO QUICK QUESTIONS

Q9.1 A neutral atom with 37 protons also has 37 electrons.

Q9.2 The boron atom has $N = 5$. The fifth electron goes into one of the unoccupied orbits having the lowest energy in the second shell.

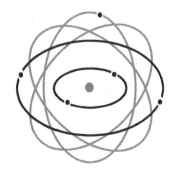

Boron (B)
$N=5$

EXERCISES AND PROBLEMS

Exercises

E9.1 Explain the various ways in which real atoms are different from a model analogous to the solar system (planets orbiting the Sun). What behaviors of electrons are similar to and/or different from human-scale objects such as baseballs or planets?

E9.2 Explain the concept of a particle in physics. Give examples of physical entities that can be considered particles and examples of physical entities that cannot.

E9.3 An element has 33 electrons. How many protons does it have, assuming the atom is overall neutral? What element is this? What is the mass of one atom of this element? What can you say about the number of neutrons the atom has?

E9.4 If extra electrons are added to or removed from a neutral atom, it is then called an ion. If two of the electrons are removed from the atom in E9.3, what will be the net charge of the remaining ion?

E9.5 If two ions of the type described in E9.4 come near each other, will there be a repulsive or an attractive force between them?

E9.6 What is wrong in this picture of an atom?

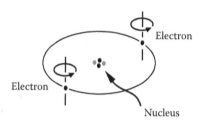

E9.7 Go online to http://nobelprize.org and transcribe the citation for Niels Bohr's Nobel Prize. From his biography there, to what three countries did he flee during the Nazi occupation of his home country, Denmark?

E9.8 A basketball is rotating (spinning) at 3 revolutions per second around the axis shown as a black line. In 5 sec, how many black seam lines pass a fixed point indicated by the pointing finger? What is the frequency of seam lines per second passing the pointing finger? Discuss a possible analogy between the basketball and the electron wave in an atom.

E9.9 Think of three everyday examples of physical quantities (not mentioned in the text) that are quantized and three quantities that are continuous.

E9.10 Think of an amusing analogy to illustrate the Exclusion Principle that could be used for teaching the concept to middle school students.

E9.11 (a) Draw the occupied orbits (as circles or ellipses) for the oxygen atom, showing the distinctions between shell radii and the direction of spin of each electron.
(b) Draw the occupied orbits for the scandium (Sc) atom.

E9.12 (a) Draw the energy-level diagrams for the following elements: carbon (C), nitrogen (N), oxygen (O), sodium (Na), and aluminum (Al).
(b) What prevents the highest-energy electron in a sodium atom from being in the second energy shell?

E9.13 There exists a minimum energy level in an atom, below which an electron's energy cannot be. Explain why this fact is incompatible with the classical theory based on Newton's and Maxwell's theories. *Hint*: An oscillating electric charge always emits or radiates EM waves. What would this energy emission do to the orbit of an electron according to classical concepts?

E9.14 How is the concept of an electron state related to the concepts of orbit and of spin?

E9.15 (a) Use and cite some of the quantum principles to explain in some detail why solid beryllium (Be) is a metal and why it conducts current when a small voltage is applied. Draw sketches of the energy bands and explain where (in energy, not location) the electrons are in these bands and the effect of an applied voltage.
(b) Likewise, use some of the quantum principles, plus a drawing of the bands, to explain in some detail why solid Argon (Ar) is an insulator.

E9.16 The following solids have band gap values given in parentheses. Determine whether each solid is a conductor, a semiconductor, or an insulator.

(a) SiO_2 (1.44 aJ) [glass]
(b) C (0.86 aJ) [diamond]
(c) Si_3N_4 (0.80 aJ) [silicon nitride]
(d) GaAs (0.23 aJ) [gallium arsenide]
(e) Si (0.18 aJ) [silicon]
(f) InN (0.11 aJ) [indium nitride]
(g) Ge (0.10 aJ) [germanium]
(h) Al (0.0 aJ) [aluminum]
(i) Cu (0.0 aJ) [copper]

Problems

P9.1 Jakob Balmer's observations of the spectrum of light emitted by a hot hydrogen gas led him to a practical formula for calculating the frequencies of light corresponding to the spectral lines observed.

$$f_{Balmer} = (3.28992 \times 10^{15} \text{ Hz}) \times \left(\frac{1}{4} - \frac{1}{n^2} \right)$$

The symbol n can equal any integer greater than or equal to 3; that is, $n = 3,4,5,...$

Use Balmer's formula to calculate the frequency and the wavelength (in vacuum) of each of the spectral lines corresponding to $n = 3,4,5$. What part of the visible spectrum (i.e., what color) does each line fall into? Check your results against the spectrum of hydrogen shown in In-Depth Look 9.1.

P9.2 The figure below shows two graphs of the emission spectrum of light versus wavelength from a hot metal object at temperature 5,000K. One is the experimentally observed brightness or intensity of light colors. The other is an incorrect prediction based on Newton's and Maxwell's theories. From the discussion in the chapter, determine which curve is which, and explain your reasoning.

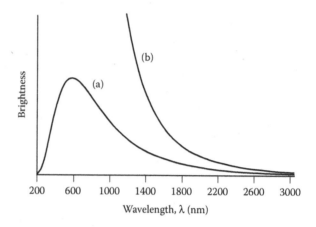

P9.3 Find a heavy rope or flexible chain and with one hand, hold it dangling down toward the ground. Whirl your hand around in a horizontal, circular motion to make the rope or chain whirl around. Observe that at certain rotation frequencies, distinct patterns occur on the rope. Describe in words and sketches at least three of these patterns and how they depend on the frequency at which you rotate the upper end. How is this analogous to quantization of electron waves?

P9.4 Bohr's theory, which we have not derived here, gives a formula for the energies of the electron shells in the H atom:

$$E_n = R \times \left(1 - \frac{1}{n^2}\right), \quad R = 2.1799 \times 10^{-18}\,\text{J}$$

R is a constant, and n is an integer labeling each energy shell. The lowest-energy shell (with energy defined to be zero) has $n = 1$, the second-lowest-energy shell has $n = 2$, etc. Calculate the energies of the first four energy shells in H, and give your answers in units of Joules and in attojoules.

P9.5 (a) Table 9.2 indicates that gallium (Ga) is a metal. Explain why.
(b) At very low temperature, argon (Ar) becomes a crystalline solid. Explain what would be the electrical conductivity (high, medium, or low) for solid argon and why.

P9.6 Not every crystal can be easily classified as a conductor, semiconductor, or insulator only on the basis of its atomic number and the number of electrons occupying its highest occupied shell. For example, diamond is a form of carbon, with four electrons in its outer shell, in which there is space in each atom for more electrons. Nevertheless, diamond is an insulator. Explain this behavior of diamond. *Hint*: Study the explanation in this chapter for why silicon crystals are semiconductors. There is a similar effect in diamond, only much stronger. Each carbon atom is bonded to four other carbons.

P9.7 The polonium (Po) atom has atomic number 84, and a mass 3.472×10^{-25}kg. The density of a particular crystal form, alpha-polonium, is 9,230 kg/m^3. The structure of this crystal is the simplest possible—the simple cubic lattice, shown below at left. In this arrangement, the atoms are stacked vertically and horizontally in square patterns. Assuming the atoms are tightly packed, so they are "touching," we conclude that each atom's diameter equals the distance between two adjacent atoms' centers, shown by the bold bars in the magnified drawings. The centers of any eight adjacent atoms form a cube, whose side length equals the single atom's diameter. Each atom can be circumscribed by one of these cubes. From this information, calculate the diameter of a polonium atom. Give your answer in both meters and in nanometers.

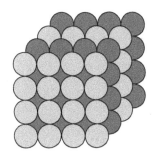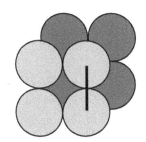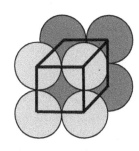

P9.8 At room temperature, a typical electron in a solid has an average energy of 0.004 aJ. In a silicon crystal, the band gap is 0.18 aJ. Explain, in light of these numbers but without any calculation, how room-temperature silicon (Si) can act as a weak conductor.

P9.9 (a) Use the Bohr formula given in P9.4 to write the formula for the difference of energies of the shell labeled by integer n_1 and another shell labeled by integer n_2.

   (b) Compare your result to Balmer's formula given in P9.1. What does this comparison tell you about the relation between the energy difference between two levels and the frequencies of light emitted by the H atom?

10

Semiconductor Physics: Transistors and Circuits

The general aim of the ... research program on semiconductors initiated at the Bell Telephone Laboratories in early 1946 ... was to obtain as complete an understanding as possible of semiconductor phenomena, not in empirical terms, but on the basis of atomic theory.

John Bardeen
(Nobel Lecture, 1956)

Eight-inch Si wafer containing hundreds of integrated circuits, each containing millions of transistors.

John Bardeen, Walter Brattain, and William Shockley (left to right), the inventors of the first practical transistor at AT&T Bell Laboratories in 1947. (With permission of Lucent Technologies Inc./Bell Labs.)

10.1 SILICON, TRANSISTORS, AND COMPUTERS

As carbon is the building block for living things, silicon (Si) is the building block for information technology. Si-based semiconductors are the basis of most modern electronic circuits. It is abundant (from sand), and it can be purified to an extremely high degree. *Semiconductor* devices can be designed to carry out switching and gate operations. That is, semiconductors can be turned into conductors when needed, and then turned back into insulators, as in turning a water faucet on and off. This chapter explains how this is done. Other technologically important semiconductors are germanium, gallium arsenide, gallium phosphide, and indium arsenide. These have special uses, such as making lasers and light-emitting diodes, discussed in Chapter 12.

The understanding of semiconductors, and Si especially, in the 1940s led to a rapid development of **solid-state electronics**—the basis of most computer and Internet technology. *Solid-state* refers to the use of crystals as electronic components in circuits. This development culminated in the development in 1947 of the first practical semiconductor transistor by John Bardeen, Walter Brattain, and William Shockley at AT&T Bell Laboratories. Although other scientists before Bardeen, Brattain, and Shockley had accomplished closely related work, these three were recognized widely as having made the big breakthrough, and were awarded the Nobel Prize in 1956 for the invention of the transistor.

As Bardeen said in his Nobel lecture, the immediate goal of the research he and his team were doing was to understand the basic physics of semiconductor crystals, on the basis of the theories of atomic physics that had been developed not long before. They also paid close attention to the practical implications of their newly found physics understanding, and, using what they had learned, built the world's first transistor.

It was immediately clear to physicists and engineers that transistors had important applications as electronic switches and as power amplifiers. Nevertheless, it took 11 years from the invention of the transistor to the making of the first miniaturized, or integrated, transistor circuit. This was accomplished by Jack Kilby at Texas Instruments (TI) in 1958 and started the digital revolution. Computer processors and memory are based on **integrated circuits** (ICs). ICs contain hundreds of millions of transistors packed into an area less than the size of a small coin.

In this chapter, we will explore how semiconductor crystals are modified by adding impurities, and how this leads to transistor action—the ability to control an electric voltage by applying a second, smaller voltage. This enables electronic switching and amplification and other operations useful in communication technology.

10.2   CONTROLLING THE CONDUCTIVITY OF SILICON

As we learned in the previous chapter, pure Si crystals are not good conductors. On the other hand, various elements can be put into an otherwise pure Si crystal to alter its properties. This process is called **doping**, and the added element is called an **impurity** or a **dopant**. In a typical doped crystal, there is about one "impurity atom" for every 1 million Si atoms. From what we have learned about energy bands and the principles of quantum theory, we can understand how doping changes the electrical conductivity of the crystal.

Figure 10.1 shows a schematic diagram of a crystal of pure Si, with each Si atom having four outer-shell *electrons*. Each Si atom is surrounded by eight outer-shell electrons—four from itself and one each from its four nearest neighbors. The crystal is neutral.

Recall (Chapter 9, Section 9.7) that at very low temperature, a pure Si crystal cannot conduct electricity because its energy bands are all either empty or completely filled. At room temperature, there is very small conductivity, but let us ignore that for now. Also shown in Figure 10.1 is a Si crystal with a very small amount of phosphorous (P) added as an impurity. The P atoms replace a small fraction of the Si atoms, which would otherwise be at those locations. Each P atom has five outer electrons (see Figure 9.19). This means that surrounding each P atom that has replaced a Si atom there is one extra electron that would not be present in a pure Si crystal. This electron is indicated

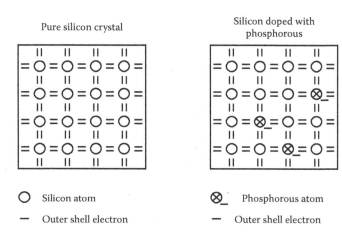

FIGURE 10.1   Pure Si crystal (on left) with four outer-shell electrons (–) for each Si atom (○). On the right is a Si crystal that is weakly doped with P atoms (⊗), having five outer-shell electrons. This creates a semiconductor.

in the drawing as an extra (–) symbol. Note that the crystal is still neutral because the P atom also has one excess proton in its nucleus, as compared with Si.

Where does the P atom's excess electron go? To answer this we must discuss both its location and its energy. Recall (Chapter 9, Section 9.7) that in a pure Si crystal there is a small energy band gap just above the highest filled energy shell. The excess electron from the P atom cannot be in the same energy band as the other electrons because that band is completely filled, and the Exclusion Principle forbids adding one more, because this would require two electrons being in the same state. So, the excess electron has no choice but to go into the next higher energy band—the conduction band, which was empty in the pure Si crystal (at low temperature). This is shown in **Figure 10.2** as the black dot in the conduction band, just above the small gap. Si that is doped with P is called an **n-type semiconductor** because there are extra electrons, which carry negative (n) charge, compared to an undoped crystal. The crystal is, nevertheless, still neutral.

The excess electrons in a phosphorous-doped (n-type) Si crystal make the crystal highly conductive. Because the excess electrons are in a partly empty energy band, they are mobile; that is, able to move. When a low voltage is applied between two ends of the crystal, the electrons in the partly filled band can easily be accelerated, gradually raising them in energy into unoccupied energy states, as indicated by the small arrow in Figure 10.2. This acceleration corresponds to electrons moving through the crystal, as shown in **Figure 10.3**. Thus, a phosphorous-doped Si crystal is a good conductor.

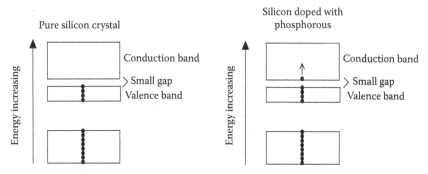

FIGURE 10.2   Energy bands and electron occupation for a pure Si crystal (left) and for a phosphorous-doped Si crystal (right). The temperature is assumed to be very low. The excess electrons from the doping go into a partly empty band, making them able to conduct electrical current. This is an n-type semiconductor.

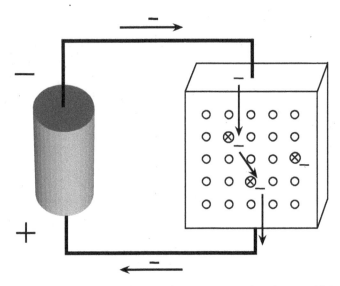

FIGURE 10.3 A phosphorous-doped Si crystal is an n-type semiconductor, which conducts current when a voltage is applied.

Notice that for every electron that moves into the crystal from the wire, one electron leaves the crystal at the other side. The arrows shown inside the crystal indicate the coordinated replacing of each electron by the electron coming behind it. Therefore, there is no charge buildup, and the crystal remains neutral.

A simple model for electron conduction in an n-type semiconductor is a ski lift with every seat occupied, taking people down a mountain (see **Figure 10.4**). As one person enters a lift seat at the top of the mountain, another person leaves a seat at the bottom. In a doped semiconductor, the partly filled band acts like the ski lift: as one electron enters one side of the crystal to occupy an energy level in the partly filled band, another electron leaves the crystal at the other side.

Doping with P atoms is not the only way to make Si conductive. By instead adding boron (B) atoms to a pure Si crystal, we also can make a Si crystal conductive. A boron atom has three outer electrons, as shown earlier in Figure 9.19. So, when a B atom replaces a Si atom in the crystal, it contributes one fewer electron in its outer shell than a Si atom would. This results in the absence of one outer-shell electron in the region near the B atom, shown as the small empty box in **Figure 10.5**. This small empty region is called a *hole*. A hole is a place in a crystal where an electron normally would be, but is not. Boron doping creates a crystal that is deficient in outer-shell electrons. This is called a *p-type semiconductor*, because a lack of negative charge appears like

FIGURE 10.4 Electrons traveling in a P-doped Si crystal are analogous to people traveling down a ski lift.

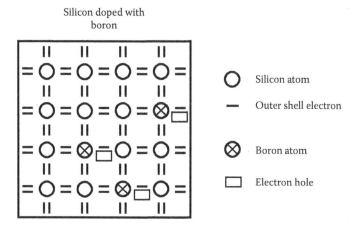

FIGURE 10.5 Si crystal doped with boron, which leaves a "hole" in the electron outer shell near the boron atoms. This creates a p-type semiconductor.

a positive charge. Again, as in the case of doping with P atoms, the crystal as a whole is still neutral, because the boron atom has one fewer proton (+ charge) in its nucleus than does a Si atom.

THINK AGAIN

Do not confuse the symbol "p" in the term *p-type* (positive-type) with the symbol P, which means the P (or phosphorous) atom.

How does this absence of an electron, or presence of a hole, affect the electrical conductivity? There is now a partly empty energy band, with a hole in it, as shown by the small empty rectangle in **Figure 10.6**. The presence of holes in the crystal's energy bands allows current to flow easily.

There is an electron in Figure 10.6 that has energy slightly below the energy of the empty hole. If a voltage is applied to the crystal, this electron can be accelerated. As it accelerates, it gains energy and moves up into the energy level that the hole formerly occupied. This in turn leaves a hole at a lower energy, as shown in **Figure 10.7**. The acceleration of electrons into holes allows electron current to flow through a p-type doped crystal. Note that if there were no energy hole, no electrons could accelerate under the influence of a voltage, because there would be no higher available energy levels for them to go into.

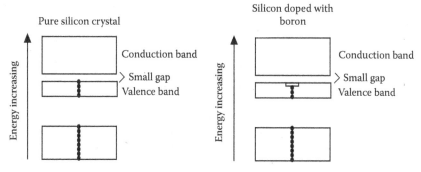

FIGURE 10.6 Energy bands and electron occupation for pure Si crystal (left) and boron-doped (p-type) Si crystal (right). Low temperature is assumed. The deficit of electrons from the doping creates a partly empty (not filled) band, making the crystal able to conduct electrical current. This is a p-type semiconductor.

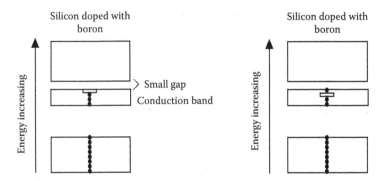

FIGURE 10.7   Electron occupation for boron-doped Si crystal (left), showing a hole in the highest-energy occupied band. Under the influence of an applied voltage, an electron gains energy and moves into the empty hole, causing a new hole just below the energy of the former hole.

10.3   P-N JUNCTIONS AND DIODES

The first solid-state devices used in electronics were diodes, used as radio receivers. A *diode* is similar to a one-way turnstile for electrons. Electrons can flow in one direction through a diode, but not in the other direction, as in **Figure 10.8**.

The earliest diodes were constructed by making contact between the end of a fine metal wire and a piece of semiconductor. Later, diodes were greatly improved by using all semiconductor materials in their construction. They play an important role in modern electronic circuits. Understanding diode operation will prepare us to understand the operation of transistors—the most important component in modern computer circuits.

A diode is made of two joined semiconductor crystals—one p-type and the other n-type, as in **Figure 10.9**. The region just around the contact is called a *p-n junction*. The figure shows the two pieces before and after being joined. In the n-type piece, excess electrons, from doping, are shown as minus symbols (–), and in the p-type piece, holes are shown by the box symbols. All other symbols such as the Si, P, and B symbols are not shown, nor are the Si outer-shell electrons. Remember that all of these are present in the crystals, making each crystal neutral (zero total charge) before they are joined.

After joining, the crystals lose their neutral property in a small region around the junction. Charge builds up as some of the excess electrons flow from the n-type piece into the p-type piece, where they occupy former holes. Why do they flow that way? There are no external forces on these electrons initially to make them flow one way or the other, but they wander, or diffuse. This motion is a natural consequence of their random jiggling motion due to thermal energy at room temperature. Both pieces are initially good conductors because they are doped, so electrons can move through them.

Diffusion is the random wandering of particles throughout some region. You can think of electrons as being like salmon in a large lake, wandering randomly from place to place, as in **Figure 10.10**. If the lake has no outlet, the salmon will spread out more or

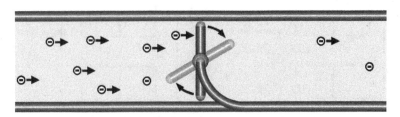

FIGURE 10.8   A good analogy for a diode is a one-way turnstile. A bar can rotate clockwise, but not counterclockwise, allowing electrons (or people) to pass only from left to right.

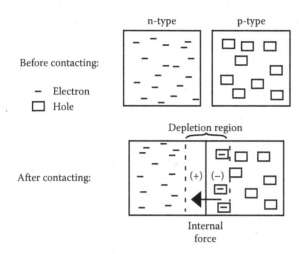

FIGURE 10.9   A p-n junction before and after contacting. A few electrons diffuse into the p-type side until the internal force on electrons, caused by the resulting charge imbalance, holds back any further net charge drift. The joined crystals comprise a diode.

less evenly throughout the lake. The average density of salmon stays about the same in each area. Then, if a channel is opened up into a second, empty lake, salmon will tend to wander at random into that second lake until they are spread out roughly evenly in both lakes. After some time, there will be on average as many salmon leaving the first lake as entering it, so that the number of salmon in each lake stays roughly constant.

Electrons are charged negatively, so if many electrons diffuse into the p-type semiconductor crystal, then the part of that crystal nearest the junction will become charged negatively, shown by the (–) symbol in Figure 10.9. Then the part of the n-type crystal nearest the junction (initially neutral), from which the electrons have departed, will become charged positively, shown by the (+) symbol. There is a thin region (about 10 μm in length) around the junction where the n-type piece is positively charged and the p-type piece is negatively charged. This region is called the ***depletion region***. In regions away from the depletion region, the crystals are neutral, as usual.

The result of the diffusion of electrons is that in the depletion region the two pieces of crystal act as a battery, with one side of this region becoming positive and the other side negative. We know that the negatively charged region (the p-type side) will repel electrons, and the positively charged side (the n-type side) will attract electrons. This effect is called the internal battery created at the junction of the p-type and n-type crystals. This internal battery creates a force on the electrons that pushes them toward the n-type side. In Figure 10.9, the arrow indicates this internal force on electrons. This causes the

QUICK QUESTION 10.1

If you put a drop of food coloring into still water, diffusion takes place. Explain.

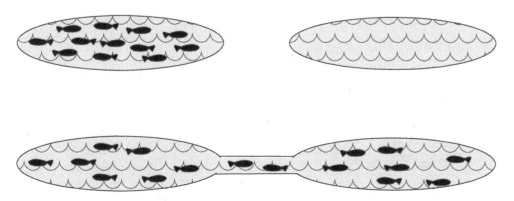

FIGURE 10.10   Salmon swimming between two lakes as an analogy of diffusing electrons.

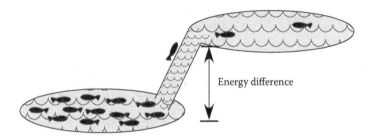

FIGURE 10.11   Salmon swimming in two lakes connected by a waterfall. Fast fish can go up. Slow fish cannot. Any fish can go down.

potential energy of electrons to be higher on the p-side of the diode. That is, it takes work to push an electron "uphill" from the n-side to the p-side, against the force.

In our salmon analogy, the effect of the internal battery is like having a waterfall between the two lakes, as in **Figure 10.11**. The higher lake is a region of higher potential energy. To reach the higher lake, salmon have to swim swiftly, energetically jumping up the waterfall. Once they reach the higher lake, they can stay there with no effort. But, if they happen to wander to the top of the falls, they may fall down into the lower lake. The waterfall is an impediment only in the upward direction. If the waterfall is high, then there are few salmon in the lower lake energetic enough to make the jump into the higher lake. On the other hand, any salmon in the higher lake can easily fall into the lower lake, regardless of how much energy it has. After some time, there will be many slow-moving salmon in the lower lake, swimming around but not able to make the jump, and a few fast salmon waiting to make the jump. In the upper lake, there will be very few salmon. Therefore, the rate of fish going down the falls is small. There will be roughly equal numbers going up and down, so that the number in each lake will not be changing. There is no net flow of fish because the tendency for upward diffusion is balanced by the tendency for downward travel between the lakes.

If we were to disturb the balance between upward diffusion of the salmon and the downward force associated with the waterfall, then a net flow, or "current," of salmon would result. Raising or lowering the height of the upper lake relative to the lower lake would disturb this balance. **Figure 10.12a** again shows the two lakes with relative

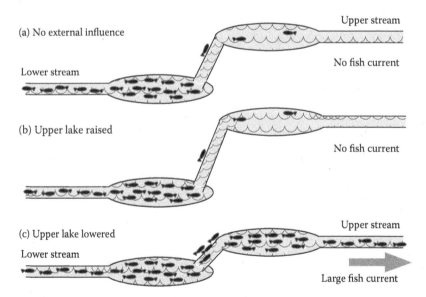

FIGURE 10.12   Analogy for the operation of a diode. Lowering the height of the upper lake allows large numbers of not-so-energetic salmon to leap up the falls and swim away in the stream to the right.

height such that there is a balance between upward diffusion and the downward force, so there is zero net fish exchange between lakes. In this case, there is zero net fish flow in the two streams going in and out of the lakes. If the upper lake is then raised to a higher level than before, as in Figure 10.12b, fish will have an even harder time jumping up the waterfall, so there is still zero or very little net fish current. On the other hand, if the upper lake were dropped to a lower level than before, as in Figure 10.12c, there is a dramatic effect: now a very large number of salmon, moving at moderate speeds, can make it up the falls, whereas they were not able to make it up when the lakes were at their original heights. Because the rate of salmon going down the falls is not much affected by lowering the upper lake, there is now a large net rate of salmon going from the lower lake into the higher lake. This results in a large net number of fish traveling from the lower stream to the upper stream. There is now a large net "current" of fish moving from left to right in the figure.

The same happens with electrons moving in a diode. In Figure 10.9, the force from the internal diode battery (likened to the waterfall) holds the slow, low-energy electrons back from diffusing "up" to the p-type side from the n-type side, where the potential energy is higher. Only a few energetic electrons can do so, depending on the temperature. Recall the discussion around Figure 9.28. Any electron in the p-type side can easily diffuse back into the n-type side. A balance gets established between electron diffusion and the internal-battery force caused by those same electrons. A brief time after joining the semiconductor crystals, these processes come into equilibrium, and after that time there is no further net flow of electrons.

If one disturbs this delicate balance between upward diffusion and the force from the internal battery, then current flows. This can be done by connecting an external battery to the two ends of the joined crystal, as in **Figure 10.13.** If, as in part (a), we connect the (–) side of the battery to the p-type side of the diode, the battery gives an increased potential energy to electrons at the p-side, analogous to raising the height of the upper lake in the salmon example. This slows down the already small rate of diffusion from the n-type to the p-type side, but does not affect the diffusion rate in the other direction. This leads to a very small net electron current from the (–) side of the battery to its (+) side. This current is so small that we will say it is zero for simplicity of discussion. We refer to this manner of attaching the battery as being in the backward direction.

On the other hand, if, as in Figure 10.13b, we connect the (+) side of the battery to the p-type side of the diode, then the battery decreases the potential energy of electrons

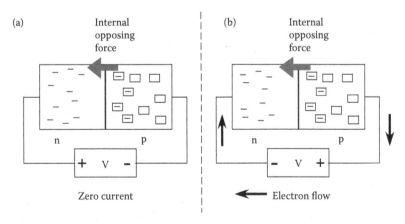

FIGURE 10.13 In a diode, the internal force opposes electron flow. An external battery adds a force on electrons pushing from the wire at the (-) side of the battery toward the wire at the (+) side of the battery. Only if the (+) side of the battery is connected to the p-type side of the diode will the battery add a force on electrons that counteracts the internal force and allows electrons to flow.

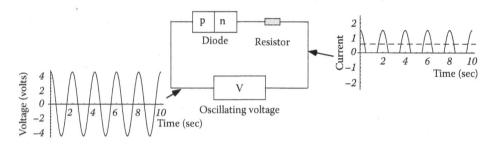

FIGURE 10.14   A p-n junction diode acting as a voltage rectifier. When the oscillating source V (the signal) creates a plus voltage on the left side of the diode, the diode conducts current from left to right, sending current through the resistor. In contrast, when the source creates minus voltage on the left side, the diode blocks the current flow through the resistor.

at the p-side, analogous to lowering the height of the upper lake in the salmon example. This allows a much larger number of the moderate-energy electrons to diffuse from the n-type side to the p-type side. This results in a large net electron flow from the (–) side of the battery to its (+) side. This explains why a diode acts as a one-way turnstile for electrons. To summarize, current flows through a diode only if the battery's plus voltage is connected to the p-type side and the minus voltage is connected to the n-type side. We call this applying voltage in the forward direction.

THINK AGAIN

Keep in mind the distinct meanings of the terms *p-type* (positive-type) and *n-type* (negative-type), which refer to types of crystals, and *plus* (+) and *minus* (-), which refer to polarities of voltages produced by batteries.

10.3.1   Rectifying an Alternating Signal

A diode acts like a one-way turnstile for electrons. The conductivity in the forward direction is large, whereas the conductivity in the backward direction is nearly zero. An application of a diode is as a ***rectifier***—a device that converts an oscillating signal (containing plus and minus voltages) into a positive-only signal. Rectifiers are a key component in radio receivers, discussed further in Real-World Example 10.1.

Figure 10.14 shows a circuit with an oscillating signal voltage, V, connected to a diode and a ***resistor*** in series. When the voltage is positive, as shown, current flows through the diode and the resistor. When the voltage is negative, the current equals zero.

**REAL-WORLD EXAMPLE 10.1: A SIMPLE CRYSTAL
AM RADIO RECEIVER**

If the oscillating voltage shown in Figure 10.14 is replaced by an antenna, and the resistor is replaced by an audio speaker, a simple radio receiver is created, as shown in **Figure 10.15**. The other connection of the speaker is attached to a metal wire that is inserted into the ground or earth outside (or a piece of indoor plumbing that is connected to the earth). The antenna is a long metal wire, inside of which a time-varying current is produced by an incoming radio wave. The incoming wave labeled (a) in **Figure 10.16**, is a carrier

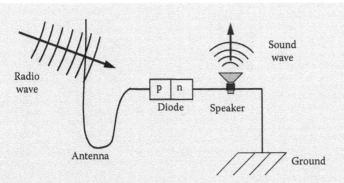

FIGURE 10.15   A simple crystal radio receiver, made of an antenna, a diode, and a small speaker.

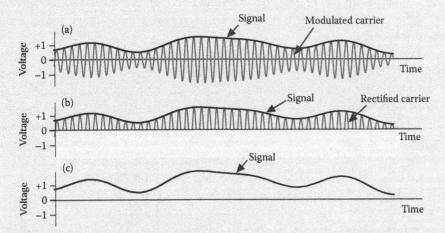

FIGURE 10.16   (a) The incoming radio wave is a carrier wave modulated by an audio signal. (b) After the diode rectifies the incoming signal, only positive voltages remain. (c) The slow response of the speaker cone follows the original audio signal.

wave (e.g., at 650 kHz), the **amplitude** of which is modulated by an audio signal. For details about modulation, see Chapter 8, Section 8.2. The diode rectifies the oscillating voltage induced in the wire by the radio wave to produce the wave shown in part Figure 10.16b. The electromagnet in the speaker is connected to the output voltage of the diode, which rectifies the incoming signal voltage. The rectified voltage creates a current in the speaker coil, which drives the motion of the speaker cone. The magnet and cone cannot rapidly respond to the kHz carrier wave of the radio signal; instead, they move in proportion to the slower-changing amplitude of the rectified signal. Without the diode, the slow response would average to zero, and the cone would not move. Changes of the amplitude of the radio wave cause corresponding changes in the amplitude of the rectified voltage driving the speaker. This modulates the cone position, creating a sound wave, labeled (c) in Figure 10.16. No battery or other external power source is needed, because the radio wave itself can deliver enough power to drive a small speaker such as an earphone.

10.4   TRANSISTORS

Now we can use our understanding of diodes to understand transistors—one of the most important devices in computer circuitry. A **transistor** is a semiconductor device that acts as a controllable valve for electrons, as a valve or faucet does for water. As in **Figure 10.17**, if a source of water is kept at a higher pressure than the drain, then when

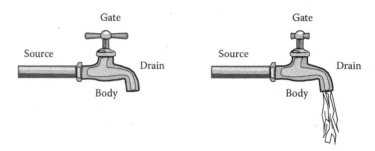

FIGURE 10.17 A transistor acts as a valve for electrons.

the control knob is activated, opening the valve, water flows from the source through the drain. If the drain has higher pressure than the source, backward flow will occur. For transistors, the control knob is called the control gate and voltage is analogous to the water pressure.

Transistors provide the low-power voltage-controlled switches needed to build practical logic circuits for use in computers. The first transistors to be put into practical use were bipolar transistors—composed of a pair of back-to-back diodes. Bipolar transistors—discussed at the end of this chapter—were initially used in semiconductor-based computer circuits, but they use a lot of power to operate. This generates heat in the circuit, and the need to remove this heat limits the speed at which computer chips can be operated reliably. Bipolar transistors were replaced in the late 1970s by the **field-effect transistor**, or **FET** (pronounced f-e-t, or simply "fet"). FETs are preferred for use in computer logic circuits because they use very little electrical power when they switch.

In a **FET**, the flow of electrons is controlled by an electric field that is applied to a narrow region in a semiconductor device. This narrow region, called an electron channel, conducts electrons only when exposed to the effect of an electric field; hence the name field-effect transistor. A schematic of a FET is shown in **Figure 10.18**. The FET contains two p-n junctions similar to those in Figure 10.9. Two n-type regions are contacted to a p-type region, and there is an **insulator** region, labeled i, adjacent to the p-region. The insulator is silicon dioxide, SiO_2, or *oxide* for short. The two n-type regions

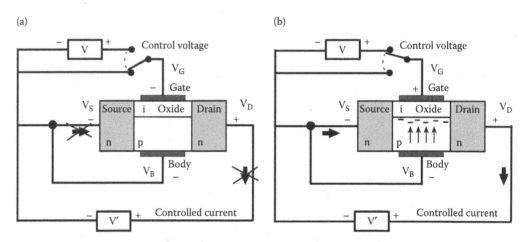

FIGURE 10.18 An n-FET. An insulating oxide layer, labeled i, separates the gate from the body. (a) With no voltage applied to the gate, no current can flow from the source to the drain. (b) A positive voltage V applied to the gate creates a conducting n-type channel through the p-type semiconductor, allowing electron current to flow from source to drain and through the battery V′.

are called the source and the drain, reminiscent of water plumbing. In addition, there are two conducting plates (metal or another conducting material[1]). One conducting plate, called the control gate, contacts the insulator region. The other conducting plate, called the body, contacts the p-region. The body is connected to the source through a wire, and to the minus sides of both batteries. This structure is called an n-channel metal-oxide-semiconductor (MOS) device.

Recall that a p-type semiconductor has a deficit of electrons in its outer electron shells; that is, it has electron holes (absences) in its outer shells. An n-type semiconductor has an excess of electrons in its outer shells. At each p-n junction in the FET structure, there is a thin depletion region where all holes are filled by diffusing electrons. Because the p-n junction on the left side of the insulator has zero voltage across it, according to our understanding of diode action, no current flows through this junction. The p-n junction on the right side has a voltage applied to it in the backward direction, so no current flows through it as well. When the control switch in the figure is in the down position, as in Figure 10.18a, the control gate and the body are at the same voltage; that is, there is zero voltage between them. There is no current flowing anywhere in this case.

When the control switch is moved to the up position, as shown in Figure 10.18b, a positive voltage appears at the gate, relative to the body. The electric field from the battery labeled V pushes electrons from its minus side, through the body, into the p-type region, and onto the lower surface of the oxide insulating region. The electrons cannot pass into the insulator, so a negative charge builds up near its surface, shown by minus signs. If enough electrons build up in the region of the p-type semiconductor near the surface, it becomes n-type. This occurs if all of the holes in this region are filled and some extra electrons arrive as well. These extra electrons can move under the influence of an applied voltage. Now there will be a continuous conducting path of n-type semiconductor material between the source and the drain, and current flows. The direction of flow is shown with the bold arrows. The FET shown in Figure 10.18 is called an *n-channel FET*, or *n-FET*, because an n-type channel is created for electrons to flow.

To make the description of FET operation more clear, **Figure 10.19** separately shows the two parts of the circuit: the controlling circuit and the controlled circuit. The voltage labeled V in the controlling circuit controls current in the controlled circuit. Notice that no current flows between the controlling circuit and the controlled circuit. This makes the FET highly efficient—almost no current is used in the controlling action.

We can describe the n-FET operation using a simplified drawing showing the FET as a controlled switch, as illustrated in **Figure 10.20**. The four voltages labeled V_G, V_B, V_S, and V_D are the voltages at the control gate, body, source, and drain. Notice that for the example given in the figures above, the body voltage is always the same as the source voltage because they are connected by a wire; that is, $V_B = V_S$.

Rule for n-FET operation: If the control gate voltage is significantly greater than the body voltage, then the n-FET is ON. Otherwise, the n-FET is OFF.

For a typical FET, "significantly greater than" means greater by about 0.5 volts (V). We can state the condition for an n-FET to be ON as: "V_G is greater than V_B by at least +0.5 V." The ON state is analogous to a conducting switch being connected: current can flow between the source and drain.

---

[1] The earlier-used aluminum metal was later replaced by heavily doped polycrystalline silicon (polysilicon).

Controlling circuit

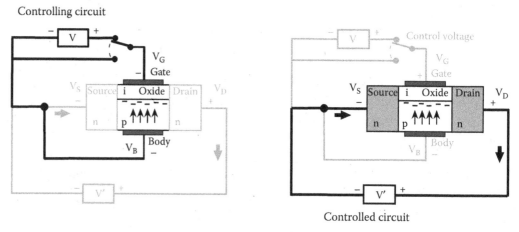

FIGURE 10.19   Highlighted controlling circuit and controlled circuit of an n-FET.

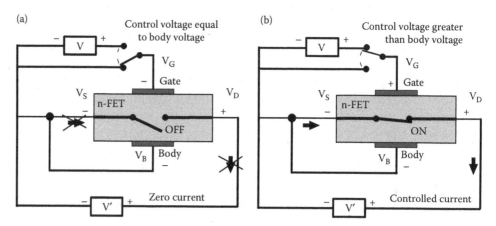

FIGURE 10.20   An n-FET represented as a controlled switch. (a) The control voltage is roughly equal to the body voltage, and the FET acts like a switch that is nonconducting, or OFF. (b) If the control voltage is at least 0.5 V more positive than the body voltage, the FET acts as a switch that is conducting, or ON.

Voltages can be defined in a relative manner (just as energy is relative). The n-FET will be ON if gate and body voltages are both positive, or if both are negative, or if the gate is positive and the body is negative, as long as the gate voltage is significantly greater than the voltage on the body. For example, the n-FET is ON if $V_G = 5$ V and $V_B = 0$ V. In addition, it is ON if $V_G = 0$ V and $V_B = -5$ V. (Note that in terms of negative numbers, -5 is less than 0.) Finally, it is ON if $V_G = 5$ V and $V_B = -5$ V. It is OFF if, for example, $V_G = 0.4$ V and $V_B = 0$ V.

Another type of FET is the *p-channel FET*, shown in **Figure 10.21**, in which the n- and p- regions of Figure 10.20 are interchanged. The battery polarities are also reversed. A p-FET will become conducting if the voltage applied to the gate is negative relative to the body, as in Figure 10.21b. When the gate is more negative than the body, electrons are pushed away from the surface of the oxide insulator, leaving holes in that part of the n-type region. The small empty boxes depict these holes. This creates a p-type channel through the n-type region; thus the name p-FET. The presence of the holes allows electrons to flow from the drain to the source, under a force exerted by the V' battery.

Rule for p-FET operation: If the gate voltage is less than the body voltage by at least 0.5 V (i.e., V_G is less than $V_B - 0.5$ V), then the p-FET is conducting, or ON.

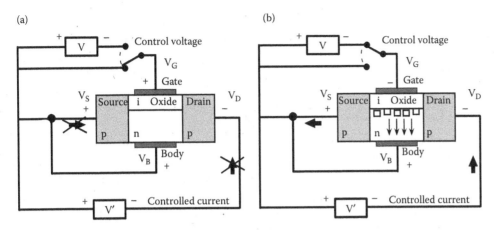

FIGURE 10.21   A p-FET. (a) With the gate voltage equal to the body voltage, no current flows from the source to the drain. (b) A negative voltage *V* applied to the gate relative to the body creates a conducting p-type channel through the n-type semiconductor, allowing electron current to flow from drain to source.

The ON state is analogous to an electrical switch being closed: current can flow between the source and drain in either direction, according to the sign of the applied voltage. Otherwise, the p-FET is OFF. This is illustrated in terms of controlled switches in **Figure 10.22**.

The p-FET can be ON if both gate and body voltages are positive, or if both are negative, or if the gate is negative and the body is positive, as long as the gate voltage is significantly less than the body voltage. For example, the p-FET is ON if $V_G = 0$ V and $V_B = 5$ V. In addition, it is ON if $V_G = -5$ V and $V_B = 0$ V. Finally, the p-FET is ON if $V_G = -5$ V and $V_B = 5$ V.

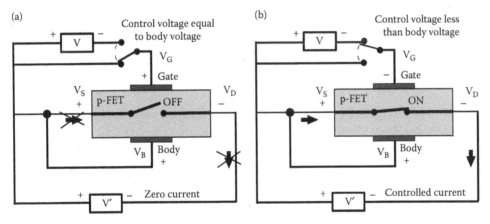

FIGURE 10.22   A p-FET represented as a controlled switch. (a) The control voltage is equal to the body voltage and the FET is OFF. (b) If the control voltage is at least 0.5 V less than the body voltage, then the FET is ON.

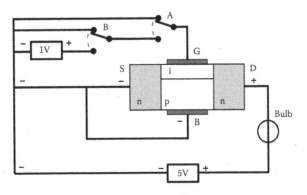

QUICK QUESTION 10.2

An n-type FET is connected to two mechanical push-ON switches, A and B, as well as two batteries and a light bulb, as shown to the left. If the switches are logic inputs, and the bulb is the logic output, what logic operation does this circuit perform? (See Chapter 6, Section 6.4.)

To summarize, we call an n-FET a "high-ON switch." It turns ON (conducting) when the control voltage to its gate is high relative to the body, and remains OFF otherwise. In contrast, a p-FET is a "low-ON switch." It turns ON (conducting) when the control voltage to its gate is low relative to the body and remains OFF otherwise.

10.5   CMOS COMPUTER LOGIC

Computer logic circuits are made using FETs. The FET implementation of the NOT logic gate is shown in **Figure 10.23**. A p-FET and an n-FET are connected in series as shown, with the two drains (D) connected to one another. The source (S) of the p-FET is held at a fixed high voltage, V_{HIGH}, which is typically +5 V, whereas the source of the n-FET is held at a low voltage, V_{LOW}, typically 0 V. The logic input is the input voltage, V_{IN}. The input voltage is sent to both of the FET control gates (G). The logic output is the voltage, V_{OUT}, taken from the common drain voltage.

The operation of the NOT gate can be understood from the diagrams. Recall that an n-FET is a high-ON switch and a p-FET is a low-ON switch. When the input voltage is high (+5 V), the n-FET is ON and the p-FET is OFF. This means that the output voltage is connected through the n-FET to the low voltage, $V_{LOW} = 0$ V, and the output is low. In contrast, when the input voltage is low (0 V), the n-FET is OFF and the p-FET is ON. This means that the output voltage is connected through the p-FET to the high voltage, $V_{HIGH} = +5$ V, so the output is high. In each case, the output voltage is the opposite of the input voltage. This is a logical NOT operation.

As usual, we use voltage values to represent logic values. Let us define a logical zero (0) to correspond to 0 V, and a logical one (1) to correspond to +5 V. The logic table for the circuit in Figure 10.23 is shown below. It corresponds to the logic table for the NOT operation.

V_{IN}	n-FET	p-FET	V_{OUT}	A	OUT
+5 V	ON	OFF	0 V	1	0
0 V	OFF	ON	+5 V	0	1

(NOT)

The type of circuit in Figure 10.23 is called **CMOS** (pronounced *cee-moss*), for *complementary metal-oxide semiconductor*. *Complementary* reflects the fact that the two types of FETS in the circuit act in complementary ways—when one is ON, the other

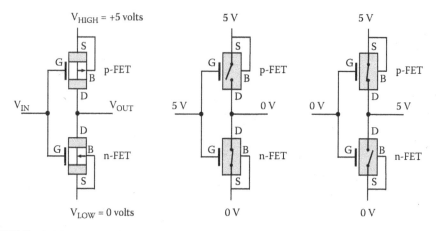

FIGURE 10.23   NOT gate constructed with an n-FET in series with a p-FET. The symbols S, G, D, and B label the source, control gate, drain, and body of each FET, respectively. When the input voltage is high (+5 V), the output is low (0 V), and vice versa.

is OFF. The advent of CMOS circuits in the 1970s was a breakthrough for computing capability, especially for microcomputers, which were introduced in the 1980s, and which later evolved into desktops and laptops. The use of back-to-back complementary FETs minimizes power consumption, because current flows only during the brief time of the switching operation. By the year 2006, the energy needed for a single switching operation reached as little as one femtojoule (10^{-15} J or fJ). After the switching takes place, no current is needed in the circuit to hold a constant output voltage indefinitely. This contrasts with the older circuit types, which used bipolar transistors and lots of current and power.

THINK AGAIN

Do not confuse the terms *logic gate* and transistor *control gate*. This is an unfortunate double usage of the word *gate*. A logic gate is any device that performs a logic operation. A control gate refers specifically to the input side of a FET.

Recall from Chapter 6, Section 6.4 that the NOR logic operation can be used to construct any other logic gate. This makes the NOR logic gate especially useful. The logic table for the NOR operation is:

NOR		
A	B	OUT
0	0	1
0	1	0
1	0	0
1	1	0

We can construct the NOR logic gate using CMOS transistors, as shown in **Figure 10.24**. The circuit consists of two n-FETs and two p-FETs, numbered 1 through 4.

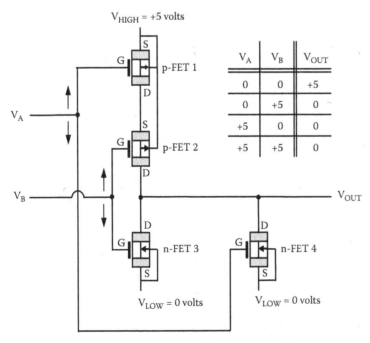

V_A	V_B	V_{OUT}
0	0	+5
0	+5	0
+5	0	0
+5	+5	0

FIGURE 10.24   NOR gate constructed with two n-FETs and two p-FETs.

The logic inputs are the voltages V_A and V_B. Input V_A is sent to the control gate of p-FET-1 and simultaneously to n-FET-4. Input V_B is sent to the control gate of p-FET-2 and simultaneously to n-FET-3. The bodies of FET-1 and FET-2 are connected to the constant +5 voltage source. The p-FET-1 is ON only if V_A is less than +5 V, whereas p-FET-2 is ON only if V_B is less than +5 V. The bodies of FETs 3 and 4 are connected directly to V_{LOW}. Therefore, n-FET-3 is ON only if V_B is greater than 0, whereas n-FET-4 is ON only if V_A is greater than 0.

Because there are two inputs, there are four possible combinations of input values, listed in **Table 10.1**. The only condition that leads to high (+5 V) output V_{OUT} is the case that both inputs are low (0 V), as shown in **Figure 10.25a**. In this case, FET-3 and FET-4 are OFF, and FET-1 and FET-2 are ON. This means that the output is connected through FET-1 and FET-2 to V_{HIGH}, and disconnected from V_{LOW}. For any other combination of input voltages (e.g., Figure 10.25b), at least one of FET-1 or FET-2 is OFF, disconnecting the output from V_{HIGH}, and at least one of FET-3 or FET-4 is ON, connecting the output to V_{LOW}.

Complex logic circuits can be implemented by combining NOR gates, as discussed in Chapter 6, Section 6.4.2. Many modern computer circuits are constructed in just this way. Instead of building several types of electronic circuits—one for each logic operation (NOT, AND, OR)—it turns out to be more efficient and economical to use only one type of electronic gate to do the tasks of the others.

We can also implement the NAND logic operation by using CMOS, as shown in **Figure 10.26**. The only combination of inputs that gives $V_{OUT} = 0$ is the case that

TABLE 10.1
Logic Table for FET NOR Gate

V_A	V_B	FET 1	FET 2	FET 3	FET 4	V_{OUT}
0	0	ON	ON	OFF	OFF	+5
0	+5	ON	OFF	ON	OFF	0
+5	0	OFF	ON	OFF	ON	0
+5	+5	OFF	OFF	ON	ON	0

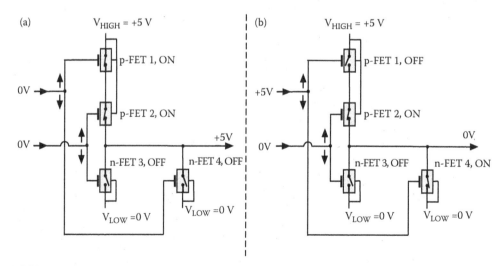

FIGURE 10.25 NOR gate constructed with two n-FETs and two p-FETs. Part (a) shows the only input condition for which the output equals +5 V, part (b) shows one of the other three possible input conditions.

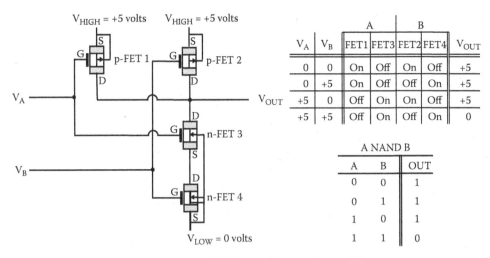

The table in the figure:

V_A	V_B	FET1	FET3	FET2	FET4	V_{OUT}
		A	A	B	B	
0	0	On	Off	On	Off	+5
0	+5	On	Off	Off	On	+5
+5	0	Off	On	On	Off	+5
+5	+5	Off	On	Off	On	0

A NAND B

A	B	OUT
0	0	1
0	1	1
1	0	1
1	1	0

FIGURE 10.26   NAND gate constructed with two n-FETs and two p-FETs.

FETs 3 and 4 are both ON, and FETs 1 and 2 are both OFF. This occurs only when both inputs equal +5 V. In every other case, the point in the circuit connected to V_{OUT} is connected to one or the other of the +5 V voltage sources (and is automatically disconnected from V_{LOW}.

IN-DEPTH LOOK 10.1: WATER-EFFECT TRANSISTORS

A whimsical analogy using the pressure of water in pipes can help us understand the operation of the FET. For fun, I call this the *water-effect transistor*, or *WET*. Recall, from Chapter 5, Section 5.7, that pressure in a water tank is analogous to the voltage in a circuit. Pressure in water refers to the potential to create a fast, powerful stream of water if a small rupture is opened in the tank's wall. Pressure does not refer to the actual flow of water, only the potential to create a flow if given the chance. A FET is analogous to a pressure-activated water valve. There are two possible types of such a valve. I will call these high-ON types and low-ON types. A high-ON valve turns ON (open) when its control line pressure is high, and remains OFF (closed) otherwise. A low-ON valve turns ON when its control line is low, and remains OFF otherwise.

Models for such valves are shown in **Figure 10.27**. As in the semiconductor FET, there are source, drain, control gate, and body regions. Each region is a hollow pipe or channel where water can flow. The pressures in the body and in the source are kept equal through a small connecting pipe. For the High-ON valve, high water pressure at the gate pushes down a piston if the gate pressure is significantly greater than the pressure in the body. This opens a path in the pipe for water flow between the source and drain. If the gate and body pressures are equal, a spring keeps the piston in the up position, closing the valve.

For the low-ON valve, the piston is held down by a spring in the OFF position when the gate pressure is higher than or equal to the body pressure. When the gate pressure drops significantly below the body pressure, the body pressure pushes the piston up, opening the valve, turning it ON. This allows water flow, as indicated by the double-headed arrow.

In summary, the high-ON valve is ON only if the gate pressure is significantly greater than the source (body) pressure, while the low-ON valve is ON only if the gate pressure is significantly lower than the (body) source pressure. The high-ON valve is analogous to the n-FET, and the low-ON valve is analogous to the p-FET.

By combining both types of water valves, we could construct a NOT logic gate. **Figure 10.28** shows a high-ON valve and a low-ON valve—controlled by a common line—and two large constant-pressure tanks. The two tanks are connected to pumps (not shown) that maintain the tank pressures—one high pressure and one low pressure. Because the control lines of the two valves are connected, they have the same pressure.

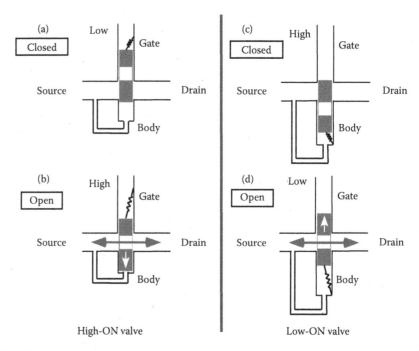

FIGURE 10.27 The water-effect transistor, or WET. The shaded regions are solid and block the flow of water. A high-ON valve is closed when the gate control pressure is lower than or equal to the body pressure (a), and is open when the gate pressure is significantly higher than the body pressure (b). A low-ON valve is closed when the gate pressure is higher than or equal to the body pressure (c), and is open when the gate pressure is significantly lower than the body pressure (d). Double-headed arrows indicate flow of water, which can occur in either direction.

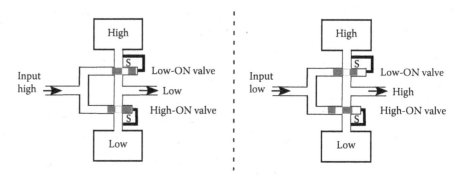

FIGURE 10.28 NOT logic gate constructed with a high-ON valve in series with a low-ON valve. The symbol S labels the source for each valve. When input pressure is high, output is low, and vice versa.

When one valve is open, the other is closed, as can be seen by considering two cases. When the input pressure is high, the output pipe is connected only to the low-pressure tank, and so it is low, as shown in the left figure. When the input pressure is low, the output pipe is connected only to the high-pressure tank, and so it is high. Bit values can be represented as 1 (high), and 0 (low). Thus, this arrangement acts as a NOT logic gate.

Notice that the only time that water flows is during or just after the switching from high to low or from low to high. If we think of water as a resource, then this is an efficient system, because it consumes little water. It operates mostly by virtue of pressure, not flow.

If several such water valves were combined in the proper manner, it would be possible to construct any type of logic gate. By using many such valves, one could construct a rudimentary computer that could carry out, for example, addition of two several-bit numbers.

10.6 MINIATURIZATION, INTEGRATED CIRCUITS, AND PHOTOLITHOGRAPHY

Where ... the ENIAC [computer] is equipped with 18,000 vacuum tubes and weighs 30 tons, computers in the future may have 1,000 vacuum tubes and perhaps weigh just 1½ tons.

–Popular Mechanics
(March 1949)

The transistor is the most important component in computer circuits. The first semiconductor double-junction transistor, which was constructed in 1947 by Bardeen, Brattain, and Shockley at AT&T Bell Laboratories, was a crude and large affair by today's standards. It was a *bipolar transistor*, which operates on different principles than the FET, as described in In-Depth Look 10.2. **Figure 10.29** is a photograph of their first transistor, measuring about 1 cm on a side. Texas Instruments introduced the first commercially available transistor in 1954. Since then, the size of transistors has shrunk by more than 10,000 times. This allows circuits to be fantastically miniaturized, so that hundreds of millions of logic gates fit into less than a square centimeter (about one-sixth of a square inch). By the year 2006, the FET gate had been reduced in width to only 37 nanometers (nm), and the oxide layer that insulates the gate was only 1 nm thick. To get a feeling for these numbers, imagine that the entire metropolitan area of Greater Los Angeles could be miniaturized and placed into an area the size of a small coin, with all of its freeways, streets, buildings, cars, and people. That gives an idea of the complexity of a modern computer chip. The historical trend of miniaturization (Moore's law) that began at least as early as 1980 indicates that the number of components that can be put onto a semiconductor circuit of a fixed size doubles every 18 months. This trend is expected to continue at least until 2020.

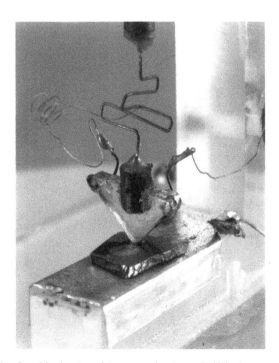

FIGURE 10.29 The first bipolar transistor was about one-half inch across. (With permission of AT&T Archives and History Center.)

Modern computer processors and memory circuits are based on ***integrated circuits*** (***ICs***). An IC is a circuit in which all elements are fabricated on a single semiconductor crystal. The first IC was made by at TI in 1958 Jack Kilby and is pictured in **Figure 10.30**. It contained one double-junction transistor, three resistors, and one capacitor. These elements were connected using external wires, so the circuit is not considered monolithic ("single stone"). The monolithic IC was proposed in 1959 at Fairchild Semiconductor company by Robert Noyce, who suggested using the lithography process to fabricate wires as thin metal strips layered directly on the surface of a crystal. Noyce, a few years later, co-founded Intel Corporation with Gordon Moore. Intel made the first single-chip microprocessor—the Intel 4004—in 1971. This chip, shown in **Figure 10.31**, measured 1/8 inch by 1/16 inch and held 2,300 MOS transistors. It was as powerful in computing capability as the earlier Eniac computer, which contained 18,000 vacuum tubes and occupied 3,000 cubic feet.

FIGURE 10.30   The first IC, built by Jack Kirby. The main part of the device is a bar of germanium (Ge) measuring 7/16 × 1/16 in., with protruding wires and glued to a glass slide. (Courtesy of Texas Instruments.)

FIGURE 10.31   Photo of the first single-chip microprocessor—the Intel 4004—introduced in 1971. (With permission of Intel Corp.)

The process used to create millions of transistors on a piece of Si is analogous to a method known to artists—lithography. Lithography is a technique for transferring a pattern of ink from a flat surface, such as stone, to a piece of paper. The stone can be reused many times to create many nearly identical artworks. Lithography was discovered in 1798 in Germany, and was named from the Greek *lithos*, "stone." This process yields a one-to-one transfer of the image. **Photolithography** is a high-tech form of lithography developed for transferring images onto a Si crystal. Jules Andrus at Bell Laboratories first used photolithography to create semiconductor devices in 1957. In the following sections, we describe the basic steps of making ICs.

10.6.1   Silicon Crystal Preparation

A nearly perfect Si crystal is needed to begin the making of an IC. The method used for making such crystals was described in Real-World Example 4.1. After a thin Si crystal wafer, called a *substrate*, is polished, it is exposed to water vapor (H_2O) at high temperature (1,000°C). This causes *oxidation*, meaning that oxygen atoms penetrate the upper layer of crystal surface and chemically bond with the Si to create a very thin layer of a type of SiO_2. A typical film thickness is 500 nm. This silica form of SiO_2 is noncrystalline (amorphous, or random) and is an electrical insulator.

A second useful type of thin film that can be applied to the surface is a thin film of metal, which serves as a conductor. When patterned into thin strips, metal films act as wiring connecting different components on the surface. A commonly used metal for this purpose is aluminum. Copper has better conductivity properties than does aluminum, but is harder to work with in fabricating reliable circuits. Other types of thin films are often used as conductors, such as polycrystalline Si that has been heavily doped.

10.6.2   Lithography for Fabricating a p-n Junction

Because the transistor is composed of two p-n junctions (diodes), we will use the fabrication of a single p-n junction as our basic example of photolithography. The overall principles of fabrication are similar in all cases, including far more complex designs. Each technique consists of a series of distinct steps, summarized as:

1. Thin-film deposition
2. Photoresist deposition
3. Photo-exposure through a patterned mask
4. Chemical etching to remove unwanted portions of the photoresist
5. Etching away of the exposed portions of the thin film
6. Removal of remaining photoresist
7. Introduction of dopants
8. Metal film deposition to form conducting contacts (circuit wires)

Some of these steps are carried out in separate machines, with the wafer being moved from machine to machine either by hand or by robots.

Figure 10.32 illustrates each of these steps. Starting with a phosphorous-doped n-type Si crystal as substrate (a), a SiO_2 thin film is created on the surface (b) by oxidation, as mentioned above. A thin layer of a liquid polymer (plastic) material is spread evenly on this oxide surface, as shown in (c). Upon heating, the polymer dries into a thin film. This polymer is called a photoresist, and has the property that when exposed to ultraviolet (UV) light, the exposed area undergoes a chemical change and becomes stronger and resistant to being dissolved by certain chemicals called developers. This is analogous to processing chemicals used in photography.

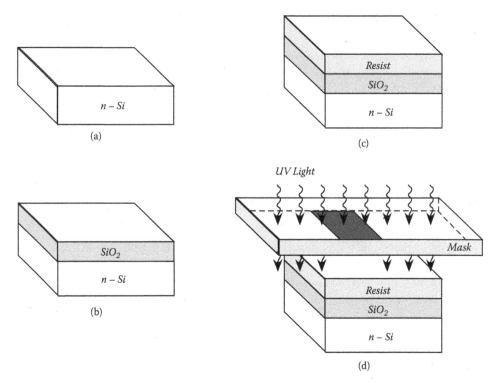

FIGURE 10.32 Fabrication of a p-n junction (a–d). In part (d), light is blocked in the dark-shaded region of the mask. (Adapted from S.M. Sze, *Semiconductor Devices, Physics and Technology,* Wiley, 2002. With permission.)

Some readers will be familiar with the idea of etching glass or metal for artistic purposes. A pattern of an acid-resistant substance (analogous to a photoresist) is painted on a glass surface. Then the whole surface is treated with an acid solution, which etches, or removes, the unprotected parts of the glass surface to a certain depth, which depends on the time of exposure. To stop the etching process, the acid is washed away using water. The pattern created in the etched glass is the negative of that pattern originally made by the photoresist. Silica (SiO_2) is a type of glass, and therefore is susceptible to etching by acid.

A similar process is used in lithography for making a p-n junction, except that the features to be created are far smaller than in the art example. Therefore, the manner for creating the pattern in the photoresist must be capable of much finer control. The method used is to first create a tiny pattern in a metal mask. This mask is a separate thin piece of metal, which is cut or machined very finely using a focused beam of electrons. The metal mask, with its patterns of holes, is placed in contact with the polymer photoresist layer on the SiO_2 layer on the substrate, as in Figure 10.32d. UV light is shone through the openings in the metal mask and into the photoresist just below it. The light causes the photoresist to harden (polymerize) only in the areas directly below the openings in the mask. In Figure 10.32d, the mask is simply a solid bar, shown shaded. The width of the bar may be 500 nm or so. Light does not penetrate this bar and so the resist there is not hardened.

In the next step, a chemical developer is spread over the surface, where it dissolves away the unhardened photoresist in the center region, leaving the pattern of photoresist shown in **Figure 10.33a.** Then a strong acid solution is applied to the surface, where it etches into the SiO_2 (silica glass) thin film, leaving the pattern shown in Figure 10.33b. Next, the remaining photoresist is removed by using a second, stronger developer chemical, leaving the result in Figure 10.33c.

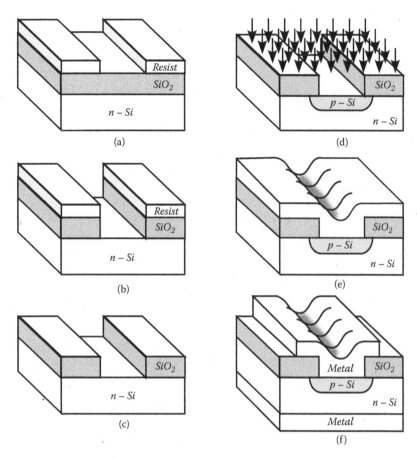

FIGURE 10.33 Fabrication of a p-n junction (e–j). (Adapted from S. M. Sze, *Semiconductor Devices, Physics and Technology,* Wiley, 2002. With permission.)

To create a p-n junction, we need, in addition to the n-type Si crystal, a region of p-type Si crystal. This is created in the region of the substrate that is now not covered by any SiO_2. This small portion of the substrate is converted from n-type to p-type by bombarding the wafer with B (boron) atoms. These enter the crystal by diffusion and, with their deficiency of electrons (compared to Si), create electron holes. This p-type material is created only in the region labeled p-Si in Figure 10.33d.

Now the p-n junction is complete, except for a means of connecting it in an electronic circuit. Rather than connecting external wires for this purpose, thin layers of aluminum metal are deposited on the top and bottom sides on the wafer, as in Figure 10.33e and f. The final diode device is shown in Figure 10.33f. Current will flow if a battery is connected with its plus side to the p-type Si (the top of the device) and the battery's minus side to the n-type (bottom). If connected in the opposite manner, no current (or very little) will flow.

When realizing the complexity of creating just one junction by the many steps described above, it is mind boggling to contemplate creating an entire IC containing millions of diodes, transistors, and other components. Fortunately, all of these components can be fabricated simultaneously, rather than one at a time. A very challenging step in this process of creating an IC is the design and electron-beam cutting of the needed masks. This is a specialty unto itself, and we will not discuss it in detail. The important point is that a single set of masks can be used repeatedly to create thousands of nearly identical circuits. This mass production capability is the essence of lithography.

IN-DEPTH LOOK 10.2: BIPOLAR TRANSISTORS

The same property that makes FETs ideal for logic circuits—low power consumption—makes them less useful as power amplifiers. Power amplifiers are used in audio and music systems such as guitar amplifiers. They are also used to boost signal levels in long-distance digital communications systems—discussed in later chapters.

The earliest and still most common transistor-based power amplifier is the bipolar junction transistor. In contrast to the FET, which is a voltage-controlled valve, a bipolar junction transistor is a current-controlled valve. It is a double diode structure, the main electron current of which can be controlled by a second, smaller current. Bipolar junction transistors come in two types—npn and pnp. An npn transistor is made by contacting a very thin (0.1–5 μm) p-type crystal, called the base, between two larger n-type crystals, called the emitter and collector, as in **Figure 10.34**. This forms two p-n junctions, back to back, labeled 1 and 2. Electron diffusion causes the two n-type crystals each to develop a positive charge near the junctions, indicated in the figure by (+). The p-type crystal develops a negative charge, indicated by (–). As in a diode, these charge buildups lead to forces on electrons in the directions shown by the large arrows at the top of the diagram. These two forces (interpreted as *internal batteries*) push electrons in two directions away from the center of the base region. This is like having a potential energy "ridge" at the middle of the base. Electrons can "fall" in either direction away from this ridge.

First, consider the case that the switch S is in the down position (shown as dashed), so the battery A is irrelevant. The conducting switch keeps the base at the same voltage (potential energy) as the emitter. For this setting of the switch, junction 1 has no external voltage, so no current flows through it. Furthermore, battery B provides a backward external voltage across junction 2 (+ sign to the n-type side), so net current does not flow through junction 2. Therefore the bulb is not lit.

Now consider what happens after the switch S is moved to the up position, connecting battery A across junction 1. Assume that the voltage V_A is much smaller than voltage V_B. Now battery A provides a forward external voltage to junction 1 (+ sign to the p-type side), counteracting the internal force that tries to hold electrons from passing through junction 1 from left to right in the diagram. Because of this counteracting force, junction 1 is now conducting, and this allows electrons to flow around the circuit labeled "small current." A few electrons move per second from battery A, through the emitter, into the base and into the (+) side of battery A.

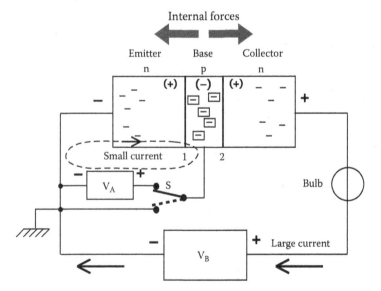

FIGURE 10.34 An npn bipolar transistor. With switch S down, junction 2 blocks current from flowing in the large outer loop. When S is up, as shown, the small current created by battery A in the inner circuit loop enables a large flow of electrons in the outer loop, driven by battery B.

There are also electrons that can be provided by battery B, whose (–) side is trying to push them into the emitter and pull them out through the collector. Battery B provides a much larger voltage than battery A, and provides far more electrons per second than battery A does. As electrons enter the base from the emitter on their way to the (+) side of battery B, many of them find themselves on the right side of the potential energy ridge, at the middle of the base. They are rapidly swept into the collector (hence its name) before they ever reach the wire leading to battery A. This opens up the possibility for another electron to enter the base from the emitter and take its place. The consequence is that a large current flows in the outer loop in the diagram, enabled by the much smaller current flowing in the inner loop. This large current lights the bulb.

A pnp transistor has the same structure as a npn transistor, except that the n-type and p-type regions are interchanged. In terms of the fish analogy used earlier to understand diode action, it is like having a small pool on a ridge of a mountaintop, with water falling on both sides, as in **Figure 10.35**. If a fish on top moves slightly to the right, it falls down on the right, whereas if instead it moves to the left, it falls down on the left. If the left lake is raised, it gives a helping boost to fish trying to go up the falls from left to right. Now many not-so-energetic fish can make it up to the small pool at the top of the ridge, at which point most of them are swept down the falls on the right side into the collector lake.

Bipolar junction transistors were used in early computer logic circuits because this technology was developed earlier than the FET technology. The disadvantage of these transistors for logic circuitry is that they carry considerable current, and this leads to heating of the circuit. Recall from Chapter 5, Section 5.7 that all conducting materials have some resistance to electron current, and this resistance is similar to friction, leading to heat. When building a computer that contains millions of transistors, heating at each transistor is the last thing we want. In fact, one of the major limitations to the number of transistors that can be fabricated on a computer chip is the amount of heat generated, because if the circuit gets too hot, the transistors cease to operate properly. The high temperature creates too many electrons in the upper, conducting energy band, and their sensitivity to the emitter-base current is lessened. For this reason, FETs are preferred and are used almost exclusively in modern computers.

Before the invention of the semiconductor-junction transistor in 1947, signal amplification was done primarily using vacuum tubes. These are familiar today mostly for their use in guitar amplifiers and some high-end stereo systems. Tube amplifiers were used by AT&T for long-distance telephone signals, but they were unreliable, bulky, and expensive. After it was found during the Second World War that semiconductors such as Si and germanium could be used as receivers for microwaves, the leaders of AT&T decided to

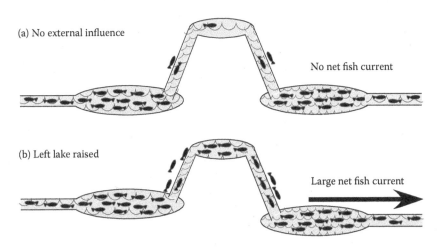

FIGURE 10.35 Fish analogy for transistor action.

research whether semiconductors could also be used as signal amplifiers. Although there had been several precursor inventions along these lines elsewhere, the work at AT&T provided the understanding to make semiconductors into a practical technology.

The team assigned to the project consisted of Shockley, Brattain, and Bardeen. After some initial ideas of Shockley's, and failures in the laboratory, Brattain and Bardeen finally were successful in creating the first pnp point-contact transistor. With Bardeens' theoretical insight and Brattain's expert laboratory technique, the two devised and made the device shown earlier in the photograph in Figure 10.29. As shown in the schematic in **Figure 10.36**, it consisted of a 1-centimeter triangular block made of an insulating material with two pieces of gold metal foil on the surface, separated by a 0.05 millimeter gap at the tip of the triangle. This tiny gap region was gently pressed onto the surface of a crystal of doped n-type germanium (Ge) semiconductor. With gold metal in contact with Ge, which had an excess of electrons, some of the electrons diffused into the metal, which is conducting and initially had no charge imbalance. The electron diffusion stopped when the charge separation created a strong enough internal electric force to prevent any further separation of charge. This diffusion caused small regions under the metal contacts to have a deficit of electrons, and thus they were p-type semiconductors.

Figure 10.36c shows the gap and the small regions below the metal contacts that became p-type because of the electron diffusion into the metal. There is an n-type region between two p-type regions. A pnp double junction transistor was created. Compare to Figure 10.34, which shows an npn transistor. The gold metal on the left side connects to the emitter (p-type region), the gold metal on the right side connects to the collector (p-type region), and the metal at the bottom connects to the base (n-type). When Brattain connected wires with negative voltage to the base and collector, and a wire with positive voltage to the emitter, he found that a small current flowing through the emitter-base loop created a large current flowing through the emitter-collector loop. This was the first example of amplification by a pnp transistor.

The three shared the 1956 Nobel Prize for their contributions to physics. The race was on to use this breakthrough, and many others that quickly followed, for improving communication systems—telephone, radio, and television—as well as computing machines.

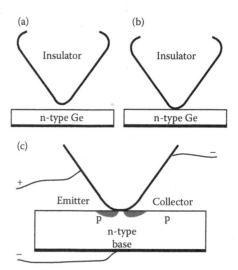

FIGURE 10.36   The first bipolar transistor, shown earlier in Figure 10.29. (a) Triangular insulator material with gold metal foil strips (thick lines) on edges. (b) Making contact with an n-type germanium crystal. (c) Close-up of contact region, showing resulting p-type regions shaded. Metal wires are connected with plus or minus voltages as shown.

SUMMARY AND LOOK FORWARD

In this chapter, we studied the important physics principles governing semiconductor behavior that were discovered in the 1940s and 1950s. Knowledge of this area of physics allowed scientists and engineers in the 1960s and 1970s to create electronic devices that are essential to the operation of computers and the Internet—miniaturized logic and switching circuits.

The keys to understanding semiconductor physics are as follows:

1. Doping—The addition of impurity atoms into a pure Si crystal to make it conducting. Dopants such as P (phosphorous), which bring extra electrons into the crystal, make an n-type crystal, whereas dopants such as B (boron), which are electron-deficient, create electron holes and make a p-type crystal.
2. Diffusion—Electrons move freely and randomly through a conducting material (a doped semiconductor or a metal).
3. p-n junction—If two materials having different concentrations of excess electrons are joined, diffusion of electrons takes place, and a separation of charge occurs. This charge separation causes an internal electric field to build up, which prevents the diffusion from causing any further charge separation.

On the basis of these principles, useful electronic components were created.

1. Diode—A p-n junction, in which current can flow only if the applied voltage is more positive at the p-type side of the diode than at the n-type side. This can be used to rectify a signal (i.e., remove its negative components) leaving a positive-only signal.
2. Transistor—The workhorse of the Information Age is the transistor, in particular the MOS FET, or MOSFET. A voltage applied between the gate and the body of the transistor turns the transistor ON, connecting the source and drain through a conducting channel. By using one n-FET and one p-FET in a complementary arrangement, voltage switching can be achieved with very little current flow. Lower current allows circuits to use a minimum of power and therefore not overheat.

We have reached an important milestone in our understanding of the physics behind computer operation. From the principles of quantum physics and electromagnetism, we are able to understand the workings of transistors. The CMOS-based electronic circuits we have studied can be used to construct any logic operation desired. In Chapter 6, we discussed how logic operations are used to accomplish any computer information-processing task. We are close to understanding all of the elements needed for the operation of computer hardware, but one element has not been covered—computer memory. We will discuss that in Chapter 11. Another application of semiconductors is based on how they emit or absorb light, which is the basis of photocells, photodetectors, light-emitting diodes (LEDs), and lasers. We will study the physics of semiconductors and light in the next chapter.

SOCIAL IMPACTS: LABELING EVERY OBJECT IN THE WORLD

Unlike a UPC bar code, Electronic Product Code provides for the unique identification of any physical object in the world.

Steve Meloan
(*Sun Developer Network,* 2003)

Most people think of a personal computer or a PDA as things connected to the network. But here we are connecting trees, race cars, and astronauts to the network. In the future, everything of value will be on the network in one form of another. And once they're on the network, we can aggregate data from those diverse devices, and then deliver that data to equally diverse devices.

John Fowler
(*Software CTO of Sun Microsystems*)

Radio-Frequency ID is about to take on a life of its own.

"RFID Goes Mainstream: Alien Technology Whitepaper"
(2007)

Supermarkets use black-and-white printed bar codes to identify items for pricing and inventory. To read its code, a worker moves it close to an optical reader. Imagine a new system, in which tiny microchips, smaller than an ant's head, are attached or inserted into every item in a store, and could be read electronically from a distance of several meters. Now imagine that these chips could be implanted under the skin of dogs, cats, children, husbands, wives, and parents. Finally, imagine a system where every person, pet, car, bike, skateboard, guitar, compact disk, book, pair of pants, gun, bullet, flag, etc., had not only an electronic "bar code" identifying the type of item, but also a unique bar code identifying every individual item. The basis of such a system is already developed and is being implemented increasingly in the commercial world. As with any new technology, such a system has wonderfully positive potential, and frightening potential for abuse.

Before discussing the social issues, let us briefly consider how the technology works. First, how many items could be labeled with unique codes? Using a binary number having N bits to label each item, we can create 2^N unique codes (see Chapter 2). For example, with $N = 4$, we can create $2^4 = 2 \times 2 \times 2 \times 2 = 16$ unique codes: 0000, 0001, 0010, etc., up to 1111. If, instead, we used 33 bits to create each number, we could create $2^{33} = 8,589,934,592$ unique codes, one for every human on Earth. For example, your personal code could be

$$011000111000100001110010000101010.$$

If we wanted to expand this labeling scheme to every single object on Earth, this could be done, at least in theory. If we used 90 bits to create them, we could create $2^{90} = 1.2 \times 10^{27}$ unique codes. These are enough codes to label uniquely every cubic millimeter of material (including rocks, dirt, and ocean) within 1 kilometer of the Earth's surface.

The technology used for labeling is radio frequency identification (RFID) tags. Each tag contains a tiny antenna, which is used for receiving and transmitting, and a Si chip used for data storage. A tag reading device, or reader, transmits a low-power radio signal to the tag, which responds by broadcasting a message back, containing information stored in the tag's memory. There are two types of RFID tags. Passive tags have no battery and must receive power from the radio waves sent by the reader. Therefore, these tags work only at close range, up to a couple of meters. Active tags contain their own battery for power and can therefore be read from greater distances.

Two significant commercial developments in this area are the Electronic Product Code (EPC) Network and the subdermal RFID implant. The EPC Network is based on inserting tags into each item produced by a company so that it can be tracked during manufacturing, distribution, sales, and (potentially) after being purchased by a customer. Subdermal RFID implants are inserted under the skin, often in the back of the upper arm, or triceps muscle. These can be used either to monitor people, such as Alzheimer's patients who might wander off and get lost, or workers in banks or other high-security positions, where their identity can be easily verified. Note that a person carrying a typical implanted chip cannot be tracked from a distance more than a few hundred meters.

Great benefits can result from using these labeling and tracking technologies. Companies can put resources when and where they are needed and potentially save costs, as well as do a better job of conserving energy. The worldwide market size for RFID products and services was about $4.5 billion in 2008. People with health risks can "get chipped," and then if they end up unconscious in an emergency ward, doctors can identify them by their code and look up their medical records in an online database. Companies can make more money by better identifying what types of products are bought where and in what combinations. On the lighter side, news agencies reported that the Baja Beach Club in Barcelona, Spain is the first business to offer RFID subdermal implants in customers to give them access to VIP areas and provide an easy payment option. On a darker note, some people might be in favor of implanting chips under the skin of sex offenders or other criminals, although no program of this kind is in force as of 2008.

It is easy to see why some people are concerned about potential abuses of this technology. It could raise the specter of a dictatorial government (called Big Brother by George Orwell [1]) monitoring our every move [2]. There have been studies that suggest subdermal implants might cause cancer in laboratory mice [3,4]. Those standing to benefit financially from the industry point to counter reports arguing that the mice studies should not be seen as conclusive [5]. There are concerns about people being forced to accept chip implants by governments (security workers or "troublemakers") or companies (employees). Wisconsin, North Dakota, and California have passed laws prohibiting employers and others from forcing anyone to have an RFID device implanted under their skin [5]. There are concerns that a person's RFID implant could be read by others without permission. There are also privacy concerns about RFID chips placed into passports, driver's licenses, or student identification cards. Also, hackers could possibly alter the data in someone else's implanted chip, for their own purposes.

So which is it: *Brave New World*, or *Better-Living-Through-Technology World*? Or both? Or neither?

REFERENCES

1. Orwell, George. *Nineteen-Eighty-Four*. London: Harcourt, Brace & Co., 1949.
2. Albrecht, Katherine, and Liz McIntyre. *Spychips: How Major Corporations and Government Plan to Track Your Every Move with RFID*. New York: Plume, 2006.
3. Albrecht, Katherine. "Microchip-Induced Tumors in Laboratory Rodents and Dogs: A Review of the Literature 1990–2006." CASPIAN Consumer Privacy, November 19, 2007. http://www.antichips.com/cancer.

4. Le Calvez, Sophie, Marie-France Perron-Lepage, and Roger Burnett. "Subcutaneous Microchip-Associated Tumours in B6C3F1 Mice: A Retrospective Study to Attempt to Determine their Histogenesis," *Experimental and Toxicologic Pathology* 57 (2006): 255.

5. Wustenberg, William. "Effective Carcinogenicity Assessment of Permanent Implantable Medical Devices: Lessons from 60 Years of Research Comparing Rodents with Other Species." AlterNetMD Consulting, Farmington, MN, 2007.

SOURCES FOR SOCIAL IMPACTS

How Stuff Works. http://electronics.howstuffworks.com/rfid1.htm.
RFID Journal. http://www.rfidjournal.com.

SUGGESTED READING

See the general physics references given at the end of Chapter 1.

Bierman, Alan W. *Great Ideas in Computer Science—A Gentle Introduction.* Cambridge, MA: MIT Press, 1997.

Brinkman, William, Douglas Haggan, and William Troutman. "A History of the Invention of the Transistor and Where It Will Lead Us," *IEEE Journal of Solid-State Circuits* 32 (1997). http://www.sscs.org/AdCom/transistorhistory.pdf.

Hepher, M. "The Photoresist Story," *Journal of Photographic Science* 12 (1964).

Riordan, Michael, and Lillian Hoddeson. *Crystal Fire.* New York: W. W. Norton & Company, 1997.

Seitz, Frederick, and Norman G. Einspruch. *Electronic Genie: The Tangled History of Silicon.* Urbana: University of Illinois Press, 1998.

"Transistorized!" ScienCentral, Inc. and The American Institute of Physics, 1997. http://www.pbs.org.

For instructors:

There are few references with elementary discussions of FET operation. Books that are helpful:

Ferry, David K., and Jonathon P. Bird. *Electronic Materials and Devices.* San Diego: Academic, 2001 (see pp. 249–282).

Melissinos, Adrian C. *Principles of Modern Technology.* Cambridge, UK: Cambridge, 1990. (Sections 1.7–1.9 describe FET operation, including depletion FETs. Section 2.7 describes DRAM.)

Neamen, Donald. *An Introduction to Semiconductor Devices.* New York: McGraw-Hill, 2006.

Pierret, Robert F. "Field Effect Devices," in *Modular Series on Solid State Devices*, edited by R. F. Pierret and G. W. Neudeck, Vol. IV. Reading, MA: Addison-Wesley, 1983. (Sections 2.1, 2.2, and 5.1 give qualitative discussions of FET operation.)

Sze, S.M. *Semiconductor Devices, Physics and Technology.* Hoboken, NJ: Wiley, 2002.

Taur, Yaun, and Tak H. Ning. *Fundamentals of VLSI Technology.* Cambridge, UK: Cambridge, 1998 (Sections 2.3.1, 2.3.2, 5.1 and Chapter 3 Introduction).

KEY TERMS

Amplitude

Bipolar transistor

CMOS

Depletion region

Diffusion

Diode

Dopant or impurity

Doping

Electron

Field-effect transistor (FET)

Hole

Insulator

Integrated circuit (ICs)

Logic gate

n-channel FET, or n-FET

n-type semiconductor

p-channel FET, or p-FET

p-n junction

p-type semiconductor
Photolithography
Rectifier
Resistor
Solid-state electronics
Transistor

ANSWERS TO QUICK QUESTIONS

Q10.1 The molecules of dye feel random forces exerted on them from the nearby water molecules colliding with them. Each collision pushes a given dye molecule in some random direction. After many such collisions, different dye molecules will find themselves at different locations in the water. As a whole, the dye molecules eventually spread out to fill the water more or less uniformly.

Q10.2 If both switches are pushed ON (logic value = 1), the FET's gate becomes more positive than its body, and current flows through the source and drain, lighting the bulb. This performs the AND logic operation.

EXERCISES AND PROBLEMS

Exercises

E10.1 For some elements, when several atoms of the same element are brought together, they form a crystal. The energy-level diagram of the allowed energies of electrons for an individual atom changes dramatically when this occurs. Below are three energy-level diagrams for three different crystals. Label each diagram as a conductor, insulator, or semiconductor and explain what distinguishes them. (In the pictures here, dots indicate full or partially full bands, but the particular numbers of dots do not have any significance.)

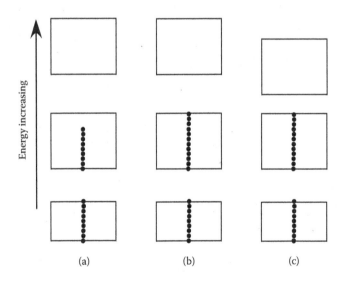

E10.2 Explain why the electrical conductivity of a pure Si crystal changes when it is warmed from 80°C below freezing to room temperature (20°C). Does the conductivity increase or decrease? How does the resistance change?

E10.3 If the band gap in Si were equal to 1.8 aJ instead of 0.18 aJ, would this make it a better or worse conductor at room temperature? Why? *Hint:* You may need to look at some information given in the previous chapter.

E10.4 In 2006, hyperpure Si cost as much as \$3/g (about \$100/oz). One cubic centimeter of Si has a mass of 2.3 g, meaning that the cost per cubic centimeter equals about \$7.00. Consider a cylinder (boule) of hyperpure Si that is 20 in. (51 cm) in length and has diameter of 8 in. (20 cm). The boule is sliced into 510 1-mm-thick disks or wafers, and 100 ICs are made on each wafer.

   (a) What is the volume (in cm³) and cost of the Si for each wafer? *Hint*: The volume of a wafer equals its thickness times its surface area, and the surface area equals πr^2, where r is the radius (one-half the diameter) of the wafer.
   (c) What is the total cost of the boule?
   (d) What is the Si cost per IC?

E10.5 (a) Explain how adding small amounts of P to a pure Si crystal greatly increases its conductivity.
   (b) Predict, using the Periodic Table, at least one other element besides P that would likely have a similar effect when doped into Si.
   (c) Explain how adding small amounts of B to a pure Si crystal greatly increases its conductivity. On the basis of the Periodic Table, predict at least one other element besides B that would likely have a similar effect when doped into Si.

E10.6 Invent and explain a situation from everyday life that is analogous to the statement, "The presence of electron holes allows current to flow easily in a crystal that would otherwise be nonconducting."

E10.7 Use the following analogy to explain how a semiconductor diode works to allow electrical current to flow in one direction, but not in the other. Two boxes, called box A and box B, sit on a table and contain a few hundred small marbles each. One box (box A) is continuously shaken side to side randomly, so the marbles in it never stop rolling around, bouncing off the walls and each other. A small opening is cut into the side of each box, down to the level of the table, and a piece of cardboard is folded into a rectangular channel, allowing marbles to freely move from one box into the other. Continue to develop this analogy to illustrate diode operation. (Read the explanation in terms of swimming fish.) What do you need to do with box B to make the diode analogy complete? *Hint*: Think about potential energy and the two lakes in the fish analogy.

E10.8 A fun model for a p-n junction can made by imagining a rock concert, where the area in front of the stage is separated into two zones by a curtain. The zone to the left is the mosh pit—a region filled with closely packed people (moshers), with zero space between them. The mosh pit is "doped" with a few crowd surfers, who have enough energy to move above the heads of the moshers. (An amusing animation of crowd surfing is at http://members.aol.com/rik0lar/moshing/mosh.htm.) In the zone to the right of the curtain, the closeness of standing people is not great enough to allow crowd surfing—a surfer would fall to the floor between people. Elaborate this model further to explain what happens when the curtain is suddenly opened. Invent a mechanism analogous to the internal battery that develops at a p-n junction, which prevents the charge separation from increasing indefinitely.

E10.9 (a) Extend the mosh-pit analogy discussed in the previous exercise to model the operation of an n-FET. As a joke, I call this a MOSHFET. Consider the insulator region to be represented by the stage, which is too high for crowd surfers to reach.
   (b) Do the same for the p-FET.

E10.10 (a) When a material is classified as a conductor, insulator, or semiconductor, what is meant by these classifications in terms of the electrical current flow?

(b) Explain why a ceramic object (made of randomly packed crystallites) might not be a good conductor.

(c) In an n-type semiconductor, it is the excess electrons (which are in the higher-energy conducting band) that are free to move. We say such electrons are mobile. In a p-type semiconductor, in which energy band do the mobile electrons reside? Make an energy-band drawing to explain your answer.

E10.11 (a) In a FET, does the voltage between the source and drain determine whether it is ON or OFF? Explain.

(b) For each of the following cases, would an n-FET be ON or OFF? (i) $V_G = 0$ V and $V_B = -5$ V; (ii) $V_G = 0.3$ V and $V_B = 0$ V; (iii) $V_G = 1$ V and $V_B = 0$ V; (iv) $V_G = 0$ V and $V_B = -5$ V; (v) $V_G = 0$ V and $V_B = -1$ V.

(c) For each of the following cases, would a p-FET be ON or OFF? (i) $V_G = 0$ V and $V_B = 1$ V; (ii) $V_G = 4$ V and $V_B = 5$ V; (iii) $V_G = -1$ V and $V_B = 0$ V; (iv) $V_G = 0$ V and $V_B = -5$ V; (v) $V_G = 0$ V and $V_B = -0.3$ V.

E10.12 A rectangular IC chip has sides of length 1 cm and 2 cm. This chip is completely covered with transistors each having dimensions 1×1 μm. Calculate the maximum number of such transistors that can fit on this chip.

E10.13 Using state-of-the-art transistors in 2006, a single switching operation could be achieved using as little as 1 fJ (10^{-15} J). The smallest switching time achieved was 10 ps.

(a) What power does this correspond to?
(b) How many such switching operations can be performed in 1 sec?

E10.14 Do some research online and/or in the book references given at the end of this chapter to learn more about the history of science leading to the development of semiconductor electronic devices. Present the history in the form of a timeline, showing 10 to 20 major events occurring between about 1900 and 1960. Provide commentary on each event. Give references.

Problems

P10.1 A pure Si crystal contains about 5×10^{22} atoms/cm³.

(a) How many protons per cubic centimeter does a Si crystal contain?
(b) How many electrons per cubic centimeter does a Si crystal contain?
(c) In a pure Si crystal at temperature 200°C (very hot), there are about 1.5×10^{14} electrons/cm³ in the conducting band. Which would be more conducting—a pure Si crystal at 200°C, or a Si crystal doped with one part per million of phosphorus at room temperature (about 22°C)?

P10.2 The discussion of p-n junctions in connection with Figure 10.9 assumes that some of the excess electrons in the n-type region diffuse the to the p-type region; that is, only electrons in the highest energy band diffuse. This is a simplification of the actual situation in real crystals. Redraw Figure 10.9 allowing also for holes to move (diffuse) from the p-type region to the n-type region. When we say a hole diffuses, this means that some other physical objects move to fill in a hole, leaving a new hole nearby. In this case, what physical object actually moves, and which energy band is this object in?

P10.3 A certain p-n junction is made of Si doped with P at a density 5×10^{16} atoms/cm³ and Si doped with B at a density 5×10^{16} atoms/cm³. The depletion region is 10 μm thick and 100×100 nm in area.

(a) What is the volume of the depletion region, in cubic centimeters?
(b) How many excess electrons from the n-type side have diffused over to the p-type side?

P10.4 In the four drawings below, indicate which light bulbs (indicated by ovals) would light.

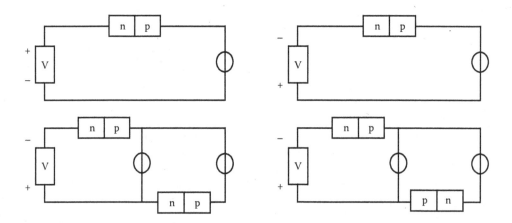

P10.5 A diode and a resistor in series are connected to an oscillating voltage, as shown below. The graph of voltage indicates the voltage at the left side of the voltage source (V) relative to the voltage at the right side. Make graphs of the current in the resistor and in the diode, assuming that current flowing left to right corresponds to a positive current, and that the maximum current in each equals 2 amp.

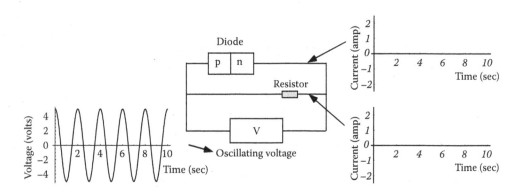

P10.6 Recall from Chapter 6 that NOR logic gates can be used to construct any other logic gate. By combining NOR gates of the form shown in Figure 10.24, how many n-FETs and p-FETs are needed to construct each of the following logic devices?

(a) AND
(b) OR
(c) half-adder
(d) full-adder

P10.7 Construct a diagram showing how the NOR logic operation could be implemented using the whimsical water-effect, or WET devices, described in In-Depth Look 10.1.

P10.8 Prepare the voltage (logic) tables for voltage inputs and outputs for the following CMOS circuit. Fill in all of the blanks, and explain in words the overall operation. *Hint*: This is a NAND gate.

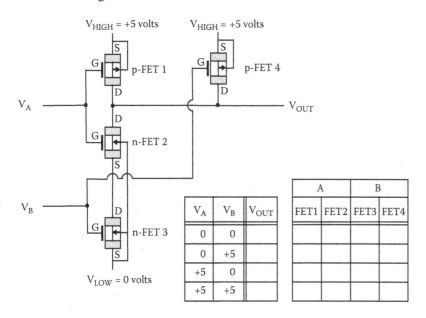

V_A	V_B	V_{OUT}	FET1	FET2	FET3	FET4
0	0					
0	+5					
+5	0					
+5	+5					

P10.9 Work out the voltage (logic) tables for voltage inputs and outputs for the following CMOS circuit. Fill in all of the blanks in the logic table below and explain in words the overall operation. *Hint*: This gate operation can be summarized as: [NOT($V_C = +5$)] AND [NOT($V_A = +5$) OR NOT($V_B = +5$)]

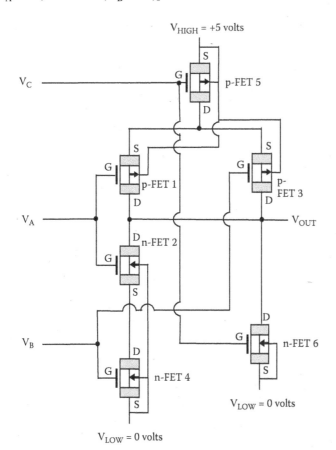

					A		B		C	
V_A	V_B	V_C	V_{out}		p-1	n-2	p-3	n-4	p-5	n-6
0	0	0								
0	0	+5								
0	+5	0								
0	+5	+5								
+5	0	0								
+5	0	+5								
+5	+5	0								
+5	+5	+5								

P10.10 (This problem depends on In-Depth Look 10.2)

(a) Describe the operation of a bipolar-junction transistor that has a thin n-doped region sandwiched between two p-doped regions. This is a pnp transistor. Draw a circuit diagram showing how the batteries should be wired (including their polarities; i.e., location of plus and minus terminals) to obtain the same valvelike action discussed in this chapter for npn transistors.

(b) Draw the corresponding fish-transistor diagram, analogous to the diagram in Figure 10.35 in In-Depth Look 10.2.

P10.11 (This problem depends on In-Depth Look 10.2)

Further develop the boxes and marbles analogy in E10.7 to explain how a bipolar semiconductor transistor works.

11

Digital Memory and Computers

A device with two stable positions, such as a relay or a flip-flop circuit, can store one bit of information.

Claude Shannon
(1948)

Using a laptop computer in the park at the University of Oregon.

Programmers hard-wiring a program for the ENIAC, the first large-scale, general-purpose computer, built in 1945 at the University of Pennsylvania. (With permission of the Army Research Laboratory Technical Library. Courtesy of Mike Muuss.)

11.1 PHYSICS, MEMORY, AND COMPUTERS

Now we are at the point where we can put it all together and see how a computer operates. All computers have several features in common: (1) means for a user to enter logical bit values, (2) means to store those values, (3) an arrangement of logical operations (a *program*) to manipulate the bit values, and (4) means to send the logical results of the program out to the user. In Chapter 2, Sections 2.8 and 2.9, we discussed binary numbers and the concept of information. In Real-World Example 5.1, we discussed random-access memory (RAM), which is based on an array of capacitors. In Chapter 6, we discussed the concepts of logic, binary arithmetic, rudimentary electronic switches, and relays for implementing logic operations. In Chapter 8, Sections 8.2 and 8.3, we discussed analog and digital signals and how digital signals are stored and transmitted as binary information. In Chapter 10, Section 10.5, we discussed how transistors are used to perform logic operations much faster and less expensively than could be accomplished using electromechanical switches and relays.

From our discussions in the chapters mentioned above, we can state the following two underlying principles of computing:

PHYSICAL PRINCIPLES OF COMPUTING (I, II)

(i)   Computers require that physical devices be used for representing and storing logic bit values. Each object can be in one of two states, or configurations, representing logic values 1 or 0.
(ii) Computers require physical switches (usually electronic) for performing logic operations.

These imply that the ultimate limits of computing machinery are subject to the laws of physics. Computing does not exist in an abstract, purely mathematical realm. It exists in the physical world.

In this chapter, we will see that in order to make electric circuits that can store bit values (i.e., act as memory) we need to consider logic circuits that are somewhat different from those we studied in Chapter 6. We will also see that the physical devices we use for short-term memory are different from those we use for long-term memory. We will discuss the hierarchy of memory types used in computers and see how this leads to efficient use of hardware. Finally, we will discuss how the components of a computer are organized, and how information is processed through these components.

11.2   SEQUENTIAL LOGIC FOR COMPUTER MEMORY

In this text we will not delve deeply into how logic instructions are carried out in a computer. Rather, we will focus on the physics underlying some of the important components. Especially important is implementation of computer memory. What is memory? In your brain, memory is a physical record of some earlier event. How can we build electronic circuits that have this property? We will see that the underlying physics of each type of memory technology determines whether it is best suited for high-speed, temporary memory or for low-speed, permanent memory, or for something in between.

Switches and logic gates are likely candidates for building memory, but there is a catch. The circuits that we studied in Chapter 6, called **combinational logic circuits**, cannot function as memory devices. This is because their output values are determined completely by the present values of their inputs. They are not influenced by the values that the inputs had previously. Recall the rules for combinational logic circuits:

Rule 1. It is forbidden to combine two wires into one wire.
Rule 2. It is allowed to split a wire into two or more wires.
Rule 3. An output wire from one gate can be used as an input to a different gate.
Rule 4. No output of a gate can eventually feed back into the input of the same gate.

To make **memory circuits**, we must change rule 4; that is, we must allow the output of a gate to feed back into the input of the same gate. This creates a self-reinforcing situation, which continually reminds the gate of what its input was at an earlier time. It puts the gate into a kind of logic loop that it has a hard time getting out of. (It is a little like when you cannot get an irritating song out of your head.) Logic circuits that have feedback are called **sequential logic**, because their outputs depend on the sequence in which the inputs change from one value to another.

Rule 4′ for sequential logic: By permitting a logic gate's output to feed back into
its input, a logic circuit can be designed to store information.

Consider some examples of sequential logic:

Feedback Example #1: NOT Loop

Figure 11.1 shows two NOT gates wired into a feedback loop. The lower output Q is
"fed back" to the input of the upper NOT gate, and vice versa. This *feedback* circuit
has two stable states (configurations). It is stable if $Q = 1$ and $Q' = 0$. It is also stable if
$Q = 0$ and $Q' = 1$. If we could find a way to cause the circuit to switch between these
two stable configurations, we could use it as a memory to store a single bit value. The
problem with this circuit is that it has no inputs, so it cannot be switched.

We can draw the NOT feedback loop in a more suggestive way, as in **Figure 11.2**.
This might remind you of two cats chasing each others' tail.

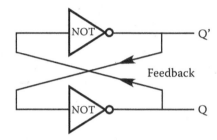

FIGURE 11.1   NOT loop feedback circuit.

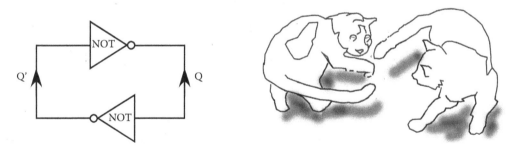

FIGURE 11.2   NOT loop feedback circuit as mutual tail chasing.

Feedback Example #2: One-Time Latch

Next, consider the circuit in **Figure 11.3**. The output Q is fed back to the input B, with
the arrows showing the direction of travel of the bit values. The other input, A, can
equal 0 or 1. What are the possible stable configurations of inputs and outputs? Suppose

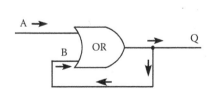

Sequential Logic History			
Event Order	A	B	Q
1	0	0	0
2	1	1	1
3	0	1	1
4	1	1	1

FIGURE 11.3   Feedback circuit, forming a one-time latch. After the input A is set to 1, flipping Q to
1, it cannot be reset to $Q = 0$.

that A = 0 and B = 0. Then the output Q is 0, and that is consistent with B = 0. So, this is a possible stable configuration.

Assume that initially the circuit is in this configuration: A = B = 0, with Q = 0. In the sequential logic history shown, this top row is called "event 1." The sequential logic table is similar to a story (a history). It should be read starting at the top row, and progressing down one row at a time (as frames on a movie film strip). If at some later time (event 2) the value of A is changed to 1, then Q goes to 1, causing B to equal 1 as well. We say the Q value has been set. This is also a stable configuration. In fact, it is so stable that even if, at a later time (event 3), A is changed back to 0, the Q value remains at 1 (and so does B). The value of Q is locked and cannot be changed after this. This latch is not resetable. It is called a ***one-time latch*** because it can operate only once. Notice that A = 1, B = 0 is a configuration that can never be reached. This circuit certainly has memory. It "knows" whether or not it has been set. It would be nice, though, to have a circuit that can be set to Q = 1, and then reset back to Q = 0.

QUICK QUESTION 11.1

For the circuit below, the A value initially equals 1, giving Q = 1, in a stable configuration. Then you change the input value to A = 0. Explain why this does not lead to a stable configuation.

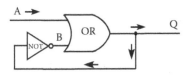

Sequential Logic History

Event Order	A	B	Q
1	1	0	1
2	0	?	?

THINK AGAIN

Not all input values will lead to stable configurations. Consider the example in the following Quick Question.

11.2.1 The Set-Reset Latch

Let us design a latch that can be reset; that is, one that can be set to Q = 1, then later set back to 0, and so on. This type of latch is called a ***set-reset latch***, or ***S-R latch***. It can be constructed by combining two feedback circuits, as in **Figure 11.4**. The logic value S is the *set* input, the logic value R is the *reset* input, and Q and Q′ are the outputs.

In such a circuit, the output values are not given uniquely by the present values of the inputs, but depend on their past history. The behavior of a S-R latch is summarized as follows. First, consider two possible cases:

(a) If R = 0 and S = 1, then Q = 1, and Q′ = 0. That is, Q = S (and Q′ = NOT S).
(b) If R = 1 and S = 0, then Q = 0, and Q′ = 1. That is Q = S (and Q′ = NOT S).

In each of these two cases, we say that Q follows S. Next, start from either of the above cases, then change S or R so that both R = 1 and S = 1. Then Q holds the previous value of S. The input R = S = 0 is to be avoided, because this leads, in some cases, to unstable values. (If we have R = S = 0, and then we simultaneously change both R and S to 1, then the Q and Q′ values switch forever between 1 and 0.)

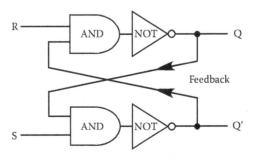

FIGURE 11.4 Set-reset (S-R) latch. S is the *set* input, R is the *reset* input, and Q and Q′ are the outputs.

Let us see how this behavior comes about. There are four useful logic states:

$$R = 0, S = 1 \quad \text{with} \quad Q = 1, Q' = 0$$
$$R = 1, S = 0 \quad \text{with} \quad Q = 0, Q' = 1$$
$$R = 1, S = 1 \quad \text{with} \quad Q = 0, Q' = 1$$
$$R = 1, S = 1 \quad \text{with} \quad Q = 1, Q' = 0$$

With the first row of values, the upper AND gate sees $R = 0$ as an input, so it puts out a 0, regardless of the Q' value. The output of this AND gate is changed to 1 by the NOT gate, producing $Q = 1$. The Q value feeds back to the input of the lower AND gate, making its output 1, which is changed to a 0, making $Q' = 0$. Note that in this case $Q = S$. A similar argument holds for the second row of values in the list, again leading to $Q = S$. In these two cases, Q follows S.

With the third row of values, the upper AND gate sees two 1s as its inputs (assuming $Q' = 1$), so it puts out a 1, which gets inverted to a 0 by the NOT gate, producing $Q = 0$. The Q value feeds back to the input of the lower AND gate, making its output 0, which gets changed to 1, making $Q' = 1$. The Q' value feeds back to the input of the upper AND gate, and we see that all is consistent. A similar argument holds for the fourth row of values in the list.

There is an interesting property of the third and fourth rows. They have the same inputs, but different outputs! This is clearly not a combinational logic circuit, whose outputs depend only on the present values of the inputs. The S-R latch circuit depends on prior history. For example, say the configuration is presently as in the first row, with $R = 0, S = 1, Q = 1$. Now change R to 1, so the configuration is as in the fourth row (not the third row!). Q has not changed. Its value is held to its previous value.

This circuit works as memory, but there is a potential problem. If both R and S equal 0, then the circuit would give $Q = Q' = 0$, and erase the earlier history. Worse yet, if we have $R = S = 0$, and then we simultaneously change both R and S to 1, then the Q and Q' values switch erratically. We need a way to avoid the $R = S = 0$ state.

The $R = S = 0$ configuration can be avoided by placing a push-off switch at each input, with a logical value 1 to the left of each switch, as in **Figure 11.5**. Only one switch can be pushed at a time (as indicated by the single finger) to create a 0 input, and when let go it pops back to the 1 value. This type of switch acts as a NOT gate. Next, we will consider how to implement this idea by using only logic gates, rather than switches and a finger.

11.2.2 The Enabled Data Latch, or D-Latch

The circuit in **Figure 11.6** is an *enabled data latch* (or *D-latch*, or a *D flip-flop*), which solves the problem in the S-R latch by automatically preventing the $R = 0$, $S = 0$ input from occurring. The right-hand side of the circuit is again the S-R latch. But now we

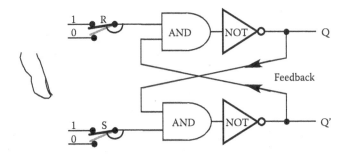

FIGURE 11.5 Set-reset (S-R) latch. S is the *set* input, R is the *reset* input, and Q and Q' are the outputs. Only one switch can be pushed open at a time.

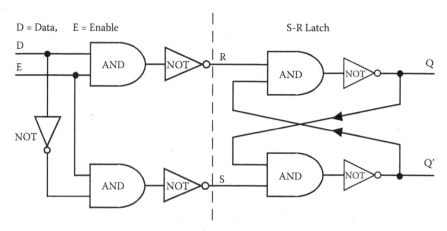

FIGURE 11.6 Enabled D-latch. When the enable E = 1, then Q equals the value of the data D. When E = 0, the value of Q cannot be changed.

send the *Data*—the value we wish to store in memory—into the D line, and we have a new input called *Enable*, which goes into the E line. The idea is that only when E = 1 can the stored value of Q be changed. When E = 1, the value of Q will follow D (i.e., Q = D). Then when E is changed to 0, the value of Q becomes frozen, or protected. Any change of D afterward will not affect Q, as long as E = 0.

The input part of the circuit prevents R = S = 0, because this could happen only if both of the AND gates on the left-hand side had zero outputs. That is impossible because of the NOT gate between the D input and the lower AND gate.

To see how the enabling operation works, consider this example sequence, shown in the sequential logic history table.

1. E = 0. Both AND gates on the left have outputs = 0, which are inverted by the NOT gates to give R = 1, S = 1. This causes the S-R latch to hold the present value of Q (in this case Q = 1), as we saw in our earlier discussion.
2. Changing D from 0 to 1 does nothing to the value of Q. (Still Q = 1.)
3. Now change E to 1. This causes R to change to 0. This is the configuration of the latch that causes Q to follow S. It is not hard to see that S equals D in this case. Therefore, Q follows D; that is, the memory location Q equals the value of the data D that we are trying to store.
4. Next, change D to 0. Q follows D, so Q = 0.
5. Next change E back to 0. R changes to 1, and now the Q value (Q = 0) is again protected against further changes in D.

This circuit makes a stable read-write memory. It remembers for as long as we hold E = 0. The only way to make a further change in Q is to first change E to 1.

QUICK QUESTION 11.2

Fill in the remaining blank entries in the Sequential Logic History for Enabled D-Latch.

Sequential Logic History for Enabled D-Latch

Event Order	E	D	R	S	Q	Q'
1	0	0	1	1	1	0
2	0	1	1	1	1	0
3	1	1	0	1	1	0
4	1	0	1	0	0	1
5	0	0	1	1	0	1
6	0	1				
7	1	1				
8	0	1				

11.3 STATIC RANDOM-ACCESS MEMORY

An important type of electronic memory is *RAM*, or *random-access memory*. The term *random access* means that any memory location can be accessed without having to go through all memory locations one after the other. (A telephone book is random-access, in that you can open it right to a person's name without reading the book cover-to-cover. An example of a memory that is not random-access is a video magnetic tape, because to reach a certain data location you must wind through the tape.) There are two common types of RAM. Let us first discuss *static RAM*, called *SRAM*. (RAM is pronounced "ram," as in the animal, and SRAM is pronounced "es-ram".) SRAM is very fast in its operation, but is bulky and expensive, compared with some other types. Within a computer, SRAM is used only for limited purposes for which speed is more important than cost. A typical amount of SRAM in a desktop computer processor is about 5 megabytes (MB).

SRAM consists of many enabled D-latches. Each D-latch is called a memory *cell*, and can store one bit of information. The D-latches are constructed using transistors to make a circuit analogous to that shown in Figure 11.6. It is easier to understand the transistor-based circuit if we first redraw Figure 11.6, using the definition of a NAND gate, which is an AND followed by a NOT. This gives the D-latch diagram shown in **Figure 11.7**. As we learned in Chapter 10, Section 10.5, each NAND gate can be built using four *field-effect transistors* (FETs). Therefore, to implement the D-latch as shown in Figure 11.7 requires 18 FETs, including two for implementing the NOT gate. Circuit designers have actually found more compact circuits than this, which require only six FETs per cell. This technique is described in In-Depth Look 11.1, but is not essential to our discussion of computers.

The benefits of SRAM built using FETs are: (1) data can be stored and retrieved fast (nanoseconds per bit), (2) storage of bits is maintained as long as a voltage is applied to the gates of the FETs, and (3) very little electrical power is used while storing bits—values are stored as long as power is maintained in the circuit.

How many SRAM memory cells can be put on a silicon wafer with an area of 1 square centimeter? One square centimeter (cm^2) equals 1×1 cm, or 1×10^{-4} meters squared (m^2). Each SRAM cell is comprised of six transistors. Assuming that one transistor occupies an area of 100×100 nm (or 1×10^{-14} m^2), one SRAM cell occupies an area of about 6×10^{-14} m^2. The number of SRAM cells that can be put in 1 cm^2 is therefore

$$\frac{1 \times 10^{-4} \, m^2}{6 \times 10^{-14} \, m^2} = 0.17 \times 10^{10} = 1.7 \times 10^9 \text{ cells}$$

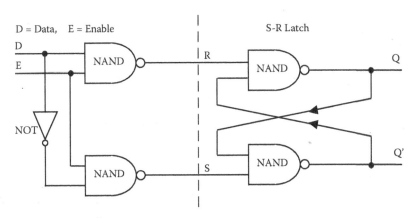

FIGURE 11.7 Enabled D-latch implemented using NAND gates.

Each SRAM cell stores one bit, so 1 cm² of SRAM can store about 1.7 billion, or 1.7 Gbits. Because 8 bits makes a byte (B), this equals about 0.2 gigabytes (GB), or 200 megabytes (MB) of SRAM.

IN-DEPTH LOOK 11.1: SRAM WITH SIX TRANSISTORS

The most compact circuit for implementing an SRAM cell is based on the simple NOT feedback loop shown in Figure 11.2. A logic circuit with *Read* and *Write* lines added is shown in **Figure 11.8**. If the *Write Enable* switch is CLOSED, then a *Data* value, 1 or 0, is passed into the loop. If the *Write Enable* switch is OPEN, as shown, then the *Data* value presently in the loop is stored indefinitely. If the *Read Enable* switch is CLOSED, the stored value is read out along the *Read Data* line.

The *complementary metal-oxide semiconductor (CMOS)* electronic circuit uses FETs to implement the logic for SRAM cells, as in **Figure 11.9**. When the *Write Enable* voltage is set to high, the *Write Data* voltage level is applied to the input line of the NOT gate on the right (made from two FETs). Then the FET output V_{OUT} loops around and drives the input of the second NOT gate, storing the *Data* value. To read the stored *Data* value, the *Read Enable* voltage is set to high, then the stored value is sent out along the *Read Data* line.

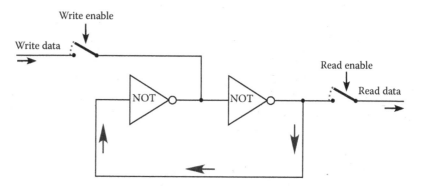

FIGURE 11.8   Enabled NOT loop feedback circuit, shown as logical implementation.

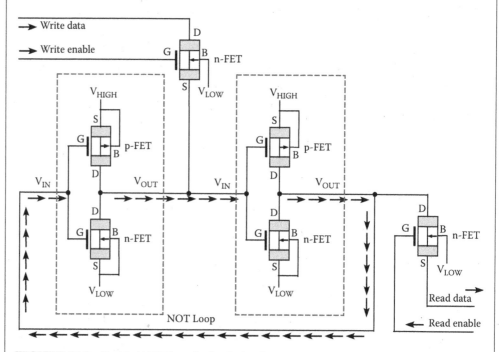

FIGURE 11.9   Enabled NOT loop feedback circuit, shown in its CMOS electronic implementation. Each circuit enclosed by a dashed box is a NOT circuit.

11.4 DYNAMIC RANDOM-ACCESS MEMORY

DRAM, or *dynamic random-access memory*, is a type of electronic memory constructed using *capacitors*, which are small, fast, and cheap. Capacitors, common components in electronic circuits, were described in Chapter 5, Section 5.6. A capacitor is made by placing two small metal plates very close together, with a thin insulating material between them. The insulator prevents electrons from flowing between the plates.

The circuit for a single DRAM memory cell is shown in **Figure 11.10**. The FET transistor's source, drain, and gate are labeled S, D, and G, respectively. The symbols with the cross-hatching represent the grounds; all of these grounds are connected through wires to a common point in the circuit (not shown). The bit-storage capacitor is labeled C. The devices outside of the dashed box are located on the circuit board, outside the memory chip. The switch labeled *Enable* controls whether or not the capacitor's charge can be changed. Applying a voltage using the switch labeled *Select* changes the charge.

Say that we want to write a data value (1 or 0) onto the capacitor. First, set the *Select* switch to the *Write* position (as shown). Then close the *Enable* switch. (All switches are actually transistors but are shown as mechanical switches for simplicity.) Then apply a data voltage value by closing (or not) the *Write* switch. If the *Write* switch is OPEN, nothing happens—the capacitor remains uncharged. But if the *Write* switch is CLOSED, then charge flows onto the capacitor from the battery V. (Recall that the grounds are all connected.) Finally, the *Enable* switch is opened, preventing the capacitor's charge from any further changes.

Next, consider how we can read a data value (1 or 0) previously stored on the capacitor. First, set the *Select* switch to the *Read* position. Then close the *Enable* switch. If charge is present on the capacitor, it will flow out along the *Read* line, representing a logical bit value of 1. If no charge is present, no charge will flow, representing a logical bit value of 0.

A DRAM chip contains many memory cells, each consisting of one capacitor and one transistor. The layout inside a typical DRAM memory chip is shown in **Figure 11.11**. A similar layout was diagramed in Real-World Example 5.1, which you should review. The memory cells are arranged in a two-dimensional pattern and are addressed through a grid of wires—one set of wires to send *Enable* commands, and the other set to send or receive data. Cells are arranged in rows, labeled 1, 2, 3, etc., in the figure. There is a collection of *Enable* switches, arranged at the tops of columns labeled A, B, C, etc. Each *Enable* switch controls one column of cells: when that switch is CLOSED, every cell in the column below it is enabled. In the configuration shown, only one *Enable* switch, labeled "B," is CLOSED. To select which individual cell will

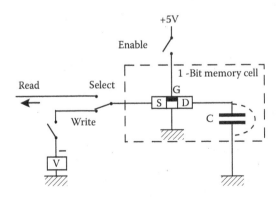

FIGURE 11.10 A single DRAM cell contains a capacitor and a FET transistor.

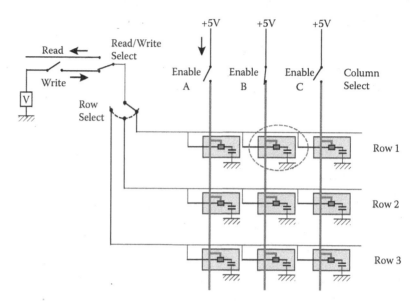

FIGURE 11.11   Inside a DRAM chip. Only nine of the many millions of cells are shown here.

be accessed, a *Row Select* switch is used to connect the data *Read* or *Write* lines to only one row of cells. In the example shown, Row 1 is selected. Therefore, only one cell is both enabled and selected; this cell is indicated by a dashed oval. Now, data can be read or written at this individual cell, leaving all other cells unaffected. To address other cells, we would use different combinations of *Enable* and *Row Select* switch settings.

DRAM chips can contain over 1 GB of data. They can be made very compact, as illustrated in **Figure 11.12**, which shows a state-of-the-art (in 2006) compact DRAM for use in cell phones.

How many DRAM memory cells can be put on a silicon wafer that is 1×1 cm, or 10^{-4} m^2 in area? Each cell requires only one transistor and one capacitor. Assuming that one transistor occupies an area of 100×100 nm (or 1×10^{-14} m^2), the number of transistors that can be put there is $1 \times 10^{10} = 10$ billion. The capacitors can be placed

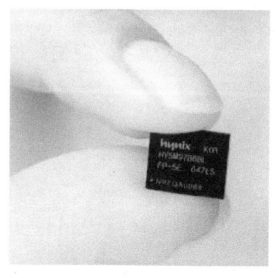

FIGURE 11.12   A DRAM chip for cell phones, containing 512 megabits of storage, which can be read at a rate of 12 Gbits per second. (Courtesy of Hynix Semiconductor, Ichon, Korea.)

above the transistors, so they do not occupy additional space in the layer containing the transistors. Therefore, the number of memory cells, or bits stored, is 10 billion, which equals around 1.25 GB.

DRAM is inexpensive, but there is a practical problem with it. Any capacitor has some small leakage of charge through it or around it, as indicated in Figure 11.10 by the dashed curve. After roughly a millisecond, the charge on the capacitor leaks off, returning the bit value prematurely to 0. The way to overcome this leakage is to restore, or rewrite, the bit value every one-tenth of a millisecond. That is, the data value must be read and rewritten to every memory cell ten times every millisecond. If we are using 10 MB of DRAM for word processing, this means there needs to be $10,000 \times 10,000,000 = 1 \times 10^{11}$ writing operations every second! These operations are done automatically in the brief periods of time while the computer would otherwise be sitting idle.

11.5   NONVOLATILE MEMORY

DRAM, discussed above, is called *volatile memory* because the charge leaks from it after a short time. SRAM is also volatile, because it loses its data when the power is turned off. A memory that holds its bit values even after the power is turned off is called **nonvolatile memory**. Electronic nonvolatile memory, based on a so-called *floating gate*, was invented in 1967 by D. Kahng and S. M. Sze while at Bell Laboratories.

Nonvolatile random-access memory, or NVRAM, is based on a special type of capacitor that has almost no charge leakage. Once charged, such a capacitor can hold its charge for over 10 years. Logic bit values are represented by the amount of charge on the capacitor. A high amount of charge represents a 0, and a small amount of charge represents a 1.

Uses of nonvolatile memory include read-only memory (ROM, PROM, EPROM, etc.), and so-called *Flash* memory. The former are used, for example, in digital cameras and cell phones to store the software programs used for wireless communication. Pocket-sized Flash memories—the size of a pack of chewing gum—can now hold over 5 GB.

Figure 11.13 shows a metal-oxide-semiconductor (MOS) transistor that is similar to the basic FET in Figure 10.18 with the addition of an isolated island of silicon, called a floating gate. This floating gate is surrounded by undoped insulating-type silicon and is the location of long-term charge storage. Because it is insulated all around, once a charge is placed onto the floating gate, it can remain there for years. The tricky part, from a physics point of view, is putting the charge into this capacitor in the first place. We will discuss this later.

How is the presence or absence of charge on the floating gate detected or read out? We use the fact that charge on the floating gate causes a change in the operating properties of the MOS transistor. Recall that the transistor will conduct current between the source (n-type Si) and drain (p-type Si) if a proper voltage is applied to the control gate.

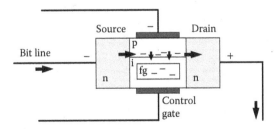

FIGURE 11.13   Nonvolatile memory cell. To write a bit value 1, negative charge is placed onto a floating gate (fg) surrounded by undoped insulating-type silicon (i). When the floating gate is negatively charged, it repels electrons from the source, decreasing the current flow from source to drain.

To create a current, the control voltage must have the proper polarity (plus or minus) and it must be of a sufficiently large value. The presence of charge on the floating gate changes the electrical properties, so that a larger voltage is needed on the control gate to produce current between source and drain. This occurs because the negative charges on the floating gate repel the excess electrons at the interface between the conducting p-type Si and the insulating i-type Si, making the conducting channel smaller and therefore capable of carrying less current.

The procedure for detecting the presence or absence of charge is as follows: Smoothly vary the control voltage from zero to a large value. If the transistor becomes conducting at the usual value expected for a FET, then we know that the floating gate is uncharged; that is, the logic value equals 1. If the transistor does not become conducting at the usual value, but does become conducting at a higher control voltage, then we know that the floating gate is charged; that is, the logic value is 0.

A Flash memory chip contains millions or billions of floating-gate memory cells, arranged in an array as in the DRAM shown in Figure 11.12. Again, a grid comprised of control and bit wires is used to write a bit at any cell of choice. We also need to be able to erase the memory to reuse (rewrite) it. This is done by exposing the entire memory, or a block of locations in the memory, to a high negative voltage at all of the control gates. This drives any charge present on the floating gate off of the floating gate and into the source-drain transistor circuit. With the memory now cleared (1s everywhere), we can write fresh data onto those memory locations of our choice. To write a 1-bit value onto a nonvolatile memory location, an uncommonly high voltage (approximately 13 V) is applied to the floating gate through the bit line. This drives charge through the insulator onto the floating gate, where it is stored. Later, to read out the memory cell, the process can be reversed.

IN-DEPTH LOOK 11.2: QUANTUM TUNNELING

The question, from a physics point of view, is how does charge move to and from the floating gate, given the fact that it is electrically insulated all around. To understand this, we must return to the quantum properties of electrons. Because the floating gate is insulated all around, charge cannot flow to or from it in the usual way that we think of when we visualize an electron as a small particle. Nevertheless, the quantum properties of electrons do allow them to pass through very thin regions of insulator material. This process is called *quantum tunneling* and is illustrated in **Figure 11.14**.

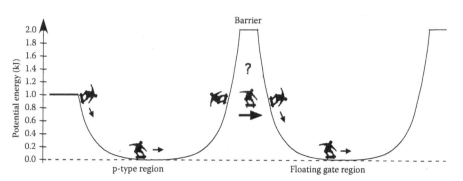

FIGURE 11.14 Quantum tunneling through an energy barrier allows charge to transfer from the p-type Si region to the floating gate in a Flash memory cell. This is analogous to a skateboarder apparently passing through an energy barrier that is too high to surmount, given his or her energy.

Recall that in Chapter 9, we stated, "A classical particle has a definite location in space and a definite speed." A particle, in the classical sense understood by Newton, cannot penetrate into a region that would require the particle to have more energy than it possesses. This would be similar to a skateboarder rolling toward a ramp that is higher than he has the energy to surmount, and mysteriously coming out through the other side, without ever having gone over the top. This behavior—impossible for a person—is possible for an electron, because of its quantum nature.

In Chapter 9, we stated, as part of Quantum Physics Principle (i), that an electron "cannot be said to have a definite location and a definite speed. Although an electron has mass—a particle-like property— it also has wavelike properties, with the amplitude of the electron wave determining the likelihood of finding the electron in a certain region." This tells us that for an electron the very concept of position is ill defined. Perhaps, then, it is not too surprising that an electron can "pass through" a region in which it has insufficient energy to exist. To make a long story short, if a strong force is applied to an electron, pushing it against a very thin energy barrier, it can pass "through" the barrier although it has insufficient energy to pass "over" the barrier. This process of quantum tunneling would seem to violate the fundamental physical law of energy conservation. Yet it does not, because it is impossible to observe the electron in the disallowed region without perturbing it enough to make us uncertain about just how much energy it actually has. This is a feature of quantum physics—measurements interact strongly with the object we are trying to measure.

The likelihood of an electron "tunneling" through a barrier is greatest for very thin barriers. In nonvolatile memory, the barrier is the insulator material separating the floating gate from the p-type semiconductor. This insulator layer is extremely thin (10 nm), making the tunneling possible.

Quantum tunneling is responsible for the movement of electrons through the insulating layer in a nonvolatile memory device. This is remarkable to me. Each time I use a Flash memory device, I cause quantum tunneling of electrons. For example, to safeguard against losing the above paragraphs as I write this text, I just copied them to a Flash keydrive (USB keychain) memory, plugged into my laptop computer. These Flash memory devices are about the size of a pack of chewing gum and are available with memory capacity over 8 GB and data read/write rates up to 480 Mbits per second. It is remarkable that such a common device operates according to principles that are so counterintuitive. However, the success of science does not rest on its principles being intuitive—only that they be logical and self-consistent. The mathematics of quantum theory can describe tunneling quite nicely and predict precisely the behavior of electrons that enable the Flash memory to operate.

11.6   MAGNETIC TAPE AND HARD DISK MEMORY

To achieve permanent, nonvolatile storage of large amounts of data, digital bits can be recorded onto magnetic media such as tape and hard disks. These contain metals such as iron or cobalt that can be magnetized by the presence of a nearby magnetic field, as discussed in In-Depth Look 5.4, which you should now read. A magnetic tape passing over the write head of a tape recording unit is shown in **Figure 11.15**. The head is comprised of an electromagnet whose ends (poles) are brought near to each other by a bend in an iron rod or core. When current passes through the coil, which is wrapped around the rod, the ends of the core become magnetic poles with a small gap between them. Before the electromagnet is activated, the tiny magnetic regions in the tape point in random directions, as shown in In-Depth Look 5.4. The electromagnet induces the magnetic region in the tape just above the gap to align with its north pole pointing toward the south pole of the core; this alignment represents a bit value, say 0. If the

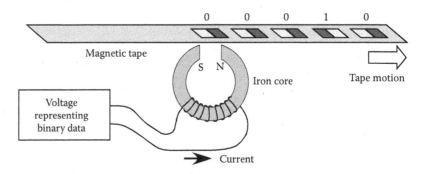

FIGURE 11.15 A magnetic tape recording head. Current flows in the electromagnet, inducing magnetized regions in a tape.

current is made to flow in the opposite direction in the coil, the tape region becomes aligned in the other direction, representing a bit value of 1.

To read out the values of bits that are stored on a tape, the process is reversed. As the tape moves above the head, a time-varying voltage is created in the wire coil around the core. A positive current is read as a 1, whereas a negative current is read as a 0. This voltage is measured and stored in RAM.

Magnetic tapes are also used for **analog** recording and play back. In this case, the data are not represented as fully magnetized discrete regions, as in digital recorders. Rather, the strength of the magnetization of each small region of tape is proportional (i.e., analogous) to the strength of the sound wave being recorded. During readout of an analog recording, the measured voltage is sent directly to a circuit that amplifies it to a level strong enough to drive a loud speaker.

The operation of a magnetic hard disk drive is similar to that of the magnetic tape described above. Magnetic metal particles are embedded in the surface of the disk, which spins at high speed, just under the poles of a miniaturized electromagnet called the read-write head, as illustrated in **Figure 11.16**. The electromagnet is placed at the end of a small moving arm, which scans in and out from the center to the edges of the disk to record and read data at different locations on the disk. Typical hard disks can store about 100 GB in an area of around 100 cm^2. Data can be read or written at rates of approximately 10–100 MB per second. The mechanical aspects of a hard-drive were discussed in Real-World Example 3.1.

11.7 OPTICAL COMPACT DISC MEMORY

Another permanent data storage device is the optical compact disc, or CD. Although slower in access time than a magnetic hard disk, CDs can store large amounts of data and are economical to produce. The digital data are stored as indentations, or **pits**, arranged in a continuous, spiral-shaped track on the metallic surface of a CD, as

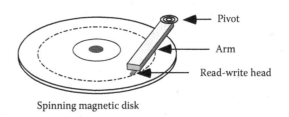

FIGURE 11.16 Magnetic hard drive.

illustrated in **Figure 11.17**. Such CDs are mass-produced by imprinting a master CD onto plastic blanks and then coating the surface with a thin metal layer.

On standard audio CDs, the distance (pitch) between adjacent lines of the spiral track is 1.6 μm. Individual pits are circular areas about 0.5 μm wide. The pits are read as a tightly focused laser beam is reflected from the surface of the CD. The reflected beam is detected by a semiconductor light detector (see **Figure 11.18**). From areas called *land*, where there is no pit, a stronger laser back reflection is detected, but when the laser beam encounters a pit, the back reflection is weaker, making the pit detectable. As the disc is rotated, the laser spot passes over the pits at a constant speed of 1.25 meters per second (m/sec) (for a first-generation "times-1" speed CD). The laser starts reading near the center of the disc and progresses toward the outer edge of the disc. As the laser spot moves outward, the disc rotation rate slows to keep the laser spot moving at a constant speed over the pits.

If the minimum distance between pits were chosen to be 1 μm, then each pit would take up an area of $1\,\mu m \times 1.6\,\mu m = 1.6\,\mu m^2$. The total area used for storing data on a CD is about 100 cm². Because 1 cm equals 10^4 μm, this area can be stated as $100 \times 10^8\,\mu m^2$,

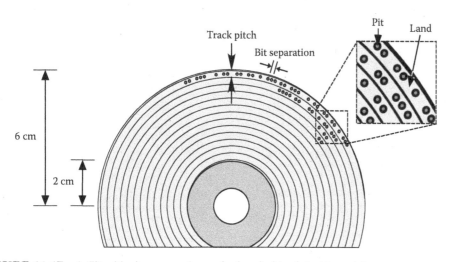

FIGURE 11.17   A CD with pits arranged on a single spiral track (not to scale).

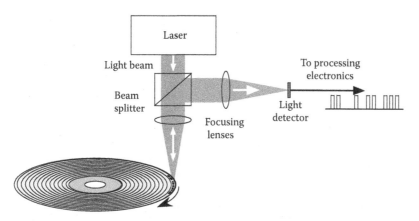

FIGURE 11.18   A laser beam that is focused onto the surface of a CD strongly reflects from a land region, creating a strong electrical signal in the photodetector, corresponding to a logical 1 value. When the laser spot is located over a pit on the disc, the reflected laser power is weakened, corresponding to a logical 0 value.

or 1×10^{10} μm^2. Using this storage scheme, the number of bits (each corresponding to one pit or land) that could be stored on this surface area is

$$\frac{area\,of\,disc}{area\,of\,pit\,region} = \frac{1 \times 10^{10}\ \text{μm}^2}{1.6\ \text{μm}^2} = 6.25 \times 10^9$$

This is 6.25 billion bits, or 6.25 Gbits, which equals about 0.8 GB (Gbyte).

CD technology actually uses a more efficient storage scheme than the one just described, in which each pit or land independently represents one bit value. In the more efficient scheme, as illustrated in **Figure 11.19**, the pits are partially merged together along the track, making some pits elongated. A data-encoding scheme called *edge encoding* is used. This is shown in **Figure 11.20**. Each bit value of 1 is represented by a transition from pit to land or by a transition from land to pit. That is, any change of height of the CD's surface represents a 1. Each bit value of 0 is represented by no change of surface height during a given time interval. As illustrated, a sequence of 0s is represented sometimes by an elongated pit, and sometimes by a constant segment of land.

When the disc rotates, the laser spot scans the land and pits. Because the spot is larger than the width of the pits, there is always some reflection of the laser light (from

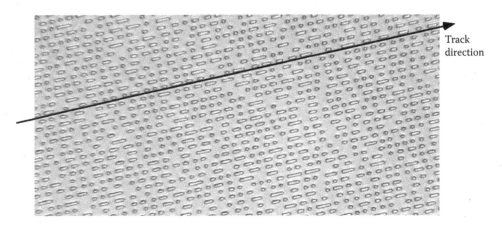

FIGURE 11.19   Microscope photo of the surface of a CD recording of music, showing elongated pits surrounded by flat regions. As the disc rotates, the focused laser spot moves over the surface parallel to the line drawn. (Courtesy Jerry Gleason, University of Oregon.)

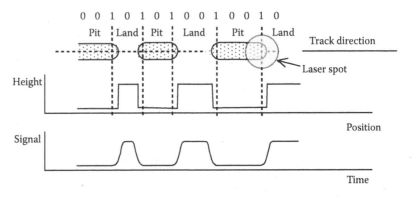

FIGURE 11.20   A series of pits and lands on a CD, with the height profile shown. The signal is created when the laser spot scans over the pits.

the illuminated land area). The larger the fraction of the laser spot that is over a pit, the smaller the reflected signal. This leads to the signal (reflected laser power) varying in time as shown in the lower part of the figure. Whenever the signal suddenly increases or decreases, the readout electronics recognizes this event as a bit value 1. Wherever the reflected power does not change over a certain time interval, this is recognized as a bit value 0. Clever coding schemes allow the avoidance of two or more consecutive 1s in the data, which would be difficult to read using the laser.

Using this type of edge encoding allows a distance of just 0.3 μm between bits on the track. This scheme allows placing about 3 times more bits in the length of the track than in the simpler scheme described earlier. Thereby about $3 \times 6.25 \times 10^9 \approx 2 \times 10^{10}$ bits, or about 20 Gbits, can be stored on one optical disc. The number of bytes stored is about $2.0 \times 10^{10} \div 8(\text{bits per byte}) = 2.5 \times 10^9$, or 2.5 GB.

A typical audio CD holds 800 MB or 0.8 GB or 6.4 Gbits of data. For a stereo music recording, how many samples does this correspond to? Recall our discussion of sampling in Chapter 8. Each sample corresponds to a number between 0 V and, say, 5 V. The larger the number of divisions we break this voltage interval into, the higher the precision we will have, and the higher the fidelity of the recording. The standard is to break the voltage interval into 65,536 distinct values. This can be represented digitally in binary using 16 bits for each sample, because $2^{16} = 65,536$. To store stereo music, with left and right channels, we need two samples for each sampling time. So, ideally, we would need 32 bits for each sample. So far, we have assumed that everything is perfect, with no errors or noise entering the system. In real systems, extra bits are needed for the encoding and for so-called error-correction schemes that are used to make the reconstructed music free of noise. This requires around twice as many bits per sample as estimated above, so we need 64 bits of stored information for each sample of the music signal. This means we can store 6.4 Gbits ÷ (64 bits/sample) = 1×10^8 samples.

Standard music CDs use a sampling rate of 44,100 samples per second. This allows faithful sound reproduction for audio frequencies up to 22 kHz, as explained in Chapter 8, Section 8.6. At this rate, a CD can store data representing a time duration of 1×10^8 samples ÷ (44,100 samples/sec) = 2,270 seconds. This estimate equals about 38 minutes, which is close to the 60 minutes we know typical music CDs can hold.

DVDs, or digital video discs, use similar sizes for pits and land regions on the disc surface, but a more efficient encoding scheme, which allows storage of up to 4.7 GB, or 37.6 Gbits.

The upper limit for the density of bit storage on an optical disc is set by the physics of waves. As we learned in Chapter 7, light is a wave. This limits the smallness of the light spot produced by a focused laser beam. A good rule of thumb is that a beam of light with *wavelength* equal to λ can be focused to a spot as small as one-half of λ, but no smaller. The smaller the laser spot is, the closer together we can place the pits and land regions, thereby increasing the storage density. Standard audio and data CDs use near-infrared laser light with wavelength equal to $\lambda = 780$ nm for reading pits. This means that the smallest spot that this light can be focused to is about 780 nm, or a little smaller. This allows the pits to be spaced about 300 nm apart and still be readable, using edge encoding.

To store more bits on a given-sized disc, as needed for video, we need to decrease the distance between pits. This can be done only if the light spot can also be made correspondingly smaller. Regular DVDs use light with a wavelength 650 nm, slightly less than used for CDs, allowing slightly higher data density. For high-definition video, manufacturers introduced optical disc players that use blue light, with a wavelength 405 nm. This allows the focused spot diameter to be about two times smaller than it is with the red lasers used for CDs or DVDs (see **Figure 11.21**). This means the spot area

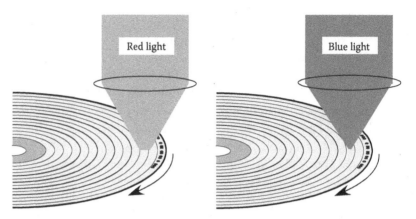

FIGURE 11.21   CDs and DVDs use near-infrared or red light, which cannot be focused as tightly as the blue light used for high-definition videodiscs.

is four times smaller than for CDs, allowing about four times the number of bits stored on one disc. A standard commercial system is "Blu-Ray," storing 25 GB on a disc.

11.8   ERROR IMMUNITY OF DIGITAL DATA

Representing data digitally using bits has a great advantage. It is immune to errors and noise. In a perfect system, all 1-bit voltages would be equal, and all 0-bit voltages would be equal. But in practice, voltage variations occur as a result of noise and other imperfections. Fortunately, when using a technique called voltage thresholding, it does not matter if there are small variations in the voltages from bit to bit. For example, if a small portion of the surface of a CD becomes dirty, and the reflected light beam is a little weaker than expected, the processing electronics have no problem distinguishing a 1 bit from a 0 bit. This is illustrated in **Figure 11.22**, which shows the signal voltage coming from the CD reader. In this example, every 1-millisecond (msec) time interval corresponds to a bit. Any voltage above a chosen threshold value of 3 V is considered to be a logical 1 value, and any voltage less than the threshold value is considered to be a logical 0 value. Thresholding can be performed automatically using a circuit called a comparator.

We can now state another physical principle of computing:

PHYSICAL PRINCIPLE OF COMPUTING (III)

Digital voltage thresholding can lead to error-free computer operation. This is not attainable using analog techniques.

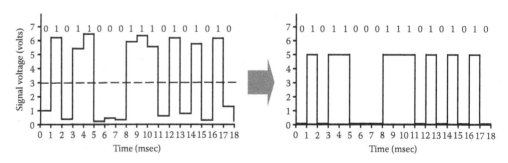

FIGURE 11.22   Thresholding is used to "clean up" a digital voltage signal and restore it to its intended form. In the example, any voltage above 3 V is interpreted as a bit value 1 and is restored to 5 V, whereas any signal less than 3 V is interpreted as a bit value 0 and is restored to 0 V.

11.9   THE STRUCTURE OF A COMPUTER

Finally, we are in a position to discuss a computer as a whole system. **Figure 11.23** is a schematic of the general layout, or architecture, of a computer. In the figure, each solid connecting line represents a group of wires. A *memory* stores the data as binary values, 0 or 1. These data values are denoted A, B, C, etc. Each is stored in a known memory location, which has an "address," numbered 1, 2, 3, etc. As an analogy, think of the memory addresses as being like the house addresses on a street. You may think of each number A, B, C... as being the resident of each house. The idea is to choose two numbers—say A and C—on which we wish to perform a logic operation, such as addition. By creating a computer program, we can inform the instruction controller which operation we wish to perform. Carrying out this logic operation requires a sequence of instructions, sent to the various parts of the computer using the control bus, which is a set of wires. After the program begins, the instruction controller sends a message to the memory, telling it which two addresses to access. Number A is sent across the data bus into a special memory location called register 1, and number C is sent to register 2. The instruction controller then instructs the logic operation unit (which could be an adder) to carry out its operation on the two numbers sitting in registers 1 and 2. The result of the operation (say the sum A + C) is placed in register 2. (This replaces the number C that was previously stored there.) Then the instruction controller sends the value now sitting in register 2 to a particular address in the memory, say address 4 (where it overwrites any number, D, sitting there already). All of these instructions must be carried out in a synchronized manner. To enable this, an electronic digital *clock* provides timing pulses to all circuit components. The set of elements enclosed by the dashed line in Figure 11.23 make up the *central processor unit*, or *CPU*.

The mathematician John von Neumann is the scientist whose name is most often associated with the origination (or at least popularization) of this general architecture. In 1946, he collaborated on these ideas with John Mauchly and J. Presper Eckert at the University of Pennsylvania—the designers and builders of the first large-scale, general-purpose electronic computer, named ENIAC (for Electronic Numerical Integrator and Computer). See the photo of ENIAC at the beginning of this chapter.

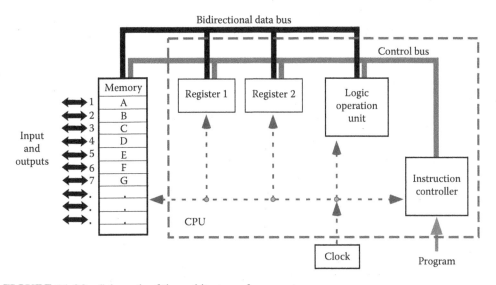

FIGURE 11.23   Schematic of the architecture of a computer.

11.10   HIERARCHY OF COMPUTER MEMORY

Computers use a hierarchy of different types of memory devices, each with its own capabilities and limitations, as shown in **Figure 11.24**. The layers in the hierarchy communicate with neighboring layers. The hierarchy is shown as a pyramid, with the lower layers having larger storage capacity, but slower access. The highest layers have small amounts of very fast-access memory.

At the top are the registers, located inside the CPU, in which the logic or arithmetic is carried out. Next is the *cache* memory, which is made of fast SRAM based on transistor logic. It is located physically close to the registers, and used to store data that is used most often by the registers. Next is the main memory, made of DRAM, which is based on storing charge in capacitors. This is a very large, fairly fast, inexpensive form of memory. Next is the hard disk, which stores massive amounts of data magnetically, but is rather slow in its access time. At the lowest level are optical disks, magnetic tape, and Flash memory, which are often located outside of the computer housing. Flash memory is becoming large enough and affordable enough that it can replace hard disks in many applications.

This memory hierarchy provides the best performance for the lowest cost. The great effectiveness of using the memory hierarchy can be expressed as a principle.

PHYSICAL PRINCIPLE OF COMPUTING (IV)

Utility and efficiency are improved by structuring computer memory as a hierarchy, ranging from large amounts of slow, long-term memory to small amounts of fast, short-term memory.

11.11   HEAT-IMPOSED LIMITS OF COMPUTERS

A computer is a machine, and as such, it generates heat. Of course, there is the heat generated by friction in the hard drive's motor and heat from the lamp that lights the screen. More fundamentally, any electronic circuit that performs logic operations generates some heat. You might have noticed that a laptop gets hotter when it is running programs. There are fundamental physics reasons why a faster-running computer circuit must get hotter than a slower-running one. The physics behind this is analogous to the friction-caused heating that occurs in any machine, whether mechanical or electrical. We can summarize this as a principle.

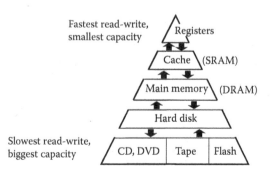

FIGURE 11.24   The hierarchy of computer memory. The higher levels are physically nearer to the CPU, are faster for reading and writing, and have smaller storage capacity.

The rate of thermal-energy removal from a computer's processor and memory circuits places limits on computational speed and overall capability.

The need to remove thermal energy from any electric circuit also limits the density of memory cells that can be placed in a memory chip. If too many cells are placed close together, the chip will overheat and fail.

A typical desktop computer uses about 200 watts (W) of power when running. Internet server companies such as Google, Yahoo!, and Microsoft need the equivalent of hundreds of thousands of such computers in a single location to operate their extensive web searching and data storing facilities; 250,000 such computers together use 50,000,000 W (50 MW) of power, comparable to reported values for a single server "farm." Much of this power is transformed into heat, requiring massive cooling capabilities. A typical household air-conditioning unit might provide a cooling power of 1,000 W. A large server-farm building would therefore require the equivalent of 50,000 such air conditioners operating.

11.12 REPRESENTING INFORMATION IN COMPUTERS USING CODES

How does a computer store this sentence: "May the Force be with you." A computer stores only bits: 1s and 0s. We need a way to translate our sentence into the computer's binary language. To do this, we create a code. A code is a rule for converting a series of symbols—which have meaning in some representation—into a different, corresponding series of symbols. A code is made up of code words, each of which has an assigned significance. Each code word is made up of several characters, selected from some "alphabet." In binary, the alphabet consists of only the symbols 1 and 0. For example, Table 2.2 in Section 2.7 is an example of a look-up table, showing the correspondence between the decimal numbers 0, 1, 2, 3, 4, 5, 6, 7, 8, 9, 10, 11, etc. and the code words (e.g., 1001) composed only of ones and zeros that are used in the binary representation.

Another simple example is International Morse Code. Morse Code is based on a binary alphabet, made of only two characters, called "dot" (•) and "dash" (–). Each of the 26 letters in the ordinary Latin alphabet is represented by a combination of between one and three dots or dashes. For example, the letter S is represented by three consecutive dots (• • •), and the letter O is represented by three consecutive dashes (– – –). Separate letters are separated by a longer space. The code for each letter can be found in a look-up table. These can be combined into meaningful combinations, such as (• • – – – • • •), which is the international distress signal: $S \, O \, S$.

In addition to the code words, we also need a set of rules, called a ***protocol***, used for reading the received series of characters and reconstructing the intended message. In Chapter 8, we discussed the idea of a protocol in the context of interpreting a series of binary numbers received from a radio broadcast. As another example, in Morse Code, each group of symbols representing one Latin letter is followed by a pause, to let the receiver know that one letter has been sent.

11.12.1 ASCII Code

In 1968, the American National Standards Institute (ANSI) adopted the American Standard Code for Information Interchange (ASCII). ASCII is the standard code used to represent the Latin alphabet A–Z and the Arabic (base-ten) numerals 0–9 in binary. It uses an eight-bit binary word to represent each character. This makes it possible to store and transmit Latin characters using digital methods. **Table 11.1** gives the binary translation of each character in the alphabet. Notice that zeros are added to the left where needed to create an eight-bit word for each character. This avoids a more complicated protocol, which would be needed if different characters were represented by code words of varying length. Because each word contains the same number of symbols, we do not need a pause or some other special symbol to separate the bits making up different words: they can all be run together. For example, the word "Hi" is encoded as 0100100001101001.

QUICK QUESTION 11.3

What is the ASCII code for the English capitalized word "Force"?

TABLE 11.1
Part of the Look-Up Table for Binary ASCII Code

A	B	C	D	E	F	G
01000000	01000010	01000011	01000100	01000101	01000110	01000111
H	I	J	K	L	M	N
01001000	01001001	01001010	01001011	01001100	01001101	01001110
O	P	Q	R	S	T	U
01001111	01010000	01010001	01010010	01010011	01010100	01010101
V	W	X	Y	Z		
01010110	01010111	01011000	01011001	01011010		
a	b	c	d	e	f	g
01100001	01100010	01100011	01100100	01100101	01100110	01100111
h	i	j	k	l	m	n
01101000	01101001	01101010	01101011	01101100	01101101	01101110
o	p	q	r	s	t	u
01101111	01110000	01110001	01110010	01110011	01110100	01110101
v	w	x	y	z		
01110110	01110111	01111000	01111001	01111010		

11.13 CODING IMAGES

How can images be stored as binary numbers? The simplest type of image is one composed of only black or white box-shaped regions, or **pixels**. The word *pixel* means *picture element*. An example of such an image is shown in **Figure 11.25**. The obvious way to encode this image as binary numbers is to make a rule, or protocol, in which we start

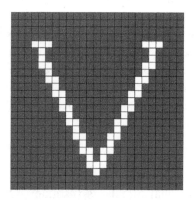

FIGURE 11.25   Image containing the letter V.

at the upper left-hand corner, and move pixel by pixel to the right along the top row. When reaching the end of the top row, move to the far left pixel of the second row, etc. At each pixel location, record whether the pixel is black (coded by 0) or white (coded by 1). A single one-bit number codes each of the 625 pixels in the image.

To encode a picture that contains eight different shades of gray, we could use a three-bit number to represent the level of brightness at each pixel. For example, 000 could represent black, 001 could represent the next brightest shade of gray, and so on, up to 111 for representing white. A gray-scale picture of a shark is shown in **Figure 11.26** using this type of coding. Under high magnification, you can see the individual pixels and the eight shades of gray used. Likewise, color images can be represented in binary form by assigning, for example, a 16-bit number to represent one of 65,536 different shades and colors. An example of a color image monitor was given in Real-World Example 2.1.

FIGURE 11.26 A grayscale picture of a shark, showing different levels of magnification.

11.14 DATA COMPRESSION

Data compression is a method of representing and storing a message by using a lesser amount of information than was initially used in creating the message. This allows greater efficiency in using the computer's memory. Common examples are mp3 files for music and .zip files for data. An original file is compressed using one of these formats, and then it can be stored using less space in the memory than originally.

It is common to apply data compression to images. A black-and-white image, or picture, can be represented in the binary world by a list of binary digits (bits), each indicating whether a given pixel (picture element) is black (bit value = 0) or white (bit value = 1). Consider the two images (25 × 25 blocks) in **Figure 11.27**. Before reading further, please guess the correct answer to the question, "Which image, A or B, contains more information?"

On first look, it might appear that image A contains more information, because it contains a recognizable shape—the letter V. This shape has some meaning to many people, whereas image B apparently contains just random spots. But this conclusion

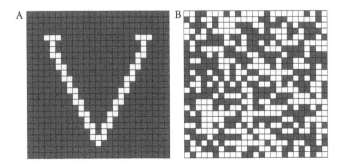

FIGURE 11.27 Image A contains the letter V, whereas image B is random. Which appears to you to contain more information?

is misleading. Recall from Chapter 2 that the technical definition of information content in a message or image is, "How many yes/no questions would I have to answer to specify completely for another person the full content of the message?" This number equals the amount of information, in bits, contained in the message. According to this definition, image B contains more information that image A.

Another way to analyze the information content in an image is to ask, "How efficiently can I compress the data making up the image?" If, indeed, image B is completely random, with no hidden pattern, then it cannot be compressed at all. An example of a pattern that is completely random is one generated by flipping a coin 625 times and recording black for heads and white for tails. If we want to send this image intact, then we need to send one bit for each pixel, of which there are 625.

On the other hand, image A is not completely random, and so it can be significantly compressed. A simple way to do this is to recognize that there are large regions consisting of continuous black. So, it might be efficient to just describe each location where the black begins and ends. An algorithm could be: Scan the image in this fashion. Start at the upper left corner and move right 25 times (the number of pixels in a row), then jump down one row and back to the left side, and repeat. While scanning, make a list giving the length of the first continuous black region ($4 \times 25 + 3 = 103$ pixels), followed by the length of the following white region (3 pixels), followed by the length of the next following black region (13 pixels), etc. This method is called run-length encoding. Image A is represented, using run-length encoding, by the sequence:

103, 3, 13, 3, 7, 1, 15, 1, 8, 1, 15, 1, 8, 2, 13, 2, 9, 1, 13, 1, 10, 2,11, 2, 11, 1, 11, 1, 12, 2, 9, 2, 13, 1, 9, 1, 14, 2, 7, 2, 15, 1, 7, 1, 16, 2, 5, 2, 17, 1, 5, 1, 18, 2, 3, 2, 19, 1, 3, 1, 20, 2, 1, 2, 21, 1, 1, 1, 22, 3, 23, 1, 62

The number of entries in this list is 72. No number in the list is larger than 127, meaning that we can use 7-bit binary numbers to represent each entry. For example, 103 (decimal) = 1100111 (binary), and 4 (decimal) = 0000100 (binary). The total number of bits needed to represent the 72 entries is $72 \times 7 = 504$. This number of bits is smaller than the 625 bits that are needed to represent the image without using compression.

A general relation between information content and compressibility is:

When comparing two images containing the same number of pixels, the image that contains the least information is the most compressible.

Mathematicians have developed compression algorithms that are even more efficient than the one discussed here, and these are widely used for storing and transmitting files on the Internet. They all rest on the idea that in most messages or images, certain combinations of bit values occur more frequently than others. This repetition of bit patterns is called redundancy. It occurs commonly in the written text of most languages, as well as in most common types of images. For example, in images, groups of adjacent black pixels occur frequently, and, in text, certain letter combinations (such as *t-h-e* or *q-u*) occur frequently.

SUMMARY AND LOOK FORWARD

From a physicist's point of view it seems clear that the capabilities of computing will always be dictated by the physics of the real-world objects from which a computer is made. The laws of nature determine how information can be made to flow during a computation.

In this chapter, we brought together many of the concepts that we learned in the previous chapters to see how a computer works at the basic, hardware level. A computer is a machine that uses electro-magneto-mechanical components. We learned how information is stored within computer memory and how computers are organized. Sequential logic for computer memory operates by permitting a logic gate's output to feed back into its input. Other types of memory do not make direct use of sequential logic, but represent bit values directly in physical objects, such as charge on a capacitor, the magnetization of a small region on a disk, or a pit in a CD's surface. Storing data in binary form, along with thresholding, has the advantage that it is immune to small errors in the physical signal used to represent the bit values. We also considered how information—whether it is numerical, textual, or pictorial—can be stored most conveniently or most efficiently by encoding the data using an appropriate code.

The next major subject that we will tackle is that of communication networks, including the Internet. The physics behind lasers and optical fibers plays a leading role in the development of the Internet.

SUGGESTED READING

Conceptual basics and history of computer development:
Pohl, Ira, and Alan Shaw. *The Nature of Computation*. Rockville, MD: Computer Science Press, 1981 (Section 5.1).

Excellent, simple discussion of logic and memory circuits:
Maxfield, Clive. *Bebop to the Boolean Boogie*. Amsterdam: Newnes, 2003 (Chapters 11 and 15).

More complete discussion of logic and memory circuits:
Murdocca, Miles J. and Vincent P. Heuring. *Principles of Computer Architecture*. Upper Saddle River, NJ: Prentice Hall, 2000 (Chapter 7 and Appendix A: Digital Logic).

A simple discussion of computer hardware and architecture:
Rizzo, John, and K. Daniel Clark. *How the Mac Works*. Indianapolis, IN: Macmillan, 2000.

More advanced texts covering digital electronics, computer architecture, and memory:
Kang, Sung-Mo, and Yusuf Leblebici. *CMOS Integrated Circuits, Analysis and Design*. Boston: McGraw-Hill, 2003 (Chapter 10 describes SRAM and DRAM).
Melissinos, A.C. *Principles of Modern Technology*. Cambridge, UK: Cambridge, 1990.

History of microelectronics:
Seitz, Frederick, and Norman G. Einspruch. *Electronic Genie*. Urbana: University of Illinois Press, 1998.

KEY TERMS

Analog
Cache
Capacitor
Cell
Combinational logic circuit
CMOS (complementary metal-oxide semiconductor)
CPU (central processing unit)
DRAM (dynamic RAM)
Enabled data latch, or *D-latch*
Feedback
Field-effect transistor (FET)
Land
Memory
Memory circuit
Nonvolatile memory
One-time latch

Pit
Pixel (picture element)
Protocol
RAM (random-access memory)
SRAM (static RAM)
Sequential logic
Set-Reset latch (S-R latch)
Wavelength

ANSWERS TO QUICK QUESTIONS

Q11.1 Initially B equals 0. After you change A to 0, Q becomes 0, giving B = 1, which switches Q to 1, which changes B to 0, etc. The Q value jumps back and forth between 0 and 1 forever.

Q11.2 The Enabled D-latch in Figure 11.6 follows the history:

	Sequential Logic History for Enabled D-Latch					
Event Order	E	D	R	S	Q	Q′
5	0	0	1	1	0	1
6	0	1	1	1	0	1
7	1	1	0	1	1	0
8	0	1	1	1	1	0

Q11.3 The ASCII code for the English word "Force" is 0100011001101111011100100011 0001101100101. Because this is hard to read, here is the same code with commas added to help you read it:

01000110, 01101111, 01110010, 01100011, 01100101.

The commas are not actually present in the ASCII representation.

EXERCISES AND PROBLEMS

Exercises

E11.1 Explain the logical behavior of the following electromagnetic-relay circuit. Give two examples of stable configurations, indicating the states of relay R1, relay R2, switch S1, switch S2, lamp Q, and lamp Q′. Compare the operation of this circuit to the NOT Loop logic diagram in Figure 11.1. Also see Chapter 6, Section 6.6.

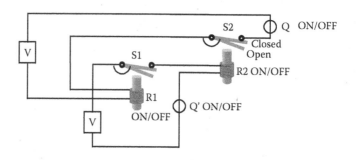

E11.2 Design an electronic circuit, using switches and electromagnetic relays (current-driven electromagnets), batteries, and light bulbs to implement the one-time latch shown in Figure 11.3. See E11.1 above and Chapter 6, Sections 6.4 and 6.6.

E11.3 A nine-cell DRAM array is shown in Figure 11.11. For simplicity, assume charge does not decay from the capacitors. Assume that, initially, several of the cells are charged (bit value = 1), but you do not know which. Give the series of operations needed to achieve the following sequence: (a) Read the bit value (0 or 1) of the center cell at Row 2, Column B. (b) Erase all cells. (c) Write a 1 bit on the cell at Row 3, Column C.

E11.4 In a Flash memory, the region called the floating gate is where the electrons are stored to indicate a bit value 1. This region is completely surrounded by a near-perfect electrical insulator. Why is an insulator used here?

E11.5 A magnetic tape, as shown in Figure 11.15, contains tiny iron particles. In light of the discussion in In-Depth Look 5.4, explain in words and pictures what happens when a bit value is written and later erased on the tape.

E11.6 List the logical bit values of this series of data, in as far as you are able, using 1, 0, or indeterminate. Explain.

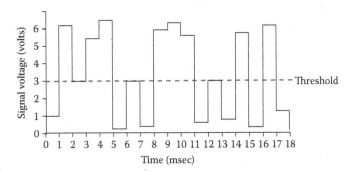

E11.7 Compact optical discs are rated by the maximum data-transfer rates. A times-1-speed CD rotates at 200–530 rpm and has a data transfer rate of 150 kB/sec. A times-2-speed CD rotates at 400–1,060 rpm and has a data transfer rate of 300 kB/sec. (a) Assuming this trend holds, what is expected for the rotation rate and data rate of a 8× CD? (b) Explain why the rotation rates specified are not constant (e.g., 400–1,060 rpm). (c) Why is the ratio of the fastest rotation rate to the slowest rotation rate equal to about 2.5? *Hint*: The circumference of a circle with radius R equals $2\pi \times R$, with $\pi = 3.14$.

E11.8 Find a description of International Morse Code on the Internet or in an encyclopedia and use it to encode the sentence, "What hath God wrought." This is the first message that Samuel F.B. Morse sent on his telegraph line installed between Baltimore and Washington DC in around 1838. Explain what protocol you are using.

E11.9 Find a description of read-write optical disks, or CD-RW, on the Internet or in an encyclopedia, and write a description of the physics behind their ability to be written and erased by the user. There are various types; concentrate on the Magneto-optical drive, which works by the Magneto-optical, or Magneto-optic, effect.

Problems

P11.1 How many memory cells of DRAM can be located on a silicon wafer 3×4 mm in area? Assume that the capacitor is much smaller than the transistor, and that one transistor occupies 100×100 nm in area. How much memory is this, in Gbits and in GB?

P11.2 This is an alternative logic diagram that implements the enabled latch for memory. Verify and explain in detail that it has the proper logical behavior. Give an example of a history table, indicating values of E, D, D', F, G, H, I, Q, Q', to prove your argument.

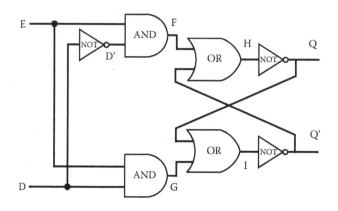

P11.3 The capacitor in a DRAM cell is charged with 9,000,000 electrons, indicating a logic value equal to 1. Following this charging event, the electrons begin leaking from the electron, degrading the information being stored. Every 1 msec, two-thirds of the previously remaining electrons leak from the capacitor. That is, after every 1 msec, the remaining number of electrons is multiplied by a factor of one-third, or 1/3. Draw a graph of the number of electrons on the capacitor at these times: 0, 1, 2, 3, 4, 5, 6, 7, 8, 9, and 10 msec.

P11.4 Shown below is an electromagnetic relay circuit that implements the set-reset latch, analogous to that shown in Figure 11.5. Verify and explain in detail that it has the proper logical behavior. Give an example of a history table, indicating the states of the inputs S and R, the relays R1 and R2, and the lamps Q, Q'. In this example, a logical 1 is defined as pushing or pulling a switch down (toward the bottom of the page). See Chapter 6, Sections 6.4 and 6.6.

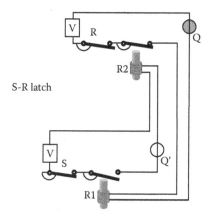

P11.5 The distance between bit locations (pit or land) along the track on a CD is approximately 300 nm. The length along the track for one revolution of the disc nearest the center is approximately 140 mm. The length along the track for one revolution of the disc nearest the outer edge is approximately 380 mm. When the laser begins reading data, it starts nearest the center, and the disc rotates at a certain rate, given in units of revolutions per minute (rpm). By the time the laser gets to the outer edge, the disc has slowed to a smaller rpm, so that at all times the laser spot passes over the surface of the disc at a constant speed of 1.25 m/sec (for a 1× speed CD).

(a) From this information, calculate the disc rotation rate in rpm when the laser is nearest the center.
(b) When it is at the outer edge.
(c) Calculate the number of kilobytes that the laser reads per second.

P11.6 If all of the 700 MB of data on a music CD were arranged in a straight line instead of in a spiral, how long would that straight line be? Assume a bit separation equal to 300 nm. If you traveled next to the line at a speed 1.25 m/sec, how long would it take you travel its whole length?

P11.7 Compare the storage densities (bits/cm^2) for a CD and a magnetic hard disk. *Hint*: Assume both have a usable area of 100 cm^2, the CD stores 800 MB, and the hard disk stores 80 GB.

P11.8 Estimate the size of the magnetic bit-storing domains on an analog music tape. *Hint*: For a tape recording to have high fidelity, the voltage signal produced when the tape is played back must contain frequencies up to a maximum of approximately 15 kHz. This means the strength of the magnetization changes about 15,000 times/sec as the head moves past it. You can estimate the speed of the tape at the head by knowing that the diameter of the spool of tape (when it is half wound) is approximately 1.5 in., or approximately 35 mm. An observation on a stereo system shows that the tape spool makes one revolution approximately every 2.5 sec.

P11.9 As described in the text, ASCII is the standard code used to represent English characters in binary. It uses one 8-bit word to represent each character. This makes it possible to store and transmit English by using digital methods. Table 11.1 gives the binary equivalent of each letter in the alphabet. How many distinct symbols can be represented using the 8-bit ASCII code? (Not all are shown in the table.) Use the table to encode the sentence, "What's up?" in ASCII. You need to do some research (library or online is okay) to find the ASCII codes for the punctuation in this sentence, because it is not given in Table 11.1.

P11.10 To code an image and transmit it digitally, each pixel (box) in the low-resolution image below has one of four possible darkness values (grayscale values) 0, 1, 2, or 3. Your task is to make up a coding scheme, and make up a protocol, so that you could write on paper a consecutive list of binary numbers, without commas or other separators, and give it to someone along with information about the protocol, such that they could reconstruct the image exactly.

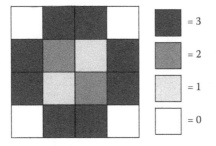

P11.11 Shown here is an electromagnetic relay circuit that implements the enabled latch analogous to that shown in Figure 11.6. Verify and explain in detail that it has the proper logical behavior. Give an example of a history table, indicating values of E, D, R, S, Q, and Q′, to prove your argument. In this example, a logical 1 is defined as pushing or pulling a switch down. See Chapter 6, Sections 6.4 and 6.6. Analyze this circuit in steps:

(i)   Initially E = 0 (open), so both R2 and R6 are not activated. Therefore changing D does nothing to relays R3 and R7, so bulbs Q and Q′ cannot change regardless of how D is set. Assume that Q = 1 (ON) and Q′ = 0 (OFF).
(ii) Now change E to E = 1 (closed). Set D = 1.
(iii) Now change E back to 0 (open).
(iv) Set D = 0.
(v)   Set D = 1.

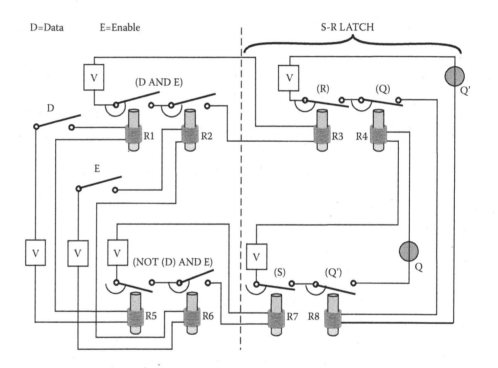

P11.12 An SRAM memory device can be constructed with two NOT gates, as shown in Figure 11.1. The diagram below shows an implementation of this using water-pressure-activated valves. There are two types of valves—high-ON valves and high-OFF valves—which respond to the pressure in externally applied control pipes. A high-ON valve is conducting only when the control pressure is high, whereas a low-ON valve is conducting only when the control is low, as illustrated. The four rectangular tanks have constant pressures, as labeled. There are write-control and read-control gates. When they are both OFF (closed), as in the drawing, the circuit holds at the location B whatever value (high or low) was entered earlier through the *write* bit pipe. To write the stored bit value, open the *write* control valve, connecting the *write bit pipe*. Give the pressures at each location, labeled A through F, for the following sequence:

(i) The bit value at B is initially high, and the write control and read control are initially low, so valves 1 and 6 are closed.
(ii) Then the read control is changed to high; what pressure value is read out in pipe G?

(iii) The read control is then changed back to low. Next the bit pipe A is set to low, then the write control valve 1 is opened, and the write control is then closed.

(iv) Finally, the read control is changed to high; what pressure value is read out in pipe G? Is the bit value that is read out the same or the opposite of the original bit value stored?

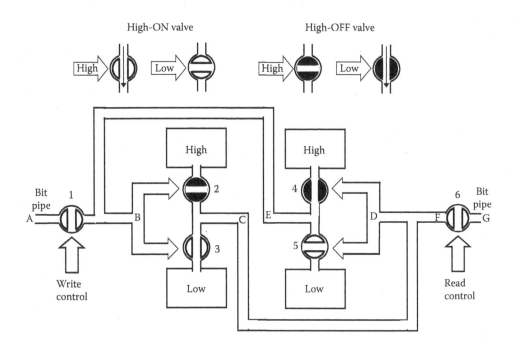

12

Photons: Light Detectors and Light-Emitting Diodes

According to the assumption considered here, when a light ray starting from a point is propagated, the energy is not continuously distributed over an ever increasing volume, but it consists of a finite number of energy quanta, localized in space, which move without being divided and which can be absorbed or emitted only as a whole.

Albert Einstein
(1905)

Large color video display containing millions of LEDs at the University of Colorado football stadium. (Courtesy Grandwell Video.)

Max Planck in 1901, a year after he revolutionized physics by proposing quantum theory. (Courtesy of the Max-Planck Society, Berlin.)

12.1 LIGHT, PHYSICS, AND TECHNOLOGY

Vision is one of our main senses, and most of the information we receive from the world around us comes to us via *light*. But what, really, is light? That question occupied philosophers and scientists for thousands of years, and still does. In Chapter 7, following Maxwell's line of reasoning, we came to a firm conclusion, or at least we thought we did, that light is a *wave* of electromagnetic fields—the entities that transmit electric and magnetic forces from one place to another. However, there is a caveat: some properties of light cannot be understood in terms of a classical wave picture. This was a huge surprise to scientists around 1900, because they thought that Maxwell's earlier concept of light as a wave had put an end to any misunderstanding about the nature of light. This is a good example of an important fact about science: Every theory in physics, no matter how good at explaining known phenomena, is always subject to being revised if further experimentation shows new phenomena that are not explained by the then-current theory.

Two key players in this story were Max Planck and Albert Einstein, who around 1900–1905 realized that certain experimental observations could not be explained by assuming the model of light as a smooth wave. Rather, they postulated that light carries energy in the form of tiny lumps or bundles. We now call these energy bundles by the name *light quanta,* a name emphasizing that these are quantities that can be counted. An alternative name is ***photons***. "Photo" means light, whereas "-on" reminds us of a particle such as a proton or electron.

In this chapter, we will begin to see how the discovery of the quantum properties of light revolutionized our ability to create and control light using modern light sources such as lasers. For example, in optical-fiber communication systems, light created by lasers travels through optical fibers and is detected at the end by semiconductor devices called ***light receivers***, or ***photodetectors***. ***Light-emitting diodes (LEDs)*** and lasers are "everywhere" in consumer electronics nowadays, from CD and DVD players to magnetic hard disks and monitors and displays. The chapter-opening photograph shows another everyday application of advanced light sources—arrays of LEDs used for making very large video displays for sporting events.

12.2   THE QUANTUM NATURE OF LIGHT—PHOTONS

During the 19th century, physicists successfully used a set of principles, which we now call the principles of classical physics, to understand the motion of everyday objects influenced by gravity or by electric and magnetic forces. In this view, ***electrons***—the small entities in ***atoms***—are thought of as particles (having definite mass, position, and speed), whereas light is thought of as waves of the electromagnetic field. In Chapter 9, we learned that electrons are not really particles in the classical sense—rather, they are spread out and have some wavelike properties as they move in orbits within an atom. The quantum behavior of electrons determines the properties of solids and the operation of ***semiconductor*** devices such as ***transistors***.

If a light wave enters a medium made of atoms, some of the energy in the incident light wave may be transferred to the atoms, thereby decreasing the power in the light wave. We call this ***absorption*** of light. On the other hand, if there is already some energy stored in the atoms, then some of the stored energy may be converted into light energy. We call this ***emission*** of light by atoms. Emission is the reverse of absorption.

Experiments carried out around 1900 showed that light cannot be understood completely as a wave in the classical sense. Physicists found that when the energy carried by a light wave is absorbed by a material object, the energy that is being deposited in the object arrives in small, indivisible lumps or bundles. In Chapter 9, we discussed the nature of atoms, using Niels Bohr's planetary orbit model. In the chapter-opening quote there, Bohr stated his postulate, or principle, that an electron can change from being in a higher-energy orbit to being in a lower-energy orbit by emitting one "quantum of radiation." This idea corresponds to the "lumpiness" of light, that is, the fact that a single atom gives off a small bundle of light energy when the electron in the atom loses a discrete, quantized, amount of energy. Therefore, we conclude that light is not simply an ***electromagnetic wave*** in the classical sense. Neither electrons nor light can be described purely as particles or waves. They are something else, which we refer to as quantum objects, which are not familiar in our common-sense experiences.

Although it is not easy to gain a completely correct intuitive picture of light, by using the principles of **quantum theory** we can predict and understand all we need to know to design and build electronic and optical devices. If this were not so, you would not have your computer, your compact disc (CD) player, or your MP3 player.

The first quantum principle relating to the nature of light is:

QUANTUM PRINCIPLE (IV):

Planck's law: When light of a single color travels through space, it has wavelike properties (wavelength λ, frequency f, wave speed c) and wavelike behaviors (refraction, diffraction, interference). But when energy is transferred to or from this light wave, the transfer occurs in discrete bundles of energy called photons, each having energy equal to the light's frequency times a constant:

$$E_f = h \cdot f \ \text{(with } h = 6.6 \times 10^{-34}\,\text{J} \cdot \text{sec/photon)}$$

The symbol h is called **Planck's constant**, after the German physicist Max Planck, who proposed this formula in 1900, ushering in the era of quantum physics. This constant is one of the fundamental constants of physics and is crucial to our understanding of nature. The units of h are (joule · seconds per photon). Recall that joule is a unit of energy.

As an example, red light has a wavelength equal to about $\lambda = 700$ *nm*, and so its frequency is

$$f = \frac{c}{\lambda} = \frac{3 \times 10^8 \ \text{m/sec}}{700 \times 10^{-9} \text{m}} = 4.3 \times 10^{14}/\text{sec} = 4.3 \times 10^{14}\,\text{Hz}.$$

Using Planck's law, we can compute the amount of energy contained in one red photon, as

$$E_f = h \cdot f = (6.6 \times 10^{-34}\,\text{J} \cdot \text{sec/photon}) \cdot (4.3 \times 10^{14}/\text{sec}) = 2.9 \times 10^{-19}\,\text{J/photon}$$

The above is a very small quantity of energy, 0.29 attojoules (aJ). Recall that 1 aJ = 10^{-18} joules (J). To make one joule (the amount of energy emitted by a 1-watt light bulb in one second) requires 3×10^{18} red photons, because $1 \div (2.9 \times 10^{-19}) \approx 3 \times 10^{18}$. This is 3 billion-billion photons.

We can make a cartoon-like representation of light, in an attempt to capture some of its odd-seeming quantum properties. We can imagine that light is made of many photons traveling along, or guided by, some kind of wave, called a guiding wave, as shown in **Figure 12.1**. Each photon is drawn as a fuzzy bundle of energy to remind us

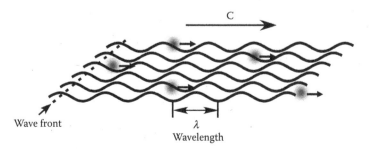

FIGURE 12.1 Photons traveling along with imagined guiding wave. Light is lumpy or grainy, containing little bundles of energy called photons, which should not be thought of as particles.

that photons are not really pointlike particles. The photons travel at the same speed as the wave, that is, c in vacuum. All of the energy in the wave is carried by the photons, and each photon has energy equal to that given by Planck's formula. The photons, of which there are an enormous number, are spread randomly throughout the wave. This means that when they arrive at a detector and are absorbed, the absorption events occur at random (i.e., unpredictable) locations and random times.

THINK AGAIN

Scientists don't really believe that the picture of photons traveling with a guiding wave is realistic, but it is perhaps the best we can do to explain the needed concepts without using advanced mathematics.

12.3   POWER AND ENERGY IN LIGHT

We know that when light strikes an object, the object is warmed. The rate at which such an object is heated depends on the power in the light beam. Recall that power is the rate at which energy is converted from one form to another or transferred from one place to another. Also recall that energy is conserved; that is, energy cannot be created or destroyed, but only converted between different forms, including mechanical, electrical, chemical, and thermal energy.

The units of power are joules per sec, or J/sec. A power of 1 watt (W) equals 1 J of energy delivered per second. (1 W = 1 J/sec) This means that the total energy E delivered by an energy source with power P in a time interval of duration t is given by $E = P \cdot t$. That is,

$$Energy = Power \cdot time$$

$$Joules = Watts \cdot Seconds$$

According to the quantum picture of light, each photon in a light wave with frequency f carries a bundle of energy equal to $E_f = h \cdot f$. All of the photons travel at the **speed of light**, which in vacuum equals c. In a dense medium with refractive index equal to n, the photons travel with speed c/n, which is less than c. The light beam power striking the surface of an object is the energy carried per photon multiplied by the average number of photons striking the surface per second. The average number of photons per second is called the **photon rate**, and is denoted by the symbol R_f. The word *rate* describes how often a process occurs. *Rate* does not mean the same as *speed*. An example of a rate occurs when a basketball player participates in a three-point shooting contest. Each player has 60 sec to shoot as many balls as possible. Only balls passing through the hoop count. Say the player shoots 30 balls in 60 seconds, 17 of which pass through the hoop. The total rate of shooting balls is 30 balls/minute (or 0.5 balls/sec), and the average rate of successful shooting is 17/minute (or 0.28/sec).

The power in a light wave with frequency f is equal to the energy per photon times the photon rate in the beam:

$$P(\text{J/sec}) = E_f\,(\text{J/photon}) \cdot R_f\,(\text{photons/sec})$$

$$P = E_f \cdot R_f$$

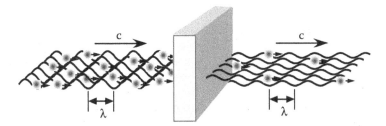

FIGURE 12.2 Photons traveling along with guide wave enter a medium, where some photons are absorbed. The amplitude of the light wave is greater where the number of photons is greater.

For example, consider a red lamp emitting light whose wavelength is $\lambda = 700$ nanometers (nm), meaning that each photon carries an energy $E_f = 2.9 \times 10^{-19}$ J/photon. If the lamp emits 3.4×10^{19} photons per second, then the power is

$$P = E_f \cdot R_f = \left(2.9 \times 10^{-19} \, \frac{J}{photon} \right) \cdot \left(3.4 \times 10^{19} \, \frac{photon}{sec} \right) = 10 \, \frac{J}{sec} = 10 \text{ W}$$

As another example, if the same lamp emits a power of 100 W, then the photon rate is

$$R_f = \frac{P}{E_f} = \frac{\left(\dfrac{100 \, J}{sec} \right)}{\left(\dfrac{2.9 \times 10^{-19} \, J}{photon} \right)} = 3.4 \times 10^{20} \text{ photon/sec}$$

How is the photon description of a light beam related to its wave properties? In the classical picture of waves, a higher power corresponds to a higher *amplitude* of the wave. In the quantum description of light, a higher power corresponds to a greater number of photons per second traveling along with the guiding wave. This is illustrated in **Figure 12.2**, which shows a light wave traveling from left to right, passing through a medium that absorbs some of the light energy. Light entering the medium has an initial wave amplitude, indicated by the height of each oscillation in the wave. It also has an initial density of photons (number of photons per cubic meter). As the wave passes through the medium, some of the photons are absorbed. When a photon is absorbed, it ceases to exist, and its energy is given up to the atoms in the medium. Therefore, the wave exiting the medium has a lower density of photons, as shown in the drawing. The exiting wave carries less power, so its wave amplitude is lower than the amplitude of the entering wave.

Again, let us emphasize that photons are not particles (they do not have definite location), but they do seem to have some particle-like properties in that each carries a certain amount of energy. They can be thought of as fuzzy bundles of light energy, which cease to exist when absorbed by a medium.

12.4 ABSORPTION OF LIGHT BY ATOMS AND CRYSTALS (OR "HOW EINSTEIN GOT HIS NOBEL PRIZE")

How can we predict whether or not light of a given wavelength will be absorbed by a particular medium? This question is important for designing light detectors, which are used in many information technologies (CD players, optical fiber communication,

QUICK QUESTION 12.1

A lamp emits light with wavelength $\lambda = 700$ nm (photon energy $E_f = 2.9 \times 10^{-19}$ J), and light with wavelength $\lambda = 350$ nm (photon energy $E_f = 3.8 \times 10^{-19}$ J). If the lamp emits 1×10^{19} photons of each color per second, what is the total power emitted?

etc.). Albert Einstein received the Nobel Prize in physics for providing the answer to this question—not for his famous theory of relativity! He explained this in terms of the then-new quantum ideas, including photons. We can summarize his idea in terms of a new quantum principle.

QUANTUM PRINCIPLE (v)

Absorption: If an electron in an atom is initially in a state (orbit) with lower energy equal to E_L, then it can absorb a photon of light having frequency f (with energy $E_f = h \cdot f$), and "jump" to a higher-energy state with energy E_H, only if the photon energy matches the difference of energies between the two electron states; that is, only if

$$h \cdot f = E_H - E_L$$

This rule is consistent with the principles of conservation of energy and Planck's law. The energy gained by the electron equals the energy given up by the light wave when it loses one photon. In this process, the photon disappears. We illustrate this in **Figure 12.3**. This shows an energy-level diagram for an atom. Recall that an electron in an atom is capable of having only certain discrete energy values. The energies (E_L and E_H) of the two states of interest are indicated by tick marks on the vertical energy axis. A photon, labeled by its energy $h \cdot f$, is shown as a wavy arrow impinging on the energy levels. The vertical arrow is the "jump process," in which the electron suddenly gains energy as the photon disappears. The white dot shows the energy the electron had before the jump. The cartoon on the right illustrates a person, representing the electron, jumping up the energy staircase after gaining just the right amount of energy. We call this type of jump, which is caused by photon absorption, an upward ***quantum jump***. If the incoming photon energy is less than that needed for the electron to make the jump, then no jump occurs.

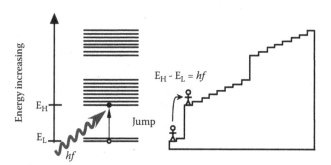

FIGURE 12.3 A photon (wavy arrow) with energy $h \cdot f$ is absorbed by an atom, causing an electron in the atom to make an upward quantum jump to a higher-energy state.

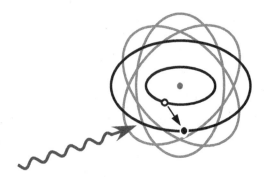

FIGURE 12.4 A photon (wavy arrow) excites an electron in a low-energy atomic orbit to a high-energy orbit.

An alternative way to visualize the absorption of photons by atoms is in terms of the planetary model of the atom, as in **Figure 12.4**. Initially the electron is in a small-radius, low-energy orbit. When a photon having the proper energy is absorbed, the electron jumps to a large-radius orbit with higher energy.

12.4.1   Absorption of Light by Crystals

The absorption of light in *crystals* is similar to absorption by single atoms. The main difference is that the electron energy levels are crowded together into nearly continuous bands, as shown in **Figure 12.5**. The energy difference between the top of the filled band and the bottom of the unfilled band is called the energy of the *band gap*. An electron can have energy anywhere within the range of a band, as long as there is not another electron already occupying that energy level (*Exclusion Principle*). If the incoming photon has enough energy to cause an electron to make a jump from a lower-energy band to a higher-energy band, then such a jump may occur, and the photon disappears. To cause such a jump, the photon energy must equal or exceed the energy of the gap, E_{GAP}. The figure shows an incoming photon, with energy slightly greater than the minimum energy needed to cause the electron to jump. The small, empty rectangle shows the energy the electron had before the jump (this is now a *hole*). The minimum frequency that a photon must have to induce an electron to jump to a higher level is

$$f = \frac{E_{GAP}}{h}$$

If the photon's frequency is less than E_{GAP}/h, then no jump occurs.

As an example, the energy needed for an electron to jump the band gap in a silicon crystal at room temperature is $E_{GAP} = 0.18$ aJ. What is the lowest light frequency that can cause such a jump? Using Planck's law, we find the frequency of a photon for which the energy equals silicon's band-gap energy:

$$f = \frac{E_{GAP}}{h} = \frac{0.18 \text{ aJ}}{h} = \frac{0.18 \times 10^{-18} \text{ J/photon}}{6.6 \times 10^{-34} \text{ J} \cdot \text{sec/photon}} = 2.7 \times 10^{14} \frac{1}{\text{sec}} = 2.7 \times 10^{14} \text{ Hz}$$

We can calculate the wavelength for such a photon:

$$\lambda = \frac{c}{f} = \frac{3 \times 10^{8} \text{ m/sec}}{2.7 \times 10^{14}/\text{sec}} = 1.1 \times 10^{-6} \text{ m} = 1.1 \mu\text{m} = 1,100 \text{ nm}$$

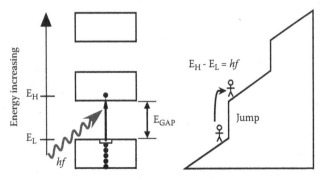

FIGURE 12.5   A photon (wavy arrow) with energy $h \cdot f$ is absorbed by a crystal, causing an electron in the crystal to make an upward quantum jump to a higher-energy state. This creates a hole in the lower-energy band.

The calculated wavelength corresponds to infrared light. Light with wavelength larger than 1,100 nm will not be absorbed by a silicon crystal, because the photons in such a light have too little energy to cause an electron to jump. In contrast, light in the visible range of the spectrum, red through blue (wavelength 400–800 nm), has photons with energy sufficiently large to cause such jumps. Therefore, silicon crystals absorb visible light.

IN-DEPTH LOOK 12.1: INABILITY OF CONSTANT VOLTAGE TO ACCELERATE ELECTRONS IN AN INSULATOR

In Chapter 9, we described an electrical *insulator* as a material that does not allow electrons to flow through it when a steady (or slowly changing) voltage is applied across it. We learned that in an insulator the highest occupied energy band is completely occupied by electrons, and that there is a large gap between the highest filled band and the next, empty band. By large, we mean about 0.5 aJ or larger. When a small voltage is applied across an insulator, no electrons can gradually accelerate, because there are no empty energy levels immediately above the highest occupied energy level. The Exclusion Principle "blocks" the electrons from accelerating in a completely filled band. Therefore no electrons can move or flow.

This explanation, although correct, leaves open a puzzle. If a moderate-sized voltage is applied across an insulator crystal, you might expect that an electron could jump from the filled band to the empty conducting band, thereby allowing current to flow. Energy would seem to be conserved by this jump process if the voltage were high enough to provide the needed energy to the electron. Why is such a voltage-induced jump not observed in insulator crystals?

The answer lies in the properties of electromagnetic waves, described by Quantum Principle (v) above. To see this, consider an insulator crystal with a band-gap energy equal to $E_{GAP} = 0.5$ aJ. Using Planck's law, we find the frequency of a single photon for which the energy equals the band-gap energy:

$$f = \frac{E_f}{h} = \frac{0.5 \times 10^{-18}\,\text{J/photon}}{6.6 \times 10^{-34}\,\text{J} \cdot \text{sec/photon}} = 7.6 \times 10^{14}\,\text{Hz} = 760\,\text{THz}$$

This means that electromagnetic radiation with frequency less than 760 THz (terahertz) cannot be absorbed by the electrons in the crystal, because there are no energy levels for the electron to jump to. As shown in **Figure 12.6**, ultraviolet (UV) radiation with frequency greater than 760 THz can cause such a jump. Visible light has frequencies roughly in the range 375–750 THz, so it is not capable of exciting electrons to the conducting band. Radio waves have even lower frequencies than visible light, so they are not able to cause conduction either.

When, in Chapter 9, we discussed applying a voltage across an insulator crystal, we had in mind a voltage that is constant, or perhaps changing a few thousand times per second. That is, we had in mind electromagnetic oscillations with extremely low frequency,

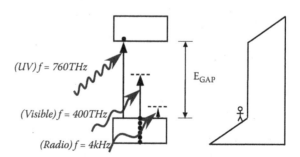

FIGURE 12.6   For a typical insulator crystal with band-gap energy equal to 0.5 aJ, light can be absorbed only if its frequency exceeds 760 THz.

perhaps even zero frequency ($f = 0$). According to Quantum Principle (v), such low-frequency electromagnetic oscillations cannot excite an electron to jump a high-energy gap. With $f = 0$, we certainly cannot satisfy the requirement that $h \cdot f$ is greater than E_{GAP}, which is needed for the electron to make an energy jump. This explains why a constant voltage cannot accelerate an electron in an insulator.

12.4.2   Absorption of Light by Metals: The Photoelectric Effect

Typically, most of the visible light that strikes a shiny, polished piece of metal is reflected from the surface. Think of the reflection you can see of your face in the surface of a shiny stainless-steel cooking pan. Although most of the light is reflected, a small amount of the light energy striking the metal can be absorbed by electrons near the surface of the metal.

If the frequency of the light is high enough (in the UV), a curious thing occurs, as illustrated by the energy-level diagram in **Figure 12.7**. An electron can gain enough energy by absorbing a photon that it can escape from the metal altogether. The stick person moving off to the right illustrates this in the figure. The electron is ejected from the metal's surface into the air. The minimum energy needed to eject the electron from the metal is called the threshold energy and is denoted by E_{THR}. The minimum frequency that light must have to cause an electron to be ejected is therefore:

$$f = \frac{E_{THR}}{h}$$

That is, the light's frequency f must be greater than E_{THR}/h to eject an electron. As an example, the metal zinc has threshold energy equal to 0.69 aJ. This means that to eject an electron, light must have a frequency greater than

$$f = \frac{E_{THR}}{h} = \frac{0.69 \, \text{aJ}}{h} = \frac{0.69 \times 10^{-18} \, \text{J/photon}}{6.6 \times 10^{-34} \, \text{J} \cdot \text{sec/photon}} = 1.04 \times 10^{15} \, \text{Hz}$$

The ejection of electrons from metal by light is called the ***photoelectric effect***, and was first studied in detail by Phillip Lenard, a Hungarian-German physicist, in 1902. A few years later, in 1905, Albert Einstein offered the explanation that we used in the above paragraphs to explain the effect. For this, he was awarded the Nobel Prize.

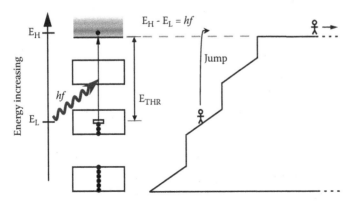

FIGURE 12.7   Photoelectric effect: A photon with high enough energy $h \cdot f$ is absorbed by a metal, causing an electron in the metal to make an upward quantum jump to a sufficiently high energy that it escapes the metal.

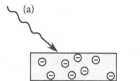
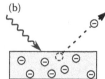
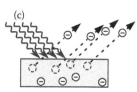

FIGURE 12.8 Photoelectric effect. (a) A photon with low frequency (long wavelength) cannot eject an electron from the surface of a metal. (b) A photon with high enough frequency and short enough wavelength can do so. (c) If the frequency of light is high enough, then increasing the incident photon rate increases the rate of ejecting electrons.

Lenard made several key observations, which allowed Einstein to put forth the explanation that we have discussed:

1. If the light's frequency was below a certain threshold value (which depended on the type of metal used), then no electrons were ejected from the metal (see **Figure 12.8a**). In this case, increasing the light's power had no effect—still no electrons were ejected.
2. If the light's frequency was above the threshold value, then electrons were ejected from the surface at some rate (see Figure 12.8b). In this case, doubling the light power striking the metal surface doubled the rate, that is, the number of electrons ejected per second (see Figure 12.8c).
3. Further increasing the light's frequency above the threshold frequency caused the ejected electrons to have increased kinetic energy (indicated by their speed) after they were ejected.

QUICK QUESTION 12.2

A copper ball has pieces of string glued at various points to its surface, as shown below. If the ball is charged positive or negative, the strings stand on end. This can happen if a plastic rod is rubbed with fur and then touched to the ball. If UV radiation with frequency 1.13×10^{15} Hz is shone on the ball, the strings do not change position. If UV with frequency 1.14×10^{15} Hz is shone on the ball, the strings flop down, as shown. From this observation, estimate the photoelectric threshold energy (in aJ) for copper.

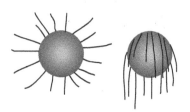

Let's think about why these observations caused Einstein to conclude that light is grainy or lumpy, that is, contains what we call photons. Consider, for a moment, the possibility that light does not contain photons, but is a smooth oscillation of electromagnetic fields, as was thought by most scientists at the time. Such smooth waves would carry energy, and this energy might be partially deposited in the metal gradually as the wave strikes it. One might think that eventually there would be enough light energy deposited in the metal to eject an electron, even if the light's frequency were below the threshold value. But this was not observed. In contrast, if light contains photons—each carrying energy equal to $h \cdot f$, and if each photon is indivisible and acts alone on an electron, then it is easy to see why increasing the light power or simply waiting a long time will not suffice to see electrons being ejected. In this case, each photon by itself simply does not have enough energy to eject any electron. Increasing the light power, or the rate of photons arriving, does not help. There would simply be more ineffective photons striking the metal each second.

The third of Lenard's observations is consistent with the following explanation: An incoming photon must give up all or none of its energy to an electron. If the photon has more than enough energy to eject an electron, then when it gives up its energy, there is excess energy, which goes into kinetic energy of the outgoing electron. This makes it move faster away from the metal.

Table 12.1 lists the threshold energies for ejection of electrons from the surface of various metals by light or other electromagnetic radiation.

TABLE 12.1

Photoelectric Threshold Energies for Various Metals

Aluminum	Zinc	Iron	Copper	Silver	Nickel	Gold
Al	Zn	Fe	Cu	Ag	Ni	Au
0.66 aJ	0.69 aJ	0.72 aJ	0.75 aJ	0.75 aJ	0.80 aJ	0.82 aJ

THINK AGAIN

The explanation just given—that increasing the number of photons striking a material is not necessarily sufficient to cause electrons to be ejected—rests on an assumption. It assumes that each photon acts separately and individually on an electron. Although some rare exceptions to this individualistic behavior are known, for our purposes we will consider this to be an accurate description.

REAL-WORLD EXAMPLE 12.1: SEMICONDUCTOR LIGHT DETECTORS

A photodetector is a device that produces electrical current when exposed to light; that is, it detects the presence of light. The photodetectors used in solar-energy applications, CD players, and in optical communication systems are made with semiconductor *diodes*. Depending on the application being considered, different semiconductor materials are used to make the diode. Recall from Chapter 10, Section 10.3 that a diode consists of an n-type doped semiconductor in contact with a p-type semiconductor, as shown in **Figure 12.9**. The n-type region contains excess electrons contributed by its *dopant* atoms, whereas in the p-type material there is a deficit of electrons (i.e, there are holes) in the outer shells of the atoms making up the crystal. At the junction between n-type and p-type materials, there is a *depletion region*, where excess electrons have diffused from the n-type into the p-type material, creating an imbalance of charge only in this small region. The excess electrons from the n-side are now trapped in the (former) holes in the p-side. The charge imbalance in the depletion region creates an internal battery at the junction, but no charge is flowing, because the electron diffusion is countered by the force created by the internal battery (shown as the large gray arrow in the junction).

In Figure 12.9a, a battery applies voltage across the diode in the backward direction; that is, minus polarity is connected to the diode's p-type side. In this case, no current flows. In Figure 12.9b, with the battery voltage still in the backward direction, light illuminates the depletion region of the diode. If the photons in the light have energy $h \cdot f$ that is greater than the band-gap energy of the crystal E_{GAP}, then electrons in the p-side of the depletion region can jump out of the holes in which they are trapped. These electrons are elevated in energy to the upper conducting band, as shown in the energy-level diagram in **Figure 12.10**.

The liberated electrons will feel a force caused by the internal battery. Because they are in the conduction band, they can be accelerated across the junction, then move through the n-side. Then they are pulled by the battery's voltage through the wire leading to the plus side of the battery. By monitoring how much current flows, we can determine how much light power is striking the detector. When a diode is used in this manner, it is called a *photodiode*.

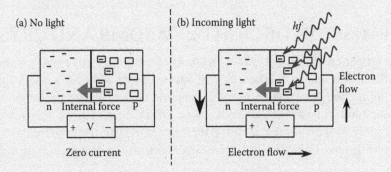

FIGURE 12.9 A semiconductor diode used to detect light.

To design and build useful photodetectors on the basis of these physics principles, we need to be able to predict whether or not a given type of crystal will detect light of a given color or wavelength. This depends on the size of the **energy gap** between the lower and upper bands, as shown in Figure 12.10. For example, a silicon photodiode has a band-gap energy equal to $E_{GAP} = 0.18$ aJ. As we calculated above, any photon whose frequency is greater than 2.7×10^{14} Hz can cause an electron to jump into the conduction band in such a crystal. This means that a silicon photodiode can detect light only if it has frequency greater than 2.7×10^{14} Hz, for example, near-infrared or visible light.

Table 12.2 lists the band-gap energies of several common semiconductor crystals, and the corresponding minimum light frequency (or, equivalently, maximum wavelength in vacuum) that each crystal can detect. For each crystal, the relation between energy and frequency is calculated using Planck's law.

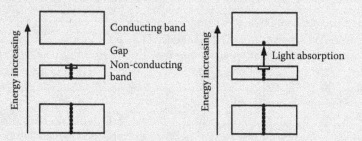

FIGURE 12.10 Light is absorbed by a semiconductor diode if its frequency is high enough so that photons in the light have sufficient energy to cause an electron in the diode to jump to a higher-energy band.

TABLE 12.2
Band-Gap Energies for Various Semiconductor Crystals

Material	Band-Gap Energy (aJ)*	Minimum Frequency (Hz)	Maximum Wavelength (nm)
Indium arsenide (InAs)	0.0577	0.874×10^{14}	3,430
Germanium (Ge)	0.106	1.61×10^{14}	1,870
Silicon (Si)	0.178	2.69×10^{14}	1,110
Gallium arsenide (GaAs)	0.227	3.45×10^{14}	870
Gallium phosphide (GaP)	0.362	5.49×10^{14}	547
Gallium nitride (GaN)	0.550	8.33×10^{14}	360

* In some texts, energy is given in units called electron-volts, or eV. The equivalence is: 0.160 aJ = 1 eV.

12.5   EMISSION OF LIGHT BY ATOMS AND CRYSTALS

Under the proper conditions, electrons in atoms and crystals will emit light. For example, the metal filament in a conventional, incandescent light bulb emits light when heated to high enough temperature by the electric current passing through it. This is easily understood in terms of electromagnetic waves created when the free electrons move randomly as a consequence of the thermal energy. In the remainder of this section, we will discuss light emission by atoms and semiconductor crystals, rather than by metals.

Emission by atoms is the reverse of absorption. To understand this, we can adapt the above-cited Quantum Principle (v). If an electron is in an atomic orbit with high energy, it can lose energy by emitting a photon, and drop down, or jump, to an unoccupied lower-energy orbit. This process, illustrated in **Figure 12.11**, is called *spontaneous emission*. It is spontaneous because there is no outside stimulus causing this process to take place; it happens on its own.

We can summarize this behavior by extending Quantum Principle (v) as follows:

QUANTUM PRINCIPLE (V, PART 2)

Spontaneous emission: If an electron in an atom is initially in a higher-energy state (orbit) with energy E_H, then it can emit a light photon having frequency f (with energy $E_f = h \cdot f$) and jump to a previously unoccupied state with lower energy E_L. The emitted photon energy matches the difference of energies between the two atomic states. That is,

$$h \cdot f = E_H - E_L$$

We can illustrate spontaneous emission with a picture of the atomic orbits, as in **Figure 12.12**. An electron is initially in a large-radius orbit, meaning it has high energy. Then it spontaneously drops to a smaller, lower-energy orbit, by emitting a photon. Spontaneous emission is random, that is, not strictly predictable. A photon can be emitted at any time, and in any direction.

Spontaneous emission can be seen by simply looking at a fluorescent neon lamp, often used in storefront windows. The tube of such a lamp contains neon atomic gas, through which an electric current is passed. As electrons move through the gas, they collide with atoms and give energy to electrons that are in the neon atoms. This "excites" the neon-atom electrons, causing them to jump to higher-energy states. Then these electrons spontaneously return to their lower-energy states, emitting photons of various frequencies (or colors). This process converts electrical energy into light energy.

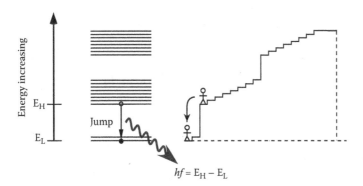

FIGURE 12.11   Spontaneous emission: An electron initially with energy E_H in an atom can emit a photon and drop to a lower-energy state E_L, emitting a photon with frequency f.

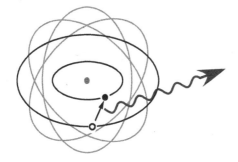

FIGURE 12.12   An electron in a high-energy orbit emits a photon and "drops" to a lower-energy orbit.

In Chapter 9, we discussed the emission spectrum of light emitted by atomic lamps. In Figure 9.8, we showed the sharp-line spectrum of the light emitted by an atomic lamp. And in In-Depth Look 9.1 we discussed a mathematical formula that can be used to predict the frequency of light emitted in each line of the spectrum. We said there that the observation of these sharp lines was evidence that the quantum theory is correct, but we deferred a detailed discussion of this point until the present discussion. Now we can see that by using Planck's law for the relation between light frequency and photon energy, we can explain why the atomic emission spectrum contains only sharp lines. It is because the energies of the electrons in an atom are discrete, or quantized, and therefore give off photons of specific frequencies or colors. This provides a major argument in favor of the quantum theory being correct. It also helps us understand how to design light detectors for specific purposes, as we will discuss next.

12.5.1   Emission of Light by Crystals

Electrons in crystals can emit light in a manner similar to electrons in atoms. **Figure 12.13** shows an electron initially in a higher-energy band. If there is no hole (empty energy level) in the lower band, then the electron cannot jump down into that band, according to the Exclusion Principle. If there *is* a hole in the lower band, then the electron can jump from a higher-energy band to the lower-energy band, emitting a photon. The emitted photon's frequency is again given by Planck's law.

12.5.2   The Rate of Spontaneous Emission: Exponential Decay

How rapidly does spontaneous emission occur? The more rapidly that atoms emit photons, the brighter an atomic lamp will be. The word *rate* is used to make quantitative the concept of "how rapidly" a process occurs. (This is not the same as *speed*, which tells how fast something moves.) The rate of spontaneous emission equals the number of atoms per second that jump spontaneously from their higher-energy states to a lower-energy state by emitting a photon. An important principle of quantum physics specifies the rate of spontaneous emission from a collection of atoms. Let us say that there are a total of N atoms, and that the number of those atoms that have an electron in the higher-energy orbit with energy E_H equals N_H. For example, let us say the total number is $N = 3 \times 10^{12}$. If one-third of those atoms have an electron in the orbit with energy E_H, then $N_H = 1 \times 10^{12}$. The rate of spontaneous emission, the process whereby an atom in a higher-energy state emits a photon and jumps to a lower-energy state, is determined by the following principle.

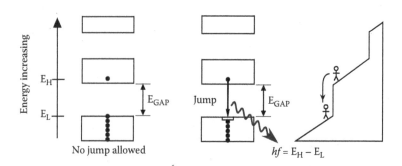

FIGURE 12.13   An electron can jump from a higher-energy band to a lower-energy band and emit a photon (wavy arrow) with energy $h \cdot f$ only if an electron hole (empty state) is present in the lower band.

QUANTUM PRINCIPLE (VI)

Rate of Spontaneous Emission: In a sample of many atoms, the *rate of spontaneous emission* ($R_{SP.EM.}$) from a higher-energy state to a lower-energy state is proportional to the number of atoms N_H remaining in the *higher* state, and is determined by the formula: $R_{SP.EM.} \propto N_H$.

Spontaneous emission causes exponential decay of the number of atoms in the higher-energy state. Exponential decay is best explained by an example involving something more familiar than atoms. Say you have 10,000 eye droppers, each containing a drop of water that can leak out spontaneously, as in **Figure 12.14**. In a time interval of 1 sec, each dropper has a 10% chance to release its drop. Start with all droppers full. Then, you would expect that in the first second, one-tenth of 10,000 droppers, (i.e., 1,000 droppers) would release a drop, leaving 9,000 full droppers. In the next 1-sec interval, you would expect that one-tenth of 9,000 (i.e., 900) droppers would release a drop, leaving 8,100 full droppers. In the third 1-sec interval, you would expect that one-tenth of 8,100 (i.e., 810) droppers would release a drop, leaving 7,290 full droppers. After 18 1-sec intervals, the number of full droppers is about 1,351, so in the next 1-sec interval, there will be about 135 drops released. The important observation to make is that the number of drops released in any 1-sec interval is a fixed fraction (one-tenth) of the currently remaining number of full droppers.

Spontaneous emission of photons from atoms is analogous to the eye dropper example. Say you have a collection of 1,048,576 atoms, all with an electron in a higher-energy state. Let us say that in any 1-sec interval about one-half of the remaining

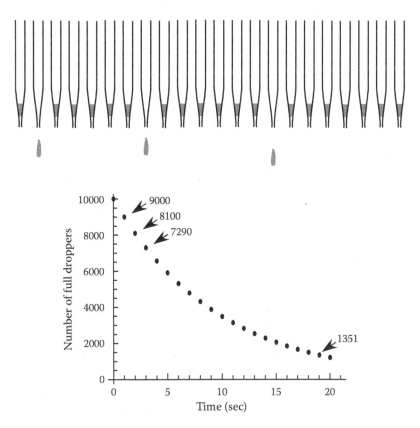

FIGURE 12.14 Graph of the number of remaining full eye droppers. Beginning with 10,000, one-tenth of the remaining full droppers release their single drop in any 1-sec interval. The previous number is multiplied by 0.9 after each second elapses.

higher-energy atoms emits a photon. The numbers of atoms remaining that have not yet emitted a photon after each successive 1-sec interval will be: 524,288, 262,144, 131,072, 65,536, 32,768, 16,384, 8,192, 4,096, 2,048, 1,024, 512, 256, 128, 64, 32, 16, etc. Each number in the list is obtained by multiplying the previous number by 0.5.

THINK AGAIN

The calculated number of atoms remaining is not rigorously accurate for a particular try of the experiment because emission is random. If tries were repeated many times, and the data were averaged, the result would be more representative.

REAL-WORLD EXAMPLE 12.2: LIGHT-EMITTING DIODES

The light-emitting diode, or LED, is very useful in situations in which a highly efficient source of light is needed. To cause a diode to emit light, we need to reverse the process shown in Real-World Example 12.1; that is, we need to reverse the polarity of the battery, thereby causing current to flow through the diode junction. Please review Chapter 10, Section 10.3 to recall the explanation of diode operation in terms of our analogy using salmon in lakes.

The operation of an LED in terms of the salmon analogy is shown in **Figure 12.15**. In the figure, the height of a lake indicates the potential energy of salmon there. Now we include in our model the fact that in any material, electrons (represented by fish) can be in a lower, nonconducting energy band or in a higher, conducting energy band. These are represented by the lower lake (labeled B) and the higher lake (labeled A). A fish in lake A might "wish to" jump down to lake B. But, such a jump is prevented by the fact that the lower lake is completely filled with fish, and the Exclusion Principle prevents having too many fish in one region. Of course, real salmon are not limited in their behavior by the Exclusion Principle, but we need to use this rule here to make the analogy more useful. As the more energetic fish jump up the falls from lake A to the raised lake C, many of them now find that they are allowed to jump down to the lake D below. This is allowed because lake D is relatively empty. Most fish travel from the left stream into lake A, then rise to lake C and immediately jump down to lake D, and exit to the right in the stream having the lower potential energy. When a salmon jumps down to lake D, it gains kinetic energy as it accelerates downward. This kinetic energy is converted into heat when a salmon hits the water in lake D. In an LED, we want the downward-jumping electrons to emit light, rather than generate heat, as in the salmon example.

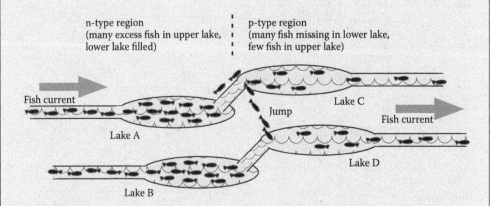

FIGURE 12.15 Salmon analogy for a LED. As energetic salmon reach the upper lake C, they can jump down to the lower lake D.

Light emission by a diode is shown in **Figure 12.16**. In Figure 12.16a, there is no externally applied voltage, so no current flows in the diode. The n-type region contains excess electrons, and the p-type region contains electron holes. Recall that diffusion of electrons takes place from the n-type region into the p-type region, filling some of the holes and creating an opposing force internal to the diode. This opposing force was represented as a waterfall in our salmon analogy. In Figure 12.16b, the switch is closed, applying the external voltage, so electrons flow in the direction shown by black arrows. As electrons are pulled out of holes in the depletion region, and new electrons flood into this region from the n-type side, many of the new electrons "fall" into holes, giving up energy as they do so (like the salmon falling or jumping to the lower lake). The energy given up by the electrons can be in the form of heat or light.

The jump process can be better visualized by using energy-level diagrams, as in **Figure 12.17**. The three parts of the figure correspond to the three regions in the diode. The left-hand-side of the figure shows the n-type region, with its excess of electrons in the conducting band. The right-hand-side of the figure shows the p-type region, with its holes in the valence band. The center region shows the depletion region, with an electron falling into a hole, emitting a photon. That electron then travels via the empty holes in the p-type region toward the plus terminal of the battery.

Why does light emission occur only in the depletion region, and not the n-type or p-type regions? The quantum Exclusion Principle prevents emission in these two regions

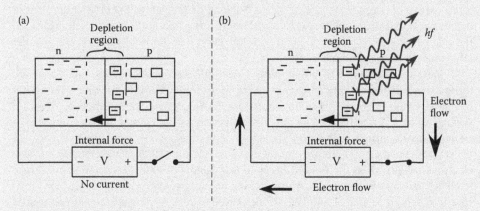

FIGURE 12.16 (a) Diode with no externally applied voltage emits no light. (b) When the switch is closed, current flows and light is emitted from the diode junction.

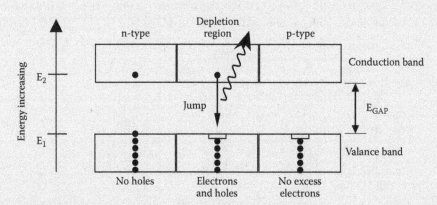

FIGURE 12.17 Energy bands of the three regions in a diode. Excess electrons pass through the n-type region into the depletion region, where they fall or jump to the lower-energy band, which contains empty holes.

as follows: In the n-type region, there are excess electrons, but no empty holes in its lower-energy band. This prevents an electron in the higher band from jumping to the lower band. In the p-type region, there are plenty of holes for electrons to fall into, but there are no excess electrons in its higher-energy band. Therefore, no jumps take place.

The type of semiconductor material used to make the diode determines the color (frequency) of light that it emits and the energy efficiency of the light source. Choosing the proper material can lead to highly efficient conversion of electrical energy into light energy. The conversion process will be efficient if most of the energy given up during the electron jumps is converted into light rather than into heat. Silicon is not a good material for making an LED. For subtle reasons that we will not explore, the rate of spontaneous emission in silicon is nearly zero. Instead of silicon, other semiconductors are used for making LEDs. A semiconductor commonly used is gallium arsenide (GaAs), a crystal made of a mixture of gallium (Ga) and arsenic (As) atoms. This is a pure semiconductor crystal, like pure silicon. Therefore, in its pure form it does not conduct electrical current well.

As when using silicon, the conductivity of a GaAs crystal can be controlled by adding impurities, or dopants, to create an excess of electrons in the conducting band, or a deficit of electrons in the originally filled band. Adding tellurium (Te) creates excess electrons in the higher, conducting band of GaAs, making the crystal an n-type conductor. Adding zinc (Zn) creates a deficit of electrons in the lower band, which creates a p-type conductor. These two doped forms of GaAs are joined at a junction to make a light-emitting diode, which operates as described above. Such an LED made with GaAs emits light in the near-infrared region, with a wavelength around 0.90 μm (900 nm), which is close to the calculated wavelength corresponding to the energy of the band gap in GaAs (see the table in Real-World Example 12.1).

We are familiar with LEDs that emit visible light—red, yellow, blue, etc. These are made using a different semiconductor crystal, for example, GaAlAs, in which aluminum (Al) atoms replace some of the Ga atoms in a GaAs crystal. The addition of Al is not doping; rather it is making a new kind of pure semiconductor crystal, which still needs to be doped by impurities such as Te. The presence of the Al atoms increases the energy of the crystal's band gap. Therefore, the energy of the photons emitted is increased, because electrons usually sit at the bottom of the upper band and emit to the top of the lower band, as shown in Figure 12.13. An increasing photon energy means a decreasing wavelength of light, as seen in Table 12.2. If the crystal has 1% Al atoms, its band-gap energy corresponds to photons emitted with wavelength around 0.88 μm, whereas for 35% Al, the emitted wavelength is 0.75 μm, as seen in **Figure 12.18**. Both of these are deep red (near-infrared).

To make an LED that emits colors other than red, it is necessary to use a different type of semiconductor altogether. For the wavelength range 0.6–0.7 μm, we can use GaAsP crystals, with various percentages of phosphorous (P) atoms. For blue

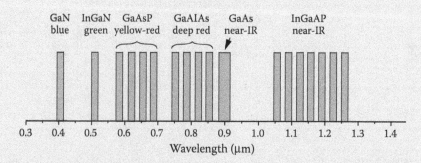

FIGURE 12.18   Using different semiconductor crystals with different band-gap energies allows making LEDs that emit different colors (wavelengths) of light.

wavelengths around 0.40 μm, we can use GaN crystals, which contain nitrogen (N) atoms. For green wavelengths around 0.52 μm, we can use InGaN crystals, which contain indium (In) atoms. To make LEDs that produce white light, a blue LED is surrounded by a white-light-emitting phosphor material, similar to those used in fluorescent lamps.

LEDs have remarkable properties compared with incandescent light bulbs. They are more efficient (10 times greater for red light) than incandescents, using less electrical power to generate the same light power. An LED can operate continuously for up to 100,000 hours—over 10 years. LEDs are very small and generate very little heat, so they can be used in very small devices, such as cell phones.

SUMMARY AND LOOK FORWARD

In this chapter, we discussed how atoms and crystals create light. This is the basis of lasers and LEDs. Atoms and crystals can also absorb light and thereby be used as light detectors or signal receivers in optical communication systems.

Planck's law tells us that light is emitted or absorbed by atoms in the form of small bundles of energy called photons. These are not particles in the classical (Newton's) sense; nevertheless, the concept helps us tally the amount of energy transferred to or from atoms. The energy of a given photon is proportional to the light's frequency, and the constant of proportionality is called Planck's constant.

If an electron in an atom is initially in some state (orbit), it can absorb a photon of light and "jump" to a higher-energy state if the energy of the photon matches the difference of energies between the two electron states. Similarly, electrons in crystals can absorb light if the energy of the photons matches the difference of energies between the two electron states in the crystal. If an electron in an atom is initially in some higher-energy state (orbit), it can spontaneously emit a photon of light and jump to a lower-energy state. The rate of such spontaneous emission events is proportional to the number of atoms that have an electron in the higher-energy state. The emitted photon energy matches the difference of energies between the two atomic states.

The colors of light emitted or absorbed by particular atoms or crystals depend on the energies of the electron states in that material. For example, in crystals with larger energy differences (gaps) between bands, the light emitted is "bluer" than the "redder" light emitted by crystals with smaller band-gap energies. This fact controls the colors emitted by LEDs made of various semiconductor materials. The LED is the precursor of the semiconductor laser—one of the most important technologies behind the Internet, and the topic of a later chapter.

Although light seems to have a curious combination of wavelike and "bundle-like" properties, physicists do have a mathematical model for light—the quantum theory—which works well for all practical purposes. In this chapter, we discussed some aspects of this model and saw how we can use it to predict the behaviors of atoms and crystals when exposed to light. Nevertheless, we still do not have a simple, clear, and intuitive model for light itself. Let us end our discussion of photons, also called *light quanta*, with a mildly surprising quote from Einstein himself

All the fifty years of conscious brooding have brought me no closer to the answer of the question: What are light quanta?

Albert Einstein
(1951)

SOCIAL IMPACTS: LIGHTING THE DARKNESS (EFFICIENTLY)

LEDs have the potential to replace most conventional light sources within the next 15 years.

William Cassarly
(*Optical Research Associates,* 2008)

Energy efficiency has become one of the paramount issues of our time. Population growth, the physical limits and side effects of natural resource extraction, and complications due to impacts on climate change have led people to look for ways to drastically reduce energy use per capita. A major, even primordial, use of energy by humans is for lighting the darkness. The discovery of fire allowed early people a degree of control over the natural elements and allowed them to see after sunset, heat dwellings, and cook food. In those days, lighting, heating and cooking made up nearly 100% of people's technological energy use. Even today, they comprise the bulk of a home's energy use. Of the total electric power delivered to homes, businesses, and factories throughout the world, no less than 25% is used for lighting alone [1]. Unfortunately, nearly 2 billion people have no access to electricity in the home [2]. The good news is that efficient lighting based on LEDs is changing that, in part because of humanitarian efforts [3].

Efficient energy use means converting energy from one form into another form nearly completely, without undue amounts of other unwanted forms being created as byproducts. For example, in the summer in Arizona, people would prefer to have indoor light bulbs that convert nearly 100% of electrical energy into visible light, rather than bulbs that generate lots of heat as well. In winter in Minnesota, this would be less of a concern, because the heat serves a useful purpose. For outdoor lighting, any amount of heat generated by a light bulb is a waste of electrical energy, which is generated at some cost by the power companies.

People have used oil-burning lamps since ancient times. The modern era of lighting began in 1879, when Thomas Edison in the United States, and Joseph Swan in Britain, invented electric lights based on filaments of various materials that glowed brightly when electric current passed through them. A very hot filament emits nearly white light, which matches well with people's viewing habits, which are based on our evolving under the Sun. These bulbs burned out after 40 hours of use. Lewis Latimer, an African-American engineer and draftsman, invented the first incandescent long-lasting bulbs. Latimer used specially designed carbon filaments enclosed in evacuated glass

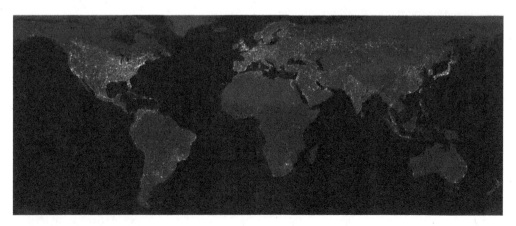

FIGURE 12.19   An image showing the distribution of lighting use worldwide. (Craig Mayhew and Robert Simmon. (Courtesy of NASA Goddard Space Flight Center.)

bulbs. The incandescent bulbs we use today are similar, with the carbon filaments being replaced by tungsten ones for even longer lifetimes. Bringing electric lights into homes had a big impact on social aspects of life [3–5]. Unfortunately, such incandescent bulbs are only about 5% efficient, meaning that 95% of the electrical energy used is wasted as heat.

The next step in lighting technology was the invention of fluorescent bulbs. In such bulbs, electric current passes through a gas of metal atoms, such as mercury or sodium, exciting the electrons in the atoms to higher energy states. These electrons lose energy by emitting light, a process that generates little wasted heat. A main challenge with mercury lamps was to create light that is nearly white colored, rather than ultraviolet. For this purpose, fluorescent paint (phosphor) is applied to the inside of the tube, which absorbs UV and emits visible light.

The efficiency of a lamp is usually given in terms of its effectiveness in aiding human vision. Because human eye receptors are most efficient for green light, a standard unit of lighting brightness is in terms of a unit called a lumen, which is equivalent to 1/683 of a watt of light power at a wavelength of 555 nm; that is, green light with power equal to 1 W is considered to equal 683 lumens. Efficiency is then defined as lumens of light generated for each watt of electrical power used to operate it; that is, efficiency is given as lumens per watt, abbreviated lpw. An incandescent bulb has an efficiency of approximately 10–15 lpw, and fluorescent bulbs have efficiency of approximately 50–100 lpw. A yellow sodium lamp, at 200 lpw, is still the most efficient bulb available, but is not the best in terms of color. Usually, engineers try to optimize lamps for lumens per watt while keeping the color as natural looking as possible.

Now roaring onto the scene is the LED, which first achieved a practical efficiency of 50 lpw around the year 2000, and can now exceed 100 lpw. Since then, manufacturing costs have decreased enough to allow LEDs to serve many markets. Early LEDs emitted light of a single color, but now white-light LEDs are common. White-LED flashlights were introduced around 2005. Current engineering goals are to meet or exceed the 200-lpw mark set by the sodium lamp at a reasonable cost.

FIGURE 12.20 American engineer Lewis Latimer, the inventor of the first long-lasting incandescent light bulb, while employed by the U.S. Electric Lighting Company, a competitor of Thomas Edison's company, which Latimer later joined. Latimer also assisted in setting up light bulb factories and installing Maxim lighting systems in New York City, Philadelphia, Montreal, and London. (Courtesy of The Queens Borough Public Library, Long Island, Latimer Family Papers.)

Issues to Ponder

1. Australia announced it will introduce higher lighting efficiency standards, which will likely result in the banning of incandescent bulbs for most uses. South Africa, California (USA), and Ontario (Canada) are considering similar steps.
2. Will increased use of cost-effective lighting result in less or more use of energy resources worldwide? If more, where will that energy come from?
3. Incandescent light bulbs contain no mercury. Compact fluorescent lamps (CFL), which are offered as more efficient alternatives, do contain mercury. Mercury is a toxic metal that is hazardous to the environment. On the other hand, coal-burning electrical power plants emit lots of mercury into the atmosphere, so the more efficient bulbs, which use less electrical power, might be better overall [6,7].

REFERENCES

1. Lister, Graeme. "Keeping the Lights Burning: The Drive for Energy Efficient Lighting," *Optics & Photonics News* 20, January (2004).
2. "LEDs: The Future of Light," Broadcast transcript, *Living on Earth*, Somerville, MA, 2007. http://www.loe.org/series/LED.php.
3. BoGo Light Project, by SunNight Solar. http://www.bogolight.com/aboutus.asp.
4. Nye, David E. *Electrifying America: Social Meanings of a New Technology, 1880–1940*. Boston: MIT Press, 1992. (Winner of the 1993 Edelstein Prize sponsored by the Society for the History of Technology).
5. Norman, Winfred Latimer, and Lily Patterson. *Lewis Latimer (Black Americans of Achievement)*. New York: Chelsea House Publications, 1993.
6. A CFL uses 75% less energy than an incandescent light bulb and lasts at least 6 times longer. A power plant will emit 10 mg of mercury to produce the electricity to run an incandescent bulb compared with only 2.4 mg of mercury to run a CFL for the same time. [From the FACT SHEET: Mercury in Compact Fluorescent Lamps (CFLs), U.S. Environmental Protection Agency.]
7. Richard, Michael Graham. "What About Mercury From Compact Fluorescents?" Gatineau, Canada on June 17, 2005, TreeHugger Radio Archives. http://www.treehugger.com/files/2005/06/what_about_merc.php.

SUGGESTED READING

Condren, S. Michael, George C. Lisensky, Arthur B. Ellis, Karen J. Nordell, Thomas F. Kuech, and Steven A. Stockman. "LEDs: New Lamps for Old and a Paradigm for Ongoing Curriculum Modernization," *Journal of Chemical Education* 78 (2001): 1033.

Craford, M. George, Nick Holonyak, Jr., and Frederick A. Kish, Jr. "In Pursuit of the Ultimate Lamp," *Scientific American* 284, no. 2, (2001): 63.

Lisensky, George C., Rona Penn, Margret J. Geselbracht, and Arthur B. Ellis. "Periodic Properties in a Family of Common Semiconductors: Experiments with Light Emitting Diodes," *Journal of Chemical Education* 69, (1992): 151.

KEY TERMS

Absorption
Amplitude
Atoms
Band gap
Crystals
Depletion region
Diode
Dopant
Electromagnetic wave

ANSWERS TO QUICK QUESTIONS

Q12.1 Total power =

$$\left(2.9 \times 10^{-19} \frac{\cancel{J}}{\cancel{photon}}\right) \cdot \left(1 \times 10^{19} \frac{\cancel{photon}}{sec}\right) + \left(3.8 \times 10^{-19} \frac{J}{\cancel{photon}}\right) \cdot \left(1 \times 10^{19} \frac{\cancel{photon}}{sec}\right) =$$

$$(2.9 + 3.8)\frac{J}{sec} = 6.7 \text{ W}$$

Q12.2 The photoelectric threshold for copper is:

$$E_{THR} = f \cdot h = (1.14 \times 10^{15} \text{ Hz}) \cdot (6.6 \times 10^{-34} \text{ J} \cdot \text{sec/photon}) = 0.75 \text{ aJ}$$

EXERCISES AND PROBLEMS

Exercises

E12.1 Light beams, A and B, are traveling side by side. Beam A has a shorter wavelength than beam B. Which beam contains photons with the larger energy? Explain, using an appropriate equation that supports your answer. What experiment could be conducted to verify your answer?

E12.2 A light pulse has a frequency of 4×10^{14} Hz, and contains 4×10^{19} photons of energy. It enters an atomic medium containing 1×10^{19} atoms, all in their lowest energy state. If one-fourth of the atoms in the medium absorb one photon from the light pulse, how much energy is contained in the pulse after it exits the medium?

E12.3 A very weak beam of green light strikes a metal plate that is charged with an excess of electrons, and it is observed that the light does not cause any electrons to leave the plate's surface. When the power of the light beam is increased by some large factor, say 1 million, still no electrons leave the plate. Explain this result and state what property of this light beam you would need to change so that electrons would leave the plate.

E12.4 A steady, very weak, beam of UV light strikes a metal plate that is charged with an excess of electrons, and it is observed that the light causes a small number of electrons each second (1,000 per second) to jump off the plate. If the power of the UV light beam is increased by a factor of 1 million, predict how many electrons per second will be ejected from the plate's surface.

E12.5 Light detectors based on semiconductors typically work best when the band-gap energy of the crystal used is equal to, or is slightly smaller than, the energy of photons in the light being detected. The table in Real-World Example 12.1 lists band-gap energies for various semiconductor crystals. From this list, choose which material would be best for detecting:

  (a) violet light with wavelength 400 nm
  (b) green light with wavelength 520 nm
  (c) UV radiation with wavelength 300 nm
  (d) red light with wavelength 630 nm

E12.6 An electrical-insulator crystal and a semiconductor crystal are at zero absolute temperature. Only one of these crystals can absorb radiation whose frequency is 9×10^{14} Hz. Which crystal is this, and why? *Hint*: In Chapter 9, we stated that the values of semiconductor band gaps fall in the range 0.01–0.5 aJ.

E12.7 Einstein, when describing the photoelectric effect, explained that increasing the number of photons striking a material is not necessarily sufficient to cause electrons to be ejected from a metal plate. His explanation rests on the assumption that each photon acts separately and individually on an electron. In some rare cases, exceptions to this individualistic behavior are known. Using drawings like that in Figure 12.7, explain how, at extremely high rates of photon impacts on the plate, an electron might be ejected from the plate, although each photon has energy less than the threshold energy for that plate. *Hint*: Generally, if an electron finds itself on one of the energy "ramps" shown in the figure, it will quickly slide back down to the lowest-energy ramp, emitting a photon as it does so. What might happen instead?

E12.8 An electron is shown in Figure 12.11 jumping to a state of lower energy and emitting a photon. Explain why there is no difference of meaning between the two ways of drawing this process shown below. Also explain why some students might initially be confused by the way the drawing is shown in Figure 12.12.

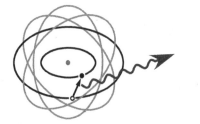 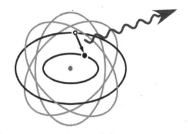

E12.9 You want to design a flashlight that emits orange light only. From the information given in Real-World Example 12.2, what semiconductor material could you use?

Problems

P12.1 A lamp emits light whose wavelength equals $\lambda = 500$ nm. What is the energy per photon? If the lamp emits 4×10^{19} photons per second, what is the total power emitted? *Hint*: Use conversion factors.

P12.2 Consider a certain LED that emits orange light when voltage from a battery is applied. We disconnect the battery and find that shining green light onto the LED causes an electric current to be created by the LED. We also find that shining red light onto the LED causes no current to be created by the LED. Explain these observations, in terms of one of the physical principles of quantum physics. Use calculations to validate your discussion. *Hints*: Consider the wavelength and frequency of typical red, orange, and green light and use the Planck relation. Also determine a reasonable estimated value for the energy of the band gap in this LED.

P12.3 An aluminum ball has many pieces of string glued to its surface, as shown below. If the ball is charged positive or negative, the strings stand on end. A plastic rod is rubbed with fur, then touched to the ball, resulting in the strings standing on end. What is the minimum frequency of electromagnetic radiation that, when shone on the ball, would cause the strings to flop down? What wavelength of light does this correspond to? Is this "light" visible to the eye? *Hint*: See the table of photoelectric threshold energies at the end of Section 12.4.

P12.4 Each "pixel" in the human eye contains three different types of light-absorbing molecules. One type absorbs blue photons, but not green or red ones. A second type can absorb blue, green, or red, but responds more strongly to green than the other molecule types. The third type can absorb red, green, or blue, but responds more strongly to red than the other molecule types. Draw three energy-level diagrams analogous to Figure 12.3, one for each type of molecule, illustrating how these color sensitivities might come about.

P12.5 Blue light with wavelength 375 nm and power 0.01 W strikes an LED, which has a band gap small enough so that every incoming light photon is absorbed and causes one electron in the LED to jump from the semiconductor's lower (nonconducting) band to its higher (conducting) band. Each electron that jumps contributes to current from the LED. How many electrons will make such a jump during each second? *Hint*: First calculate the energy of each blue photon.

P12.6 Say you have a savings account of $19,683. Let us say that after every 6-month interval, you withdraw two-thirds of the remaining balance, leaving one-third of the previous balance. How many such withdrawals would it take to reduce your balance to $1 or less? Make a graph of the balance versus time.

P12.7 Say you have a collection of 531,441 atoms, all with their electrons in the higher-energy state. Let us say that in any 1-sec interval, about two-thirds of the remaining higher-energy atoms emit a photon, leaving one-third still in the higher state. For times 0,1,2, ...12 sec, make a list of the numbers of atoms remaining that have not yet emitted a photon. Make a graph of these numbers versus time.

P12.8 The main method for dating old (but not too ancient) plant and animal materials in objects found by archeologists is radiocarbon dating. When a living organism initially grows, it absorbs from the environment a certain known fraction of carbon-14 atoms, which are carbon atoms containing six protons and eight neutrons in their nucleus. This form of carbon is unstable (i.e., it spontaneously decays) and changes into nitrogen atoms. (This decay is analogous to spontaneous decay of electrons by emission of photons, but we will not study the process by which this decay occurs.) "Ordinary," stable carbon atoms contain six protons and six or seven neutrons and do not decay. The amount of carbon-14 remaining in a material decreases by a factor of 0.5 after every 5,730-year interval since it was formed. This interval is called the half life. Therefore, by measuring the amount of carbon-14 remaining in an old material, archeologists can determine its age.

(a) The oldest objects that can be dated this way are around 63,000 years old. What fraction of original carbon-14 is still present after that length of time?

(b) An old shoe is found to contain a fraction 0.00195 of the amount that it contained when it was formed. About how old is that shoe?

P12.9 Recently many cities have switched from using incandescent lamps in traffic signal lights to using LEDs to save money. Incandescent lamps are inefficient, in part because their white emitted light has to be filtered with colored plastic to make the red, yellow, and green colors of light needed. Including the initial purchase price, compare the costs of operating a red LED lamp to a red incandescent lamp for 2 years. Assume the mechanical fixtures have the same cost, and that the red light is ON one-half of the time. Use the following data: A red-filtered incandescent lamp uses 100 W of electrical power to produce the needed brightness, whereas an LED uses 20 W to create the same brightness. Assume the cost of electrical power is 10 cents per kilowatt-hour (kW · hr). A 100-W incandescent bulb costs around $2, whereas a 20-W LED bulb costs around $50. An incandescent bulb might last 6 months before needing replacement, meaning four bulbs will be needed over two years, whereas an LED can run for over 30,000 hours, meaning no bulb replacement is needed during the 2 years.

P12.10 Recall from Chapter 9, P9.4, that Bohr's theory (which we did not derive) gives a formula for the energies of the electron shells in the hydrogen (H) atom:

$$E_n = R \times \left(1 - \frac{1}{n^2}\right), \quad R = 2.1799 \times 10^{-18} \, \text{J},$$

where R is a constant, and n is an integer labeling each energy shell. The lowest-energy shell (with energy defined to be zero) has $n = 1$, the second lowest has $n = 2$, etc. Calculate and list the energies of the first six energy shells, $n = 1$ through 6. Calculate and make a table of the differences of energies of the three groups of jumps shown in the figure here. Use the integer n to label the lower of the two orbits, and m to label the higher-energy orbit. Calculate the wavelengths of light emitted in the jumps labeled A, B, and C in the figure. The others are given in the figure, in units of nanometers, and you should verify a few of these. For this question, use the more accurate values of Planck's constant: $h = 6.626 \times 10^{-34}$ J · sec, and the speed of light: $c = 2.998 \times 10^8$ m/sec.

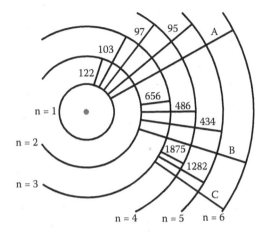

Difference of Energies

n \ m	1	2	3
2		x	x
3			x
4			
5			
6			

13

Light and Optical Fibers for the Internet

Vision perceives necessarily all the objects through supposed straight lines that spread themselves between the object and the central point of the sight.

Arab physicist Ibn al-Haytham
(Born in Basra, 1965)

A handfull of silica glass fibers could carry an almost unimaginable amount of information per second in the form of light pulses.

Donald Keck, Robert Maurer, and Peter Schultz invented low-loss optical fiber while working at Corning in 1970. They were awarded the 2000 National Medal of Technology by the President of the United States. (Courtesy of Corning Inc.)

13.1 LIGHT AS A COMMUNICATION MEDIUM

As we discussed in Chapter 5, the discovery that electricity, magnetism, and light are all aspects of the same underlying physics was one of the greatest insights in the development of science. Light and electromagnetism have long been used for communication. In ancient times, a person might have lit a fire at the top of a hill to send a simple one-bit message (YES or NO) to another person at a distant location. The advent of the telegraph around 1845 enabled people to send many bits of information across hundreds of kilometers, and later across thousands of kilometers. Then, in the twentieth century, radio transmission through air was perfected, leading eventually to modern wireless communication and cell phones. Satellites using radio transmission enabled global communication. Only in the late twentieth century did fiber-optic technology enable using light for communication over globe-spanning distances. In this chapter, we will study the physics of light traveling in various types of media, with

435

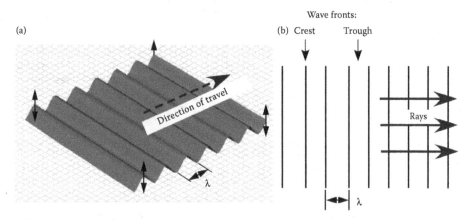

FIGURE 13.1 (a) Perspective view of a light wave, with vertical arrows showing the directions of oscillation of the electric field, and the dashed arrow showing the direction of travel of the wave. (b) "Top view," or wave-front view, of the same wave, showing the crests, and the rays.

an eye toward understanding fiber optics. We will discuss how the properties of air, water, ordinary glass, and optical fibers affect light's behavior.

Optical fiber is made of silicon, although in a glassy (noncrystalline) form rather than a crystalline form as is used for making semiconductor electronics. This means that silicon is the principal basis for long-distance communication technology as well as the basis for computers. In this sense, the Internet, which is a huge collection of silicon-based computers linked together using silicon-based optical fibers, is accurately described as *The Silicon Web*—the title of this book.

13.2 PROPAGATION, REFLECTION, AND TRANSMISSION OF LIGHT

Light is an *electromagnetic wave*, as discussed in detail in Chapter 7, Section 7.10. A simple example of light propagation occurs in air or vacuum. **Figure 13.1** shows two styles for representing a light wave. We can draw a three-dimensional picture, reminding us of the analogy of a light wave to a water wave. Or, we can draw a simpler "top view," indicating the locations of the crest wave fronts as lines. We also call this the wave-front view. Recall that the distance between crests is the *wavelength*, and is denoted λ. A wave travels, or advances, in the direction perpendicular to the *wave fronts*. We call the arrows that we draw pointing in the direction of travel *rays*. The rays are not physical, but give us a nice way to describe the direction in which the wave is advancing. For example, in the case of a laser beam, the rays point along the direction that the beam travels. As the wave travels past a fixed location, the strength of the *electric field* at that location oscillates with a *frequency* denoted by f. The frequency is related to the oscillation period T by $T = 1/f$.

Light carries *energy*, and delivers it to a surface at a certain rate, which is the *power*. If light is traveling through air and strikes the surface of a smooth, thin *medium*, such as a pane of window glass, part of the light bounces off, or reflects. This is reflection. Typically, not all of the power in the light wave reflects—some of it can be transmitted into or through the medium. We call this process transmission. These processes are illustrated in **Figure 13.2**. *Absorption* can also occur, as we discussed in Chapter 12, but in this chapter, we will discuss media that absorb so little that we can ignore this process. We call such media *transparent*.

The principle of conservation of energy, discussed in Chapter 3, Section 3.6, leads to an important property of these waves:

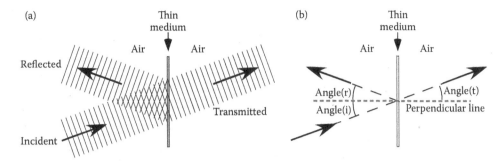

FIGURE 13.2 (a) Wave-front "top view" of light incident on a smooth, thin, transparent medium in air. Part of the incident wave is reflected, and part is transmitted. (b) The transmitted propagation angle and the reflected propagation angle both equal the incident angle.

PRINCIPLE OF LIGHT-WAVE REFLECTION AND TRANSMISSION

When a wave strikes a surface, part of the wave is reflected and part of the wave is transmitted. If the medium does not absorb any of the light, then the sum of the power transmitted and the power reflected equals the power of the incident wave.

For example, consider a light beam from a hand-held laser aimed at a windowpane. For typical glass, approximately 8% will reflect and 92% will transmit. Therefore, if the incident beam from the laser has power equal to 1 mW (milliwatt), then the reflected beam will have power equal to 0.08 mW and the transmitted beam will have power equal to 0.92 mW.

Another important property is illustrated also in Figure 13.2, relating to the angles of propagation of these waves. To define the angles, we draw an imaginary line perpendicular to the medium's surface. We can measure the angle of the incident light ray from this perpendicular line: denote this by the symbol *angle (i)*. Also denote the angles of the reflected and transmitted rays similarly, by symbols *angle (r)* and *angle (t)*. It has been known since ancient times that these three angles are equal. That the reflected angle equals the incident angle was known by the Arab physicist Ibn al-Haytham, also known as Alhazen, who was born in Basra (now Iraq) in 965 (see also the Social Impacts section in Chapter 3). In the following sections, we will see how this behavior comes about because of the properties of waves.

THINK AGAIN

If light reflects from an irregular surface, rather than a smooth one, the wave fronts in the reflected wave become irregular, and the light travels in many directions, as shown here.

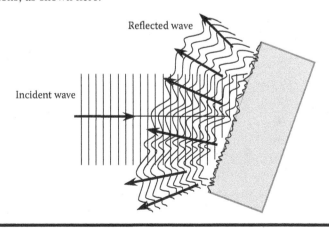

13.3   LIGHT IN TRANSPARENT MEDIA

We know that light can travel through a perfect vacuum where there is no material medium. In vacuum, the ***speed of light*** equals $c = 3 \times 10^8$ m/sec. When light enters a dense medium, such as glass, the electrons in the medium's atoms begin oscillating. Some of the energy originally stored in the light wave is transferred to the electrons in the medium, and then back to the light. This takes some time, and causes a delay, and thus a slowing of the light wave.

In a medium such as glass, which is nearly transparent, light travels without changing its strength, or brightness. It only slows down. The amount by which light slows depends on the type of medium in which it is traveling. Measurements have been done to characterize the amount by which various types of media slow light. These measurements are summarized by a parameter called the ***refractive index*** of the medium, denoted by n. The refractive index equals the ratio of the speed of light in vacuum to the speed of light in a selected medium. We denote the speed in a certain medium by c_n. The refractive index is given by:

$$\text{refractive index} = \frac{\text{speed of light in vacuum}}{\text{speed of light in medium}}$$

$$n = \frac{c}{c_n}$$

This can be stated in another useful way. The speed of light in a medium equals the speed of light in vacuum divided by the medium's refractive index; that is:

The speed of light in a medium with refractive index n is given by $c_n = \dfrac{c}{n}$.

Because the speed of light is less in a medium than in vacuum, we can think of the refractive index as the *slowing-down factor.*

Each type of medium has specific values for n. **Table 13.1** gives typical values for refractive index for visible light in several materials. The values of n actually vary slightly with the light's frequency, as we will discuss below.

For example, in air light is slowed only slightly, whereas in the semiconductor crystal gallium arsenide it slows by a factor of 3.5. That is, the speed is

$$c_{\text{gallium arsenide}} = \frac{3.0 \times 10^8 \, \text{m/sec}}{3.5} = 0.857 \times 10^8 \, \text{m/sec}$$

TABLE 13.1
Refractive Index of Media

Medium	n (Visible)
Vacuum	1
Air	1.0003
Water	1.3
Glass	1.5
Diamond	2.4
Gallium arsenide	3.5

Consider in more detail what happens when light encounters the electrons in the atoms making up the very first layer of the medium's surface. When the light hits those electrons, oscillating electric and magnetic forces are exerted on them. As a result, the electrons begin oscillating at the same frequency at which the light wave oscillates. We know from Chapter 7 that an oscillating electron will emit light. These electrons, which are oscillating as a consequence of light striking them, emit some light of their own. This emitted light has the same frequency as the electron's oscillation, which is the same as the frequency of the incident light. This emitted light travels in the same direction as the incident light and adds to the original wave. From this, we conclude:

The frequency of a light wave is unchanged when it enters a transparent medium, although the speed of the wave changes.

Let us see what this fact implies about the light's wavelength. Recall that the wavelength equals the distance the wave travels in one period T (one full oscillation cycle). Let us introduce the symbol λ_n for the wavelength in the medium. The equation relating wavelength to speed and period is: *wavelength = speed* multiplied by *time between peaks*, or

$$\lambda_n = c_n \cdot T$$

The fact that the frequency remains unchanged when the wave enters the medium implies that the period also remains unchanged (because $f = 1/T$). The equation $\lambda_n = c_n \cdot T$ implies that the wavelength must change if the speed changes. The wavelength becomes shorter in a dense medium than it was in the surrounding vacuum (or air). This is because in a fixed time interval the wave travels a shorter distance in the medium than it would in vacuum. We conclude that the wavelength inside a medium equals the wavelength in vacuum divided by n. That is,

The wavelength of light inside a medium with refractive index equal to n is: $\lambda_n = \dfrac{\lambda}{n}$

We illustrate this behavior by drawing waves inside and outside of the medium, as in **Figure 13.3**. Here we consider a medium for which $n = 2$. The wave arrives from the left, travels to the right, and enters the medium. The wave speed slows for the light inside of the medium, then speeds up again when the wave exits the medium. Remember that the frequency is the same everywhere. From the relation $c_n = c/n$, we deduce that the speed inside the medium is one-half of the speed outside. In this case, the wavelength inside is one-half as long as the wavelength outside. As the wave moves from left to right, a crest outside of the medium travels a distance λ in the same time (T) that a crest on the inside travels a distance $\lambda/2$.

13.4   REFRACTION OF LIGHT AT A BOUNDARY

The wave nature of light produces an interesting effect: when light passes from one medium into another having a different refractive index, its direction of travel can change, depending on the angle at which the incoming wave travels. The bending of the wave fronts and propagation direction of a wave at a boundary between mediums of different refractive index is called **refraction**. **Figure 13.4** illustrates this behavior, showing a "top view" of a light wave traveling from air into a dense medium, where its wave speed slows and its wavelength decreases. The direction of travel for each portion of the

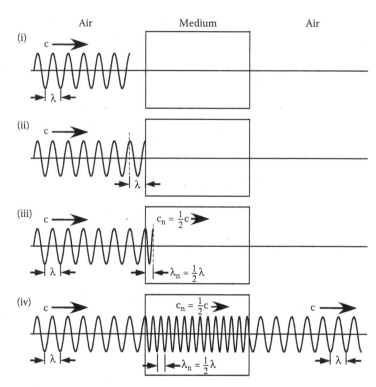

FIGURE 13.3   The first two frames show a light wave traveling from left to right in air with speed c and wavelength λ approaching a medium. The last two frames show the wave entering and passing through the medium, which has refractive index $n = 2$. In the medium, the wave travels with speed $c/2$, and its wavelength equals $\lambda/2$.

wave is shown as a bold arrow. Recall that a portion of a wave always travels in a direction perpendicular to its wave front. In the medium, each portion of the wave lags behind where it would be if it were not refracted. The wave fronts in the medium become angled relative to their original orientation. The angling of the wave fronts causes the upper and lower portions of the wave to travel in altered directions, as indicated by the arrows. As an analogy, think of a wave surfer who always travels in the direction of the steepest slope down the surface of a water wave; as the wave fronts bend, the surfer will change direction, following one of the paths of arrows shown in the figure.

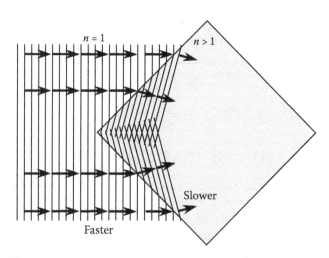

FIGURE 13.4   The shaded medium has higher refractive index than the air surrounding it. When the wave enters a higher-index medium, it refracts and the direction of travel bends relative to its original direction.

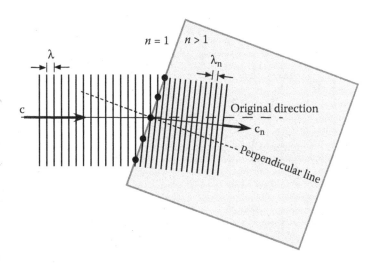

FIGURE 13.5 When the wave enters a higher-index medium from a lower-index one, the direction of travel bends toward the surface-perpendicular line relative to its original direction.

Figure 13.5 illustrates this behavior in detail, again showing a wave traveling from air into a denser medium, where its wave speed slows and its wavelength decreases. In the left side of the figure, a wave in a medium with refractive index $n = 1$ has a wavelength λ, and then enters a medium with higher index n. The wavelength λ_n in the higher-index medium is smaller than in the outside medium. The bold dots are drawn to indicate the points at the boundary where an outside wave front meets an inside wave front.

To understand how a wave behaves when it crosses the boundary between two media, we consider the following physical property: Wave fronts are continuous across the boundary between two media with different refractive indices. This is shown in the figure by the fact that every line representing a wave front outside of the denser medium meets a line representing a wave front inside the medium.

The consequence of this behavior is that inside of the medium, the wave fronts must tilt relative to their orientation outside of the medium, as shown. This means that the direction of travel tilts, or bends, when light enters a medium from another medium with a different refractive index. When a wave enters a higher-index medium from a lower-index one, the direction of travel bends toward the surface-perpendicular line, relative to its original direction.

A mechanical analogy can help to understand refraction. Consider a toy train made of a collection of rigid axles connected by compressible rods that can pivot at joints, as shown in **Figure 13.6**. The axles are initially a distance λ apart. Each axle supports two wheels, and each wheel rotates at a speed dependent on the properties of the surface below it. There are two types of surfaces—one on which wheels rotate rapidly, and one on which wheels rotate slower because of greater friction on a rolling wheel (e.g., sand). When a wheel enters the shaded region, its rotation rate is slowed. Figure 13.6a of the figure shows such a "train" of axle segments moving in a straight line, approaching a surface on which wheels rotate slower. When the first lower wheel reaches the slower region, it slows, causing the first axle to turn. In Figure 13.6b the upper wheel of the first pair reaches the slow region, as the first axle moves in the new direction in the slow region. Because of deceleration, the connecting rod is compressed, making the distance (λ') between axles less. In successive frames of the animation, additional wheels and axles enter the slower region and turn in the angled direction. The axles in this picture are analogous to wave fronts.

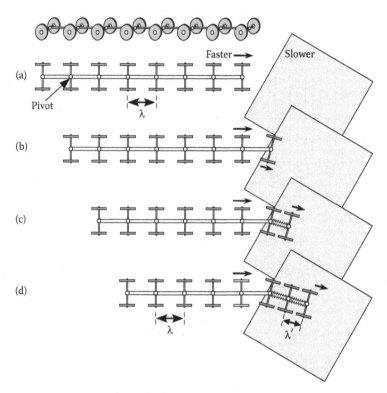

FIGURE 13.6 A "train" of wheel-and-axle segments enters a region where wheels rotate more slowly. The frames show the train at successive times.

QUICK QUESTION 13.1

Consider a transparent medium whose two surfaces are *parallel*, and which is surrounded on both sides by the same type of medium (e.g., air), as Figure 13.2a. When a light wave passes through the transparent medium and exits the other side, it travels *parallel* to the direction that it was initially traveling. This is because the refraction angle when exiting is equal but opposite to the refraction angle when entering. The amount of bending when entering and exiting depends on the ratio of the refractive indices inside and outside of the medium. Verify this conclusion by carefully drawing a continuation of the wave shown in Figure 13.5, being careful to make the wave fronts continuous at the surfaces.

On the other hand, if a wave leaves a higher-index medium and enters a lower-index medium, as in **Figure 13.7**, its wavelength increases and its speed increases. This causes the direction of travel to bend away from the surface-perpendicular line, relative to its original direction. This again results from the fact that the wave fronts are continuous when crossing the boundary, as indicated by the dark dots in Figure 13.7.

An example of refraction occurs when light enters water from the air. You are probably familiar with the way that objects seem to shift in their apparent position from their true position when you view an object that is under water. Try to explain to yourself how refraction causes this optical illusion.

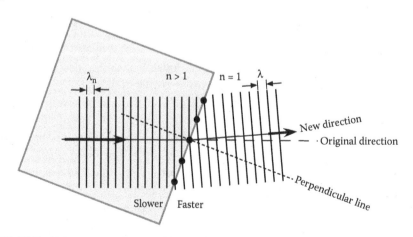

FIGURE 13.7 When a wave leaves a higher-index medium and enters a lower-index medium, the direction of travel bends away from the surface-perpendicular line, relative to its original direction.

We can summarize these observations by the following principle:

PRINCIPLE OF WAVE REFRACTION:

Because wave fronts are continuous across the boundary between two media with different refractive indices, when a wave enters a higher-index medium, the direction of travel bends toward the surface-perpendicular line, relative to its original direction. If a wave enters a lower-index medium, the direction of travel bends away from the surface-perpendicular line, relative to its original direction.

THINK AGAIN

If a light wave is incident on a medium with its wave fronts parallel to the medium's surface, the direction of wave travel does not change, because this direction is already parallel to the surface-perpendicular line.

13.5   REFLECTION OF LIGHT AT A BOUNDARY

The reflection of light at a boundary between two mediums was illustrated in Figure 13.2. Here we will see why the reflected wave travels in the direction that it does. We defined propagation angles relative to a line drawn perpendicular to the medium's surface. **Figure 13.8** shows the incident and reflected wave fronts. Both waves have the same wavelength, because they are in the same medium. The bold straight lines indicate how one incident wave front joins up with a corresponding reflected wave front. Wave fronts must be continuous at the boundary, and this causes the reflected angle to equal the incident angle.

We state this behavior for reflection as a principle:

PRINCIPLE OF REFLECTION ANGLE:

Because wave fronts are continuous at a reflecting boundary, when a wave strikes a flat surface, the reflected wave travels with an angle from the surface-perpendicular line that is equal to that of the incident wave.

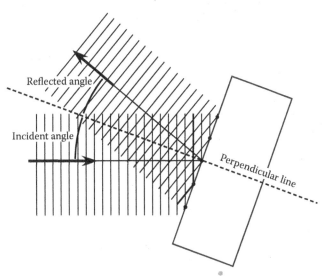

FIGURE 13.8   In reflection, the wave fronts are continuous, causing the reflected angle to equal the incident angle.

THINK AGAIN

Although *reflection* is a commonly used English word, the word *refraction* is not commonly used outside of optical physics. Note that there is no such thing as "reflaction."

13.6   TOTAL INTERNAL REFLECTION

When light is incident from a lower-index medium on a higher-index one, as shown earlier in Figure 13.5, some of the light is reflected and some is transmitted, regardless of the angle at which the incident wave travels. In the opposite situation, when light is incident from a higher-index medium onto a lower-index one, a special phenomenon can occur, called *total internal reflection*.

Figure 13.9 shows three cases of a light wave traveling initially in a higher-index medium, the only difference being the angle of the incident light relative to the surface. The wavelengths in the media are determined by the refractive index of each medium and have the same relative values in all three cases shown. In every case, the wave fronts are continuous across the boundary between the two media, as required by the principle above.

In case (a), the incident angle is not too large, so some light is transmitted, as shown, and some is reflected (wave fronts of the reflected wave are not shown).

In case (b), the incident angle equals a special value, called the *critical angle*. At this angle, the transmitted wave travels parallel to the surface between the two mediums. There is also a reflected wave.

In case (c), the incident angle is greater than the critical angle, and in this case, it is impossible to have a transmitted wave in the lower-index region. This is because the light's wavelength (λ) in the lower-index region is too lengthy to match up with the spacing (d) between incident wave fronts at the boundary. Because the wave cannot transmit with the proper wavelength and the required continuous behavior, it is totally reflected, as shown.

We summarize this behavior as a principle:

PRINCIPLE OF TOTAL INTERNAL REFLECTION:

Total internal reflection (TIR) occurs if light travels inside a medium with a higher refractive index than the surrounding medium, and if the direction of light travel makes an angle with the surface-perpendicular greater than the critical angle. Then all of the light is reflected.

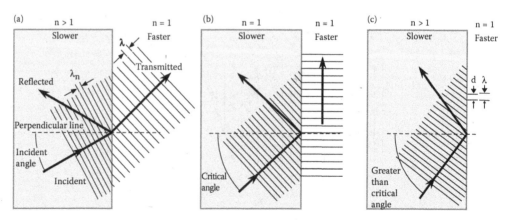

FIGURE 13.9   When light is incident from a higher-index medium on a lower-index one, some light is reflected and some is transmitted, unless the incident angle is larger than the critical angle, in which case the light is totally reflected and none is transmitted. This is total internal reflection.

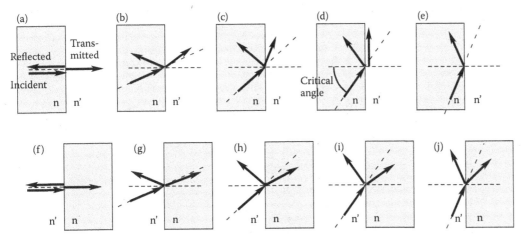

FIGURE 13.10 Ray pictures show refraction and TIR for $n > n'$. TIR occurs if the incident wave travels inside a denser medium at an angle greater than the critical angle.

In any particular case, the value of the critical angle depends on the ratio of the refractive indices in the two mediums. The TIR effect is very important for the operation of fiber optics, as we will see below.

We can replace our drawings of the wave pictures shown in the last few figures by the more convenient ray pictures, which we introduced in Figure 13.1. In the picture, only the arrow that is perpendicular to the wave fronts is shown. This arrow is called a **ray** and indicates the direction in which the wave is traveling. In **Figure 13.10**, the ray pictures for the important cases are given. In the top row, light initially travels in a higher-index medium and encounters a lower-index one. In Figure 13.10a through d some light transmits and some reflects, whereas in Figure 13.10e the light is totally reflected (TIR). Notice that in Figure 13.10a, the reflected light comes back along its original path. In the bottom row, light initially travels in a lower-index medium and encounters a higher-index one. As the incident angle from the perpendicular line is gradually increased, the amount of ray bending also increases and there is always a transmitted, as well as a reflected, wave. Notice that there is no critical angle, and therefore no TIR, when light enters a higher-n medium from a lower-n medium.

13.7 PRISMS AND SPEEDS OF DIFFERENT COLORED LIGHT

Everyone knows that a glass prism separates, or disperses, white light into a spectrum of colors. It is interesting to see how this comes about because of refraction. We first recall the physical difference between light waves of different colors. As we discussed earlier in Chapter 7, Section 7.10, red-colored and blue-colored light are distinguished in human vision by their frequencies. The frequency of red light (with wavelength in vacuum approximately 700 nm) is approximately 430 THz (terahertz), and the frequency of blue light (with wavelength in vacuum approximately 475 nm) is approximately 630 THz.

To understand prisms, we also need to realize that for a given type of glass, the refractive index n has different values for different colors, or frequencies, of light. The reason is that the electrons in the glass respond more strongly to certain frequencies than to other frequencies. In a transparent medium such as glass or water, the electrons typically respond more vigorously to blue light than to red light; that is, to light of higher frequency. This effect makes the refractive index n larger for blue light than for red light; that is, n (blue) $> n$ (red). This changing of refractive index with changing

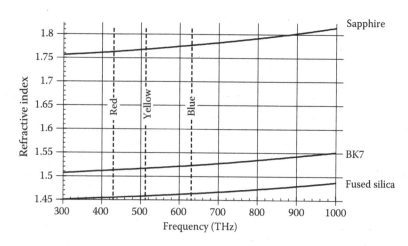

FIGURE 13.11   Refractive indices for several common transparent materials versus frequency.

light frequency is called ***dispersion***, because it leads to the dispersing of white light into its spectrum, as we explain next.

Figure 13.11 shows graphs of the values of refractive index for two types of glass (BK7 and fused silica glass) and one crystal (clear sapphire). BK7 glass is commonly used to make components for optical instruments, such as prisms, mirrors, and lenses. Fused silica is silicon dioxide, or SiO_2, and is used to make optical fibers for telecommunications. Clear sapphire is an uncolored transparent crystal (unlike the gem sapphire, which contains other elements that give it its color).

We can use these data to predict the behavior of light of various colors passing through a prism made of one of these materials. First, note that in air the refractive index has nearly constant value (1.0003) for all frequencies corresponding to visible colors of light. The refraction angles will be somewhat different when using different materials for the prism, but the color trends will be the same. That is, as shown in **Figure 13.12**, red light bends (refracts) toward the first surface-perpendicular line, shown as a dashed line, when it enters the prism from the air. When it exits the prism and goes back into the lower-index air, it bends away from the second surface-perpendicular line. Blue light behaves similarly but has larger bending angles than does red. This results from the larger n values for higher frequency. This is the case for all of the materials shown in Figure 13.11.

White light is comprised of a mixture of all colors, red through violet. So, when white light passes through a glass prism, each color exits the prism traveling in a different direction. When the surrounding medium is air, as usual, the blue light bends more

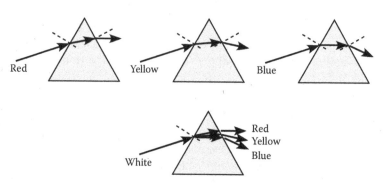

FIGURE 13.12   A glass prism refracts (bends) blue light more strongly than yellow or red light. White light is dispersed into its spectrum.

than the red. If the surrounding medium had an index higher than that of the prism, the trend would be the opposite.

13.8   LENSES AND CURVED MIRRORS

You are probably familiar with the focusing and magnifying abilities of a lens. For example, a lens can focus sunlight to a spot small enough to set a piece of paper on fire, as in **Figure 13.13**. A good way to think about a lens is as a group of prisms, arranged to refract all parallel incoming light rays to the same point, called the focal point. This is illustrated in **Figure 13.14**.

An equivalent way to illustrate the action of a focusing lens is by drawing the wave fronts, as **Figure 13.15**. You can think of each prism-like segment of the lens as bending a separate narrow wave toward the focus point. When we draw all of the wave fronts joined into a smooth wave, we conclude that the wave fronts are curved, as

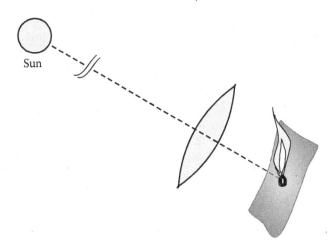

FIGURE 13.13   A lens focuses sunlight to ignite a piece of paper.

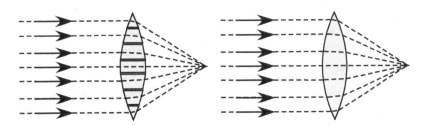

FIGURE 13.14   A lens acts like a group of prisms, arranged to refract all parallel incoming light rays to the same point.

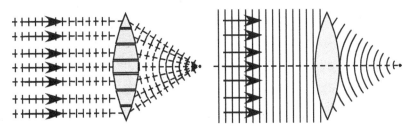

FIGURE 13.15   When a wave passes through a lens, its wave fronts curve to make circular wave fronts, all converging onto a focus point.

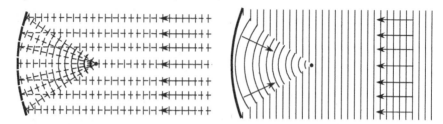

FIGURE 13.16 When a wave reflects from a curved mirror, its wave fronts curve to make circular wave fronts, all converging onto a focus point.

shown. It turns out that, because of its wave properties, light cannot focus to a perfect point. At best, the light focuses down to a small spot the diameter of which is roughly equal to the size of one wavelength of the light.

For example, blue light can be focused to a spot with diameter equal to about 400 nm, but no smaller. (Figure 4.1 in Chapter 4 shows a photograph of a single atom. Although the size of the atom is approximately 0.1 nm, the apparent size of the blurry spot seen in the photo could not be less than about 215 nm, equal to the wavelength of the ultraviolet light used to make the image.)

A similar focusing behavior occurs when light strikes a concave-shaped mirror, as in **Figure 13.16**, which shows a wave approaching the mirror from the right. Parallel rays of light reflect from different segments of the mirror and are all directed toward a common focus point. Alternatively, we can represent the focusing action of the mirror by drawing the wave fronts, which are initially straight lines and become curved after striking the mirror.

13.9 OPTICAL LOSS IN MATERIALS—THE CLARITY OF OPTICAL FIBER

Glass is made of silica—a noncrystalline form of SiO_2 called fused silica. When you look through a thin pane of ordinary window glass, it appears transparent. But a thick slab of such glass is not transparent, as shown in **Figure 13.17**. If a beam of light with power P_0 enters a block of window glass that is 1 meter thick, the light will emerge at the other end with a power one-tenth of P_0. For example, a 1 W beam will be attenuated to 0.1 W. The glass absorbs the energy that is lost. That is, the energy is deposited in the atoms making up the glass and most of it turns into heat, making the glass warmer.

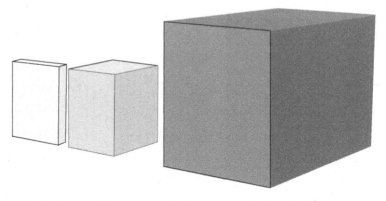

FIGURE 13.17 Ordinary window glass looks clear when a thin piece is viewed, but thick pieces look opaque.

If a light beam passes through 2 m of window glass, it is decreased by a factor $0.1 \times 0.1 = 0.01$, so only 1% is transmitted. For each meter of glass the light passes through, another factor of 0.1 is multiplied. This trend is called exponential loss. (Exponential loss is similar to exponential decay, discussed in Chapter 12, Section 12.5, but refers to a decrease with distance rather than a decrease with time.) After traveling some distance through the glass, the transmission factor that multiplies the input power is $(0.1)^M$, where M is the number of meters of glass that the light passes through. This can also be written as 10^{-M}. An example of exponential loss with distance is shown in **Figure 13.18**. This graph starts with 1 W of power at the location where the light enters the glass. As the distance into the glass increases, the power in the light decreases, as shown in the figure. At a distance equal to 1 m, the power has decreased to 0.1 W. At a distance of 2 m, the power has decreased to 0.01 W. After 20 m, the light power would be attenuated to 10^{-20} W. This is an exceedingly small number, meaning that essentially no power would transmit through 20 m of this glass.

In 1966, when Charles Kuen Kao first proposed that glass fiber could be used to transmit light for long-distance communication, he also pointed out a problem. At the time there was no known glass that could transmit light over long distances. The loss of light by absorption was too great. Kao found that the origin of the large absorption of light in glass was metal impurities such as iron, chromium, and copper. These metals had to be removed from the silica glass to create an extremely high-purity form of silica. Over the next decade, a great deal of research was carried out in an attempt to purify glass so it would have a much smaller optical loss. This would allow glass fiber to be used as a medium to transmit light large distances.

In 1970, Donald Keck, Robert Maurer, and Peter Schultz at Corning developed a silica-glass fiber with loss low enough to begin its use in optical communications. They are shown in the opening photos for this chapter with a spool of their revolutionary fiber. They achieved a fiber that transmitted 1% of power through 1 km. This means that after every 1 km, the power is multiplied by another factor of 0.01, or 10^{-2}. This was a great improvement over previously known types of glass. Again the loss is exponential with distance, but with a transmission much higher than that of window glass. After passing through 3 km of this improved glass, the power would be attenuated by a factor 10^{-6}. This is sufficient for communication over a short distance, perhaps 10 km, but not for long-distance communication.

Current-day fiber can transmit 95.5% of the initial light power after 1 km. This is specifically for a wavelength of 1,500 nm, which is near-infrared light. This means

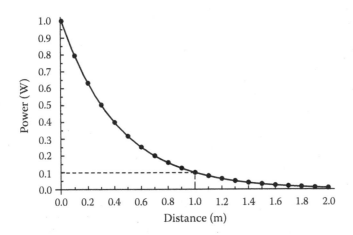

FIGURE 13.18 Exponential decrease of light power with distance in window glass.

that for each 1 km passed, a factor of 0.955 multiplies the power. The exponential loss formula is

$$P = (0.955)^N \cdot P_0 \quad (N = \text{number of km passed})$$

In this formula, the input power is represented by P_0, and the transmission factor is given by $(0.955)^N$. As an example, after passing 100 km ($N = 100$), the power transmitted is $P = (0.955)^{100} \cdot P_0 = 0.01\ P_0$. This means that 1% of the input power transmits through 100 km. This is a huge difference compared with the transmission through window glass, which is 1% through only 2 m!

The transmission factor for light traveling in glass depends on the wavelength of the light. This is easy to understand for colored glass. For example, a stained-glass window that appears red when sunlight streams through it looks that way because most of the blue and green wavelengths in the sunlight are absorbed by the atoms in the glass, but the red is not absorbed—it is transmitted. The staining of the glass refers to metal impurities that are added because they are known to preferentially absorb light of certain wavelengths. This has to do with the chemical and molecular structure in the glass, and we will not delve deeply into this. A graph of the transmission factor for 1 km of fiber is shown in **Figure 13.19**. It is seen that the highest transmission occurs for a wavelength equal to 1,500 nm. For light in the visible wavelength range (less than 800 nm),

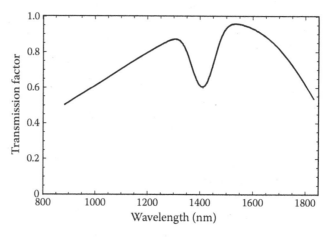

FIGURE 13.19 Transmission factor through 1 km of fiber-quality fused-silica glass, circa 1995, with a transmission dip at wavelength 1,400 nm caused by OH molecules in the glass.

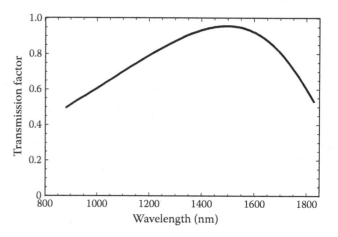

FIGURE 13.20 Optical transmission factor through 1 km in fiber-quality silica glass, circa 2001.

the transmission is low because of light scattering by residual impurities. The earliest optical-fiber communication systems used deep red light with a wavelength around 870 nm because of the availability of light-emitting diodes (LEDs) as light sources.

For light in the infrared range with wavelength greater than 1,800 nm, the transmission is low because of absorption by the SiO_2 molecules that make up the glass. Until the late 1990s, the best fiber also had a dip in its transmission around 1,400 nm because of OH molecules (oxygen bonded to hydrogen) present in the glass. This created two optimal wavelength regions for high transmission—approximately 1,300 nm and approximately 1,550 nm. These were commonly used in optical communication technology around year 2005.

The best fibers, made since 2000, have no OH absorption dip at 1,400 nm and have a transmission factor versus wavelength as shown in **Figure 13.20**. This fiber is the best that could ever be achieved using silica and allows high transmission for all wavelengths between 1,300 and 1600 nm. Later we will discuss the importance of having such a wide spectral region of high transmission.

13.10   LIGHT GUIDING

Light guiding is the basis of all fiber-optics technology and is a consequence of total internal reflection (TIR). As we discussed earlier in this chapter, TIR occurs if two conditions are met: (1) light travels inside a medium with a higher refractive index than the surrounding medium, and (2) the direction of light travel makes an angle with the surface perpendicular that is greater than the critical angle.

Figure 13.21 shows the simplest example of a light guide, also called a *wave-guide*. This is simply a solid cylinder of glass surrounded by air. The light is sent in the end nearly parallel to the sidewalls of the rod, and TIR occurs when light hits a wall.

Figure 13.22 shows practical examples of light guides that are used in communications. Light guides such as these with rectangular geometry are used in constructing semiconductor lasers. Shown on the left is a three-layered structure, made of three

FIGURE 13.21   A glass rod in air acts as a light guide. Light is guided by TIR.

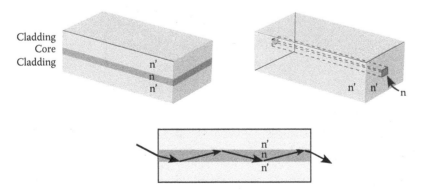

FIGURE 13.22   Optical light guides.

rectangular slabs of glass in close contact. The inner layer of glass, called the core, is narrow and has refractive index equal to n. The surrounding media, together called the cladding, have index n'. The core is made with a type of glass with a refractive index greater than that of the cladding. That is, $n > n'$. A light ray is shown (in side view) entering the core, with bending of the ray direction caused by refraction. Once inside the core, the ray strikes the inner surface between core and cladding at an angle larger than the critical angle. Therefore, the light is totally reflected at this surface. It then travels to the next surface where again it is totally reflected, etc., eventually emerging from the far end of the guide. On the right is shown a wave-guide in which the rectangular core is completely surrounded by cladding material, which better confines the light to a narrow pencil-shaped region.

13.11   OPTICAL FIBERS

An important example of a cylindrical light guide is an *optical fiber*. Such fibers are major elements in optical communication systems. **Figure 13.23** shows the structure of an optical fiber. The core is about 5–100 mm in diameter, and is made of ultrapure SiO_2 glass with a small amount (2%) of germanium dioxide (GeO_2) added. The germanium increases the refractive index slightly, giving a value around 1.461 for wavelengths in the near infrared (1,500 nm), which is best for optical communication. The surrounding cladding is made of SiO_2 glass with a small amount (8%) of B_2O_3 added. The boron increases the refractive index only slightly, giving a value of approximately 1.458, a smaller index than in the core. The cladding is about 100–200 µm in diameter. Surrounding the cladding is a protective jacket typically made of Kevlar®, a strong polymer invented by Stephanie Kwolek at DuPont Corporation around 1965. Kevlar is also used in bulletproof vests. Although optical fiber is made of glass, it is flexible because it is so thin.

The process that is used to make optical fiber is interesting. Engineers begin with a large cylindrical glass tube. They pass a gas containing certain chemical compounds through the tube. These compounds, which contain elements such as germanium, stick to the inside surfaces of the tube. Then the tube is heated until it melts and collapses to form a solid cylindrical rod of glass having a controlled concentration of impurities running along the center axis (the core). The refractive index of the center core is larger than the index of the surrounding glass. This solid rod is then clamped at both ends and heated at one end to soften the glass as a force is applied, pulling the two ends apart. As the ends separate, a thin bridge of hot glass forms between them. By further

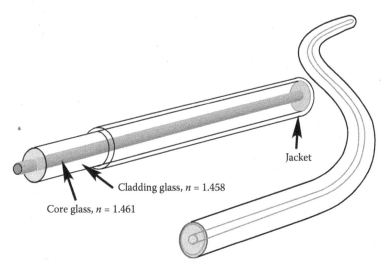

Jacket

Cladding glass, $n = 1.458$

Core glass, $n = 1.461$

FIGURE 13.23   Structure of optical fiber.

pulling on this thin bridge, a fine strand (about the thickness of a spider's web) is formed approximately 1 m in length between the two ends. As it is exposed to the air, the strand cools and hardens into a flexible glass fiber, which can be rolled onto a spool. If this process is done carefully, the fiber strand can be pulled continuously without breaking, allowing a spool to contain many kilometers of a single-strand fiber.

13.12   LIGHT PULSES IN OPTICAL FIBERS

For long-distance digital data transmission, pulses of light emitted by a laser are focused into one end of an optical fiber. The glass core guides the light pulses by TIR until they reach the other end. Ideally, the pulse would have the same duration at the end of its journey through the fiber as it had at the beginning. In practice, this is not the case. It is found that a pulse is typically longer when it exits than when it entered, as illustrated in **Figure 13.24**. This is a potential problem when using pulses to transmit digital bits, because if the pulses were originally quite close to each in time, then after broadening, they might overlap in time, confusing the receiving electronics. This leads to limitations in the number of pulses that can be sent through the fiber each second.

As an example, consider a series of light pulses representing data, traveling in a typical 1980s telecommunications fiber. As shown in **Figure 13.25**, each pulse initially

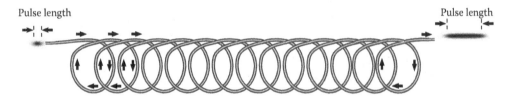

FIGURE 13.24   A short light pulse becomes longer after traveling a long distance in an optical fiber.

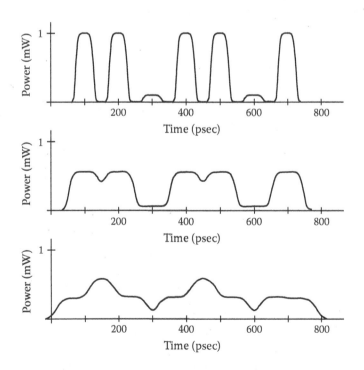

FIGURE 13.25   Pulse broadening in an optical fiber. The top graph shows the pulses injected into the fiber. Middle shows pulses after traveling 10 km in a particular fiber. Pulse broadening during longer-distance propagation progressively causes pulses to overlap.

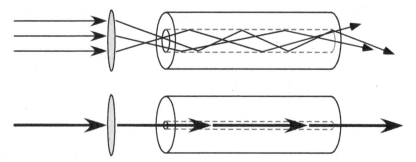

FIGURE 13.26 Multimode fiber can guide light in many paths, or modes (upper). Single-mode fiber guides light only in one mode (lower).

has a duration of 100 picoseconds (psec). After traveling 10 km in the fiber, each pulse is stretched, and now has a duration of 200 psec. The pulses now overlap, but are still distinguishable. After traveling much farther than 10 km, the pulses overlap badly, preventing accurate binary data reception.

There are two causes of this pulse broadening: the geometry of the fiber core and the material used to make the fiber. The geometrical effect is easy to understand. If the core diameter is not sufficiently small, then there is not a unique path that light rays will take. As shown in **Figure 13.26**, a light beam is focused into the fiber using a lens. The light contains rays traveling at various angles when it enters the fiber. Any ray whose angle from the surface-perpendicular line is large enough will be guided. The upper part of the figure shows an example with a "large" fiber core (e.g., 125 μm in diameter). Rays of different angles take different paths through the fiber to its end. Each of these different types of paths is called a different ***mode***. (Think of the common phrase "modes of transportation.") A fiber that can support propagation in many different modes is called a ***multimode fiber***. The distance traveled by each ray depends on which path it takes. Those rays that travel parallel to the fiber axis travel the shortest distance, and so reach the fiber end before the rays that travel at an angle from the axis. For this reason, if a 100 psec pulse enters a very long fiber, when it emerges from the end it may be 200 psec in duration. The pulse is stretched, and its shape may be altered as well. This effect is called ***mode dispersion***. (Dispersion means *breaking up* or *spreading out*.) The first optical fiber communication systems that were installed in the 1970s and 1980s used such multimode fibers, and this limited the data rate that could be achieved in these systems.

The ideal way to overcome mode dispersion is to use a fiber with a very small core (approximately 5 μm) and a very small difference between the refractive indices of the core and cladding. For such a fiber, light is guided only if it travels in a direction very close to parallel with the fiber axis. We say that light can travel in only a single mode. For this reason, such fiber is called ***single-mode fiber***. Such fibers are used in modern communication systems because there is very little pulse broadening in them.

The second source of pulse broadening is connected with the material properties of silica glass, which we reviewed in our discussion of Figure 13.11. That figure shows the dependence of the refractive index on the frequency of light in the medium. As Figure 13.11 shows, the value of refractive index n varies with changes in frequency, and therefore so does the speed, given by $c_n = c/n$. In particular, light of smaller wavelength (bluer light) travels slower than light of longer wavelength (redder light). Just as a white light beam is made up of all of the frequencies (or wavelengths) of the visible spectrum, a short pulse of light (e.g., 100 psec) is made up of a certain range of frequencies. We discussed the same idea for the case of radio waves in Chapter 8, Section 8.4. Fourier's theory established that wave pulses may be generated by adding waves with differing frequencies. Such a pulse spans a range of frequencies called the **bandwidth** of the pulse. The duration of a pulse that is synthesized by this method is given roughly by the inverse of its spectral bandwidth. That is,

$$pulse\ duration\ =\ \frac{1}{bandwidth}\ =\ \frac{1}{B}$$

For example, for a light pulse with a spectral range (bandwidth) equal to 10 GHz, the pulse duration equals $1 \div (10\ GHz)$, or 100 psec. For another example, see Figure 8.23 in Chapter 8. A 100 psec light pulse is made up of a range of wave frequencies, and the different-frequency waves travel at different speeds c_n. So, the 100-psec pulse, even if it travels in a single-mode fiber, will spread out slightly after traveling through a long fiber. We call this type of pulse broadening "material dispersion" to distinguish it from mode dispersion. It is caused by the same physical effect in glass that causes a beam of white light to be spread (dispersed) into a spectrum of different colors upon passing through a prism. A lot of engineering has gone into reducing this pulse broadening effect by designing special fiber materials and structures.

SUMMARY AND LOOK FORWARD

In this chapter, we learned that the motion of light inside transparent materials such as glass, water, or clear plastic is governed mainly by its speed in the medium. Although the speed of light in vacuum is a fixed constant, $c = 3 \times 10^8$ m/sec, the speed in a dense medium is determined by the refractive index of the medium, which is denoted by n. The speed in a medium with refractive index n equals $c_n = c/n$ and is always less than the speed in vacuum, c. When light enters one medium from a medium having a different refractive index, the frequency of the light stays the same. The wavelength of the light changes, according to $\lambda_n = \lambda/n$.

When a wave strikes a surface, part of the wave is reflected and part of the wave is transmitted. The reflected wave travels with an angle from the surface-perpendicular line that equals that of the incident wave. Wave fronts must be continuous across a boundary between two media. Consequently, to be consistent with the changing of the wavelength, a light wave will bend its direction of travel when moving between two media having different refractive indices. This bending is called refraction. When a wave enters a higher-index medium, the direction of travel bends toward the surface-perpendicular line. When a wave enters a lower-index medium, the direction of travel bends away from the surface-perpendicular line. If light travels inside a medium with a higher refractive index than the surrounding medium, and if the direction of light travel makes an angle with the surface-perpendicular line greater than the critical angle, then TIR occurs. Typically, in a dense medium, blue light travels slower than does red light, because of its stronger interaction with the atoms in the medium.

The above behaviors explain the operation of optical devices like prisms and optical fibers. When passing through a prism, blue light is bent in its direction more strongly than is red light, producing a spread-out spectrum. In an optical fiber, TIR confines light waves inside the fiber if it travels nearly parallel to the axis of the fiber; that is, with a large enough angle relative to the surface-perpendicular line. Modern fiber is made of extremely pure silica glass, so light can travel in it as far as 100 km before its power drops to less than 1% of its original value.

Our discussion in this chapter of light in dense media has prepared us to explore how lasers work, and to learn how optical-fiber communication systems operate. These topics are discussed in the following chapters.

SOCIAL IMPACTS: TOTAL IMMERSION IN A SEA OF INFORMATION

The Internet has made the world a much smaller place.

1,040 authors – Google search for "The Internet has made the world a much smaller place"
(2008)

I had a dream that the web could be ... an interactive sea of shared knowledge ... immersing us as a warm, friendly environment made of the things we and our friends have seen, heard, believe or have figured out.

Tim Berners-Lee
(*Inventor of the World Wide Web,* 1995)

We have constructed a virtual Wild West, where the masses indulge their darkest vices, pirates of all kinds troll for victims. ... And the virtual marketplace is a great place to get robbed.

Steve Maich
(2006)

Twenty-five years ago, I might have wondered about something such as the full text of a speech given by a presidential candidate during a primary season. In most cases, I felt I had so little hope of getting that information that I would not have bothered to try. Nowadays, that information is literally at my fingertips. I Google it or browse on a news company's web pages, and in a few minutes I have the speech "in hand." Now what? In former days, if I somehow had obtained the speech, I could have photocopied it and mailed it by post to my family and friends, then exchanged letters or phone calls about the contents. That is what "good" citizens should have done, but in those days we did not, because the effort was just too great. Nowadays I e-mail the speech text (or video) to my family and friends, and within a day we all hear back from many respondents with interesting analyses of the speech.

If you were born and raised in the Internet Era, it is probably hard to appreciate how revolutionary and important this change is. It would be like young fish listening to a million-year-old ancestor talking about the days when their species lived on dry land rather than in water. The young fish, being always immersed in water, would say, what "water?" Not only does the Internet make it easy for friends to communicate, it greatly simplifies communication between people who have never met and who live in very different cultural environments. It is important for me, as a scientist, to be able to learn what other scientists already have discovered through their own research. In the old days (1970s), I would read articles published in scientific journals. To read a paper written in a journal published in the Soviet Union, I would have to wait several years for it to be translated from Russian into English and then republished in the United States. It seemed exotic to read such papers, as if I were peering through a smoky tunnel into a mysterious land. It would not have occurred to me to try to communicate directly with the author of a Soviet-published article. Nowadays, I might read an online article written (in English) in Germany, which contains a mention of some unpublished research studies carried out in Russia. Using only the Russian scientist's name and institution's name, I can locate the e-mail address of that person and send an e-mail asking if he or she would be willing to share these new results with me. Then we might start up a friendly discussion, which could even lead to a research collaboration, including visits to one another's universities. Again, it is hard to convey what a big change this is.

But, is technological change always good? Does instant and wide-ranging communication make lives fuller and better, or only more complicated and stressful? In the widely read Canadian magazine *Maclean's*, Steve Maich wrote [1]:

> After 15 years and a trillion dollars of investment, just about everything we've been told about the Internet and what the information age would mean has come up short... The great multinational exchange of ideas and goodwill has devolved into a food fight.

This article set off a firestorm of defensive rebuttals from aficionados of information technology [2]. Many critics pointed out that Maich ignored the positive aspects of the Internet.

One aspect of the Internet that has potential for good as well as bad is the ease with which politically or ideologically driven groups of people can communicate or even organize from afar. In U.S. Presidential primaries, Howard Dean and Barack Obama were successful early users of the Internet for organizing grass roots support [3]. At the other extreme, there are worldwide militant or even terrorist groups who use the Internet to post information and communiqués that are useful to their causes, which in some cases are very destructive.

Not only does the Web put information at our fingertips, it can also put information about us at other peoples' fingertips. Spying is probably as old as mankind, and the Internet provides one more—powerful—means for prying into people's privacy. In some cases, this might be good, in that it could prevent a violent crime or attack by political extremists. In other cases, it might be bad, if governments or corporations violate laws or bounds of decency in their prying. On February 6, 2006, *USA Today* published an article titled "Telecoms Let NSA Spy on Calls," according to which, AT&T, MCI, and Sprint aided the National Security Agency (NSA) in eavesdropping on phone calls between people in the United States and foreign countries, without search warrants. Three days later, U.S. Senators Russell Feingold and Edward Kennedy sent letters to the heads of these companies, asking for their cooperation in a Senate investigation into the matter. This led to years of struggle between the telecoms, Congress, and the White House, with the administration arguing in favor of immunizing the telecoms against criminal charges. In 2008, Wired Blog Network reported that, "FBI headquarters officials sought to cover their informal and possibly illegal acquisition of phone records on thousands of Americans from 2003 to 2005 by issuing 11 improper, retroactive 'blanket' administrative subpoenas in 2006 to three phone companies that are under contract to the FBI." [4] Although there may or may not have been reasons for such actions, of interest here is the involvement of some of the top communications technology companies.

On the other hand, private citizens can turn the tables and use communication technology to provide news coverage of events that governments do not want covered. As I write this, the largest nonpeaceful demonstration in 20 years against Chinese authorities in Tibet is taking place, and most conventional news agencies have had their ability to cover the events restricted. Tourists or people posing as tourists have been able to make their own videos and post them on YouTube. (For example, "Tibet protests spread to neighbouring provinces," posted on March 16, 2008, with the narrator describing "teargas grenades scattering the crowd" of Tibetan monks.) The government of China has blocked Internet access to YouTube within the country, deeming it harmful.

In a different avenue, another type of information gathering, called consumer market (or marketing) research, is greatly aided by Internet technology. Marketing research is a kind of sociology, applied to the problems of group consumer behavior and the effects on such groups of advertising. Group behavior is exploited through viral marketing— the idea that promotion for a product can occur through person-to-person contacts in a social network. The Internet supports many social networks (think MySpace), and in some ways is automated. It therefore provides an ideal playground in which to try to create social viral marketing.

Although technology is all well and good for those who have it, the existence of technology does not automatically mean that everyone's lives will uniformly benefit from it. Society can (or has) become stratified into haves and have-nots, with segments of society ending up in "technology ghettos." This is referred to as the "digital divide." Not only is access to information not uniform, the opportunity to learn skills related to information technology is unevenly spread. Fiction authors, such as William Gibson (who coined the word "cyberspace") and Phillip K. Dick (who wrote the book on which the movie *Blade Runner* was based), explore this idea artfully [5,6].

The future is already here. It's just not very evenly distributed.

William Gibson
(1999)

Questions to Ponder

1. Compare the invention of the Internet with Johannes Gutenberg's invention of movable type in about 1439 (and the invention of the telegraph in the 1830s).
2. Google search: "The Internet has made the world a much smaller place."
3. Google search: "Digital divide."
4. Is it clear that providing affordable Internet access to economically disadvantaged citizens in the United States would help advance their economic standing?
5. What effect does total immersion in a sea of information have on education methods and results?

REFERENCES FOR SOCIAL IMPACTS

1. Maich, Steve. "Pornography, Gambling, Lies, Theft and Terrorism: The Internet Sucks (Where Did We Go Wrong?)" *Maclean's* 119, no. 43, 2006. http://www.macleans.ca/article.jsp?content=20061030_135406_135406 (Accessed October 30, 2006).
2. Schick, Shane. "Why *Maclean's* Scathing State-of-the-Internet Feature is so off the Mark," IT Business.ca. http://www.itbusiness.ca/it/client/en/Home/News.asp?id=40929 (Accessed October 23, 2006).
3. Trippi, Joe. *The Revolution Will Not Be Televised: Democracy, the Internet, and the Overthrow of Everything.* New York: Harper Collins, 2004.
4. Singel, Ryan. "FBI Tried to Cover Patriot Act Abuses with Flawed, Retroactive Subpoenas, Audit Finds," *Wired Blog Network* http://blog.wired.com/27bstroke6/2008/03/fbi-tried-to-co.html (Accessed March 13, 2008).
5. Gibson, William. *Virtual Light.* New York: Bantam, 1994.
6. Dick, Philip K. *Do Androids Dream of Electric Sheep?* New York: Del Rey, 1996.

SUGGESTED READING

Hecht, Jeff. *City of Light: The Story of Fiber Optics.* New York: Oxford University Press, 2003.
Hecht, Jeff. *Understanding Fiber Optics,* 5th ed. New York: Prentice Hall, 2005.
Rogers, Alan. *Understanding Optical Fiber Communications.* Boston: Artech House, 2001.

KEY TERMS

Absorption
Bandwidth

Critical angle
Dispersion
Electric field
Electromagnetic wave
Energy
Frequency
Light
Loss
Medium
Mode
Mode dispersion
Multimode fiber
Optical fiber
Power
Ray
Refraction
Refractive index
Silica
Single-mode fiber
Speed of light
Total internal reflection (TIR)
Transparent medium
Wave front
Wave-guide
Wavelength

ANSWER TO QUICK QUESTION

Q13.1 Consider a transparent medium whose two surfaces are parallel, and that is surrounded on both sides by the same type of medium (e.g., air), as in Figure 13.2a. When a light wave passes through it and exits the other side, it travels parallel to the direction that it was initially traveling. This is because the refraction angle when exiting is equal but opposite the refraction angle when entering. The amount of bending when entering and exiting depends on the ratio of the refractive indices inside and outside of the medium. This is proved by considering the continuous wave fronts shown here. The wavelength in the entering medium and in the exiting medium are equal.

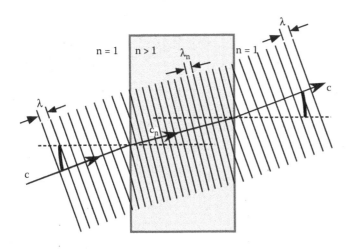

EXERCISES AND PROBLEMS

Exercises

E13.1 A person underwater holds a laser pointer 3 ft below the water's surface near one sidewall of a swimming pool. The water's surface is calm, with no ripples. She aims the beam straight up toward the water's surface. Next, she aims the beam toward a point on the water's surface a couple of feet away from her location. Then she aims the beam toward a point on the water's surface near the other side of the pool. For each case, draw a picture of the situation in side view, indicating the paths taken by the light beam. Consider both transmission and reflection and account for refraction. Include the possibility of TIR. Explain your reasoning.

E13.2 A good hands-on way to model waves in two different media is by printing wave fronts on two pieces of paper and placing one on top of the other such that both patterns can be seen together. Then, by rotating one paper relative to the other, you can find the angle between the two that allows every wave front to be continuous across the boundary between the two media. Assume one medium has a refractive index equal to 1.0, and the other has a refractive index equal to 2.0. Therefore, the wavelength in one medium is one-half of the wavelength in the other. There are two ways you can make the needed wave patterns on paper: (1) Print or photocopy the image below and cut PART A from PART B using scissors; (2) Using a pencil, a straight edge, and the printed lines on writing paper, darken every line on one, and darken every other line on the second paper.

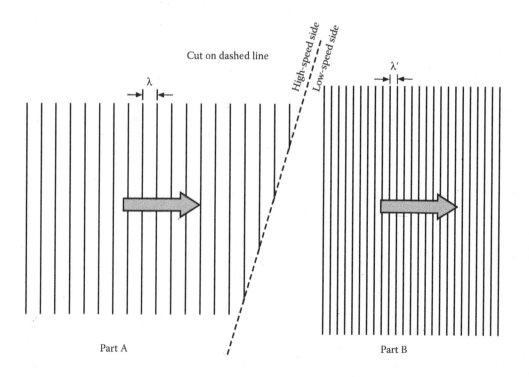

Part A   Part B

(a) Place PART B beneath PART A and rotate to make the wave fronts continuous at the boundary indicated by the dashed line. Then tape them together. Explain how the bending angle compares to the rules for this angle given in the chapter.

(b) On a separate piece of paper, draw your own PART B wave for the case that the medium in region B has a refractive index equal to 1.5. Repeat (a) and again tape the two parts together. Observe the new direction of wave travel. How does the bending angle change when the refractive index in the B region is changed in value?

E13.3 (a) Follow the same instructions as in E13.2 for the case that the wave travels from a low-speed region (refractive index = 2) into a higher-speed region (refractive index = 1). Use the diagram below. For part (b), assume that the refractive index in region B equals 1.5.

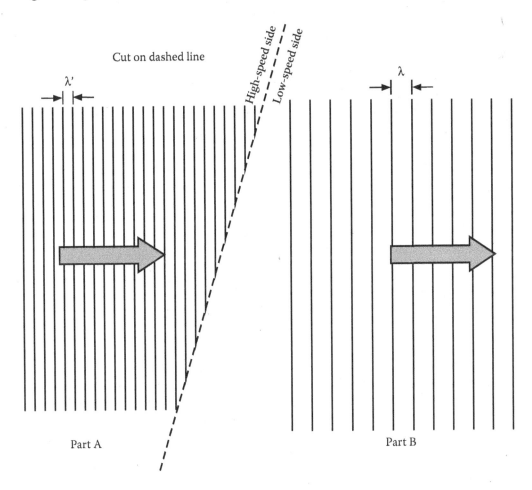

E13.4 A good way to understand the critical angle for TIR is to model waves in two different media by printing wave fronts on two pieces of paper and placing one on top of the other such that both patterns can be seen together. Then by rotating one paper relative to the other, you can find the angle between the two that allows every wave front to be continuous across the boundary between the two media. Assume one medium (B) has a refractive index equal to 1.0, and the other (A) has a refractive index equal to 2.0. Therefore, the wavelength in one medium is one-half the wavelength in the other. There are two ways you can make the needed wave patterns on paper: (1) Print or photocopy the image below and cut PART A from PART B using scissors; (2) Using a pencil, a straight edge and printed lines on writing paper, darken every line on one paper, and darken every other line on a second paper.

(a) Place PART A beneath PART B and rotate to make the wave fronts continuous at the boundary indicated by the dashed line. Then tape them together.
(b) Explain why slightly changing the angle of the incoming wave in A makes it impossible to have continuous wave fronts at the boundary.

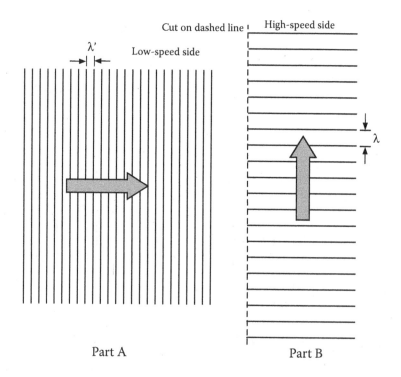

Part A   Part B

E13.5 To locate the apparent location of an object, the eye perceives the path of two or more light rays, as shown in (i) for the case of a small object in air. Referring to (ii), explain how the principle of reflection aids in explaining the apparent location of the object when reflected in a mirror. Referring to (iii), consider a small illuminated (lit up) object placed under water. The person looking at the object "sees" it to be in an apparent location, which is different than the actual location because of refraction. By analogy with (ii), draw an estimated actual location somewhere in the water, and the two light rays going from the actual location to the eyes.

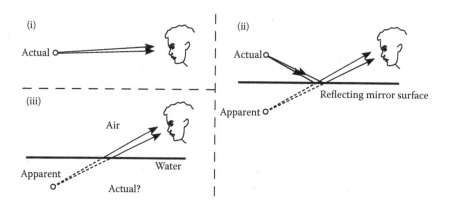

E13.6 Fill a wide, shallow container with water and insert a straight wooden stick at an angle. Surprisingly, the stick looks to be bent at the water's surface. Explain this observation, using the argument developed in E13.5. Consider four small marks painted along the stick as reference points. *Hint:* This requires thinking and drawing in three dimensions.

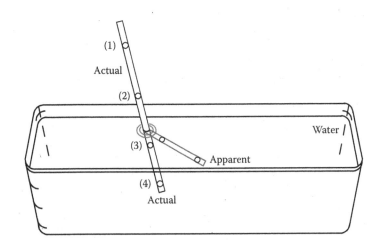

E13.7 Clear (uncolored) glass and a certain type of clear oil, called mineral oil, have nearly equal values of refractive index. Explain why a clear glass marble immersed in mineral oil is essentially invisible. Why can an illuminated, clear glass marble in water be seen by the eye?

E13.8 Why does a wide beam of white light not disperse (spread out) into a colored spectrum when it passes perpendicularly through a thin glass windowpane with parallel sides, but does disperse into a spectrum when it passes through a prism made of the same kind of glass? Answer the same question for white light passing nonperpendicularly through a thin glass windowpane. *Hint:* See Quick Study Question 13.1 and its solution. Consider not only the positions of the light rays, but also their angles when emerging from the glass.

E13.9 On a cloudy but still bright day, a person underwater looks upward and sees a bright disk of light at the surface, surrounded by darkness. Explain, using the ideas of refraction, why he does not see light at all angles from the location of his eye.

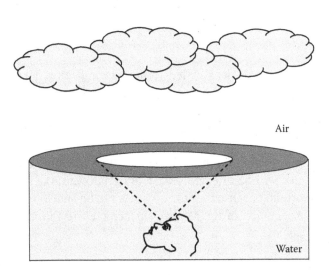

E13.10 A narrow, red light beam passes through a series of transparent glass plates, each having parallel sides, as shown. Carefully draw the path the beam will take. *Hint*: See Quick Study Question 13.1 and its solution. The beam emerges at the far end of the stack traveling parallel to the incident beam, similar to one of the dash arrows shown.

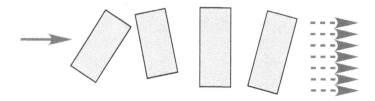

E13.11 Draw a careful picture that continues and completes the following figure showing a light wave passing through a square-shaped region, in which light travels 1.42 times more slowly than in the region around it. Assume that the incident wave exists only in the region shown by wave fronts. *Hint*: The arrows show the directions that the reflected and transmitted waves might take. The question mark prompts you to consider whether reflection at the upper tilted surface is total or partial.

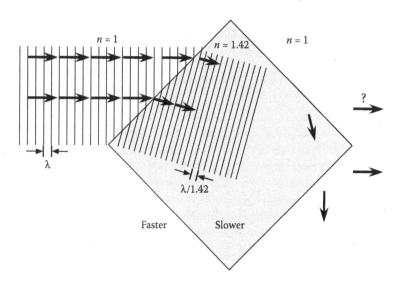

E13.12 Astronauts placed devices called right-angle reflectors (RARs) on the Moon during the Apollo missions. A RAR is actually a complex device, but for simplicity let us represent these reflectors by a nearly equivalent setup—a combination of two flat mirrors angled at exactly 90° (a right angle) from each other. Scientists can direct a laser beam from Earth toward the Moon, where the light is reflected from the RAR. Verify, by drawing several different incident rays, that any light incident on the RAR from different angles will be directed precisely back toward the source from which it came. This is important, because the scientists want to be able to see the reflected light, regardless of the orientation of the Moon's surface relative to the Earth. A single flat mirror would not achieve this.

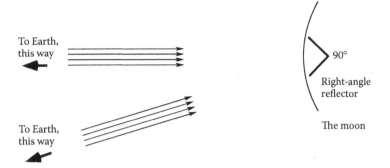

E13.13 Consider a hollow, air-filled cavity inside a solid block of glass. Draw a line showing the path of a narrow beam of red light if it enters from the left as shown. How would the colored spectrum look if white light were incident on the prism-shaped air region embedded in fused silica? Draw a figure similar to the one below for each answer.

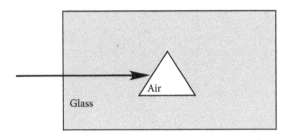

E13.14 Consider a clear glass bowl made with uniform thickness glass. Explain why this will not act as a lens. Explain how, by adding some water to the bowl, you could make it behave as a lens. A tilted wineglass also works.

E13.15 Figure 13.16 shows a light wave reflecting from a curved mirror. Explain why the reflected light-wave fronts are more strongly curved than is the surface of the mirror. *Hint*: Consider the principle of reflection.

E13.16 For visible light, the index of refraction of diamond is 2.4, compared with 1.5 for glass. Explain how this gives rise to the unique appearance of diamonds. That is, if a piece of glass is polished to have the same shape as a cut diamond (with sharp-edged, flat facets), why does it not look like a diamond to the eye?

E13.17 (a) If a thin sheet of fused-silica glass is sandwiched between two thick sheets of sapphire, will red light be wave-guided by TIR inside of the silica? *Hint*: See Figure 13.11 to estimate values of refractive index. (b) If a thin sheet of sapphire is sandwiched between two sheets of fused silica glass, will red light be wave-guided by TIR inside of the sapphire?

E13.18 You are given 1 million optical fibers, each with a diameter of 0.3 mm and length of 10 m. Your job, as a design engineer, is to invent a method for relaying an image of writing on a single thin piece of paper to another person 10 m away, without using lenses.

(a) Think of a way to do this and make a sketch illustrating your idea. Explain how it works, and what challenges there might be in constructing such a system. Do not forget that the writing needs to be illuminated, and that the relayed image must be sharp.

(b) How large of an area of writing can you relay using this method? Give your answer in square millimeters.

E13.19 You are given a clear plastic pail of water with a small hole in the sidewall near the bottom, plugged by a cork. After pulling a cork from the hole, a stream of water emerges and curves toward the floor. You are also given a laser pointer. Explain how you would demonstrate light guiding using this setup to a class of middle-school students. Give your explanation that you would offer the students.

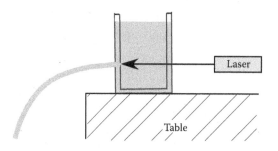

Problems

P13.1 A red light wave in air has frequency $f = 4 \times 10^{14}$ Hz. Its period is $T = 1/f = 2.5 \times 10^{-15}$ sec, and its wavelength is $\lambda = c/f = 750 \times 10^{-9}$ m $\doteq$ 750 nm.

(a) Find its frequency in units of terahertz.
(b) If this light enters fused-silica glass, what are its frequency, speed, and wavelength inside of the glass? See Figure 13.11 to estimate values of refractive index.
(c) If this same light enters sapphire, what are its frequency, speed, and wavelength?

P13.2 Current-day fiber transmits 95.5% of the initial light power a distance of 1 km (for near-infrared light). This means that for each 1 km passed, a factor of 0.955 multiplies the power. The exponential transmission formula is $P = (0.955)^N \cdot P_0$, where N = number of kilometers passed and P_0 equals the power at the input end of the fiber. Make a graph analogous to Figure 13.18 for the case that light with power of $P_0 = 1$ W enters such a fiber with a length of 100 km. Include one point on your graph for every 10 km traveled. *Hint:* After 10 km, the power equals $P = (0.955)^{10} \cdot 1$ W = 0.631 W. (If your calculator does not do exponents like this, you can multiply 0.955 by itself ten times to evaluate this. Then to obtain the next number, $(0.955)^{20}$, square the first result, $P = (0.631)^2 \cdot 1$W, etc.)

P13.3 Repeat the graphing exercise in P13.2 for a wavelength of 1,800 nm. Use the data shown in Figure 13.20.

14

Light Amplification and Lasers

The truth is, none of us who worked on the first lasers imagined how many uses there might eventually be. This illustrates a vital point that cannot be overstressed. Many of today's practical technologies result from basic science done years to decades before. The people involved, motivated mainly by curiosity, often have little idea as to where their research will lead.

Charles H. Townes

Keep in mind that nobody had ever made coherent light. One prominent physicist asserted that you could not make coherent light.

Theodore Maiman

Lasers are ubiquitous in consumer products, such as compact disc players.

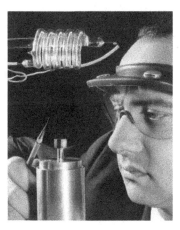

Theodore Maiman of Hughes Research Laboratories with the first ruby laser in 1960.

HOW TO USE THIS CHAPTER: If you have not read Chapters 9, 10, and 12, then you should skip the last two sections of this chapter. On the other hand, if you have studied Chapters 9, 10, and 12, then you are well prepared to study these sections, which discuss laser action in terms of quantum physics.

14.1 ATOMS AND LASERS

In 1960, the previous work of scientists resulted in a huge payoff: the laser was invented. Theodore Maiman, working at Hughes Research Laboratories (founded by Howard Hughes, of aviation fame), built the first laser using a crystal of ruby. The physics ideas behind laser operation were largely known since 1916, when Einstein published a paper explaining how atoms absorb and emit light. Nevertheless, scientists' realization that these ideas could be used to create a laser occurred only in the late 1950s, when three groups of physicists independently invented the concepts and methods for actually building lasers: the American team of Charles Townes and Arthur Schawlow, the Russian team of Nikolay Basov and Aleksandr Prokhorov, and American Gordon Gould by himself. During the intervening 44 years, a hard road of technical innovation and learning had to be traveled before Einstein's ideas could be fully recognized and put into practice. This is a good example of the process of scientific development that must be sustained over a long time for great things to happen. In this chapter, we bring together a wide range of physics topics that we learned in previous chapters (just as the scientists did in 1960) to understand how lasers work.

After the laser was invented, some technologists began dreaming of the wonderful applications that were possible. Still, it took another 30 years or so for the laser to become ubiquitous, playing key roles in thousands of applications in areas including consumer electronics, telecommunication, data storage, science, medicine, industry, entertainment, and military. One interesting example is when the Apollo 11 astronauts landed on the Moon in 1969, they placed prisms on the Moon's surface. By bouncing laser-light pulses from these prisms, scientists on Earth can determine distances between the Earth and the Moon with an accuracy of about 4 centimeters (about one part in 10 billion). Isaac Newton would have been impressed.

For our purposes, the important applications of lasers are in data storage and telecommunication. The Internet—the global communication network—uses pulses of laser light traveling in optical fibers to transmit much of its data between computers and other devices. CD and DVD players and recorders use lasers for reading and writing digital data for music, video, and computer data storage.

In this chapter, we will study the theory behind the principles of lasers and examine some specific types of lasers. In the following chapters, we will consider how lasers, radio, and electronics combine to form communication networks and the Internet.

14.2   THE UNIQUENESS OF LASER LIGHT

Lasers are the best possible sources for optical communication. Ordinary light bulbs and *light-emitting diodes (LEDs)* are not the best possible sources for this purpose. They emit light in random directions, with many frequencies, and with random wave fronts (see **Figure 14.1**). That is, each atom emits independently, with no organization or coherence between atoms. As a result, the light power is relatively low in any single direction, even for a very bright bulb. These effects make it difficult to use light from bulbs or LEDs for reliable data transmission over large distances.

Lasers, on the other hand, emit light with unique properties, which make them more suited as sources in communication systems. The unique properties of laser light are:

- *Monochromaticity*: Laser light is monochromatic—meaning single-color (or *frequency*). Rather than emitting a wide range of frequencies, or *wavelengths*, an ideal laser emits light with a single frequency.
- *Directionality*: Laser light is highly directional—this refers to the property of laser light that allows it to travel in a tight (small-diameter) beam over long distances, without spreading out too rapidly.
- *Coherence*: Laser light is coherent—this means that all of the wave fronts in a laser beam are traveling in nearly the same direction, and oscillating together in a coordinated manner. (The word *cohere* means to stick together.) Compare

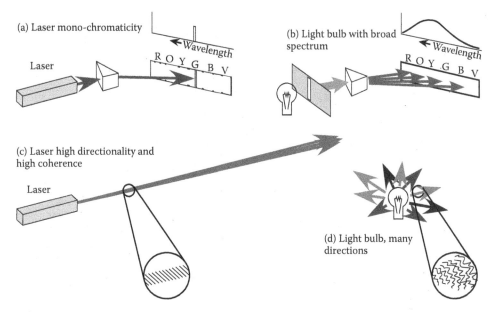

FIGURE 14.1 Comparison of lasers and light bulbs.

this to light emitted by a light bulb, in which waves travel in different directions with no coordinated relationship.

■ *Brightness*: A laser can create a very bright (powerful) beam of light having the properties of monochromaticity, directionality, and coherence.

14.3 ABSORPTION AND EMISSION OF LIGHT BY ATOMS

When light of a single color travels through space, it has wave-like properties (wavelength λ, frequency f, and wave speed c) and wave-like behaviors (**refraction**, **diffraction**, **interference**). If a light wave enters a medium made of atoms, some of the energy in the incident light wave may be converted into energy in the atoms, thereby decreasing the power in the light wave. This is **absorption** of light. On the other hand, if there is already energy stored in the atoms, then some of this energy may be converted into light energy. This is **emission** of light by atoms. Emission is the reverse of absorption. Here we will use the concepts of classical (pre-1900) physics to describe these processes. Other sections in this chapter (which are optional) describe absorption and emission using the modern quantum-physics viewpoint.

The simplest atom, hydrogen, consists of one electron moving in the vicinity of a proton. Because the electron is negatively charged and the proton is positively charged, they attract each other by the electric force. We can understand absorption and emission using a simple model for atoms. Instead of thinking of the common picture of the electron orbiting around the proton, imagine that the electron is being held near the proton by some kind of fictional spring, as in **Figure 14.2**. In this simple view, the farther the electron moves from the proton, the stronger the attractive force becomes. This leads to **simple harmonic motion** (SHM) of the electron. We know from Chapter 7, Section 7.4 that an object oscillating on a spring has a natural **resonance frequency**, which is the frequency of its motion when it is oscillating freely. We also learned that if you subject a spring to continuous driving by a periodic force, it will oscillate at a frequency equal to the driving frequency. The amplitude of an oscillator's harmonic

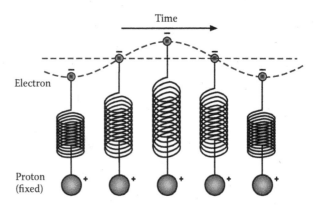

FIGURE 14.2 An atom modeled as an electron (−) and a fixed proton (+), connected by a fictitious spring. When struck by light with frequency f that obeys the resonance criterion, the electron gains energy and the light loses energy.

motion becomes maximum if the frequency of the driving force equals the natural resonance frequency of the oscillator.

In our simple electron-on-a-spring model for an atom, the natural oscillation of the electron has its own resonance frequency. An *electromagnetic* (EM) wave can drive the electron in the atom, because the electric field exerts an oscillating force on the electron. When an EM wave drives an electron, it can transfer energy to the electron. This process corresponds to absorption of light and occurs only if the light's frequency is nearly equal to the electron's resonance frequency:

$$f_{EM} \approx f_R$$

For example, the relevant electron in the neon atom has a resonance frequency f_R equal to around 475 THz. The period of this oscillation is

$$T = \frac{1}{f_R} = \frac{1}{475\,\text{THz}} = \frac{1}{475 \times 10^{12}\,\text{cyc/sec}} = 0.210 \times 10^{-14}\,\text{sec/cyc} = 2.10 \times 10^{-15}\,\text{sec}$$

This is an extremely small period for an oscillator. This means that if we want to drive this electron resonantly, in order to "excite" it to have a large-amplitude oscillation, we need to drive it at a frequency $f_{DRIVING} = 475$ THz. What type of EM wave has this frequency? The wavelength of this EM wave is found by

$$\lambda = \frac{c}{f} = \frac{3 \times 10^8\,\text{m/sec}}{475 \times 10^{12}\,\text{cyc/sec}} = 632 \times 10^{-9}\,\text{m} = 632\,\text{nm}$$

According to the EM spectrum in **Figure 7.37** in Chapter 7, this corresponds to red light.

Emission of light by atoms is the reverse of absorption. If an electron in an atom already has energy (it is oscillating), the electron can lose energy by emitting light. This process is called *spontaneous emission*. It is called spontaneous because there is no outside stimulus causing this process to happen. The emitted light frequency matches the electron's resonance frequency.

In 1916, Einstein hypothesized that there is a third process that can take place, which he called *stimulated emission*, described in the following:

FIGURE 14.3 Football player (A) steadily pushes at the resonant frequency of the child (C) on a swing. Baby (B) also tries to push, but is successful in helping only when pushing in-phase with the swinging child. If instead the baby pushes out-of-phase with the swinging child, then the baby gains energy (ouch!).

Principle of Stimulated Emission: If an electron in an atom already has energy (it is oscillating), and light whose frequency is resonant with the electron's frequency (i.e., with $f_{LIGHT} \approx f_R$) strikes the atom, the electron can lose energy by emitting light with the same frequency and direction as the incident light. This process increases the power of the light beam, and is called stimulated emission.

Stimulated emission increases the power and amplitude of the light wave. That is, it causes amplification of the light. Amplification of light by stimulated emission occurs only under special conditions. As an analogy, consider a child on a swing, being pushed by her football-player father, as in **Figure 14.3**. The father powerfully pushes the child at the natural resonance frequency of the swing, so that she oscillates with a large amplitude. A baby tries to get into the act by pushing on the child as she goes by. If the baby pushes in synchrony (in-phase) with the child's motion, the baby will add a little energy to the child's swinging. But if the baby pushes out-of-phase, then the swinging child will lose some energy, and the baby will gain that energy and be knocked backward. In this analogy, the father is the driver (or energy pump), the child is like the electron in an atom, and the baby is like the light wave. The energy in the light wave can either be decreased (absorbed) or increased (amplified), depending on the relative phase (or synchronicity) between the wave's and the electron's oscillations.

Figure 14.4 illustrates light amplification. A beam of light with low power (energy per second) travels from left to right through a medium in which many atoms have electrons that are already oscillating (and so have energy). When the light wave emerges from the medium, it has higher power and its amplitude is correspondingly larger. Correspondingly, the atoms in the medium have less energy. The word *laser* is an acronym for *light amplification by stimulated emission of radiation*.

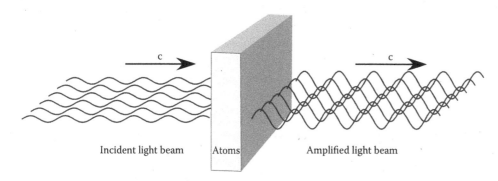

Incident light beam   Atoms   Amplified light beam

FIGURE 14.4 Light traveling in a beam enters an amplifying medium, where additional power (energy per second) is added to the beam. The amplitude of the light wave is greater after the light is amplified. The frequency is unchanged.

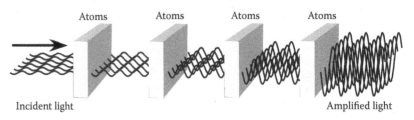

FIGURE 14.5   A light pulse traveling through a series of gain media grows exponentially in energy.

The atomic medium is called a gain medium. The factor by which the light power increases is called the **gain**, and we use the symbol G to denote it. If the power in a beam of light entering the medium is $P(in)$, and the power after the light exits the medium equals $P(out)$, then the relation between them is

$$P(out) = G \cdot P(in)$$

If G is greater than one, there is amplification of light. For example, if the incident power equals 2 watt (W), and the gain equals 1.4, then the power after the light exits the medium equals $P(out)$ = 2.8 W.

If a pulse of light passes through a sequence of gain media, as in **Figure 14.5**, there is exponential growth of the pulse's power. Upon each passing of the pulse through a gain medium, the pulse power is multiplied by G. As more passings occur, the pulse power grows exponentially. The mathematics of exponential growth was discussed in Chapter 2, Section 2.9. The power grows according to this sequence

$$P(in) \rightarrow G{\cdot}P(in) \rightarrow G^2{\cdot}P(in) \rightarrow G^3{\cdot}P(in) \rightarrow G^4{\cdot}P(in) \rightarrow \ldots$$

For example, if the gain equals $G = 1.2$, and there are four amplifiers, the sequence is

$$P(in) \rightarrow 1.2{\cdot}P(in) \rightarrow 1.2^2{\cdot}P(in) \rightarrow 1.2^3{\cdot}P(in) \rightarrow 1.2^4{\cdot}P(in)$$

$$P(in) \rightarrow 1.2{\cdot}P(in) \rightarrow 1.44{\cdot}P(in) \rightarrow 1.73{\cdot}P(in) \rightarrow 2.07{\cdot}P(in)$$

To obtain gain by using a collection of atoms, we somehow need to keep a larger number of atoms in a state of high energy (called an excited state) than there are atoms in their low-energy (non-oscillating) state. Otherwise, the unexcited atoms would tend to absorb the light more strongly than the excited atoms could amplify it, leading to no gain. Since the late 1950s, scientists have discovered ways to do this in many kinds of materials—including atoms in gases, liquids, glass, or crystals. It is beyond the scope of this text to explain the details of how to achieve the synchronization of phases between the atoms' electrons and the light wave, needed in order for amplification to occur. An alternative way to understand amplification by stimulated emission is given in Section 14.9. Let us next consider how light amplification can be used to make a laser.

14.4   LASER RESONATORS

Creating gain for light amplification is necessary for a laser to operate, but it is not sufficient. We must also enclose the gain medium in a **resonator**. The purpose of the resonator is to assist the light wave in building up its energy. A resonator is an object, usually some kind of box, that resonates a wave. A resonator is also called a **cavity**. To *resonate* does not mean to amplify. It simply means to build up the energy of the wave

FIGURE 14.6 An acoustic guitar body and a microwave oven are examples of resonators.

in a certain region. This occurs only if the frequency of the wave equals one of the special values, or resonance frequencies, of the resonator.

An example of a resonator for sound waves is the hollow body of an acoustic guitar, shown in **Figure 14.6**. Because there is no external power going to the guitar body, other than the string vibration, the body cannot amplify the sound wave. It can, however, resonate it. This means that for certain frequencies, the sound wave will be reinforced by the effect of wave interference in the air inside of the body. This leads to the buildup of sound energy. Guitar bodies are designed to have many different resonance frequencies, so they do not emphasize one note at the expense of the others. An example of a resonator for EM waves is a microwave oven. Such a microwave resonator also has many different resonance frequencies.

Resonators operate by the principles of standing waves, which we discussed in Chapter 7, In-Depth Look 7.2, which you should review. The crests and troughs of a *standing wave* do not move, but simply oscillate between positive and negative values. Such a wave occurs when two waves with the same frequency travel in opposite directions and interfere. In Chapter 7, we considered the example of water waves in a rectangular-shaped water-filled pan, which is shaken side to side. At certain frequencies of shaking, a resonant effect is observed: the waves interfere constructively, and the amplitude of the water wave becomes large. At other frequencies, called nonresonant, the waves interfere destructively, and the amplitudes of the resulting waves are small.

A simple physical example of standing waves occurs when a long rope is stretched and attached between two rigid walls and is driven by a vibrator, which applies an oscillating force near one of the rope's ends, as shown in **Figure 14.7**. You can also imagine two people holding a jump rope taut, and one of them jiggling the end. If one person oscillates an end smoothly and continuously with a constant frequency, then there is the possibility of creating a standing wave. At certain frequencies of driving, a standing wave is created in which the wave appears to move up and down; that is, in a direction perpendicular to the original stretched rope.

Each distinct form of standing wave is called a different *mode of oscillation*, or *mode* for short. A mode is created in the medium if any whole number (1, 2, 3, etc.) of half-wavelengths fits exactly between the two walls. Several examples are shown in **Figure 14.8**.

The same behavior occurs when an EM wave is trapped between two highly reflecting metal walls, as in a microwave oven. The EM wave takes on forms analogous to

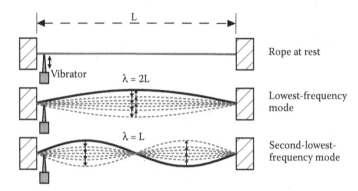

FIGURE 14.7 A taut rope stretched between rigid walls, when driven at certain frequencies, undergoes periodic oscillations in the form of standing waves.

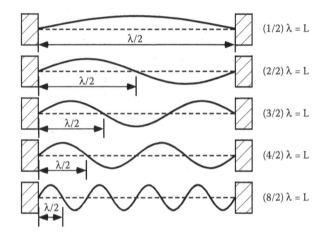

FIGURE 14.8 Each distinct form of standing wave is called a different mode of oscillation.

those shown in the mode patterns for the rope oscillations. The only resulting wavelengths of microwaves are those for which a whole number of half-wavelengths fits exactly between the two walls.

Standing waves also occur when a light wave is trapped between two highly reflecting mirrors, forming a laser resonator. Because light has very short wavelengths, a very large number of wavelengths can fit between the two mirrors. For example, if the light's wavelength is 500 nanometers (nm), as determined by the atoms' energy-level difference, then we could build the resonator so its length equals 500 mm; then exactly 1 million wavelengths would fit between the walls. This wavelength corresponds to a frequency equal to

$$f = \frac{c}{\lambda} = \frac{3 \times 10^8 \text{ m/sec}}{500 \text{ nm}} = 600 \text{ THz}$$

If we slightly increase or decrease the light's frequency from this value, the wave will not be resonant in the resonator, and therefore will not build up constructively. We say these frequencies are "not allowed" in the resonator. The fact that only certain frequencies are resonant in a laser resonator of a given length explains why lasers emit light of very pure color.

A laser resonator is like an exclusive nightclub—only certain frequencies are allowed.

IN-DEPTH LOOK 14.1: LASER RESONATOR FREQUENCIES

As explained above, in a resonator of length L, the only allowed wavelengths of light are those for which a whole number of half-wavelengths fit exactly between the two walls. If we denote the wavelength by λ, and the whole number by the symbol m ($m = 1,2,3,...$), we can state this condition as an equation.

$$m \cdot \left(\frac{\lambda}{2}\right) = L$$

We can use this equation to determine the allowed frequencies of the light in the resonator. Recall that wavelength equals the distance traveled by a wave during a time equal to one oscillation period T, which also equals one divided by frequency ($1/f$). Denoting the speed of light by c, as usual, we have $\lambda = c/f$. The condition above then gives:

$$m \cdot \left(\frac{c}{2f}\right) = L$$

Multiplying both sides by f and dividing both sides by L gives allowed resonating frequencies:

$$f = m \cdot \left(\frac{c}{2L}\right) \quad (m = 1,2,3,...)$$

14.5   HOW A LASER WORKS

A laser is a device that emits a highly directional beam of pure-colored, coherent light. To make a laser, we combine the idea of light amplification with the idea of a resonator. **Figure 14.9** shows the layout of a laser, consisting of a gain medium with an energy pump to transfer energy to the electrons in the medium's atoms and two reflective mirrors at the ends, making a resonator for light. A large circle shows an atom that contains stored energy and can contribute to light amplification. A small circle shows an atom that has no stored energy and will absorb (not amplify) light.

The resonator is made having two mirrors, one each placed at the ends of the gain medium. A mirror does not necessarily reflect 100% of the light incident on it. A mirror that reflects less than 100% of the incident light is called a partially reflecting

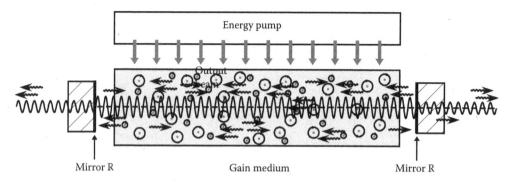

FIGURE 14.9 A laser consists of a resonator made with two reflecting mirrors, with a gain medium inside. Large circles indicate atoms containing stored energy. Small circles indicate atoms not containing stored energy.

QUICK QUESTION 14.1

Explain how "one-way" sunglasses work. That is, why can the wearer see an onlooker, but on a sunny day the onlooker cannot see the wearer's eyes? Consider that mirror reflectivities are symmetric with respect to inside and outside.

mirror. An example of such a mirror can be found in one-way sunglasses, with reflectivity equal to 0.90. This means that 90% of the light hitting either surface (inside or outside) is reflected back, and 10% is transmitted through. The **reflectivity** of a mirror equals the fraction of incident power that is reflected when a light beam hits the mirror, as illustrated in **Figure 14.10**. That is,

$$R = \frac{P(reflected)}{P(incident)}, \quad P(reflected) = R \cdot P(incident)$$

The value of the reflectivity of each mirror making up the resonator is denoted by the symbol R. The value of R is between 0 (no reflection) and 1 (total reflection).

A partially reflecting mirror is typically made by coating a thin layer of aluminum or other material onto the surface of a clear piece of glass. This thin layer is shown in Figure 14.9 as a bold line. Mirrors behave symmetrically; that is, the **reflectivity** of light is the same regardless of which side of the mirror the light is incident on.

Now we are in a position to understand how a laser works, as illustrated in **Figure 14.11**. Initially, the pump excites the electrons in atoms in the gain medium to states of

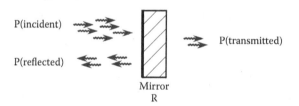

FIGURE 14.10 Light is incident on a partially reflecting mirror. A fraction R of the light power is reflected and the rest is transmitted.

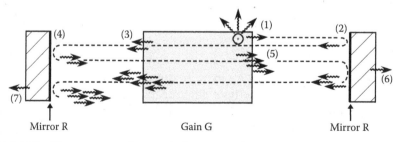

FIGURE 14.11 The sequence of events in laser operation. In this way of representing light, the number of wavy arrows indicates the power in the light beam.

higher energy. Then a sequence of events leads to laser operation (numbers correspond to numbers in parentheses in the figure).

1. A small amount of light is emitted from the gain medium by spontaneous emission, most of which travels in directions away from the mirrors.
2. Some of this emitted light travels toward one of the mirrors, where a fraction R of it is reflected back into the medium.
3. The reflected light beam returning from the mirror is amplified as it passes once through the gain medium, increasing the light power by a factor G.
4. The light travels beyond the gain medium and strikes the other mirror, where a fraction R is reflected back toward the medium.
5. The light again passes through the gain medium, being amplified by a factor G.
6. Some of the light striking the mirror transmits through it, creating an output beam from the laser. The reflected light beam returning from the mirror is amplified as it passes through the gain medium, increasing the power by a factor G.

This process is repeated over and over.

To summarize, each time the light goes through the gain medium and then reflects from a mirror, the power of the light is multiplied by the factors R and G. The mirror reflectivity R is less than 1. For laser action to occur, the gain G should be greater than one. Two cases can occur, depending on the product of R and G:

1. If $R \cdot G < 1$, there is a net loss of light on each round trip through the resonator. There is no build up of laser light. The laser is OFF. We say the laser is below threshold.
2. If $R \cdot G > 1$, there is a net increase of light on each round trip. Laser light builds up. The laser is ON. We say the laser is above threshold.

The gain G increases as you increase the pump power. A very slight change of pump power can cause the laser to go from OFF to ON. For example, consider a laser, for which the mirrors have reflectivity equal to $R = 0.90$. To have $R \cdot G > 1$, we must have the value of G be greater than 1.11. For example, if $G = 1.12$, the product of R and G equals 1.008, which is greater than 1. If instead $G = 1.10$, the product of R and G will equal 0.99, and the laser will be OFF.

An analogy to illustrate this high sensitivity to the pump power can be seen in the behavior of a public address (PA) system. A PA system consists of a microphone, an amplifier, and a speaker, as in **Figure 14.12**. You may know that if you turn up the volume (gain) on the PA's amplifier just a little too high, you will hear an uncontrolled, very loud squeal. This is caused by feedback—allowing the sound from the

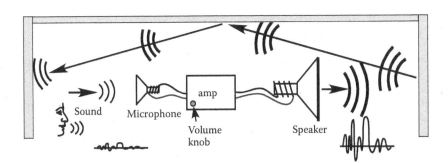

FIGURE 14.12   Feedback in a PA system. A person whispering acts analogously to spontaneous emission in a laser to start the self-oscillation of the system.

speaker's output to be picked up by the microphone, amplified, and sent into the speaker's input. **Feedback** is the sending of an amplifier's output back into its own input. In the case of a PA system, reflections of sound waves from walls and the ceiling can create a path for feedback. The room acts as a resonator. The sound energy is reflected and picked up by the microphone multiple times, being amplified each time. This effect is called self-oscillation, because the squeal that is heard often has a particular tone or frequency, although that frequency was not introduced into the microphone intentionally. Rather, the background noises (traffic, etc.) near the microphone contain many frequencies, and only certain of those efficiently travel around the feedback path, being reinforced. If the PA gain is turned down just slightly, the self-oscillation stops.

Optical feedback is needed for a laser to operate. Optical feedback is caused if, after being amplified by the gain medium, the light is sent back into the gain medium. Light is passed through an amplifier many times as a result of mirror reflections. Atoms in the amplifier medium may emit light of various colors, but the only light that experiences self-oscillation is the light having frequency that is resonant with the resonator and is proper for stimulated emission by the atoms. This is a highly selective condition and leads to the nearly pure color of laser light. Furthermore, the properties of stimulated emission guarantee that the emitted light has the same frequency and travels in the same direction as the initially present light. This means the light beam is nearly monochromatic and is highly directional. The light is coherent, with all wave fronts oscillating in phase and lining up in an orderly manner.

In a laser with $R \cdot G > 1$, the light has net gain upon each round trip in the resonator. On the other hand, the power in the light cannot grow indefinitely after the laser is switched ON. This is because the **pump source** is providing a fixed amount of power (J/sec) to the gain medium, and energy conservation prevents the laser from emitting a light beam containing more power than is provided by the pump. A little while after the energy pump source is switched on, the laser power builds up to a certain level and remains constant. We define the **efficiency** of the laser as the ratio of the laser light output power to the input pump power.

$$Efficiency = \frac{P(laser\ output)}{P(pump)}$$

For a typical red helium-neon gas-tube laser, the efficiency is approximately 0.001, or 0.1%, whereas for a semiconductor diode laser it is as high as 0.70, or 70%.

To summarize, the necessary parts of a laser are:

1. A gain medium.
2. A pump source to maintain most of the atoms in states of higher energy.
3. A light-energy resonator, made with mirrors, to provide feedback of the emitted light to the medium, while leaking out a small fraction of the light to create an output light beam.

14.6   THE HELIUM-NEON LASER

An important type of laser is the gas laser, the most common example of which is the helium-neon laser, or HeNe laser. It emits a nearly monochromatic red beam with wavelength 632 nm, and is inexpensive (about $200). HeNe lasers became commonly used in supermarket checkout scanners in the late 1980s, although now semiconductor lasers, which are more compact, have replaced them.

It is interesting to contrast a HeNe laser with a neon fluorescent lamp, which we discussed in Chapter 9, Section 9.3 concerning atomic vapor lamps. Neon lamps can be seen commonly in store windows, where the glass tube holding the red-glowing neon gas is twisted into the shapes of letters making words for advertising. Although such a lamp has the proper type of atoms—neon—with which to make a laser, there is not enough energy stored in the atoms to create gain, and there is no resonator. Therefore, the light it emits is just spontaneous emission, which is not directional and is not coherent.

HeNe lasers are constructed as in **Figure 14.13**. The mixed helium and neon gases are contained in a glass tube, which has mirrors glued directly to its ends. The example shown here has a flat, highly reflecting ($R = 0.999$) mirror glued to one end, and a second, curved, mirror (with $R = 0.99$) at the other end. Light partially reflects off the curved mirror and passes back through the gas in the tube, where it is amplified and reflected by the flat mirror. The output light beam emerges from the lower-reflectivity, curved mirror.

The three required laser elements for a typical HeNe laser are:

1. Gain medium—a mixture of helium (85%) and neon (15%) atomic gases.
2. Pumping—electrical current creates an ionized gas (a gas with some of the atoms' electrons removed from the atoms and free to move in the spaces between atoms), allowing electrical current to flow.
3. Resonator—a flat mirror and curved mirror.

One of the mirrors is curved to keep the returning light tightly focused in a narrow beam, as in **Figure 14.14**. This is needed because the inner bore of the gas tube is quite small, and the light must pass through the bore for it to be amplified sufficiently on each pass. In a HeNe gas laser, the gain factor G is quite small, approximately $G = 1.02$. This requires that the mirror reflectivity be greater than $R = 0.98$.

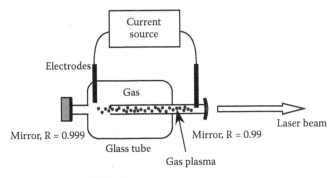

FIGURE 14.13   Helium-neon (HeNe) laser.

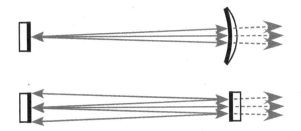

FIGURE 14.14   Laser resonators having one flat and one curved mirror serve to keep the light beam or mode confined to the axis of the laser tube. A laser resonator made with two flat mirrors will not confine the beam as it reflects back and forth—the beams spread out, as shown in the bottom panel.

When electrical current passes through the gas, the fast-moving electrons collide with the He atoms, exciting some of their electrons to higher energy. These excited He atoms then collide with Ne atoms, transferring their electron excitation to the electrons in the Ne atoms. The Ne-atom electrons then undergo stimulated emission, amplifying any passing light of the proper frequency. The electric current deposits approximately 1 W of its power in the ionized gas, and only 1 milliwatt (mW) of that deposited power goes into the emitted laser beam. The rest is dissipated as heat. That is, the efficiency of a HeNe laser is quite low—only approximately 0.1%.

IN-DEPTH LOOK 14.2: EXTREME LASER FACTS

Scientists and engineers love lasers because they offer the possibility of extreme behavior. This allows scientists to probe nature on very fine scales and to burn holes through solid objects. To find out about extreme lasers—the X-GAMES™ of LASERS—I asked some experts[1] about the biggest, "baddest," shortest, longest, slowest, fastest, and smallest lasers they knew of. These stats are from circa 2007 and are not necessarily the record holders but are important representatives of each category. The numbers in parentheses correspond to references given at the end of the chapter, where more information can be found.

Highest-energy laser pulse: 150 kilojoules (kJ) in a 10-nanosecond pulse. The National Ignition Facility (NIF) at the Lawrence Livermore National Laboratory in Livermore, California, achieved this result in 2005. The energy contained in the 150-kJ pulse is equivalent to a 1-ton automobile traveling at about 60 miles per hour. Soon, the NIF aims to achieve a pulse energy of a couple of megajoules. All of this energy will slam into a tiny pellet containing deuterium—a form of hydrogen—in an effort to induce nuclear fusion, which may serve as a futuristic source of power, as well as aiding in nuclear-weapons research [1].

Shortest laser pulse: Just under 1 femtosecond (fsec), or 10^{-15} sec. This subfemtosecond pulse was created at the Max Planck Institute for Quantum Optics in Garching, Germany, and is less than the time for a single oscillation cycle of a visible light wave. How short is this pulse? The duration of this pulse is to 10 sec as 10 sec is to the age of the Earth! To make such a short pulse, researchers have to generate light containing all colors from the visible region well into the ultraviolet region of the spectrum. Such a pulse of light looks white to the eye. This is very different from typical lasers, which emit a single color [2].

Highest instantaneous power: Just over 1 PW, or petawatt (i.e., 10^{15} W). This was first achieved at the Lawrence Livermore National Laboratory in 1996. This unimaginably high power exceeded the entire electrical generating capacity of the United States by more than 1,200 times, but was over in such a short time—440 fsec—that it produced "only" 680 J of energy. Because laser light can be focused to a very small spot, the focused energy density reached the equivalent of 30 billion J in a volume of 1 cubic centimeter, far larger than the energy density inside of stars. At such high energy densities, the electric field of the light is so strong that electrons become accelerated to almost the speed of light! [3]

Highest average power: More than 1 MW (megawatt) of continuous output power. This dangerously high level of power was first achieved by the Mid-Infrared Advanced Chemical Laser (MIRACL), at the High Energy Laser Systems Test Facility at White Sands Missile Range, New Mexico. Because the power is so high, it is operated only for seconds at a time, producing several megajoules of energy in an outburst [4].

"Baddest" laser: The U.S. Air Force's airborne laser, a close relative of the MIRACL, described above, is mounted in a modified Boeing 747-400F freighter aircraft. It produces

---

[1] Thanks to Wayne Knox, the Director of The Institute of Optics at the University of Rochester; Philip H. Bucksbaum, Director of the Stanford PULSE Center at Stanford Linear Accelerator Center (SLAC); and Roger Falcone, Director of the Advanced Light Source, Lawrence Berkeley National Laboratory.

megawatt-power pulses with a duration of seconds and energies of megajoules, for shooting down enemy tactical nuclear-weapons ballistic missiles. That is one bad laser [5].

Longest laser: 1.3 kilometers (0.8 mi). This is the length of the Linac Coherent Light Source, a "free-electron" X-ray laser, which is being built as a component of the Stanford Linear Accelerator Center. The building is so long that scientists use golf carts to travel around in it [6].

Shortest laser: Several micrometers, or millionths of a meter in length. This is the length of the resonator in vertical-cavity surface-emitting lasers, or VCSELs. These special semiconductor lasers were invented at the Tokyo Institute of Technology and are becoming important in telecommunications and other applications [7].

Longest time of laser-cycle stability: 13 sec. This is the duration that a laser at the National Institute of Standards and Technology (NIST) was able to oscillate with precisely the same frequency. That is, the laser did not gain or lose even one wave cycle over a 13-sec time, during which it oscillated 1,064,721,609,899,145 times. That is one stable laser! Atomic clocks based on stable lasers such as this are used to synchronize the Global Positioning System (GPS), which is available to consumers via satellite [8].

Most precise length measurement using a laser: About 1 attometer (i.e., 10^{-18} m). This length-measuring sensitivity was achieved by the Laser Interferometer Gravitational Wave Observatory, or LIGO. This unbelievably small length is one hundred billionth of the diameter of an atom and is measured as the change of distance between two mirrors that are separated by 4 kilometers (2.5 mi)! The purpose of LIGO is to detect the passing by Earth of a gravitational wave created by two neutron stars colliding [9].

14.7   VARIABLE-COLOR SEMICONDUCTOR LASERS

One of the workhorses of the Internet Age is the semiconductor diode laser. Such lasers provide the light that "lights the fiber" and does the reading and writing of compact discs. Semiconductor lasers are made with the same types of crystals used to make LEDs; for example, gallium arsenide (GaAs), a crystal made of a pure mixture of gallium (Ga) and arsenic (As) atoms. This is a semiconductor crystal. Semiconductor lasers can be designed to emit different colors of light, including infrared. Similar to LEDs that emit different colors of light, these are made using different semiconductor crystals; for example GaAlAs, in which aluminum (Al) replaces some of the Ga atoms in a GaAs crystal.

A semiconductor laser can have the ability to vary the color that it emits. Because frequency changes with changing color, this is also called "tuning" the laser, analogous to tuning the frequency of a guitar string. Tuning a laser allows you to vary the laser beam's frequency for use as a specific channel in a communication system. We will discuss this application in the next chapter.

To make a tunable laser, add a frequency-selective element inside of the resonator. **Figure 14.15** shows a laser with a prism placed inside of the resonator. The resonator is comprised of the 100% reflecting mirror on the bottom side of the semiconductor crystal and the partially reflecting mirror on the far side of the prism. The prism has the effect of filtering the optical feedback; that is, it restricts feedback to a narrow range of frequencies. Because the prism refracts light of different frequencies in different directions, a small range of light frequencies are sent back by the mirror into the center part of the crystal, where the gain is. Only light with the proper frequency is able to self-oscillate and create a strong coherent beam. The particular frequency that is allowed to self-oscillate is determined by the angle of the prism. Therefore, we can tune the frequency of the laser light by rotating the prism to change its angle.

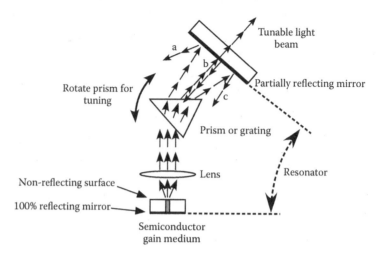

FIGURE 14.15 Frequency-tunable semiconductor lasers can be made with a semiconductor crystal having one reflecting and one nonreflecting surface, a prism, and a partially reflecting mirror on the other side of the prism.

In practice, diffraction gratings, rather than prisms, are used in most tunable semiconductor lasers, but the principle is the same for both. A diffraction grating has millions of thin lines scribed onto its surface and splits light into a spectrum of colors much as the surface of a compact disc does.

14.8 OVERCOMING LOSSES IN FIBER-OPTIC SYSTEMS

Light cannot travel forever in a medium such as a glass fiber if the medium has even a small amount of absorption. In Chapter 13, Section 13.9, we pointed out that high-quality optical fiber transmits 95.5% of the power after 1 kilometer (km), for a wavelength of 1,500 nm (near infrared). This means that for each kilometer passed, the power is multiplied by a factor 0.995. For example, after passing 100 km, the fraction of power transmitted is $(0.955)^{100} = 0.01$, or 1%. After passing 300 km, the fraction of power transmitted is $(0.955)^{300} = 10^{-6}$. This is far too small for a photodetector to pick up. Something needs to be done to increase the strength of the light pulses if they are to travel any farther than a few hundred kilometers.

Figure 14.16 shows a typical fiber-optic communication system. A semiconductor laser creates pulses of light that serve as binary bits representing some information. Lenses focus the light pulses into the end of a fiber, which transmits them to a distant location where they are detected by a photodetector. After the pulses travel a long distance in the fiber, they suffer loss and become weaker, as shown. How can we avoid this problem?

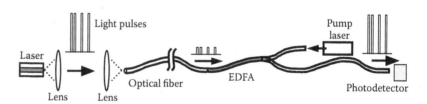

FIGURE 14.16 A semiconductor laser creates pulses of light with a wavelength of 1,550 nm, which are sent into a fiber, where they suffer loss. An erbium-doped fiber amplifier (EDFA), placed in-line with the signal-carrying fiber, boosts the power of the signal to original levels. A semiconductor pump laser provides the energy for creating gain in the EDFA.

Fortunately, stimulated emission comes to the rescue. We know that stimulated emission can amplify a light wave. To boost the light level back to its original power, an optical amplifier is placed in the fiber every 100 km or so. A convenient way to make an optical amplifier inside of an optical fiber is to put some gain-creating atoms into the glass core of the fiber and to pump these atoms using some additional energy source. A successful scheme that has been developed for doing this is called the erbium-doped fiber amplifier (EDFA), illustrated in Figure 14.16. Erbium (Er) is an element that can be doped easily into silica glass fibers. The electrons in the Er atoms are excited into higher-energy states by passing a powerful pump laser beam through the fiber in the direction opposite to that of the optical signal pulses. The pump laser light is absorbed by the Er atoms, which store energy. This causes amplification of the signal pulses that pass through the erbium-doped fiber, boosting them back up to their original power levels. The gain factor of an EDFA can be as high as $G = 1,000$.

14.9   QUANTUM PHYSICS DESCRIPTION OF LASERS

Required background: Chapters 9, 10, and 12

The discussion so far in this chapter has used the concepts of EM radiation as a wave phenomenon. Thinking of light as an EM wave, we were able to understand most of the details of laser operation. On the other hand, using the principles of quantum physics (which we learned in Chapters 9, 10, and 12) can give us a more complete and deeper understanding of the physics behind laser operation.

In Chapter 12, we learned that light has some properties that are wavelike and some properties that are in some sense particle-like (or "lumpy"). When light of a certain color travels through space, it has wavelike properties (wavelength λ, frequency f, wave speed c) and wavelike behaviors (refraction, diffraction, interference). But, when energy is transferred to or from this light wave, the transfer occurs in discrete bundles of energy called **photons**, each having energy equal to the light's frequency times Planck's constant, h. We can visualize light as in Figure 12.1, which showed photons, each with energy $E_f = h \cdot f$, being carried along with an imagined guide wave. The constant $h = 6.6 \times 10^{-34}$ J · sec/photon is Planck's constant and gives the relation between light frequency and the energy of each photon in the wave. For example, red light has a wavelength equal to approximately $\lambda = 700$ nm, and its frequency is $f = 4.3 \times 10^{14}$Hz. Using Planck's relation, we can compute the amount of energy contained in one red photon, as

$$E_f = h \cdot f = (6.6 \times 10^{-34} \text{ J} \cdot \text{sec/photon}) \cdot (4.3 \times 10^{14}/\text{sec}) = 2.9 \times 10^{-19} \text{ J/photon}$$

We learned about absorption of light by atoms in terms of the quantum description. If an electron in an atom is initially in a state (orbit) with energy equal to E_L, then it can absorb a photon of light having frequency f (with energy $E_f = h \cdot f$), and "jump" to a higher-energy state with energy E_H. This can occur only if the photon energy matches the difference of energies between the two electron states; that is, $h \cdot f = E_H - E_L$. The energy gained by the electron equals the energy given up by the light wave when it loses one photon.

We also learned about emission of light by atoms, in terms of the quantum description. If an electron is in an atomic orbit with higher energy, it can lose energy by emitting a photon and drop down, or jump, to an unoccupied lower-energy orbit. This process is called spontaneous emission. As in absorption, the emitted photon energy matches the difference of energies between the two atomic states; that is, $h \cdot f = E_H - E_L$.

In 1916, Einstein hypothesized that there is a third process that can take place, which is different from absorption or spontaneous emission. This is *stimulated emission*. He considered what happens if the electron is in a higher-energy orbital state and then light strikes it. Einstein proposed the following principle:

QUANTUM PRINCIPLE (VII)

Stimulated Emission: If an electron in an atom is initially in a higher-energy state with energy E_H, when a photon with frequency satisfying $h \cdot f = E_H - E_L$ strikes it, the atom can be induced, or "stimulated," to jump to the lower state with energy E_L. When the atom makes this jump, the original incident photon still exists, and the atom emits an additional photon with the same frequency and direction as the incident photon.

That is, stimulated emission is the emission of light by an atom initially in an excited energy state when an external light stimulus is present. The picture in **Figure 14.17** illustrates this. On the left is shown an atom with its outer electron in a higher-energy state (larger orbit). A photon with frequency f is incident, where f satisfies the energy conservation rule $h \cdot f = E_H - E_L$. The right side of the figure shows the situation after the photon passes. The atom's electron is now in the lower-energy orbit, and a second photon has been created with the same frequency and direction as the first photon.

We can illustrate the same process in terms of an energy-level diagram, as in **Figure 14.18**. Stimulated emission occurs because the incoming photon, which is associated with an EM wave, exerts forces on the atom's electron. These forces make the electron oscillate with the same frequency as that of the incoming photon. This induced motion of the electron causes it to emit light with the same frequency as that of the incoming photon, and in the same direction that the incoming photon is traveling. The emitted photon is a perfect copy of the incoming photon.

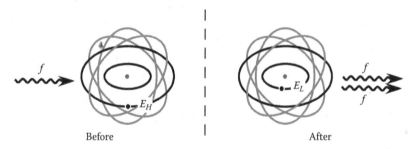

Before   After

FIGURE 14.17 The atom with its electron in a higher-energy state (orbit) is struck by a photon with frequency f, obeying the energy-conservation criterion. The electron loses energy and a second photon is emitted with the same frequency and direction as the incident photon.

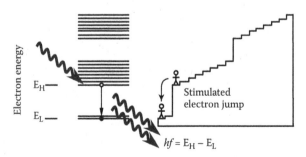

FIGURE 14.18 The electron in a higher-energy state is stimulated by a photon with frequency f, obeying the energy-conservation criterion. The electron descends to a lower energy state and a second photon is emitted with the same frequency and direction as the stimulating photon.

Stimulated emission increases the number of photons, and therefore the light power, in the beam of light; that is, it causes amplification of the light. **Figure 14.19** illustrates light amplification in terms of photons. A beam of light with low power enters a medium in which many atoms occupy higher-energy orbits. When the beam emerges from the medium, it contains a larger number of photons and its wave amplitude is correspondingly larger.

Special conditions are needed for light amplification to take place. Recall that electrons in lower-energy states can absorb light, making the light weaker, whereas electrons in higher-energy states can amplify light, making it stronger. Therefore, for amplification to dominate over absorption, there must be more atoms with electrons in the higher-energy state than there are atoms with electrons in the lower-energy state. This condition is called ***population inversion***. The word *population* refers to the number of electrons occupying a certain type of orbit, analogous to the number of people occupying a floor in a building (recall the picture in Figure 9.25). The word *inversion* reminds us that in the usual situation there are more atoms with lower energy than there are with higher energy. To make a laser, we must invert the usual situation.

We can state this conclusion in terms of a new quantum principle, involving the *rates* of the processes. We introduced the concept of rate of photon processes in the previous chapter. Say you have a medium containing N atoms. Denote the number of atoms with an electron in the higher-energy state as N_H, and denote the number in their lower state as N_L.

QUANTUM PRINCIPLE (VIII)

Rates of Absorption and Stimulated Emission: In a medium containing many atoms, the ***rate of photon absorption*** R_{ABS} is proportional to the number of atoms N_L that are in the lower state with energy E_L. The ***rate of stimulated emission*** $R_{St.Em.}$ is proportional to the number of atoms N_H that are in the higher-energy state with energy E_H. Both of these rates are also proportional to the rate R_f of photons (with the correct frequency f) that are striking the sample.

The principle says, for example, that if you double the number of lower-energy atoms, the rate of absorption will double. Likewise, if you double the number of higher-energy atoms, the rate of stimulated emission will double. Another example: Say there are 1,000 atoms in a medium, 900 of which are in the lower state. Then $N_L = 900$ and $N_H = 100$. If the rate of photons arriving at a typical atom is $R_f = 10^{19}$ photons per second, then the rate for absorption is proportional to $N_L \times R_f = 900 \times 10^{19}$, and the rate for stimulated emission is proportional to $N_H \times R_f, = 100 \times 10^{19}$.

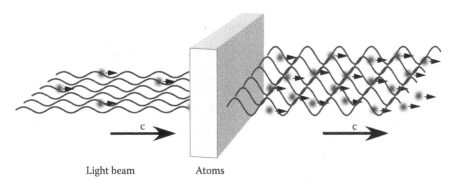

Light beam   Atoms

FIGURE 14.19 Photons traveling in a beam enter an amplifying medium, where additional photons are added to the beam. The amplitude of the light wave is greater after the light is amplified.

From this quantum principle, we derive an important rule:

For the rate of stimulated emission to exceed the rate of photon absorption, we must have N_H greater than N_L; that is, the number of atoms with an electron in the higher-energy state must be greater than the number of atoms with an electron in the lower-energy state.

14.9.1   Laser Gain

The amplification process is shown in terms of photons in **Figure 14.20**. Photons with the proper frequency for inducing stimulated emission are incident on an atomic medium containing N atoms. There is a population inversion, because there are more electrons in large-diameter orbits than in small-diameter orbits. Photons are represented by arrows. In this case, the rate of photons exiting the atomic medium is greater than the rate of photons entering it. The factor by which the number of photons increases is the gain, and we use the symbol G to denote it. If the number of photons in a pulse of light before entering the medium is $n(before)$, and the number of photons after the pulse exits the medium is $n(after)$, then the relation between them is:

$$n(after) = G \cdot n(before)$$

We can also write this as:

$$G = \frac{n(after)}{n(before)}$$

If G is greater than one, there is amplification of light. The atomic medium is called a gain medium.

14.9.2   Exponential Growth of the Number of Photons

Consider a single photon, with the proper frequency, entering a gain medium in which every atom is in the higher-energy state, as in **Figure 14.21**. At the location labeled "b", the photon stimulates an atom to create a second photon; now there are two. At location "c", each photon stimulates an atom to create a second photon; now there are four. If this happens four times, the number of photons at the end of the medium will equal $2^4 = 16$. In this case, the gain G equals 16.

The sequence of numbers of photons at successive locations follows: 1, 2, 4, 8, 16, that is, 2^0, 2^1, 2^2, 2^3, and 2^4. This is exponential growth.

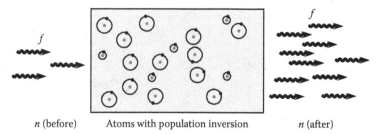

n (before)   Atoms with population inversion n (after)

FIGURE 14.20 Photons having frequency f enter a medium of atoms having a population inversion. There are more electrons in higher-energy electron orbits (drawn with larger diameter) than electrons in lower-energy orbits. Many of the incident photons cause stimulated emission, thereby increasing, or amplifying, the number of photons.

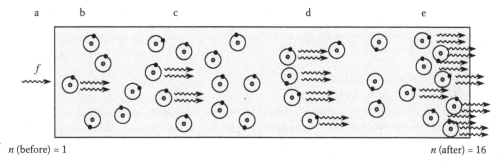

a b c d e

f

n (before) = 1 n (after) = 16

FIGURE 14.21 One photon enters a medium in which every atom is in the higher-energy state. The incident photon causes stimulated emission, amplifying the number of photons.

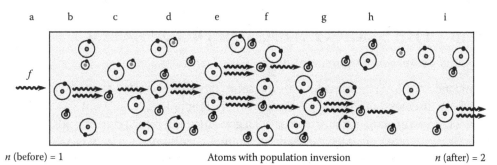

a b c d e f g h i

f

n (before) = 1 Atoms with population inversion n (after) = 2

FIGURE 14.22 One photon enters a medium that has a population inversion. There are only slightly more atoms in higher-energy states than atoms in lower-energy states. Stimulated emission and absorption take place, leading to a gain in this case equal to 2.

Next, consider a single photon with the proper frequency entering a gain medium in which only slightly more atoms are in the higher-energy state than are in the lower-energy state, as in **Figure 14.22**. At the location labeled "b", the photon stimulates a higher-state atom to create a second photon; now there are two. At location "c", a photon is absorbed into a lower-state atom, leaving one photon. At location "d", the remaining photon stimulates an atom to create a second photon; now there are again two. The process continues like this until at the end of the medium, two photons emerge. In this case, the gain G equals 2. On the other hand, if there is no population inversion (there are more atoms in their lower states) then there will be a net loss, or attenuation, of the light as it passes through the medium.

14.9.3 Gain-Medium Pumping

To obtain gain by using a collection of atoms, we need to keep a larger number of atoms in their excited (higher-energy) state than in their lower-energy state. This is one of the key points that Einstein did not seriously consider, because he did not believe it would be practically possible. There are two general points to appreciate:

1. In most naturally occurring situations, there is no population inversion—electrons tend to accumulate in the lowest-energy states available.
2. By "pumping" energy into a medium in the form of electrical, chemical, or thermal energy, it is possible to create population inversions in certain types of materials.

14.9.4   Laser Operation—Quantum Description

As explained in Section 14.5 above, to make a laser we combine the idea of light amplification with the idea of a resonator, as was shown in Figure 14.9. The description given above remains valid, when considering the quantum description, if you realize that each wavy arrow shown in Figures 14.9 through 14.11 represents a photon. For example, we can reinterpret Figure 14.10 showing light being partially reflected from a mirror in terms of photons. Denote the number of photons incident on the mirror by $n(incident)$ and the number of photons reflected by $n(reflected)$. We now say that a mirror's reflectivity equals the fraction of incident photons that is reflected when a beam hits the mirror. That is,

$$R = \frac{n(reflected)}{n(incident)}, \qquad n(reflected) = R \cdot n(incident)$$

14.10   THE SEMICONDUCTOR DIODE LASER

Required background: Chapters 9, 10, and 12

Semiconductor lasers are made with the same types of crystals used to make LEDs, which we discussed in Chapter 12, Real-World Example 12.2, which you should review now. There we explained that silicon is not a good material for making crystals that emit light. Instead, other crystals are used; for example, gallium arsenide (GaAs), a crystal made of a pure mixture of gallium (Ga) and arsenic (As) atoms. This is a semiconductor crystal, like pure silicon, and therefore, in its pure form it does not conduct electrical current well. As with silicon, the conductivity of a GaAs crystal can be controlled by adding impurities, or dopants, to create an excess of electrons in the conducting band, or a deficit of electrons in the originally filled band. Adding tellurium (Te) creates excess electrons in the higher, conducting band of GaAs, making the crystal an n-type conductor. Adding zinc (Zn) creates a deficit of electrons in the lower band, which creates a p-type conductor. To make an LED, these doped forms of GaAs are joined at a junction. Laser action can be obtained in an LED structure by adding reflecting mirrors to the two surfaces that are perpendicular to the junction, but it turns out that this creates a rather inefficient laser.

To make an efficient laser, a different type of diode is formed, called a *double junction*. As shown in **Figure 14.23**, a thin layer of pure, undoped GaAs is sandwiched between pieces of n-type and p-type GaAs semiconductors. This thin central region

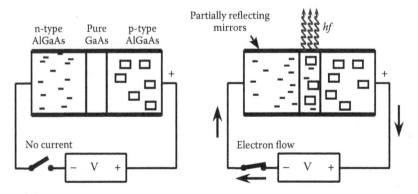

FIGURE 14.23   The semiconductor laser is made with a double junction, consisting of a thin layer of pure GaAs sandwiched between a piece of n-type GaAs and a piece of p-type GaAs. When a voltage is applied, electrons and holes flow into the central region, where an electron can "fall" into a hole, emitting a photon.

improves laser action in two ways. The region has a higher refractive index than the surroundings. Therefore, the central region acts like a wave-guide for the laser light, just as does an optical fiber. Therefore, total internal reflection keeps the laser light tightly confined within the active region, allowing it to experience the maximum amount of gain possible.

The second way in which the thin central region improves laser action is by providing a larger volume of semiconductor material in which electrons and holes are both present. In the usual single-junction structure of an LED, only a very narrow depletion region is created, so not many electrons and holes can be present in this region. In contrast, when current flows through the double-junction structure, there are electrons and holes throughout the central pure region, which can be wider than a typical depletion region in an LED. This larger volume can provide higher gain and a stronger laser output beam than can a single junction.

To make the double-junction structure into a laser, we must add two mirrors to provide optical feedback. This is done in one of two ways. As shown in Figure 14.23, we can apply a thin metal layer to the top and bottom edges of the diode structure. Thin metal layers reflect some light and transmit some light. It is sufficient to have a mirror reflectivity of around $R = 0.50$, because in a semiconductor diode the gain factor G can be quite large, say $G = 3$. The second method was shown earlier in Figure 14.15. There, we showed a partially reflecting mirror and a prism used to make the resonator. By rotating the prism, the frequency of light that is fed back into the gain medium is changed, thereby tuning the frequency of the generated laser light.

Semiconductor lasers can be made to emit different colors of light. These are made using different semiconductor crystals, for example GaAlAs, in which aluminum (Al) replaces some of the Ga atoms in a GaAs crystal. The addition of Al makes a new kind of pure semiconductor crystal, which still needs to be doped by impurities such as Te to make it conducting. The presence of the Al atoms in a formerly pure GaAs crystal increases the energy of the crystal's band gap. Therefore, the energy of the photons emitted is increased, because electrons usually sit at the bottom of the upper band and emit to the top of the lower band, as shown in Chapter 12, Figure 12.13. Increasing photon energy means a decreasing wavelength of light, and a bluer color. If the crystal has 1% Al atoms, its band gap energy corresponds to photons emitted with wavelength around 880 nm, whereas for 35% Al, the emitted wavelength is 750 nm. Both of these are deep red (near infrared). To make a laser that emits colors other than red, it is necessary to use a different type of semiconductor. For the wavelength range 600–700 nm, we can use GaAsP crystals with various percentages of phosphorous (P) atoms. For blue wavelengths of approximately 400 nm, we can use GaN crystals, which contain nitrogen (N) atoms. For green wavelengths of approximately 520 nm, we can use InGaN crystals, which contain indium (In) atoms.

SUMMARY AND LOOK FORWARD

Lasers operate according to physics principles that were developed over a period of 100 years. The first is the principle of stimulated emission, put forth by Einstein in 1916. Stimulated emission is the process in which light of the proper frequency impinges on an atom, the electron of which is in a higher-energy state of oscillation. The incoming light induces the electron to lose energy, emitting light whose frequency is the same as the frequency of the incoming light. If a medium has a sufficient number of higher-energy atoms, then stimulated emission is stronger than absorption, and amplification of a light beam can occur. The factor by which the light power is increased is called the gain.

A laser resonator is made of two mirrors facing each other. Specific light frequencies are reinforced or "resonated" by the resonator, leading to constructive interference and a buildup of light power. The specific frequencies that can be resonated depend on the length of the resonator. This frequency selectivity of the resonator contributes to the purity of the light color emitted by a laser.

The three required elements of a laser are:

1. Gain medium—an atomic medium that can absorb and emit light.
2. Pumping—a power source that can excite atoms to a high-energy state.
3. Resonator—two opposing mirrors that reflect the light emitted by the atoms.

If the pumping power is sufficiently high enough to overcome the losses of light through the partially reflecting mirrors in the resonator, the laser is above threshold. Only in this situation does the laser begin self-oscillating as a result of optical feedback. This produces a coherent beam of light.

Lasers are the best sources for optical communication because they emit light with these properties:

1. Monochromaticity—a laser emits light with nearly a single frequency.
2. High Directionality —a laser beam travels nearly in a single direction.
3. Coherence—the wave fronts in a laser beam line up parallel, making a highly organized wave.

In the next chapters, we will study the design of laser-based communication systems and discuss the physical factors that limit the rate at which we can send data using such systems.

REFERENCES FOR EXTREME LASER FACTS

1. National Ignition Facility Project, Lawrence Livermore National Laboratory. https://lasers.llnl.gov.
2. The first subfemtosecond laser pulse was created in 2004 by Ferenc Krausz and his group, at the Max Planck Institute for Quantum Optics in Garching, Germany. See *Physics Today*, October 2004, 21.
3. Perry, Michael. "The Amazing Power of the Petawatt," in *Science and Technology Review*, March 2000, 4. http://www.llnl.gov/str/MPerry.html.
4. Pike, John, and Sherman, Robert. *Mid-Infrared Advanced Chemical Laser* (MIRACL). Federation of American Scientists, Space Policy Project, 1998. http://www.fas.org/spp/military/program/asat/miracl.htm.
5. ABL YAL 1A Airborne Laser, AirForceTechnology.com, Copyright 2007, SPG Media Limited, a subsidiary of SPG Media Group PLC. http://www.airforce-technology.com/projects/abl.
6. Linac Coherent Light Source, Stanford Linear Accelerator Center, http://www-ssrl.slac.stanford.edu/lcls/index.html.
7. Iga, Kenichi. "Surface-Emitting Laser—Its Birth and Generation of New Optoelectronics Field," *IEEE Journal of Selected Topics in Quantum Electronics* 6 no. 6 (2000): 1201.
8. *Mercury Atomic Clock Keeps Time with Record Accuracy*, NIST News Release, National Institute of Standards and Technology, http://www.nist.gov/public_affairs/releases/mercury_atomic_clock.htm.
9. LIGO Caltech, MIT Collaboration. http://www.ligo.caltech.edu, http://www.ligo.mit.edu, http://www.ligo-wa.caltech.edu.

SUGGESTED READING

Hecht, Jeff. *Understanding Lasers: An Entry-Level Guide* (IEEE Press Understanding Science & Technology Series), 2nd ed. Piscataway, NJ: Wiley-IEEE Press, 1993.
Taylor, Nick. *LASER: The Inventor, the Nobel Laureate, and the Thirty-Year Patent War.* New York: Simon & Schuster, 2000.

KEY TERMS

Absorption
Brightness
Cavity
Coherence
Diffraction
Directionality
Efficiency
Electromagnetic
Emission
Feedback
Frequency
Gain, G
Interference
LASER as a device
LASER as a process
Light-emitting diode (LED)
Light resonator
Mode
Monochromaticity
Photon
Planck's constant
Population inversion
Pump source
Rate of absorption
Rate of spontaneous (or stimulated) emission
Reflectivity, R
Resonance frequency
Resonator
Simple harmonic motion (SHM)
Spontaneous emission
Standing wave
Stimulated Emission
Threshold
Wavelength

ANSWER TO QUICK QUESTION

Q14.1 The reason that highly silvered sunglasses seem to allow a person to see only "one-way" is that a person looking at the wearer sees 90% of the light reflected from the silvered surfaces; whereas this 90% of light is much greater than the 10% of the light that illuminates the wearer's eyes from the side and is transmitted through the surface of the glasses to the onlooker. Therefore, the onlooking person sees her own face reflected in the glasses and does not see the wearer's eyes. The wearer, on the other hand, sees 10% of the bright light illuminating the onlooker.

EXERCISES AND PROBLEMS

Exercises

E14.1 We can try to mimic the properties of laser light by the following scheme. Use a bright white incandescent bulb and a small hole in a distant opaque screen to create a highly directional beam of white light. Then pass the directional beam through a prism, followed by a narrow slit in a second opaque screen, to select a nearly pure color. The resulting beam is coherent, directional, and nearly single color, so the scheme works in some senses. What important aspect of laser light would not be produced by this method?

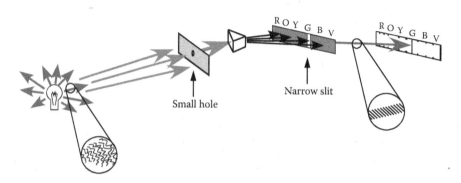

E14.2 Explain in words and pictures the differences between light absorption, spontaneous emission, and stimulated emission.

E14.3 A certain type of atom (calcium) does not absorb light whose wavelength equals 421, 422, 424, or 425 nm, but does absorb light whose wavelength equals 423 nm. Estimate the resonance frequency corresponding to the electron motion inside of this atom.

E14.4 A gas of a certain type of atom (rubidium) is contained in a hollow glass tube and has electrical current passing through it. There are no other sources of light nearby, and the atoms in this tube emit light whose wavelength equals 795 nm. Is this spontaneous or stimulated emission? Explain.

E14.5 Explain the role of an energy pump in the operation of a light amplifier.

E14.6 A beam of light with initial power equal to 1 W passes through a series of gain media, as in Figure 14.5. Each gain medium increases the beam's power by a factor equal to 2. After passing through ten such media, what will be the beam's power?

E14.7 An example of an EM resonator is the metal food compartment in a microwave oven. A microwave generator creates EM waves with a frequency equal to 2.45 GHz, corresponding to a wavelength equal to 120 mm, about 4 ¾ inches. If you were to completely cover the inside floor of the compartment with marshmallows closely-packed in a regular array, describe what you would observe after cooking for just long enough to blacken a marshmallow. (No need for quantitative answer.)

E14.8 Consider a jump rope stretched taut between two fixed walls. For this particular rope, if you oscillate an end smoothly and continuously with a constant frequency equal to 1 Hz, then a standing wave of the form shown here is created. Draw the pattern that is created if you oscillate the end with a frequency equal to 2 Hz. Answer the same for 3 Hz.

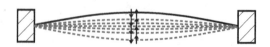

E14.9 Think of Jimi Hendrix holding his plugged-in guitar and standing face-toward a guitar amplifier speaker that is turned all the way up (preferably to "11"). Explain how, without touching the guitar strings, he could sustain a note for as long as he wanted.

E14.10 Why is it best to use at least one curved mirror to make a laser resonator rather than two flat mirrors?

E14.11 Neon fluorescent lamps emit light using the same atoms (Ne) that emit light in a helium-neon laser. The wavelengths of both are 632 nm. In what ways is the emission process similar or different in the two cases? In what ways are the properties of the emitted light similar or different in the two cases?

E14.12 In what ways are LEDs and semiconductor lasers similar and different? In what ways are the light emitted by LEDs and semiconductor lasers similar and different?

Problems

P14.1 A 10-cm-long glass rod contains erbium atoms, the electrons of which are maintained in a higher-energy state by a pump-energy source. When infrared light with frequency 2.0×10^{14} Hz passes once through the rod, it becomes amplified by a gain factor of 1.09. If such a light pulse initially contains energy 4×10^{-9} J and passes through the rod 85 times, how much energy will it then have?

P14.2 As stated in In-Depth Look 14.1, only certain frequencies of light are present in a resonator. What is the lowest possible frequency for a resonator with a length of 0.25 m? What is the wavelength of this light? What is the second lowest possible frequency, and what is its wavelength? What is the one-millionth lowest possible frequency, and what is its wavelength?

P14.3 A certain laser has two mirrors, each with reflectivity equal to 0.83. How high does the gain need to be for the laser to self-oscillate?

P14.4 A certain laser has a gain medium with gain equal to 0.9. Is there a value for the reflectivity of the two mirrors such that the laser will turn on (i.e., oscillate)? If so, give the value. If not, explain. Answer the same if the gain equals 1.9.

P14.5 A certain laser draws a power equal to 3 W from the electric wall plug. It emits a light beam having power of 0.05 W and converts the rest into heat. What is the efficiency of this laser? How much light energy is emitted in 2 sec? How much thermal energy is released in 2 sec?

P14.6 (This problem requires review of Section 14.9.) When the Apollo astronauts landed on the Moon, they placed prisms on the Moon's surface. By bouncing laser-light pulses from these prisms, scientists on Earth can very precisely determine the distance between the laser on Earth and the prisms. The best distance measurements to date have precision of 2 cm (approximately 1 in.). A recent experiment on lunar ranging[2] uses a short (100 psec) green laser pulse containing 0.1 J of energy with wavelength equal to 530 nm, which is aimed using a 3.5-m diameter telescope at the area on the Moon where the prism reflectors are located.

(a) How many photons are in this outgoing pulse? (*Hint*: Calculate the energy per photon using the Planck relation.) When the pulse hits the Moon's surface, it illuminates a circular area of the Moon's surface. Because of the wave properties of light (all waves spread upon travel) and the turbulence of Earth's atmosphere (variations of the atmosphere's refractive index distort the light's wave fronts), the laser spot is quite large when it reaches the Moon, approximately 1.8 km in diameter. (Area $= \pi \, (diameter/2)^2 = 4.1 \times 10^6$ m^2). The combined area of the group of prisms equals around 0.5 m^2.

(b) How many photons hit the combined area of the prisms? (*Hint*: Calculate the ratio of the areas of the group of prisms and the area of the laser spot on

---

[2] Apache Point Observatory Lunar Laser-Ranging Operation, at the University of California at San Diego. http://physics.ucsd.edu/~tmurphy/apollo.

the Moon.) About one-third of those photons hitting the prisms actually reflect toward the Earth, and the spot that the reflected beam makes on Earth's surface is about 15 km across. The same 3.5 m telescope used to send the original pulse out collects only about 1 in 30 million of the photons arriving at Earth.

(c) How many photons are collected? (*Hint*: The answer is less than 100,000, but it should be greater than 1, otherwise the experiment would not work.)

(d) What is the round-trip time of a photon if the distance to the Moon is 395,000 km?

(e) If the arrival time of the photon pulse can be measured to a precision of about 10 psec, what precision of measuring the Earth-to-Moon distance does this give?

P14.7 (This problem requires review of Section 14.9.) A small light bulb that emits 2,000 W of light power is placed at the center of the floor of a room with a dome overhead. Both the radius and center height of the dome are 10 m. A small hole, with a diameter of 1 mm, at the top center of the dome allows a narrow beam of light to pass through and shine toward the sky.

(a) What is the power of this narrow beam? *Hint*: One half of the lamp's emitted power uniformly illuminates the dome ceiling, the total area of which equals $(1/2)4\pi R^2 = 628$ m². The area of the hole is πr^2, where r is the radius (one-half diameter) of the hole.

(b) If you place a color filter (a prism and a slit) outside of the hole so that only a very small spread of yellow light frequencies is transmitted into the sky, such that only one part in 10^5 of light passes through this filter, what is the power of the outgoing beam?

(c) Say that the wavelength of this filtered yellow beam is 550 nm. How much energy is emitted through the hole and the filter in a time 1 μsec?

(d) How many photons are emitted in this time?

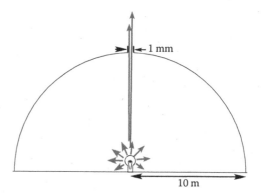

P14.8 (This problem requires review of Section 14.9.) A sample of gas contains 2×10^{13} hydrogen atoms, 1.3×10^{13} of which have their electron in the lowest (ground) energy state and 0.7×10^{13} of which have their electron in the next-to-lowest (excited) energy state. The energy difference between these states is 1.63×10^{-18} J, corresponding to a light frequency of 2.467×10^{15} Hz.

(a) If light of this frequency passes through the gas, will it be attenuated, amplified, or neither? Explain.

(b) If light of slightly different frequency passes through the gas, will it be attenuated, amplified, or neither?

(c) Answer the same two questions in the case that 0.9×10^{13} atoms have their electron in their lowest-energy state, and 1.1×10^{13} atoms have their electron in their next-to-lowest energy state.

P14.9 (This problem requires review of Section 14.9.) A sample of gas contains 2×10^{13} hydrogen atoms. In hydrogen, the energy difference between the lowest (ground) energy state, and the next-to-lowest (excited) energy state is 1.63×10^{-18} J, corresponding to a light frequency of 2.467×10^{15} Hz. Under what condition would this gas be completely transparent to light of this frequency when passing through the gas? That is, under what condition is there no net absorption or amplification? *Hint*: Consider the numbers of atoms in their lowest energy states and in their higher energy states.

P14.10 (This problem requires review of Section 14.9.) A 10-cm-long glass rod contains erbium atoms, having electrons that are maintained by a pump-energy source in a state whose energy is 1.32×10^{-19} J higher than the ground electron state. By the Planck relation, this energy difference corresponds to infrared light with frequency 2.0×10^{14} Hz. When light of this frequency passes once through the rod, it becomes amplified by a gain factor equal to 1.09. If a light pulse initially contains 3×10^{10} photons, and then passes through the rod 85 times, how many photons will it have? (This question is related to problem P14.1.)

P14.11 (This problem requires review of Section 14.9.) For the erbium-doped fiber amplifier (EDFA), the wavelength of the signal light is 1,550 nm. The pump wavelength must be less than 1,550 nm; typically it is 980 nm. Explain why the pump wavelength could not be greater than 1,550 nm. *Hint*: Think about energies of the photons and electrons.

P14.12 (This problem requires review of Section 14.9.) A 10-cm-long glass rod contains 5×10^{16} erbium atoms, all of which are initially in an excited electron state, the energy of which is 1.32×10^{-19} J higher than the ground electron state, corresponding to infrared light with frequency 2.0×10^{14} Hz. No further pumping of energy into the gain medium takes place. When light of this frequency passes once through the rod, it becomes amplified by a gain factor equal to 2.0. If a light pulse initially contains 1 nJ of energy and then passes through the rod 75 times, how much energy will it have? *Hint*: The light pulse cannot extract more energy from the rod than is initially stored there.

P14.13 (This problem requires review of Section 14.9.) Explain how a thin layer of aluminum coated on the surface of a piece of glass acts as a mirror. *Hint*: Metals have partially filled electron energy bands, as discussed in Chapter 9. Therefore, when an EM wave impinges on a metal, the electrons in the metal can accelerate and oscillate. What does an oscillating electron do?

15

Fiber-Optics Communication

Currently, communication systems are mainly handling voice traffic, and now we are talking about adding data to it. At the same time, fiber optics is offering a unique opportunity to create a transmission network like the interstate highway system; it is really capable of handling lots of traffic.

Charles Kuen Kao
(1987)

Network specialists at the University of Oregon installing optical communications fibers. (Courtesy of the University of Oregon.)

Charles Kuen Kao was among the first to propose, in 1966, that glass fiber could be used for long-distance communication by light. (Courtesy of Emilio Segrè Visual Archives, American Institute of Physics.)

15.1 BANDWIDTH AND THE PHYSICS OF WAVES

In 1966, Charles Kuen Kao, a Chinese-born scientist working in Britain, first proposed that glass fiber could be used to transmit light for long-distance communication if the attenuation or loss of light in the fiber could be made small enough. As described in Chapter 13, scientists at Corning were the first to make glass fiber with low loss, thus making Kao's idea practical. Optical fiber systems can carry huge quantities of voice and data traffic. Compared to an AM radio channel, for example, a single optical fiber theoretically can carry millions of times more data per second! In the 1990s, a rapid increase in the amount of bandwidth available to users drove an exponential growth of the Internet. This revolution in data capacity is what recently enabled Internet abilities such as on-demand video like YouTube. Our goal in this chapter is to understand why optical fiber permits such huge data rates. The answer to this question is not in the high speed of the light itself. In fact, light waves in fiber travel about 50% slower than do radio waves in air. The answer lies in the concept of *bandwidth* and the frequencies of the waves used in each case, as we shall see.

In Chapter 8, we discussed *analog* and *digital* communication. We treated the basics of analog radio, including the modulation of carrier waves to carry information. We also discussed *frequency multiplexing*, which allows a single physical medium to carry many communication *channels*. We discussed digital *sampling* of an analog signal to change it to a binary form suitable for transmission across a digital communication channel. Digital communication requires use of a chosen *protocol*—a set of rules for interpreting a list of binary numbers.

We discussed the idea of the bandwidth of a communication channel, which is the range of frequencies allocated to each channel's broadcasting. For example, each AM station is permitted to use a small range (called a band) of radio frequencies covering about 10 kHz (e.g., 1275–1285 kHz), so its bandwidth is 10 kHz. If a channel has bandwidth equal to B, the shortest radio or light pulse that can be transmitted in that channel has time duration of about $1/B$. For example, a bandwidth of 10 kHz corresponds to a pulse duration of about $1/10$ kHz $= 10^{-3}$ seconds (sec), or 1 millisecond (msec).

In this chapter, we will learn about some of the optical hardware used to make the large bandwidth of light accessible for fiber-optic communication systems. We will see that using light pulses instead of radio signals offers a huge increase in data rate in digital systems, because the bandwidth of optical systems is far greater than in radio systems. The bandwidth of a medium is determined by its physical properties, so this leads us back to the physics of waves. In Chapter 13 we studied how light travels in glass optical fibers. The key concepts are *refraction* of light and *total internal reflection*. This chapter brings together many concepts and discussions that were introduced in earlier chapters, and there are many references to those earlier chapters. Please return to these earlier sections and review them or read them for the first time. This is an opportunity to see the connections between the many topics we have discussed in this book.

15.2 OVERVIEW OF FIBER-OPTICAL COMMUNICATION SYSTEMS

Before discussing data transmission in fiber-optical systems, let us review communication systems in general. **Figure 15.1** shows the three main elements of any communication system:

■ The *transmitter*, which converts information to a physical form suitable for transmitting.

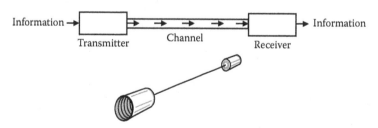

FIGURE 15.1 A communication system consists of a transmitter, a channel, and a receiver. A simple example is two tin cans and a taut string.

- The *channel*, which is a physical medium that can carry energy in a form representing the information.
- The *receiver*, which converts the energy received from the channel into a useful representation of the original information.

We discussed an example of this general scheme in Chapter 8, where we studied radio transmission systems—both analog and digital. In that case, the channel was the *electromagnetic* medium, air or vacuum (think of space communications). The transmitter converts data or audio information into physical radio waves, and the receiver converts these waves back into information, usually represented as voltages in an electronic circuit. We discussed the mathematical concept of *information* in Chapter 2, where we said that the amount of information in a signal or message equals the number of yes-no questions you would need to have answered to gain the information in the signal. Because yes-no questions have only two possible answers, the answer to a single yes-no question is a *bit*, that is one binary digit, 0 or 1. Recall that one *byte* equals eight bits. In Chapter 8, we discussed the idea of sampling an analog signal to convert it to a list of binary numbers (also called bits). We also discussed modulation of a *carrier wave*, by which we mean varying the amplitude of a wave to represent ones and zeros in the list of bits that represent the sampled signal. (You should review Sections 8.3 and 8.4.)

Since ancient times, light waves have been used for transmitting information over long distances. We learned in Chapter 7 that light waves are made of the same "stuff" as radio waves—electromagnetic fields. Oscillating electric charges create traveling waves of oscillating electric and magnetic fields. In vacuum, light waves travel with a speed $c = 3 \times 10^8$ m/sec, whereas in a dense medium light travels slower by a factor n called the *refractive index*. That is, $c_n = c \div n$, as discussed in Chapter 13, Section 13.3. For example, for glass n equals about 1.5, so the speed of light in glass is 2×10^8 m/sec.

In an optical communication system, light pulses carrying data are transmitted inside *optical fibers*, which we discussed in Chapter 13, Section 13.11. The light pulses are created using *lasers*, which we studied in Chapter 14. The elements of a basic optical communication system are shown in **Figure 15.2**. A list of bit values (ones and zeros) are sent into the transmitter in the form of voltage levels (high or low), where they control a *modulator*, which alters the power of a light beam produced by a laser. (Modulators are described below.) The laser produces a constant-power light beam, which experiences different amounts of attenuation as it passes through the modulator, depending on what bit value is being sent. The light emerging from the modulator is a series of pulses of high or low power. These pulses travel as far as 100 kilometers (km) by total internal reflection inside the core of the fiber until they reach the other end, where they are focused onto a light detector (a semiconductor photodetector), described

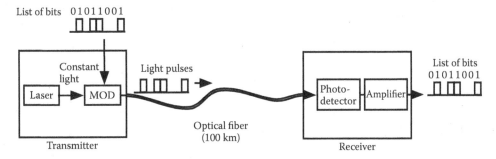

FIGURE 15.2 In a basic optical communication system, the transmitter contains a constant-power laser and a modulator, the channel is an optical fiber, and the receiver consists of a semiconductor photodetector and an electronic amplifier.

in Chapter 12 in Real-World Example 12.1. As explained there, the light entering the semiconductor crystal can elevate the energy of electrons in the crystal only if the light's frequency is high enough to cause the electron to jump from the lowest energy band to the next higher band. When this occurs, the crystal becomes electrically conducting, allowing a current to flow. A transistor amplifier increases the current of the signal to a level high enough for driving a computer memory or other device. In this way, the original list of bits can be transmitted over distances up to 100 km.

The practical distance limit of 100 km is set by the decrease of a light pulse's energy as it travels through the fiber, as discussed in Section 13.9. As we said there, for near-infrared light with wavelength equal to 1,500 nm, modern optical fiber transmits 95.5% of the light's initial power (or pulse energy) over a distance of 1 km. This means that after 100 km, the light power is decreased by a factor 0.01; that is, 1% of the initial light power transmits through 100 km. If we doubled the distance, then only 0.01% would be transmitted, which is not enough for the photodetector to detect reliably. As discussed in Section 14.8, this attenuation of the light can be compensated for by inserting laser amplifiers into the fiber every 100 km or so.

Let us estimate how many bits per second we can send using the basic system in Figure 15.2. Several factors limit this maximum data rate.

■ Limits in the transmitter arise because of limitations on how fast the modulator can vary the laser power. Typical electronic systems (in 2008) can modulate a laser beam at a rate of up to 40 gigahertz (GHz). Therefore they can generate laser pulses representing binary data at a rate of 40×10^9 bits/sec. The time duration of each pulse equals

$$\textit{bit duration} = \frac{1}{40 \times 10^9 \, \text{bits/sec}} = 25 \times 10^{-12} \, \text{sec/bit} = 25 \, \text{psec/bit}$$

This is a very brief time interval. For comparison, in the time it takes a housefly to flap its wings once (about 1 msec), the modulator can produce 40 million pulses!

■ Limits in the fiber-optical channel arise because of broadening of the light pulses. When a short light pulse enters a fiber and travels a long distance, the pulse becomes stretched or broadened in duration. As discussed in Section 13.12, this is because of two effects. The first, ***mode dispersion***, refers to the varying times required for light to pass through a length of fiber when traveling in different directions (modes) within the fiber. The second—***material dispersion***—is caused by the same effect in glass that causes a beam of white light to spread (disperse) into a spectrum of different colors upon passing through a prism. Any short pulse of light is made up of a certain range of frequencies, called its spectral bandwidth, B. The duration of a pulse is given roughly by the inverse of its spectral bandwidth. That is:

$$\textit{pulse duration} = \frac{1}{\textit{bandwidth}} = \frac{1}{B}$$

For example, for a light pulse with a spectral range (bandwidth) equal to 10 GHz, the pulse duration equals $1 \div (10 \, \text{GHz}) = 10^{-10}$ sec, or 100 picoseconds (psec). (For another example, see Figure 8.23 in Chapter 8.) A 100-psec light pulse contains a range of frequencies, and the different-frequency waves travel at slightly different speeds c_n. So, a 100-psec pulse, even if it travels in a single-mode fiber, will spread out slightly after traveling through a long fiber. This pulse broadening effect limits the number of pulses we can send during any fixed time interval. For example, if during transmission the pulses broaden to as long as 1 nanosecond (nsec; 10^{-9} sec) each, then we could not pack more than 10^9 such pulses into a 1-sec time interval; otherwise they would overlap and the bit values would become scrambled, as illustrated in Figure 13.25.

- Limits in the receiver occur because any photodetector has an upper speed at which it can operate reliably. As an example of such a response time, think of a factory assembly line where a conveyor belt moves a line of machine parts toward a worker (you). Your task is to pick each part up, perform an operation on it, and place it onto a different conveyor belt. Let us say that you can do this operation once in 3 sec. Your rate of operation completions is 0.33 Hz. As long as the parts come to you at a rate no greater than 0.33 Hz, you can process them just fine. But, if they come faster than 0.33 Hz, the parts get all jumbled together, and errors will occur. A photodetector has a certain response time, depending on how it is made. This is the time required for the detector to receive the light pulse, create a voltage pulse representing the power in the light pulse, and transmit this voltage to the next component in the electronic circuit. Very fast detectors can have a response time equal to 25 psec, corresponding to an operating rate of 40 GHz, or 40×10^9 pulses/sec. (Faster detectors exist, but they would strain the ability of the subsequent electronics to keep up.) Let us say, then, that 40 GHz is a reasonable rate to work with in practice.

If a basic fiber-optical system with a laser, a modulator, and a receiver can send about 40×10^9 pulses/sec, how can we best use this capability for communication? Consider sending telephone calls on such a system. Audio frequencies—the frequencies in speech and music—do not exceed about 20 kHz. A music compact disc (CD) is recorded using a sampling rate of 44 kHz, fast enough to be sure that no information is lost (as discussed in Chapter 8, In-Depth Look 8.1). This means that real-time sending ("streaming") of CD-quality music requires sending 44,000 pulses/sec through the fiber. This is only about one part in a million of the full capability of the fiber—40×10^9 pulses/sec. Therefore, a single fiber could carry 1 million CD-quality streaming signals simultaneously!

To see how this can be done, consider a simpler case of sending three CD signals on one fiber, as illustrated in **Figure 15.3**. Each CD player generates a stream of long voltage pulses, say 20 μsec each. In an electronic circuit called a ***multiplexer***, which is a part of the transmitter, these pulses are shortened to about 25 psec each and are interweaved as shown. The interweaving is done by delaying the streams from the signal sources by different amounts of time, then combining the streams, as shown. The pulses in the combined signal stream do not overlap in time because they are so short. They are then all sent together through the fiber in the form of optical pulses. At the receiver end of the fiber, this process is reversed. The short pulses are separated or demultiplexed, then stretched out in time to 20 μsec each, and sent to the separate users' computers or CD players. This procedure is called ***time multiplexing***.[1]

THINK AGAIN

When we say such a system is fast, we do not mean that the light travels particularly fast. We mean that the system itself has a short response time, so it can respond quickly.

The fact that such a fiber-based system can transmit data much faster (higher data rate) than metal-wire-based electronic systems or wireless radio systems is a consequence

---

[1] Called time-division multiplexing (TDM) in the technical literature.

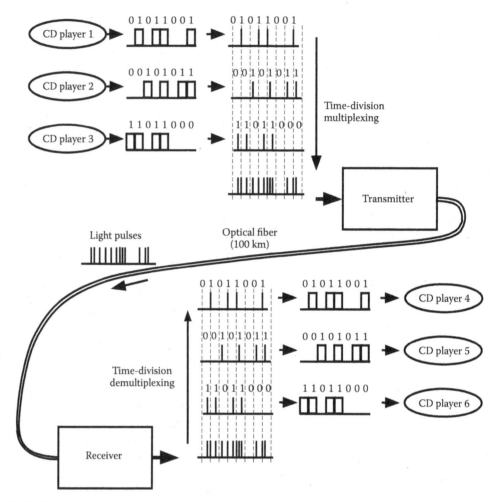

FIGURE 15.3 In a time-multiplexed optical communication system, slow (44 kHz) signals from three or more CD players are combined or multiplexed into a single faster stream (40 GHz) of shorter pulses, which are sent through the fiber. At the receiver, the shorter pulses are separated back into separate slower streams and sent to the individual users (CD players in the illustration).

of the physics of light waves in glass. It is not because of the fast speed of light itself. As we mentioned above, light actually travels slower in glass fiber than radio waves travel in air. Rather, it is because light waves have a much higher (carrier) frequency, about 10^{15} Hz, than the frequency of radio waves, approximately 10^6 to 10^9 Hz. According to our discussion above, the duration of a wave pulse is given roughly by the inverse of its spectral bandwidth. And, according to the Principle of Carrier Modulation discussed in Chapter 8, Section 8.4, we cannot modulate a carrier wave at a frequency higher than the frequency of the carrier without destroying the identity of that carrier wave. For this reason, a light wave, whose carrier frequency is much higher than that of a radio wave, can be modulated much faster than can a radio wave. This allows far more data pulses to be placed each second onto the light wave than can be put onto a radio wave. We can summarize these arguments by the following practical principle:

Principle of Single-Channel Data Rate: Because we cannot modulate a carrier wave at a frequency higher than the frequency of the carrier without destroying the identity of that carrier, a higher rate of transmitting data on a single channel can be achieved only by using a higher carrier frequency.

15.3   MODULATING A LASER BEAM WITH DATA

To impress data bit values on a laser beam in the form of pulses, you could simply turn the laser ON and OFF rapidly by varying the electrical current being sent into the semiconductor junction in the laser. Earlier fiber optical systems used this method, but suffered from an upper limit to how fast the laser could be turned ON and OFF. Because it takes around 0.2 nsec for the electrons in the junction to emit all of their photons once the current is externally turned OFF, the upper limit for modulation speed is approximately 5 Gbits/sec when using this direct modulation method. Faster methods are needed.

In the previous section, we mentioned that a modulator may be used to impress the data bit values onto the laser beam, as in Figure 15.2. In this case, the laser is operated continuously, without being switched ON and OFF. A modulator is used as a shutter—at one moment it is ON, allowing the laser's light to pass through it, and at a later instant it is OFF, blocking the light beam. A modulator is like venetian blinds on a bedroom window, which can be set to pass or block sunlight. A very fast light modulator can be made using wave interference, which we discussed in Chapter 7, Sections 7.7 and 7.11. Recall that when two waves come together in a region of space, their displacements add. If both waves simultaneously are at a crest, these crests add constructively, creating large displacement. But, if one wave is at a crest while the other is at a trough, the waves combine destructively and cancel out, leaving a small or zero displacement.

Figure 15.4 shows an optical *wave-guide*, the basics of which were discussed in Chapter 13, Section 13.10. Recall that light can be guided inside a long narrow region of a medium by total internal reflection, which occurs if two conditions are satisfied: (1) the light travels inside a medium with a higher refractive index than the surrounding medium, and (2) the light travels at an angle to the "surface perpendicular" that is greater than the critical angle of the medium. The figure shows a solid slab of a special transparent crystal with a refraction index of 2.22, in which rectangular-shaped regions with higher refractive index (2.23) are embedded. The higher-index region is

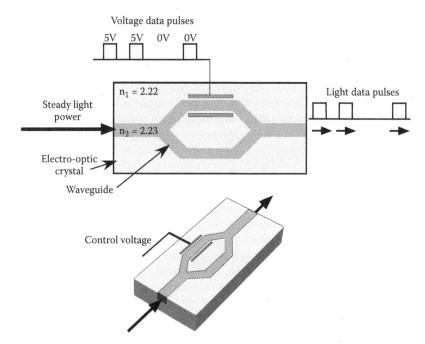

FIGURE 15.4   An optical wave-guide light modulator. Voltage pulses control transmission of laser light through the modulator, creating a series of optical data pulses.

a wave-guide. If light is sent into one end of such a wave-guide, it will travel within the wave-guide and emerge at the other end.

In the modulator, the wave-guide is split into two parallel wave-guides, splitting the light wave into two equal-power waves. After a short distance these two waves are recombined. Under normal conditions, because the two parallel wave-guides have equal lengths, the waves approaching the merging region are in phase, as shown in **Figure 15.5**. Therefore, they interfere constructively to create a strong light wave traveling in the original direction and exiting the device. On the other hand, if we could cause the refractive index in the upper wave-guide to become larger, the wave in it would slow a little, causing it to be out of phase relative to the other wave arriving at the merging region. This would cause destructive interference to occur at the merging region, creating a small light power exiting the device, as shown in the lower part of Figure 15.5. This figure shows the light scattering out of the sides of the device rather than exiting through the output wave-guide. This device is called a Mach-Zehnder interferometer.

The technique described can be accomplished using the properties of certain crystals, called electro-optic crystals. In such crystals, an example of which is lithium-niobate ($LiNbO_3$), the refractive index increases when a voltage is placed across it. This arises from the distortion of the crystal lattice by the electric force caused by the voltage. Therefore, if a voltage is applied to an electrode next to the upper wave-guide, the refractive index in the wave-guide will increase and the wave will slow, causing it to be out of phase relative to the other wave arriving at the merging region and thereby creating destructive interference.

Using this electro-optic effect to control the interference, we can use a series of data voltage pulses to create a series of corresponding light pulses, as shown in Figure 15.4. The electro-optic effect has the advantage of being very fast, with switching times much less than 1 nsec. Such fast switching is needed to modulate the laser beam at a rate of 40 GHz.

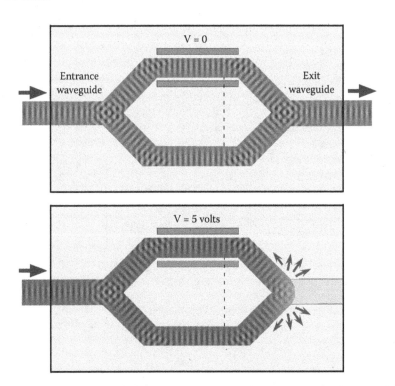

FIGURE 15.5 The operating principle of a Mach-Zehnder interferometer is illustrated. Light waves traveling in the entrance wave-guide are split into two parallel wave-guides and then interfere at the exit wave-guide. Depending on the voltage on the electrodes (0 or 5 V), the waves interfere either constructively or destructively. The dashed line shows in-phase waves (top) or out-of-phase waves (bottom).

15.4 WAVELENGTH MULTIPLEXING IN OPTICAL COMMUNICATION

You might think that the 40 Gbits/sec achievable using a single-fiber channel is such a huge data rate that there would be no need to go higher. That is not so. The economics of communication make it imperative to send as much data as fast as technically possible. So, engineers asked, what is the upper limit to the data rate in an optical-fiber system? To answer that query, let us recall from Chapter 8 that radio systems use a technique called frequency multiplexing to increase the amount of information (music or data) that can be transmitted through the air waves in a geographic region. In Chapter 8, Sections 8.4 and 8.6, we pointed out that each broadcast station is allocated a certain range of frequencies in the radio spectrum for transmitting its data. Each station uses a different value for its carrier frequency, which defines the *channel* it is using. This situation is illustrated in **Figure 15.6**, which shows many channels connecting the transmitter and receiver. Each channel is identified by a different carrier frequency. In radio, all of the channels can occupy the same air space. For example, in an AM radio system, about 25 channels typically coexist in the same air space, covering a combined bandwidth of about 1200 kHz (see Real-World Example 7.2). Theoretically, if each AM station used a bandwidth of only 20 kHz, there could be as many as 60 stations broadcasting at once, because $1{,}200 \div 20 = 60$. That is, the number of channels that can coexist in a transmission medium whose total bandwidth equals B_M is given by

$$number\ of\ channels\ =\ \frac{total\ bandwidth\ of\ medium}{bandwidth\ of\ individual\ channel}\ =\ \frac{B_M}{B}$$

In frequency-modulated (FM) radio systems, the total allocated frequency range spans a bandwidth of 20 MHz (from 88 to 108 MHz). As an example, if each station were allocated 60 kHz of bandwidth, then the number of different channels that could coexist would be

$$number\ of\ channels\ =\ \frac{B_M}{B}\ =\ \frac{20\ \mathrm{MHz}}{60\ \mathrm{kHz}}\ =\ \frac{20 \times 10^6}{60 \times 10^3}\ =\ 333$$

Given an allocated bandwidth, the data rate that can be achieved using that channel is limited. The maximum data rate for a station using a bandwidth equal to B (in Hz) is approximately B (in bits per second).[2] This is a consequence of the fact that larger bandwidth allows the creation of shorter pulses, as we discussed in the previous section. For example, an AM station with an allocated bandwidth of 20 kHz could transmit data at a rate equal to 40 kbit/sec, but no higher.

Fiber systems also benefit by using frequency multiplexing, also called **wavelength multiplexing**.[3] In a fiber system, many different frequency channels can occupy the

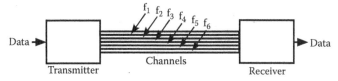

FIGURE 15.6 A frequency-multiplexed communication system transmits data over many channels simultaneously, with each channel identified by a unique carrier frequency, f_1, f_2, etc.

---

[2] The theoretical upper limit to data rate is $2B$ for a station using a channel for which the bandwidth equals B, as discussed in Chapter 8, Section 8.6.

[3] Called wavelength-division multiplexing (WDM) in the technical literature.

same fiber. How many optical channels could coexist in a fiber? An optical channel is identified by the frequency (color) of the laser light beam that is used as the carrier wave for that channel. The maximum total bandwidth that is practical for long-distance communication using an optical fiber is approximately 40 terahertz (THz; 1 THz = 10^{12}Hz). Recall the graph of the optical transmission factor versus wavelength in a fiber in Chapter 13, shown again here as **Figure 15.7**. The higher the transmission factor, the farther a light pulse can travel in the fiber before it becomes too weak to be detected. From the figure, we see that the transmission through 1 km of the best fiber is greater than 0.9 in the infrared wavelength range of 1,300–1,600 nm. The frequency corresponding to a wavelength of 1,300 nm is

$$f_1 = \frac{c}{\lambda_1} = \frac{3 \times 10^8 \, \text{m/sec}}{1300 \times 10^{-9} \, \text{m}} = 231 \times 10^{12} \, \text{Hz},$$

that is, 231 THz. The frequency corresponding to wavelength 1,600 nm is

$$f_2 = \frac{c}{\lambda_2} = \frac{3 \times 10^8 \, \text{m/sec}}{1600 \times 10^{-9} \, \text{m}} = 188 \times 10^{12} \, \text{Hz},$$

that is, 188 THz. The bandwidth B of this high-transmission range is the difference of the two frequencies, or

$$B = f_1 - f_2 = 231 \, \text{THz-}188 \, \text{THz} = 43 \, \text{THz}$$

This optical-fiber bandwidth is far larger than can be achieved using copper (coaxial) cable, which achieves a bandwidth of 300 MHz over a 10-km distance, or twisted-pair wire, which achieves a bandwidth of 3 MHz over a 1-km distance.

This total bandwidth in optical fiber, approximately 40 THz, is 20 million times larger than the total bandwidth for FM radio! From an economics point of view, we would like to use all of this bandwidth for sending data in an optical fiber. Doing this is not so easy and is not routinely done. Trying to use all of this bandwidth is like trying to drink water from a fire hose. Engineers are working to improve systems so that they can use all of the bandwidth. If they could do so, the number of channels that could be sent in a single fiber would be enormous. For example, say that each channel is

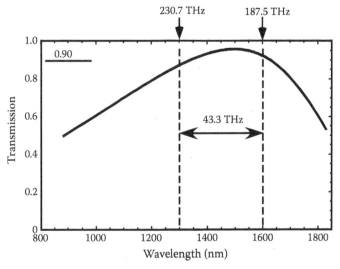

FIGURE 15.7   Optical transmission factor for 1 km of the best fiber-quality silica glass.

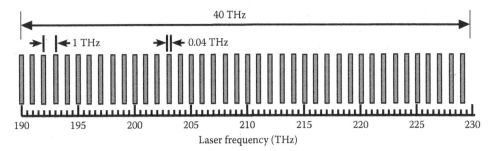

FIGURE 15.8   The spectral allocation for a frequency multiplexed optical fiber system. Forty channels, each with bandwidth 0.04 THz (40 GHz), are separated by 1 THz.

allocated a bandwidth of 40 GHz. Then 1,000 such channels could coexist in one fiber, as seen by the calculation:

$$number\ of\ channels\ =\ \frac{B_M}{B}\ =\ \frac{40\ \text{THz}}{40\ \text{GHz}}\ =\ \frac{40\times10^{12}}{40\times10^{9}}\ =\ 1,000$$

Recall that each separate channel with 40-GHz bandwidth can carry 1 million CD-quality streaming signals simultaneously. This would mean that, theoretically, on one fiber (a strand of glass just 0.2 mm thick) we could simultaneously carry 1 billion CD-quality streaming signals!

Figure 15.8 shows a more realistic frequency-multiplexing spectrum, in use commercially circa 2006. Within the 40 THz total bandwidth of the fiber, there are 40 separate channels, with carrier frequencies separated by 1 THz. The bandwidth of each channel is 0.04 THz (i.e., 40 GHz). The combined bandwidth of these 40 channels is 40×0.04 THz = 1.6 THz. This is only 4% of the bandwidth available in the fiber; the rest is wasted.

In 2007, state-of-the-art commercial systems used 160 individual channels, each with a bandwidth of 40 GHz in a fiber. The combined bandwidth of these 160 channels is 160×0.04 THz = 6.4 THz. This ideally allows a single fiber to simultaneously carry 160 million CD-quality streaming signals. However, it is not economical for companies to operate all of their fibers at these blindingly fast rates. Such high rates are used only for *long-haul channels*, connecting two major service-providing sites. In contrast, for delivering television signals to homes, slower fiber-optic connections having a single channel and data rates around 100 Mbit/sec are used.

15.5   THE VIRTUES OF LASERS FOR OPTICAL COMMUNICATION

Early optical-fiber communication systems used *light-emitted diodes (LEDs)* to create the light pulses for carrying data in fibers. Present-day systems use lasers instead. The unique properties of laser light, which make it best suited for use in fiber-optic systems, result from the properties of stimulated emission, which we discussed in Chapter 14. Recall the important properties of laser light:

■ *Monochromaticity*—A laser can be made to emit light of nearly a single color, or frequency.
■ *Directionality*—A laser emits light that travels in a tight (small-diameter) beam over long distances, without spreading out too rapidly.
■ *Coherence*—All of the wave fronts in a laser beam are traveling in nearly the same direction and oscillating together in a coordinated manner.

These have several useful consequences:

- Monochromaticity and controllability of the light's frequency are useful for wavelength multiplexing.
- Directionality (which is a consequence of the coherence) allows the light from a laser to be focused efficiently into a small, single-mode optical fiber. This allows practical data transmission over long distances.
- Coherence of laser light enables the use of techniques for multiplexing that are based on the interference of light.

15.6   HARDWARE FOR WAVELENGTH MULTIPLEXING

Here we discuss how to build a *wavelength division multiplexing (WDM)* system that can exploit the huge bandwidth provided by optical fiber. Here we use the terms wavelength multiplexing and frequency multiplexing interchangeably. Optical communication operates using a hybrid of electronic and light-based techniques. That is, computers—electronic devices—send electrical voltage pulses to a modulator to impress the data onto a light beam. The light beam travels through the fiber and is detected and processed at the far end to create an electrical signal. Let us design a hypothetical WDM system that would use *all* of the available 40 THz of fiber bandwidth. Say we want each separate channel to have a bandwidth of 100 MHz, large enough to handle most present needs. The number of such channels that could be fit (ideally) onto a single fiber would be

$$number\ of\ channels = \frac{B_M}{B} = \frac{40\ \text{THz}}{100\ \text{MHz}} = \frac{40 \times 10^{12}}{100 \times 10^{6}} = 400{,}000$$

Having this large number of channels would be great, but there is a problem. Building such a system using wavelength multiplexing would require using 400,000 lasers, each with a different frequency to provide all of the carrier waves needed. Because that is not practical, we use a combination of time multiplexing, as discussed earlier in this chapter, and frequency multiplexing. A system that does this is shown in **Figure 15.9**. This shows only 3 of up to 160 lasers that are used in a transmitter. Each laser emits

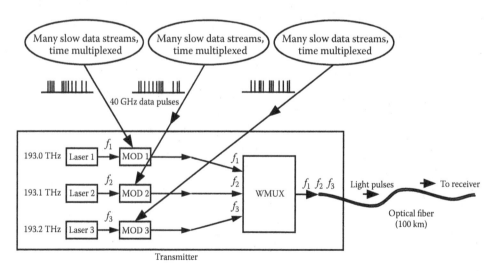

FIGURE 15.9   In a wavelength-multiplexing optical communication system, many high-speed data streams are impressed onto separate laser beams having different carrier frequencies, f_1, f_2, etc., separated by 0.1 THz. These laser beams are combined in a wavelength multiplexer (WMUX) and sent into a single fiber.

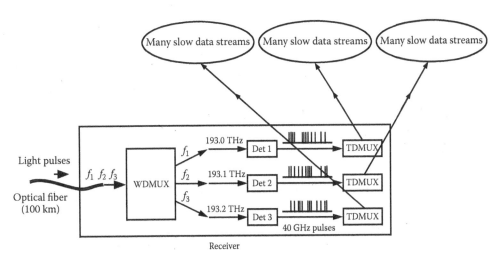

FIGURE 15.10   In a wavelength-multiplexing receiver, the different laser frequencies (wavelengths) are separated using an optical device called a wavelength demultiplexer (WDMUX) and received separately using separate photodetectors (Det). The 40-GHz stream from each detector is sent into a time demultiplexer (TDMUX), which undoes the interweaving of the pulses, and sends the results to many slower data processors, such as computers.

light with a different frequency, with a frequency difference between adjacent lasers of 100 GHz, or 0.1 THz. For example, the frequencies of the three lasers shown are 193.0, 193.1, and 193.2 THz. In terms of wavelength, these are 1,554.4, 1,553.6, and 1,552.8 nm. Semiconductor lasers having different, controlled frequencies of their output light can be made using methods discussed in Chapter 14, Section 14.7. The beams from these lasers are modulated by the 40-GHz data streams, which resulted from interweaving many slower data signals, as described earlier. Then the laser beams are combined in a device called a wavelength multiplexer (abbreviated WMUX or simply MUX) and sent together into a single fiber. At the receiving end of the fiber, the different laser frequencies are separated using an optical device called a wavelength demultiplexer (DMUX) and detected separately, as shown in **Figure 15.10**.

The WMUX and WDMUX devices can be made in various ways. A primitive concept is shown in **Figure 15.11.** In this design, which is not practical for actual systems, the different-colored light beams would be combined or separated using a simple glass prism. Recall from Chapter 13 that refraction causes light waves to bend their direction of travel when entering or leaving a denser medium, such as when going from

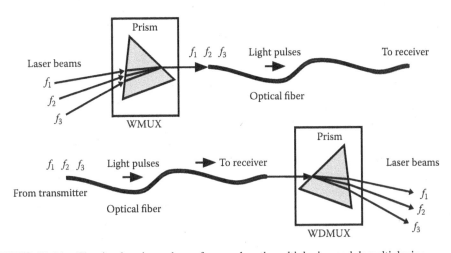

FIGURE 15.11   The simple prism scheme for wavelength multiplexing and demultiplexing.

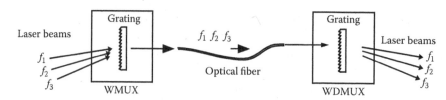

FIGURE 15.12   Using diffraction gratings for wavelength multiplexing and demultiplexing.

air to glass. Recall also (Chapter 13, Section 13.7) that blue light (higher frequency) bends more than red light (lower frequency). The spectral separating ability of a prism is limited. A WDMUX based on a prism would be effective at separating the laser beams only if the frequencies of the beams were different by at least 0.1 THz, as in the example presented here. To obtain greater separating ability, engineers use diffraction gratings or other devices based on optical interference, as shown in **Figure 15.12**. A diffraction grating is a piece of glass or plastic into which millions of straight, parallel grooves or lines have been cut or drawn. When a light beam passes through such a grooved glass plate, wave interference causes light waves of different wavelengths to deflect (diffract) into different directions.

15.7   LASER BEAM ROUTING

The schemes described above use fixed routing paths that are determined entirely by the wavelength (frequency) of each channel. But, in a flexible network we would like to be able to send any data pulse to any destination station. WDM is useful for combining channels for a long-haul connection (like putting many people into an airplane to fly across the Atlantic), but at the other end of the long haul, the packets need to be sent to specific end destinations (like the people after they deplane). We need a more powerful routing system than is provided by WDM.

A method for switching is based on *microelectrical mechanical systems* (*MEMS*). This scheme consists of building extremely tiny movable mirrors. A drawing of such a switching scheme is shown in **Figure 15.13**. A light beam emerges from the end of a fiber and passes into the entrance of another fiber located a distance of only 0.1 mm away. If the fiber ends are near enough to each other, no lens is needed to refocus the light. To activate the switch, a negative electric charge is placed onto two metal plates that make up the tiny mirror, shown as a rectangle and a dashed rectangle. By electrostatic repulsion (like charges repel), the top metal plate flips up suddenly (in approximately 10 msec) on its microhinges, reflecting the light beam into the end of a different nearby fiber.

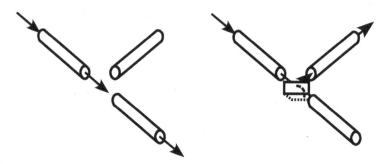

FIGURE 15.13   Laser beam routing using microelectrical mechanical systems (MEMS) occurs when a voltage is placed across a micromirror actuator, causing it to flip up and reflect the light beam into a different fiber.

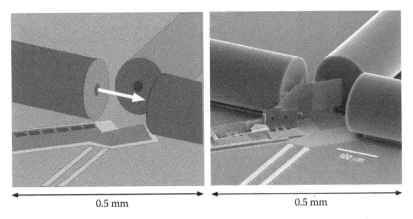

FIGURE 15.14 Left: Drawing of the micromirror (dark gray) resting horizontally in the down position. Right: Microphotograph of the actual mirror in the up position. (Courtesy of Alcatel-Lucent Bell Laboratories.)

Remarkably, the whole micromirror is only approximately 0.2 mm wide. The left panel of **Figure 15.14** shows an artist's close-up drawing of the micromirror situated in the down position, with an arrow showing the light beam path passing into one fiber. The right panel of this figure shows an actual microphotograph of the mirror in the up position, reflecting the beam into another fiber. It seems amazing that such tiny mechanical devices can be made to work, but work they do! In fact, they are the basis of certain types of digital movie projectors.

An important application of such MEMS switches is the ***add/drop multiplexer***. This makes long-haul fiber systems and short-haul fiber systems compatible. For example, there are many long-haul fibers installed underground going between San Francisco, California, and Seattle, Washington. Let us say that the government of a smaller city between San Francisco and Seattle wants to tap into one of these long-haul fibers. How can this be done? They could simply cut the fiber at their location and splice in a new fiber that feeds into their local-area network. But, such a permanent splice would reduce the flexibility of the network, which is one of the most important features of any powerful network.

An add/drop multiplexer will solve this problem. This is a switch with two input channels and two output channels, which can swap the connections between inputs and outputs. It can be built by installing a second input fiber in front of the micromirror in Figure 15.14 in the location marked by the two parallel guide rails, as in **Figure 15.15**. When the mirror is DOWN, both input beams travel across the empty gap into

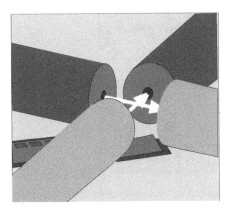

FIGURE 15.15 Two-by-two channel switch for implementing the add/drop operation in a long-haul fiber link.

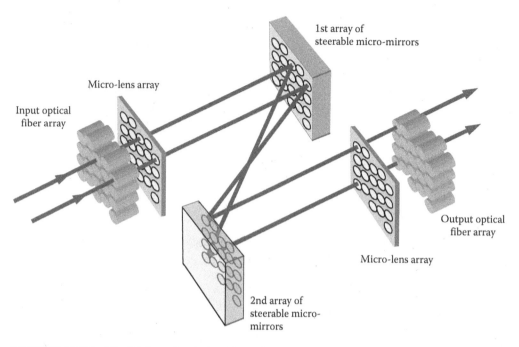

FIGURE 15.16 20×20 channel switch, which redirects the beam from any input fiber to any selected output fiber. The fibers are shown shortened for convenience.

the opposing output fibers. When the mirror is UP, the beam from each input fiber is reflected into the corresponding fiber at right angles.

Often we need the ability to switch the output of any one of many fibers to the input of any one of many fibers. This can be done using an $N \times N$ switch, illustrated in **Figure 15.16**. In this example, 20 input fibers are grouped together and their output beams are passed through an array of microlenses, the purpose of which is to image

FIGURE 15.17   Close-up photograph of an individual steerable micromirror from Lucent Technologies' LambdaRouter™, which contains 256 mirrors. The mirror is only approximately 0.2 mm across and is mounted on double-axis pivots, which are controlled electronically. (Courtesy of Alcatel-Lucent Bell Laboratories.)

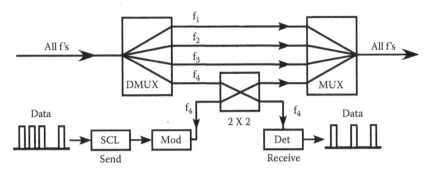

FIGURE 15.18   A frequency-multiplexing optical system, with 2×2 channel switch for implementing the add/drop operation. SCL, semiconductor laser.

the end of each fiber onto a particular steerable micromirror within an array of such mirrors. These mirrors are so small that 100 of them would fit onto the head of a pin! Each of the mirrors in the array can be independently tilted, directing the corresponding beam onto a selected micromirror within a second mirror array. Another microlens array images each mirror in the second mirror array onto a particular fiber in the output fiber array. By use of the steering mirrors, any input fiber may be connected to any output fiber.

Figure 15.17 is a close-up photograph of a single micromirror within an array of 256 mirrors, which together occupy an area less than 1 square inch. The material is entirely silicon and the fabrication process is based on photolithography and chemical etching—similar to that used for making electronic circuits for computer chips.

A summary of the above-discussed methods is shown in **Figure 15.18**, including lasers, data modulators, frequency multiplexing, and a 2×2 add/drop multiplexer. In this system, signals in four different-frequency channels travel in the same fiber, then enter a demultiplexer where they are separated into four distinct fibers. The add-drop multiplexer allows data in the f_4 channel to be removed from the fiber and detected or received, while adding new data in the f_4 channel.

SUMMARY AND LOOK FORWARD

You should now have an appreciation of how optical fibers are used to provide the extremely high data rates of modern communication systems. Two main points contribute to this: (1) because fiber optics use light, the frequency of which is very high (around 10^{14} Hz), it is possible to modulate the carrier wave at extremely high rates (theoretically up around 10^{14} bits/sec); (2) rather than trying to modulate directly at such high rates, wavelength-division multiplexing (WDM) is used, in which light signals from many lasers with different carrier frequencies are combined into a single light beam traveling in a single fiber.

For a local communication network, such as between two nearby college campuses where only moderately high data rates are needed, WDM is not necessary. Instead, a single laser is used together with time multiplexing, in which electrical data pulses carrying bits from different senders are interweaved in time before being sent to the laser-beam modulator. The modulated laser beam is then transmitted between campuses and detected; then the electrical data pulses are separated into distinct signals and sent to different computers.

In fact, if the two campuses are very near each other, there may be little or no advantage to using a laser and fiber optics. In this case, the electrical signals can be sent directly using metal wire. On the other hand, optical fiber is better than metal wire in two regards: (1) fiber can carry shorter data pulses than can wire (because it has higher bandwidth), allowing more pulses per second to be transmitted; and (2) light pulses in fiber can travel longer distances than voltage pulses can travel in wire before becoming too weak to be detected. Therefore, fiber is used rather than wire if one needs extremely high data rates or transmission over very long distance, or both. An example of this sharing of duties is the system called Gigabit Ethernet, which uses fiber for distances longer than 100 m and copper wire (called 1000 Base-T) for distances less than 100 m. This system achieves 10^9 bits per second, or 1 Gb/sec.

We discussed how to make a high-speed modulator using light interference in an optical wave-guide. We also discussed how to route optical beams from one fiber to a different fiber using micromechanical mirrors. In the next chapter we will discuss how these facts and devices influence the design of communication networks, including the Internet.

SUGGESTED READING

Introductory texts on optical communications:
Kartalopoulos, Stamatious V. *Introduction to WDM Technology*. New York: IEEE Press, 2000.
Rogers, Alan. Understanding Optical Fiber Communications. Boston, Artech House, 2001.

A readable, but more advanced, text:
Optical Switching, edited by Tarek S. El-Bawab. New York: Springer, 2006.

KEY TERMS

Add/drop multiplexer
Analog
Bandwidth
Bit
Byte
Carrier wave
Channel
Coherence
Demultiplexer (DEMUX)
Digital
Directionality
Electromagnetic
Frequency multiplexing
Information
Lasers
Light-emitting diodes (LEDs)
Microelectrical mechanical systems (MEMS)
Mode dispersion
Monochromaticity
Multiplexer (MUX)
Protocol
Receiver
Refraction
Refractive index
Sampling
Time multiplexing
Total internal reflection

Transmitter
Wave-guide
Wavelength multiplexing
Wavelength-division multiplexing (WDM)

EXERCISES AND PROBLEMS

Exercises

E15.1 Consider the tin-can-and-string system shown in Figure 15.1, assuming that it is used in the conventional manner: one person speaks into a can and the other person places a can near their ear.

(a) Identify the transmitter, channel, and receiver in this system and explain the physics of each.
(b) If you tried to use this system for communication over longer and longer distances, why and how would the system fail? (Ignore such mundane matters as the string breaking.)
(c) Explain how you could use the can system for digital communication. Would there be any advantages?

E15.2 (a) Explain the three factors that limit a basic fiber-optical system with a laser, a modulator, and a receiver (without multiplexing) to a maximum data rate of approximately 40×10^9 pulses/sec.

(b) A music CD holds approximately 800 MB of data. How much time, in seconds, would it take to transmit the data on one CD across such a system?

E15.3 Light travels 1.5 times slower in glass fiber than radio waves travel in air. Explain why, nevertheless, a light-in-fiber system can transmit far more bits per second than can a radio-wave system.

E15.4 Develop and explain an analogy for the idea of data multiplexing by considering people who write and send letters within a single town, the letters themselves, a home mailbox, a postal carrier and his or her truck, a town's post office, etc.

E15.5 Develop an analogy for the idea of time multiplexing by considering 20 high school chess players who play 20 simultaneous chess games against one chess master.

E15.6 Consider an alternative method to create a modulated light beam for optical communication using light interference from two narrow slits, as discussed in Chapter 7, Section 7.11. The setup is shown below: a light wave passes through two slits in an opaque screen and forms an interference pattern on a second screen, which has a single slit in it. A "voltage-controlled phase shifter" is inserted in front of the upper slit. The phase shifter is made of a certain transparent crystal with the property that if you apply 10 V to it, its refractive index (n) increases in value. This slows the wave that passes through the upper slit, causing it to become out of phase with the wave passing through the lower slit.

(a) Explain how this leads to modulation of the output beam.
(b) What disadvantage would this method have, compared to the wave-guide method shown in the chapter? *Hint*: Consider the power of the transmitted beam.

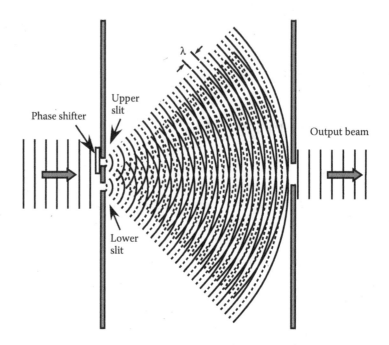

E15.7 Photodetectors, which are used as receivers for light pulses, cannot respond to a light pulse if it is too brief or is changing too rapidly in time. To simulate the idea of detector response time, consider an analogy. Ask a friend to vertically dangle a ruler or yardstick between your opposing thumb and finger, then drop it suddenly. If you can catch it, then you have a fast response. If it drops faster than your hand is capable of responding, then you cannot catch it.

(a) Try different distances from your fingers to the end of the ruler, and record at which distances you are able to catch and which you are not.
(b) Optional: Use this method to measure your response time. *Hint*: The distance that an object travels in time t after being dropped in Earth's gravity is $D = (1/2)a \cdot t^2 = 4.9 \times t^2$, where t has units of seconds (see In-Depth Look 3.1). The constant $a = 9.8$ m/sec^2 is the acceleration caused by gravity.

E15.8 An idealized design for a light modulator is shown below. We know that the angle through which a beam of light is refracted on entering a transparent crystal depends on the crystal's refractive index n. We know (Section 15.3) that we can increase the refractive index of an electro-optic crystal by applying voltage to it. In the figure as drawn, the switch is open and no voltage is applied, and the beam passes from point A to point B. Assume that when the switch is closed, the refractive index of the crystal increases enough that the light beam now hits the mirror. Make a careful drawing showing the path the light beam takes in this case.

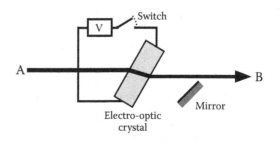

Problems

P15.1 As we studied in Chapter 13, Section 13.9, many optical fibers commonly used for communications have a transmission factor for 1 km of fiber as shown below. The most useful range of wavelengths of light is shown as the shaded region. Light in this spectral region travels through this fiber with transmission equal to 0.90 or greater.

(a) Calculate the bandwidth B of this region of the spectrum. Give the answer in hertz and in terahertz. *Hint*: Calculate the frequencies corresponding to the highest and lowest wavelengths in this band.
(b) Roughly how many bits of information could be sent per second using this band?
(c) If many laser frequencies are transmitted in this band, and adjacent laser frequencies are separated by 100 GHz (or 0.1 THz) how many laser channels could fit into this band?

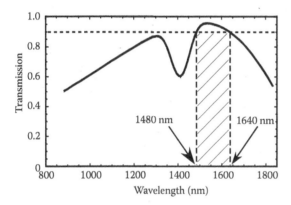

P15.2 As discussed in Section 15.3, when a voltage is placed across certain types of crystals, the refractive index of the crystal increases in value. Consider using a prism made of this type of crystal and a voltage source to steer light beams of different colors into different, controllable directions. Elaborate and describe such a system, starting with the drawing here.

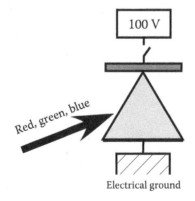

16

Communication Networks and the Internet

You can get any reliability you want—far, far greater than the reliability of the components.

Paul Baran
(2001)

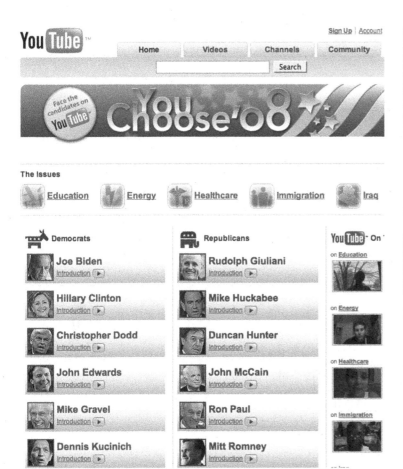

Paul Baran in the 1960s. He was one of the pioneers who invented the idea of packet switching, which is key to the operation of the Internet. (Courtesy of RAND Corporation.)

The Internet, as we see it through our computer screens, is a window to the world. For example, 2007 was the first year in which presidential debates were carried out online. (YouTube, 2007. Courtesy of Google Inc.)

16.1   THE PHYSICS BEHIND THE INTERNET

In this last chapter, we discuss how communication networks work. Of course, networks are constructed using physical materials and devices, according to the principles of physics. Before addressing the physical implementation of networks, we will discuss how computer networks are meant to function, independent of any particular physical construction. We will compare **analog** and **digital** techniques and different types of network designs. We will consider the robustness of networks in the presence of noise and breakage of various components. A most important concept is **packet switching**, which is the basis for most data transmission on the Internet. We will also discuss how wireless cell phone communications work.

This chapter is written more or less independently of the rest of the text, so that it can be read at any time during the course, including at the beginning if you wish. Some topics discussed in other parts of the text are briefly repeated here to make the present chapter self-contained (or for review, if you have already read most of the text). References are made to specific sections of the book for more detail on some of the concepts discussed.

16.2   THE GOALS OF COMPUTER COMMUNICATION NETWORKS

What is a computer network and what does it do? Computer scientists define computer networks as:

- A computer communication **network** is an interconnected collection of computers or other devices that are autonomous.
- Interconnected means they can exchange information according to some agreed upon protocol.
- A **protocol** is a set of rules for data format, voltage levels, error handling, and timing.

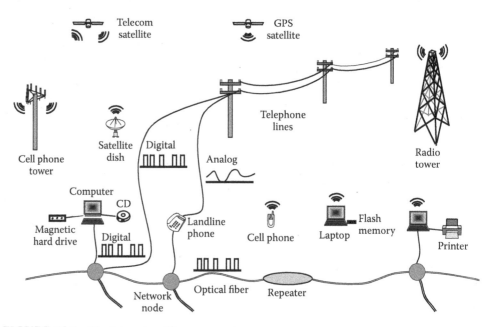

FIGURE 16.1   The Internet and its components.

- *Autonomous* means that each computer or device controls itself according to an internal set of rules.
- An *internetwork* is a network made by connecting two or more distinct networks (which might use different protocols.)
- The *Internet* is the global internetwork.

Computer networks are designed and built to achieve two main goals. The first is to make all programs, data, and other information resources available at any location. This can be accomplished if the system is set up to treat data, voice, audio, and video in the same manner, and to allow error-free data transfer. The second main goal is to provide high reliability for message transmission. Reliability is not the same as the complete absence of errors in transmitting data. It means that the network does not completely "go down," becoming inoperable, nor does it simply lose messages or fail to deliver them.

Figure 16.1 shows examples of the various types of hardware that are interconnected in the Internet (also shown in Figure 1.1). In this chapter, we will see how these elements work together to provide a communication system that looks more or less "transparent" to the end users.

16.3   NOISE IN ANALOG AND DIGITAL SYSTEMS

The goal of treating data, voice, audio, and video in the same manner was not achieved in the previous, old-style communications systems. In old-style systems, voice is transmitted in telephone lines, and video is delivered by cable. The wires used have different physical properties, suited to the different types of signals they carry. Video needs a much higher rate of information transfer than does voice-only transmission. As we saw in Chapter 8, Sections 8.4 through 8.6, a higher rate of information transfer is provided by using a medium or channel with a higher *bandwidth*. Economics plays a large role in this, too. It would be wasteful to send just voice through a dedicated cable line.

We need to ask, "To achieve the goals stated above, should we use an analog or a digital network?" Chapters 2 and 8 discuss the principles behind the two main types of communication systems—analog and digital. Here we briefly review the main points, keeping in mind the issues that are important for understanding networks. Simply put, analog means "continuously varying analogously to the input," whereas digital means "varying discretely or discontinuously." For example, the shape of the grooves in a vinyl music disk creates an analogy of the original sound wave. In contrast, the small discrete pits in the surface of a compact disc (CD) create a representation of the music based on (binary) numbers (see Chapter 8, Section 8.3).

An example of an analog network is the traditional telephone system (pre-1980 or so). It operated using the same method as the public address (PA) system described in Chapter 8, Section 8.2. As shown in **Figure 16.2**, a microphone converts sound waves from a person's voice into voltage variations, which are first amplified (increased) using

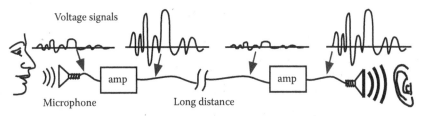

FIGURE 16.2   An analog telephone system consisting of a microphone, a long wire, two amplifiers, and a speaker.

an electronic amplifier and then transmitted on a long copper wire to another location, where it is converted back into sound. If the wire is longer than a few kilometers, then the voltage arriving at the receiving end will be much smaller than at the starting point. This decreasing of the signal strength is called attenuation, or *loss*, and is illustrated in the figure. Recall that loss, or a decrease of voltage level, is caused by the ***electrical resistance*** in the wire, as discussed in Chapter 5, Section 5.7. To summarize:

Signal transmission over long distances suffers from loss (decrease of signal strength).

As a rule of thumb, the power in an analog electrical signal in wire decreases by a transmission factor of approximately 0.1 for every kilometer traveled. This rule is valid for analog signals with frequency around 3 megahertz (MHz) in coaxial cable. After traveling 3 kilometers (km) in such a cable, the power decreases by a factor $0.1 \times 0.1 \times 0.1 = 0.001 = 10^{-3}$.

The loss of signal can be overcome by adding amplifiers to increase voltage, as also shown in Figure 16.2. Such amplifiers need to be placed in the transmission line every few kilometers or so to avoid the signal becoming too small to detect. Installing such amplifiers in undersea transmission lines is expensive and problematic.

During the transmission of a signal along a metal wire, ***noise*** corrupts the signal and distorts it. By noise, we mean unwanted, random voltage variations. A good way to hear such noise is to turn the volume on a stereo or guitar amplifier to maximum, with no signal being sent into the amplifier. The hissing sound that you will hear is caused by amplifier noise. Noise occurs in long wires in part because the signal loss process is random; you can never predict exactly how much loss will occur, and it can change rapidly with time and other factors. Furthermore, there can be external sources of noise, such as nearby lines (wires) or outside radio waves (from broadcasts or other sources) that can enter the wire and change the voltage inside of it unpredictably.

A fundamental property of amplifiers is that they always add noise to the amplified signal—there is simply no way to avoid this fact, according to our best understanding of physics. The source of the amplifier noise is in the random "jiggling" motion of electrons in the amplifier circuit, which is caused by the energy associated with heat. This idea is worth emphasizing by stating the following rule:

PRINCIPLE OF COMMUNICATION (I)

All signal amplifiers add noise.

Furthermore, the amplifier at the receiver end amplifies not only the wanted signal, but also amplifies the unwanted noise that was added during transmission.

An example of noise corrupting an analog signal is seen in **Figure 16.3**. The initial signal in Figure 16.3a is a voltage that represents a voice or music signal. After traveling a long distance in a wire, the signal strength (voltage level) is decreased, as shown in Figure 16.3b. After this signal passes through an amplifier, its voltage is restored to original levels, but now noise has been added to it, as seen by the jaggedness of the signal in Figure 16.3c. In this case, the amplified signal is distorted only slightly, making it still useful. In contrast, if the signal is being sent over such a large distance that several amplifiers are needed, the noise may become so large, as illustrated in Figure 16.3d, that the signal becomes useless. The result is that super-long telephone systems are quite noisy, meaning it is hard to hear and understand the person speaking at the other end.

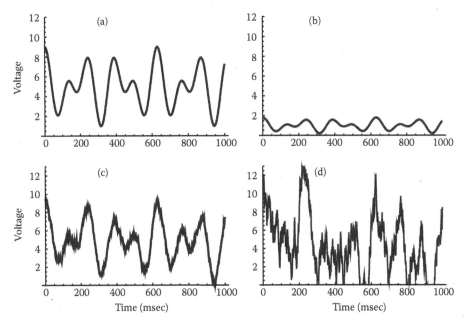

FIGURE 16.3   An analog signal consists of voltage oscillations that represent sound. A signal is sent with the original voltages as shown in (a). After the signal travels some distance in a wire, it suffers loss, and the voltages are decreased, as shown in (b). An amplifier then boosts the signal back to its original voltage levels, but noise is added, as in (c). If a large amount of noise is added, the result will be as in (d), which bears little similarity to the original signal.

As discussed in Chapter 8, digital communication systems operate on very different principles from analog systems. Before being transmitted along the wire, the signal is converted into a digital form, consisting of discrete high or low voltage values, as in **Figure 16.4**. These values represent the *zeros* and *ones* of a binary number. This conversion is done using an electronic circuit in a computer called an analog-to-digital converter (or A-to-D converter). All of the *one* bits start out their journey along the wire represented by well-specified large voltage values (e.g., 10 V). All of the *zero* bits start out with well-specified small values (e.g., 0–0.1 V). During their journey to the other end of the long transmission wire, the voltage pulses suffer loss and added noise. Along the journey, there might be electronic amplifiers, which increase the voltages to overcome the loss. These amplifiers will add noise. Finally, these digital pulses are measured at the receiving end of the wire, converted back into analog voltage form, and used by the receiver for some purpose such as driving an audio speaker to produce music.

Here we find a great difference between analog and digital systems, which has significant implications for designing communication networks. Because of noise in trans-

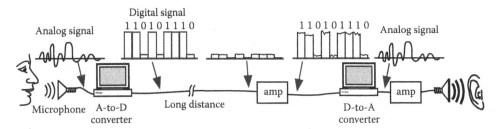

FIGURE 16.4   In a digital-electronic communication system, analog signals are converted into digital signals, consisting of discrete, high or low voltage levels that are transmitted along a wire. After loss occurs, amplifiers boost the voltages back to original levels, which are then converted back into analog form for use.

mission and amplification, the voltages received will be slightly different from the original values, as shown in Figure 16.4, but for a digital system this does not degrade or distort the information that is sent. A moderate amount of noise has little or no impact on a digital communication system. A similar conclusion was discussed for digital memory in Chapter 11, Section 11.8.

This immunity to noise is a consequence of a technique called ***thresholding***, which largely avoids the effects of noise when receiving voltage pulses. An example showing the immunity of digital communication to noise is seen in **Figure 16.5**. The original digital signal with bit values = (1, 1, 0, 0, 1, 0, 1, 0, 1) is shown in Figure 16.5a. After being transmitted over a long wire, the voltage levels are decreased, as in Figure 16.5b. Amplification boosts the voltages back up to near-original levels, but noise is added, as in Figure 16.5c. The thresholding technique involves assigning a voltage value, say 5 volts (V), as the ***threshold*** value. An electronic circuit (called a voltage comparator) is used to perform thresholding. Any detected pulse that is above 5 V is interpreted by the circuit as originally having been a 10 V pulse and is interpreted as a one. In contrast, any pulse that is detected below 5 V is interpreted as originally having been a 0 V pulse, and is interpreted as a zero. If the noise is not too large, the original signal can be restored perfectly to its original condition, as in Figure 16.5d. This capability is not possible in analog systems.

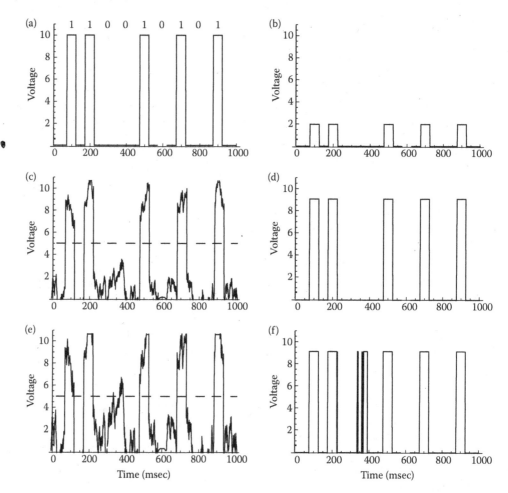

FIGURE 16.5 A digital signal consisting of nine bits is sent out with original voltages as shown in (a). After the signal travels a long distance in a wire, it suffers loss, and the voltages are decreased, as shown in (b). Amplification returns the signal to its original voltage level, but with some noise added, as in (c). The dashed line at 5 V is the threshold. Thresholding cleans up the signal and produces the result shown in (d), which is identical to the original signal. If the noise is too large, as in (e), then thresholding, as shown in (f), fails to recreate the original signal.

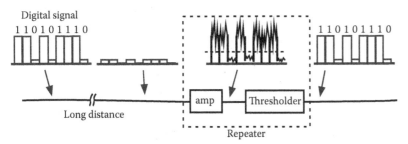

FIGURE 16.6   A digital repeater, within the dashed line, receives and amplifies a digital signal. The signal is restored in the thresholder to its original form, and a new, corrected signal is sent out.

In contrast, Figure 16.5e shows a signal that experienced very high loss and was restored using very large amplification followed by thresholding. In this case, the resulting noise is so great that one of the pulses that was originally low (representing zeros) now has voltages that go above the threshold value of 5 V. Then the comparator circuit interprets these pulses as a one, thereby making an error.

   If the distance between the sender and receiver is much larger than a few kilometers (e.g., 2,000 mi under an ocean), then simply amplifying the signal every few kilometers is not sufficient to avoid it being swamped by noise. Because every amplifier adds its own noise, which is amplified by subsequent amplifiers, the noise on the signal accumulates and becomes too large. In such a case, *digital repeaters* must be placed into the line every few kilometers or so. A ***digital repeater*** is an electronic device that receives and amplifies a digital signal, detects it using thresholding, and retransmits a fresh, corrected digital signal. A schematic of such a repeater is shown in **Figure 16.6**. Unlike in the case of analog repeaters, there is no limit to how many digital repeaters can be used, one after the other, to relay a signal across a very long distance without introducing intolerable noise. We can state these conclusions as a broad principle.

PRINCIPLE OF COMMUNICATION (II)

In digital communication systems, voltage thresholding can lead to error-free operation, which is not attainable with analog techniques.

   The two principles stated above apply for any type of communication technology. For example, in fiber-optics communication (Chapter 15), we use light pulses from lasers to represent zeros and ones. Because the light power decreases after traveling a long distance in a fiber, light amplifiers are used to boost it back up, as discussed in Chapter 14. This type of amplification also creates noise, or randomness in the light signals, in accordance with Principle (i). This noise results from the light emitted spontaneously and randomly by atoms in the amplifier, as discussed in Chapter 12, Section 12.5. Corrective thresholding is used to restore the optical signals to their original values.

16.4   CHALLENGES IN NETWORKING

Communication in computer networks takes place over various distances. **Table 16.1** gives examples of typical distances over which data need to be transmitted in different situations. Each distance range uses a different strategy for networking. When engineers think about building real communication networks, two challenges arise:

1. How can we connect a very large number of computers or other devices that are nearby?
2. How can we connect two or more computers that are very far apart?

TABLE 16.1

Typical Communication Distances

Interprocessor Distance	Processors in Same:	Example
0.1 m	Circuit board	Computer
1 m	System	
10 m	Room	
100 m	Building	Local network
1 km	Campus	
10 km	City	
100 km	Country	Long-haul network
1,000 km	Continent	
10,000 km	Planet	Interconnected long-haul networks

Source: Adapted from Tanenbaum, A.S. *Computer Networks.* New York: Prentice-Hall, 2003.

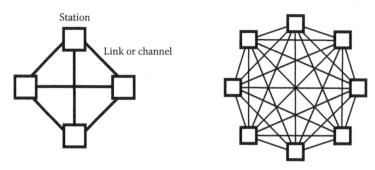

FIGURE 16.7   Fully connected networks. Each station is connected to each of the other stations by a separate link or channel through which data travel.

Figure 16.7 schematically shows two communication networks. Each box represents a *station*, meaning a device such as a computer or telephone that is used by a person for sending and receiving data. Each line in the figure represents a communication link, meaning a physical medium that can carry information. A link may be a metal wire, an optical fiber, or radio or television waves. (Note that a link may contain one or several channels—for example, a red light beam can coexist with a green light beam in a single fiber, and each color may transmit information on a different channel.) In the example shown, called a fully connected network, a separate wire (or other type of link) is used to connect every pair of stations. For a network with a very large number of stations, this fully connected arrangement becomes impractical.[1]

For connecting a very large number of stations, a better way is to use a network containing fewer links, which connect devices called *nodes*. A **node** is a computer whose job is to transfer data into and out of a network and around the network. In **Figure 16.8**, each station is connected to a node (represented by a circle), and the nodes are connected together by links to form a network.

To connect computers that are far apart, we use a long-haul link, which can be a long, dedicated wire or fiber that connects two nodes, as in **Figure 16.9**. The long-haul link would typically contain amplifiers or repeaters along its length to overcome loss and noise, as discussed in the previous section. It is common to connect smaller networks, called local-area networks, at the ends of a long-haul link, as in **Figure 16.10**.

---

[1] A simple counting exercise shows that to connect a number (equal to N) of stations together, using a separate wire for each pair, requires a number of wires equal to $(1/2)N(N - 1)$.

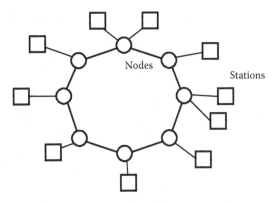

FIGURE 16.8   Stations (squares) and nodes (circles) form a network.

FIGURE 16.9   Long-haul link.

FIGURE 16.10   Local-area networks connected by a long-haul link.

16.5   BROADCASTING NETWORKS AND SWITCHING NETWORKS

How does a network having no central control keep track of which two stations are "talking" to each other? To do this, two primary schemes are used: broadcasting networks and switching networks.

In a ***broadcasting network***, all stations share a single physical medium or link for transmission. Each sender transmits a signal into the medium, and any receiver can choose to listen to it or not. The sender usually includes some identifying information in the broadcast to inform the receivers of its origin. Examples are:

- Satellite broadcasting network—The medium is comprised of radio waves traveling in the atmosphere or in space, as in **Figure 16.11**. The satellite receives from one station and rebroadcasts to all. Each station decides whether it wishes to listen.
- Cable broadcasting network—The medium is a metal wire, or cable, as in **Figure 16.12**. Each station can broadcast onto the cable and each station can listen to any message on the cable. An example is cable television.

In a ***switching network***, many separate transmission mediums are used, and each node contains physical circuit switches that are used to route messages. A circuit switch is a device that can connect an input link to one of several possible output links. A simple example is an electronic switch constructed of a movable metal bar that can be

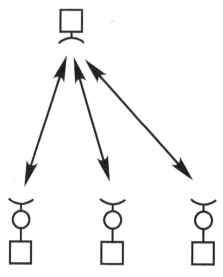

FIGURE 16.11   Satellite network.

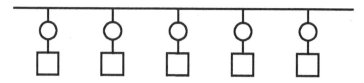

FIGURE 16.12   Cable network.

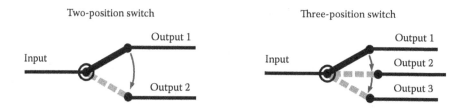

FIGURE 16.13   Circuit switches.

moved to bridge the space between an input wire and one of several possible output wires, as in **Figure 16.13**. In practice, switches are made from silicon semiconductor circuits.

There are two types of switching networks, circuit-switching networks and packet-switching networks.

16.5.1   Circuit-Switching Networks

In a *circuit-switching network*, switches connect a single, dedicated, continuous path between two stations through the nodes of the network. The path is formed by a sequence of connected links, as in **Figure 16.14**. The two stations are connected for the entire duration of the message transmission. Other stations are not connected through this dedicated link during this transmission. Circuit-switching networks can be either analog or digital.

In an analog circuit-switching network, as discussed in Section 16.3, over long distances losses and amplifier noise degrade the signal quality. This limits the allowable distance for reliable transmission. This might not be a problem for simple voice data, but it would be a big problem when transmitting, for example, financial records between

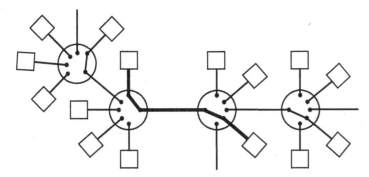

FIGURE 16.14 Circuit-switching network. Each node contains a switch that can connect various links or channels attached to that node.

banks. An example of an analog circuit-switching network is the old-style telephone system, before 1988. Properties of analog circuit-switching networks are:

1. The signal is sent through each node without delay.
2. A channel being used to send and receive data between two stations cannot be used by other stations. The other stations cannot use that channel until it becomes available. This can create network congestion.
3. Switching and amplification add noise to the signal at each node. This leads to the rule of thumb:

Ten-node rule: Because of the noise added by each switch and by each amplifier, in an analog circuit-switching network the maximum number of nodes that an analog signal can pass through before being badly degraded is approximately ten.

A schematic of the U.S. telephone circuit-switching network, prior to about 1988, is shown in **Figure 16.15**. It consisted of about 200 million telephones, served by a five-level hierarchy made up of five classes of switching centers (nodes). The ten regional centers were connected to one another through large trunk lines (groups of wires) capable of enormous data rates. The regional centers communicated through somewhat smaller lines to 67 sectional centers, which connected to 230 primary centers. The primary center served 1,300 toll centers, which served 19,000 end offices. Individual household and business telephones were hooked directly to the end offices.

Notice that to pass from a telephone station up to a regional center, then to another regional center, and finally down to a distant telephone station, required passing through

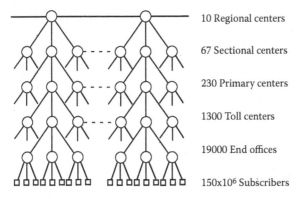

10 Regional centers

67 Sectional centers

230 Primary centers

1300 Toll centers

19000 End offices

150×10^6 Subscribers

FIGURE 16.15 U.S. telephone system, circa 1988.

about ten nodes. Thus, the ten-node rule was not violated. However, it is clear that if a trunk line broke, then large portions of the network would be cut off from one another. To overcome this, all regional centers were fully connected, as in Figure 16.7, requiring 45 trunk lines.[2] In addition, a small number of redundant (additional) lines were installed between the sectional centers. If a regional trunk line broke, some data could be routed around the break at the regional or sectional levels. Further network redundancy was installed at lower levels of the hierarchy as well.

In a digital circuit-switching network a dedicated connection is made through a series of channels between two stations. An example is the telephone system in 2004. In this type of network the data are sent using digital techniques. This does not avoid the problem of network congestion, but it does largely eliminate errors in the transmitted data. This is important when transmitting sensitive information or when much larger distances must be covered. The properties of digital circuit-switching networks are:

1. The signal is sent through each node without delay.
2. A channel being used by two stations cannot be used by other stations.
3. Switching and amplification add noise to the signal at each node.
4. Digital thresholding is used at each node to clean up the signal and prevent errors. This leads to an important property:

A digital circuit-switching network is not limited by the ten-node rule, because noise in such a system does not build up with distance.

16.5.2 Packet-Switching Networks (e.g., the Internet)

A *packet-switching* network differs from a circuit-switching network in how channels are used. No dedicated continuous channel or link is used between two stations. Instead, each digital data stream is broken into small chunks (or packets, as named by Donald Davies). A *packet* is a group of about 1,000 bits that travel in the network together. Each packet is passed from node to node, using a channel that is dedicated to this connection only for the duration of one packet's transmission between the two nodes. At every node, a packet is received, stored briefly (causing a time delay), and then, when a needed channel becomes available, the packet is sent through it to the next node. Packets are labeled with sender, destination, and packet-order information, and each packet may travel along a different route in the network. The packets may arrive at the destination node in any order. Upon arrival of all packets, they are automatically reassembled before being sent to the receiving station. This is illustrated in **Figure 16.16**. For example, a message is made up of 5,000 data pulses, which are divided into five packets. Each packet of 1,000 data pulses is denoted by a letter, A, B, C, etc. Routes for each packet are determined "on the fly" by the nodes, depending on local traffic and channel conditions.

A packet-switching network overcomes the problems of switching noise (because it is digital) and traffic congestion (because it uses packet switching). The Internet is the prime example of a packet-switching network.

Prior to around 2003, the Internet carried data, including e-mail and music files, but did not commonly carry voice or telephone signals. After 2003, sending telephone signals by using the Internet became popular. This is referred to as "voice over Internet Protocol," or "voice over IP" (most commonly referred to in its shortest form as VOIP). When you are speaking to a friend over IP, your voice is converted to digital (binary) numbers,

---

[2] $(1/2) N(N - 1) = (1/2)10(10 - 1) = 45$

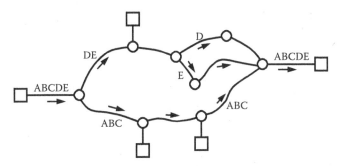

FIGURE 16.16 In a packet-switching network, a message is broken into multiple packets (A, B, C, D, E, etc.), that may travel along different routes before being reassembled at the final node.

which are separated into packets, each of which may zip around the Internet by different paths before being reassembled at a node near your friend, then converted back into your voice at your friend's computer. That is remarkable!

16.6 FAILURE-RESISTANT COMMUNICATIONS

Because packet switching is at the heart of the Internet's operation, it is interesting to learn how and why it came to be. The concept was discovered more than once by different scientists at different times, including in 1965 by Donald Davies, a researcher at the British National Physical Laboratory. The person who thought through the implications of packet switching first was Paul Baran. In the early 1960s, Baran worked at Rand Corporation, a nonprofit institution in the United States studying technical questions relating to national defense. A huge problem in the early days of nuclear weapons (and perhaps still a concern today) was "how to build a reliable command and control system." There were two goals: to avoid an accidental launch of nuclear missiles by U.S. forces; and to have a communication system that could survive, at least partly, the devastating effects of any nuclear attack by any adversary (the Soviet Union in those days). Both of these goals could be met only by a network that had very low rates of error for data transmission and a very low likelihood of breaking down and preventing communication between points on the network.

Before the advent of digital communications, Baran realized that for a message to have high reliability for delivery, it needs to be robustly "routable." This requires that switches be able to pass on data without degradation. The telephone system in the 1960s was analog and had five levels of hierarchy, as described above. After passing through ten switches, or nodes, an analog signal was degraded badly. "Analog is like making a videotape of a videotape of a videotape: The quality goes to pot," Baran said. [1] He realized that to have high reliability the system must be digital. A digital signal can go through any number of nodes without being degraded, as we discussed above. Digital packet switching:

1. Uses thresholding, which keeps error rates in digital signals extremely low.
2. Uses logical error-correction schemes, which allow one to correct residual errors (e.g., send data three times and compare results).
3. Allows transmitting messages with essentially no errors, which permits passing a packet through as many nodes as you like.
4. Provides high resistance to breakdown (see below).
5. Relieves network congestion.

A network that is robust against breakdown has many uses far beyond military ones, for example, medical. So, we should not think that the only or main reason that the first packet-switching network was built was purely for military reasons. Baran did not get to build his packet-switching network that he conceived, in part because the military initially did not believe that it would work, and possibly because the only U.S. commercial phone company at the time, AT&T, did not want to make its own analog system obsolete. In 1967 Donald Davies, a scientist working at Britain's National Physical Laboratory, independently helped convince U.S. military scientists to organize the building of the first packet-switching network—the ARPANET, which initially connected U.C. Los Angeles, U.C. Santa Barbara, Stanford Research Institute, and the University of Utah. Baran served as an unofficial consultant during the building of this network, which came online in 1969. By 1973 the network had 37 nodes, and an intercontinental link was installed between the United States and the United Kingdom. Baran later went on in his work to invent the precursors to wireless communication and DSL (digital subscriber line).

Let us study Baran's main research on packet switching in the early 1960s while at Rand. Baran emphasized the idea of *level of redundancy* in a network, which is related to the number of distinct pathways that can be used to connect any two nodes on the network. A network having one level of redundancy has just the minimum number of channels needed to connect all nodes, as in **Figure 16.17a**. A network having two levels of redundancy has about twice the number of connections needed for one level of redundancy, as in Figure 16.17b. A network having three levels of redundancy has about three times the number of connections needed for one level of redundancy, as in **Figure 16.18**. Using computer simulations, Baran showed the following important and surprising result:

PRINCIPLE OF COMMUNICATION (III)

A network with three levels of redundancy can tolerate 50–70% destruction of its nodes and still have a high probability of any two surviving nodes being able to communicate. Somewhat higher levels of redundancy can create extremely robust, survivable systems.

This general idea is illustrated by the example shown in **Figures 16.19** and **16.20**. A network containing 100 nodes is constructed with 3 levels of redundancy. If 50 or fewer of these nodes are destroyed at random, it is highly likely that all surviving nodes will be in full communication. If, instead, 60 of these nodes are destroyed at random, it is possible that all surviving nodes will still be in full communication, but it is more likely that the network will be broken up into smaller groups of communicating nodes.

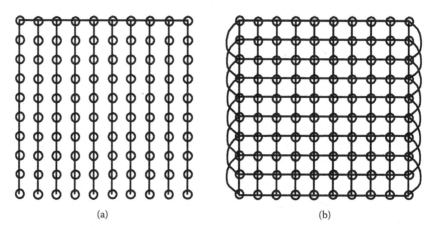

(a)   (b)

FIGURE 16.17   (a) A network with one level of redundancy. For 100 nodes, 99 channels are sufficient to connect every node. (b) A network with two levels of redundancy, 99 + 97 channels.

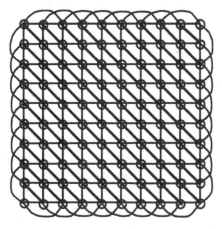

FIGURE 16.18 A network with three levels of redundancy, 99 + 97 + 97 channels.

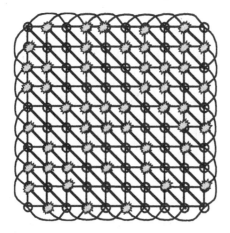

FIGURE 16.19 The result of destroying 50 randomly selected nodes. In this typical example, all surviving nodes remain in contact.

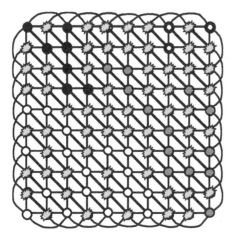

FIGURE 16.20 The result of destroying 60 randomly selected nodes. In this typical example, the surviving nodes are broken into four communicating groups (labeled using different shades or symbols), which are cut off from each other.

This example shows that a highly redundant network can withstand a loss of many of its nodes and still function from point to point with a high probability. Baran recognized that such a network could not be built using analog transmission because the noise in these networks would limit the number of nodes that a packet could pass through (the ten-node rule). In digital networks, the reliability comes from the ability of data to pass through a very large number of nodes without experiencing degradation. Only having this capability allows the use of high levels of network redundancy, and the high reliability that comes with high redundancy.

Even in the absence of catastrophic failures of the network, it is vital to have this type of robustness. Hardware can malfunction; human error can wreak havoc (think of Homer Simpson), etc. The packet-switching scheme allows us to build a reliable system with unreliable components. As Baran said in 2001, "You can get any reliability you want—far, far greater than the reliability of the components." In his early design, he decided "to have a half-second delay maximum from one point to another. Then at every one of these nodes we would keep a carbon copy of the message until we were sure it got through intact to the next node. So the carbon copy with error detection allowed repeated transmission of any message block, now called a packet." Baran still marvels at the advantages of this system that he uncovered using his computer simulations, "Packet switching had all these wonderful properties that weren't invented—they were discovered."

16.7   WIRELESS MOBILE CELL PHONE NETWORKS

One of the biggest impacts of digital communications on people's lives was the recent rapid rise of cell phone use. Cell, or cellular, phone systems are an example of wireless networks, and are especially interesting in that they are also mobile—you can carry the phone with you. The main physics idea behind cellular phone systems is the fact that the strength of radio waves falls off with increasing distance. Although this may be a disadvantage if you want to communicate directly by radio over a long distance, it has a great benefit: separate stations broadcasting radio waves will not interfere with each other if they are far enough apart. This is illustrated in **Figure 16.21**. The broadcasting station labeled S1 sends out radio waves that are easily picked up by people at locations labeled A and B. These waves decrease in strength with increasing distance, so that people at C, D, or E cannot detect them. Likewise, the station S2 sends out waves that can be detected at D and E, but not at A, B, or C. The advantage of this situation is that each broadcasting station can use the same set of frequencies to broadcast to a local group of listeners, without interfering with other groups of listeners. In this simple example, which avoids interference between stations S1 and S2, a person at C is not able to communicate. This problem can be remedied by using a slightly more involved strategy, described next.

The implementation of a cell phone system is illustrated in **Figure 16.22**. It is based on the idea that each broadcasting station has a limited range. The countryside or a city is separated into small geographical areas, called cells, represented in the figure by hexagons. Each *cell* contains one broadcasting and receiving station, called a base

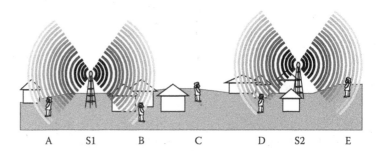

FIGURE 16.21   Radio wave propagation covering a small region around each broadcasting station (cell phone tower). This creates isolated regions within which cell phone users can communicate only with the nearest tower.

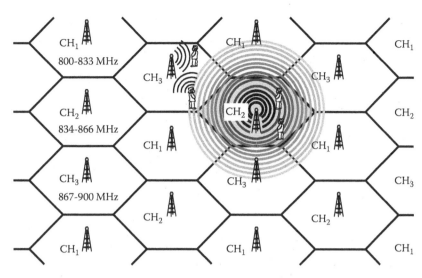

FIGURE 16.22 A large region is divided into regions called cells, within which a specific range of frequencies is used for radio communication. No two adjacent cells use the same frequency range.

station, which can be one of three types, labeled CH_1, CH_2, and CH_3. Each cell is surrounded by six cells, which are of different types from its own type. Each type uses a different range of frequencies or channels for broadcasting radio waves. For example, CH_1-type stations could use channels in the range 800–833 MHz; CH_2-type stations could use channels in the range 834–866 MHz; and CH_3-type stations could use channels in the range 867–900 MHz, as shown in **Figure 16.23**. Because these frequency ranges are not overlapping, this allows some overlapping of the geographical areas being reached by each station. For example, in the figure, one station (a CH_2-type) is shown emitting waves that partly leak into the nearest adjacent cells, but no farther. Every telephone communicating with this station is using frequencies in the CH_2-type frequency range of 834–866 MHz. The key point is that each cell adjacent to this broadcasting cell is either CH_1-type or CH_3-type. Therefore no phone users in those cells are receiving radio waves in the CH_2-type frequency range. They are using the other frequency ranges, CH_1-type or CH_3-type. We conclude that by using three separate broadcasting frequency ranges, we can cover the countryside with adjacent but not interfering cells, in which users can receive and send signals with their own local station without suffering interference from other nearby stations.

Every base station in a given area is connected by high-speed data links to a Mobile Telephone Switching Office, or MTSO, as shown in **Figure 16.24**. These links (cables or satellite channels) allow a base station to simultaneously handle many phone connections. Each MTSO is connected to both the Public Telephone Network for voice,

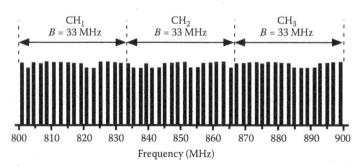

FIGURE 16.23 An example showing the full spectral region used for cell phone communication (800–900 MHz) broken into three separate regions, only one of which is active within a given cell.

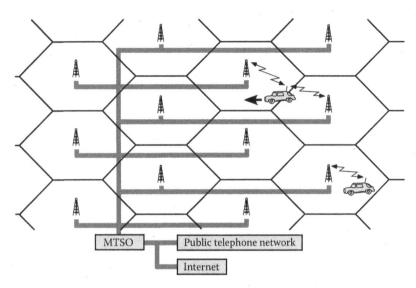

FIGURE 16.24 Every base station within a certain area is connected by high-speed links to a Mobile Telephone Switching Office, or MTSO, which is connected to the public phone network and to the Internet. When a user passes from one cell to another, the MTSO instructs one base station to hand off the user's connection to the next base station.

and the Internet for other data. The MTSO controls the operation of each base station in its area. An important operation is the handing-off of any phone connection between base stations when a moving phone user crosses a boundary between cells. This gives rise to the name *Mobile Phone System*. As the user moves between cells, her phone must quickly reconfigure itself to start using a different frequency range, for example from CH_2-type to CH_3-type. This explains why calls are sometimes dropped when a user is traveling. If she enters a new cell in which all frequency channels are already being used, the MTSO has no choice but to order that call be disconnected.

This provides an example of an internetwork, which is a network comprised of different types of networks connected together. The mobile phone system, the public telephone network, and the Internet are entirely separate networks and operate using different hardware types and protocols for handling data. Yet to a user they behave as a single communication network.

How many phone users can be accommodated within a single cell at any given time? This depends on the particulars of each system, so we can discuss only a representative example. Say that each single-user channel has a bandwidth (the width of the user's spectral region) equal to 100 kHz. For example, one channel might occupy a narrow portion between 80.0 and 80.1 MHz of the spectrum shown in Figure 16.23. If the base station in that cell can handle broadcasting frequencies between 800 and 833 MHz, then the number of single-user channels that can be accommodated equals the ratio of the total frequency range to the frequency range of a single channel:

$$\frac{833 \text{ MHz-800 MHz}}{100 \text{ kHz}} = \frac{33 \text{ MHz}}{100 \text{ kHz}} = \frac{33 \times 10^6 \text{Hz}}{100 \times 10^3 \text{Hz}} = 330$$

This relatively small number explains why a densely populated city needs many cell-phone towers per square mile.

In practice, the manner in which the spectral ranges are allocated to different users is far more complicated than described above. For example, the frequency being used by a single user might change in time according to some set pattern. Or, the band-width being used by a user might change in time according to what the user is doing at the moment—just talking requires little bandwidth, whereas downloading an MP3 file requires larger bandwidth. The bandwidth detected by a user might be much larger than the 100 kHz needed, but the user's phone may reject most of the data it receives and select the correct data needed by that user. These topics are discussed in the advanced suggested reading at the chapter's end.

16.8   PROPAGATION OF WIRELESS WAVES IN TERRAIN

As an example of using physics theory in cell phone systems, consider the complicated problem of designing the optimum shapes of the cells in a particular city. In the absence of any obstructions, radio signals tend to travel in straight lines from their source. When radio signals strike a building wall, they are partially reflected and partially absorbed. The reflected part is useful, because it may carry the signal farther by, for example, reflecting many times from building fronts as it travels down a narrow street. Also, radio signals, being waves, experience *diffraction*, or bending around sharp corners. (Think of an ocean wave being bent around a jetty.)

In open space, radio-wave strength decreases with increasing distance as the wave spreads out. In a city, you might have noticed that the strength of your cell phone connection can vary during a call. This decrease of signal is called fading, and is caused by several phenomena, including shadowing, reflection, guiding, diffraction, and wave interference. These are illustrated in **Figure 16.25**. Shadowing is simple—a large building blocks the line of sight between you and the cell tower, decreasing the signal. Shadowing is the direct loss of signal that occurs when a wave encounters a solid object. Reflection from a hard, solid object redirects a wave in a different direction. Guiding occurs when a wave traveling between two reflecting surfaces is guided a long distance without losing strength. Diffraction occurs at edges of hard objects, bending the wave around it. As you drive through the city, these effects are interacting and are changing with time, causing fading.

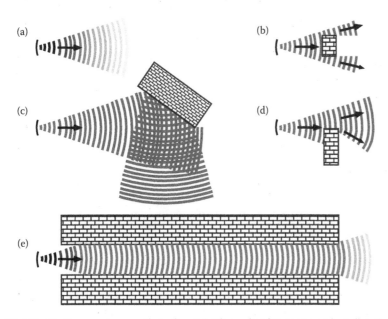

FIGURE 16.25   Radio wave propagation, shown both as circular waves and as directional arrows, called rays. (a) In open space, (b) shadowing, (c) reflection, (d) diffraction, and (e) guiding.

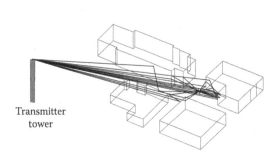

Transmitter
tower

FIGURE 16.26 Computer simulation of radio wave propagation around buildings in downtown Eugene, Oregon. (Courtesy of EDX Wireless LLC, Eugene, OR.) Aerial photo of the same buildings. (Courtesy of Lane Council of Governments.)

Reflections, shadowing, guiding, and diffraction all affect radio transmission in complex environments such as cities. The radio signal arrives at the receiver by multiple paths and at different times. Fading can result from the waves taking different paths and interfering destructively at the location of your cell phone. To model these effects, powerful computer programs are used to simulate hundreds or thousands of radio signal paths, keeping track of the time of arrival of waves at a certain location after traveling along each path. An example of such a simulation is shown in **Figure 16.26**. Here the directions of the radio-wave paths are indicated as straight lines, or rays, rather than as circular waves as in the above figure.

For example, a radio wave can be guided down a long, narrow street by sequential reflections from the buildings lining the street, as seen in the computer simulation in **Figure 16.27**. The lighter areas show the regions where the signal received from the

FIGURE 16.27 Computer simulation of radio wave propagation in Paris, France. The transmitting antenna's location is indicated by the arrow at the upper left, on the Avenue des Champs-Élysées. Lighter areas show where the signal from the antenna is predicted to be stronger. (Courtesy of AWE Communications, Germany.)

FIGURE 16.28 Computer simulation of radio wave propagation in a city, including elevation of the terrain, as well as reflection, diffraction, and attenuation by buildings, trees, and parks. Lighter areas show where the signal is predicted to be stronger. (Courtesy of AWE Communications, Germany.)

antenna is predicted to be stronger. The waves are strongly guided along streets, and diffraction causes them to bend around corners and penetrate into side streets. Trees and parks cause attenuation or loss of radio waves, as the wave energy is absorbed or scattered by the small variations is the shapes of leaves, etc. Remarkably, computer simulations can handle all of these effects, while modeling the elevation variations of the terrain in and around a city. An example of such a simulation is shown in **Figure 16.28**. Such modeling is useful for helping to determine how many cell phone towers are needed and where to locate them. It should be noted, however, that simulations over a large area of a city may not be truly accurate, because the complexity of wave propagation in complex environments pushes computer modeling beyond its practical limits. Therefore, wireless engineers rely on a combination of modeling, physical measurements in the actual environment, and intuition based on experience.

The physics of wave propagation imposes trade-offs for best performance when using wireless technology to send data. Two essential goals are: (1) reaching every user within a complex environment, such as a city; and (2) providing the highest possible rate of sending data. These two goals work against each other because of the physics of wave diffraction. In earlier cell phone systems, relatively low frequencies around 800 MHz were used. These waves have a relatively long wavelength: $\lambda = c/f = (3 \times 10^8 \text{ m/sec})/(800 \times 10^6/\text{sec}) = 0.375$ m, about the distance from your elbow to the tip of your fingers. Radio waves of this wavelength easily bend around corners by diffraction, thereby reaching all of the users in a cell. Modern cell phone systems use higher wave frequencies to provide higher bandwidth and therefore higher rates of sending data. Popular frequencies are 2.4, 3.5, and 5.8 GHz. This allows users to send and receive packet data and thereby communicate directly with the entire Internet through, for example, a hand-held web browser device. The disadvantage of using a higher frequency is that it corresponds to a shorter wavelength. For example, at 5.8 GHz the wavelength is $\lambda = c/f = (3 \times 10^8 \text{ m/sec})/(5.8 \times 10^9/\text{sec}) = 0.052$ m, about the length of your little finger. Such short-wavelength radio waves do not easily bend around corners, creating shadowing and the inability of users in the shadows to connect to the network. That is, a user would need to be able to see the cell tower with her own eyes for her phone to communicate with it. This makes sense from a physics point of view, because the light that she can see with her eyes is simply another form of electromagnetic wave (as is radio), but with an even higher frequency, approximately 500,000 GHz (500 THz). We know from experience that visible light does not bend much around corners—rather, it makes sharp shadows.

SUMMARY AND LOOK FORWARD

Computer networks are designed to achieve two main goals—to make all programs, data, and other information resources available at any location; and to provide high reliability for message transmission. Analog and digital networks operate on different principles, but both require amplification of signals for long-distance communication. The "price" we pay for using amplification is that noise is added to the signal, distorting it. Digital techniques, based on thresholding, can overcome this added noise.

There are two main types of networks: broadcasting and switching. In a broadcast network all stations share a single channel.

In switching networks, many channels are used, and nodes perform switching of connections between channels. Switching networks come in two types: circuit switching and packet switching, each of which has advantages and disadvantages. In a circuit-switching network, each channel is dedicated temporarily to a single connection, and signals are sent through nodes without delay. Circuit-switching networks are further divided into two kinds:

1. Analog circuit switching
   Advantages: simple technology; fast, when working
   Disadvantages: noise added at each node; ten node rule; not robust to node failure; network congestion
2. Digital circuit switching
   Advantages: noise is removed; no ten node rule; high redundancy is possible; robust to node failure
   Disadvantages: network congestion

Digital techniques have the great advantage that, because of the absence of data errors, data can pass through any number of switches without degrading. This allows a high level of redundancy to be built into the network, bringing extremely high resistance to breakdown of connectedness in case some of the links become damaged or inoperable.

In packet-switching networks, which are digital only, data are broken into small packets, which are sent separately. Packets are delayed at nodes and may take different routes before being reassembled. Advantages of packet-switching networks are less network congestion; noise is removed; no ten-node rule; high redundancy is possible; and robustness to node failure. Disadvantages include that it is complex and expensive.

Wireless cell phone networks operate by combining broadcasting with clever use of regional separation and frequency-spectrum separation. This allows full coverage as users move between different areas (cells). Sophisticated physics and engineering are behind your ability to talk to friends and family while on the move.

REFERENCES

1. "Founding Father, an Interview with Paul Baran," *Wired Magazine*, March 2001, 145–153.

SUGGESTED READING

For everyone:
"Founding Father, an Interview with Paul Baran," *Wired*, March 2001, 144–153.
Naughton, John. *A Brief History of the Future*. Woodstock: Overlook Press, 1999.

For advanced students and instructors:
Baran, Paul. "On Distributed Communications Networks," *IEEE Transactions on Communications Systems* CS-12 (1964) 1–9. http://www.rand.org/publications/RM/RM3420.

Goldsmith, Andrea. *Wireless Communications.* New York: Cambridge University Press, 2005. Chapters 1 and 15.

Stallings, William. *Data and Computer Communications,* 2nd ed., New York: Macmillan, 1988.

Tanenbaum, Andrew S. *Computer Networks.* Englewood Cliffs, NJ: Prentice-Hall, 1981. 4. Chapters 1 and 7.

KEY TERMS

Analog
Autonomous
Bandwidth
Broadcasting network
Cell
Channel
Circuit switching
Digital
Digital repeater
Electrical resistance
Internet
Internetwork
Level of redundancy (network)
Link
Loss
Network
Node
Noise
Packet
Packet switching
Protocol
Station
Switching network
Threshold
Thresholding

EXERCISES AND PROBLEMS

Exercises

E16.1 Imagine you are an entrepreneur whose goal is to set up a new radio music station. Give arguments in favor of using the Internet to distribute the signal, rather than using traditional analog radio-wave broadcasting. Next, take the viewpoint of a critic who argues the reverse case.

E16.2 Describe the physical factors that fundamentally limit the useful range of metal-wire-based communication systems in the case of no repeaters (intermediary amplifiers) and of using repeaters.

E16.3 A digital voltage signal is shown below:

The protocol associated with this signal is given by the following set of rules.

(a) The threshold is 5 V. Above 5 V the bit value is considered a "1" and below 5 V the signal is considered a "0".

(b) The signal is to be read starting at time t = 0.

(c) The signal pulses are each 1 msec in duration.
(d) There is no time between adjacent signal pulses.

Determine the list of binary digits (bits) associated with this signal.

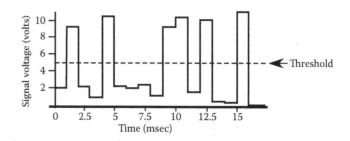

E16.4 The two main types of networks are broadcasting and switching networks. Give at least three real-world examples of each type of network. Describe at least two major differences between the two types of networks.

E16.5 The "old telephone system" is an example of an analog circuit-switching network. Consider a phone system in which your voice is no longer represented by an analog signal, but rather a digital signal. This is an example of a digital circuit-switching network. As is characteristic of all circuit-switching networks, there must be a single, continuous communication channel dedicated to data transmission, and for two people to carry on a conversation, we would also want there to be essentially no delay in the signal at the nodes in the network.

  (a) Think of at least two advantages this type of network has over its analog version.
  (b) Which phone network is more reliable and why?
  (c) What disadvantages does a digital circuit-switching network have over a digital packet-switching network?

E16.6 (a) Consider a cellular phone system that uses equal-area square cells in a checkerboard pattern, as shown. If two cells come together at a boundary so they touch along a line, or at a corner so they touch at only a point, they are considered "adjacent." What is the minimum number of different-type base stations needed to avoid any two adjacent cells using the same station type?
  (b) Think of a pattern of square cells that requires fewer distinct types of base stations than in the case of the checkerboard pattern.

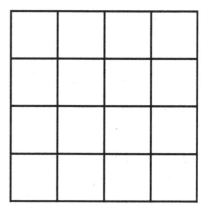

E16.7 Consider a cellular system that uses the irregular cells shown below rather than the hexagonal cells illustrated in the chapter. If two cells come together at a boundary so they touch along a line, or at a corner so they touch at only a point, they are considered "adjacent." What is the minimum number of different-type base stations needed to avoid any two adjacent cells using the same station type? *Hint*: Make several copies and try to "color" the map using only three different colors. Then try it with four, five, etc.

E16.8 (a) Consider a city occupying an area 10×10 km (6.2×6.2 mi), divided into square cells of 1 km^2 each. If each cell can accommodate 330 mobile phone users, how many users in total can talk at the same time throughout the whole city?

(b) If you drive straight through this city at a speed 30 km/hr, how long does it take you to travel through one cell? Assume you travel the shortest distance through each cell.

E16.9 Radio waves refract or bend and tend to follow the curvature of the Earth. This bending is caused by the fact that the speed of a radio wave is slower in air with higher density than in lower-density air. The density of air steadily decreases with increasing elevation above the Earth's surface. Explain, using carefully drawn pictures, how this air density behavior causes the radio "horizon" (reach of radio waves from a source near the ground) to be beyond the horizon that you can see with your eyes.

E16.10 Write a mini-essay about why and how digital, packet-switching networks solve some practical problems that circuit-switching networks suffer from. That is, what are the advantages of packet switching and how do they come about?

Problems

P16.1 The two main types of wire used for transmitting electronic signals are twisted-wire pair and coaxial cable, pictured here.

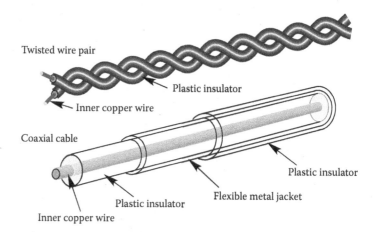

In twisted pair, the power in an electrical signal with frequency around 1 MHz decreases by a factor of 10^{-5} for every kilometer traveled. In coaxial cable, the power in an electrical signal with frequency around 1 MHz decreases by a factor of about 0.63 for every kilometer traveled. If a 1-MHz signal with 1 W of power enters a wire or cable, calculate the power remaining in these cases:

(a) After traveling 1 km in coaxial cable. (Answer: 0.63 W)
(b) After traveling 1 km in twisted pair.
(c) After traveling 3 km in coaxial cable.
(d) After traveling 3 km in twisted pair.
(e) After traveling 100 km in coaxial cable.

P16.2 (a) A digital communication system uses voltage of 10 V to represent a bit value of 1, and 0 V to represent a bit value of 0. In this system a 10-V pulse travels on a long cable, after which its value drops to 0.001 V. Then it is amplified by a factor of 10^4, bringing it back up near 10 V, but in the process noise is added. You can simulate the noise this way: toss a coin ten times and list the results. For each heads you toss, add 2 V of noise, and for each tails you toss, subtract 2 V of noise. What is your resulting voltage? Assuming a threshold of 5 V for distinguishing *zero* bits from *one* bit, does the noise cause the bit value to change?

(b) Repeat the exercise in part (a).

(c) Once more, repeat the exercise in part (a). Comment.

P16.3 (a) In a network containing seven nodes, determine, by making a diagram and counting links, exactly how many links are needed to make a fully connected network.

(b) In a network containing seven nodes, how many links are needed to make a non-fully connected network with one level of redundancy?

(c) In a network containing seven nodes, how many links are needed to make a non-fully connected network with three levels of redundancy?

(d) In a fully connected network containing 100 nodes, calculate how many links are needed to make a fully connected network, using the formula in footnote 1 on p. 526.

P16.4 The power of a radio wave in empty space decreases proportionally to the square of the distance from the emitting antenna. For every doubling of the distance, the power decreases by a factor of four. This follows simply from geometry. Consider a small satellite emitting a power of 1,000 W in Earth orbit, where there are no obstructions blocking the waves. Imagine a sphere with a radius of 2 m, centered on the satellite. The power striking the inside of the sphere is uniformly distributed on the inner surface of the sphere, which has an area equal to 50.3 m^2. This is calculated with the formula for the surface area of a sphere, $4\pi R^2$, where R is the sphere's radius. The power per square meter striking any small portion of the sphere's surface is (1,000 W)/50.3 m^2 = 19.9 W/m^2.

(a) Consider a sphere with radius 1 km centered on the satellite. Find the power per square meter striking any small portion of this sphere's surface.
(b) What is the ratio of the power per square meter in the case of the larger sphere relative to the smaller sphere?
(c) Reconcile your answer in (b) with the rule "the strength of a radio wave in empty space falls off proportionally to the square of the distance from the emitting antenna."

P16.5 (a) Make a few photocopies of the network of 100 nodes interconnected with three levels of redundancy in Figure 16.18. Use the following procedure (or some other unbiased method) to "randomly" destroy 60 nodes: Make a vertical list of 60 digits. Each digit is chosen randomly from 1, 2, 3, 4, 5, 6, 7, 8, 9, 10. (You can toss a pair of dice, or invent some other way to make random numbers.) To the right of that list, make another different list of 60 random digits. Now you have 60 pairs of digits. If you find duplicate pairs in your list, cross those out and replace by new numbers until you have 60 unique pairs. Each pair identifies one node to destroy—the first digit in a pair labels the row (counting horizontally) and the second labels the column (counting vertically) of a node to destroy. After destruction of 60 nodes, color-code the groups of still-interconnected nodes. Use a different color for each group. Make a list of the number of nodes in each group. This is like the simulation that Paul Baran did in the early 1960s when he first discovered the robustness of networks.
(b) Now consider the same network as in part (a), and add one more level of redundancy. You can do this by adding diagonal connecting lines in the other direction, plus 16 additional connections along the edges. Make a separate figure showing this. Again destroy the same 60 nodes as you did above, color-code and count interconnected nodes.
(c) Optional—for case (a) above, what is the average number of interconnected nodes per interconnected group? Answer the same for case (b).

P16.6 For a network containing N nodes, it takes (N/2)(N−1) wires to directly connect every node to every other node. In the 100-node network you worked with in P16.5, this would require 50 × 99, or 4,950, wires to make all pair-wise connections. Compare this to the number of wires needed to achieve the three levels of redundancy used in P16.5. Comment on the efficiency of the two methods for achieving a certain level of robustness against node failure.

P16.7 (a) Explain how a digital packet-switching network can outperform a digital circuit-switching network.

(b) Explain how a digital circuit-switching network can outperform an analog circuit-switching network.

Summary of Scientific Notation, Units, and Physical Constants

I. SCIENTIFIC NOTATION

Positive Powers-of-Ten Notation

1	$= 10^0$	one
10	$= 10^1$	ten
100	$= 10^2$	ten-squared
1,000	$= 10^3$	ten-cubed
10,000	$= 10^4$	ten to the fourth power
100,000	$= 10^5$	ten to the fifth power
1 (and n zeros)	$= 10^n$	ten to the nth power

Negative Powers-of-Ten Notation

0.1	$= 10^{-1}$	one-tenth
0.01	$= 10^{-2}$	one-hundredth
0.001	$= 10^{-3}$	one-thousandth
0.0001	$= 10^{-4}$	ten to the minus 4 power
0.00001	$= 10^{-5}$	ten to the minus 5 power
0.(n - 1 zeros)1	$= 10^{-n}$	ten to the minus n power

To multiply, add the exponents:

$$100,000 \times 1,000 = 10^5 \times 10^3 = 10^{5+3} = 10^8 = 10,00,00,000$$

To divide, subtract the exponents:

$$\frac{100,000}{1,000} = \frac{10^5}{10^3} = 10^{5-3} = 10^2 = 100$$

If the result is less than one, the exponent is negative:

$$\frac{1,000}{100,000} = \frac{10^3}{10^5} = 10^{3-5} = 10^{-2} = 0.01$$

If the numbers have prefactors, break up the numbers and operate:

$$200,000 \times 6,000 = (2 \times 100,000) \times (6 \times 1,000)$$
$$= 2 \times 6 \times 10^8 = 12 \times 10^8 = 1.2 \times 10^9$$

II. METRIC SYSTEM AND UNITS

Metric Prefix Abbreviations

10^{12}	10^9	10^6	10^3	10^{-2}	10^{-3}	10^{-6}	10^{-9}	10^{-12}	10^{-15}	10^{-18}
T	G	M	k	c	m	μ	n	p	f	a
tera	giga	mega	kilo	centi	milli	micro	nano	pico	femto	atto

Metric Units
Length: meter (m); time: second (sec or s); mass: kilogram (kg).

Units for Length, Position, or Distance
We measure length, position, or distance in the metric units of, for example, millimeters (mm), meters (m), or kilometers (km). The equivalences are:

Length		
1 kilometer	= 1 km	= 10^3 m = 0.621 mile
1 meter	= 1 m	= 39.37 inch (about 1 yard)
1 centimeter	= 1 cm	= 0.01 m = (1/100) m = 10^{-2} m = 0.394 inch
1 millimeter	= 1 mm	= 0.001 m = 10^{-3} m (thickness of a dime)
1 micrometer	= 1 μm	= 10^{-6} m (size of bacterium)
1 nanometer	= 1 nm	= 10^{-9} m (size of large molecule)

Units for Time
We measure time in units of seconds, abbreviated sec or simply s.

Time		
1 second	= 1 sec	(about one heartbeat)
1 millisecond	= 1 msec	= 0.001 sec = 10^{-3} sec (flap of housefly wing)
1 microsecond	= 1 μsec	= 10^{-6} sec (high-speed strobe light flash)
1 nanosecond	= 1 nsec	= 10^{-9} sec (time for light to travel 1 foot)
1 picosecond	= 1 psec	= 10^{-12} sec (time for one vibration of a molecule)

Units for Mass
The standard unit for mass in the SI system is kilograms (kg).

Mass		
1 kilogram	= 1 kg	= mass for which Earth's gravity force at sea level equals about 2.2 pounds
1 gram	= 1 g	= 10^{-3} kg (one paper clip)
1 milligram	= 1 mg	= 10^{-6} kg = 10^{-3} g (mosquito)
1 microgram	= 1 μg	= 10^{-9} kg = 10^{-6} g (small grain of sand)

Units for Acceleration
Acceleration is given in units of meters per second squared, or m/sec^2. This can also be written as m · sec^{-2}, where the dot (·) means multiplication.

Units for Force

In the SI system, we measure the strength of forces in units of newtons, abbreviated N. One newton is the amount of force needed to accelerate an object with a mass of 1 kg at a steady rate of rate 1 m/sec^2. That is,

$$1\,N = 1\,kg \cdot m/sec^2, \text{ or } 1\,N = 1\,kg \cdot m \cdot sec^{-2}$$

For example, to accelerate a 4-kg bowling ball at a rate 3 m/sec^2, you would need to apply a force equal to $F = m \cdot a = 4\,kg \cdot 3\,m/sec^2 = 12\,kg \cdot m/sec^2 = 12\,N$. Alternatively, we could ask how much acceleration we would achieve for this bowling ball if we applied a given amount of force; for example, 20 N. The acceleration in this case would be

$$acceleration = a = \frac{20\,N}{4\,kg} = \frac{20\,kg \cdot m/sec^2}{4\,kg} = 5\,m/sec^2$$

Units for Energy

The unit of energy is the joule (J). One joule (1 J) is the amount of mechanical energy needed to generate a force equal to 1 N and to use it to push an object a distance of 1 m. This means $1\,J = 1\,N \cdot 1\,m$, or $J = N \cdot m$.

Units for Power

Power equals energy delivered or converted per second. A power of one watt (abbreviated W) equals 1 J of energy delivered per second. (1 W = 1 J/sec).

Units for Electric Charge

The amount of electric charge on one proton is denoted by e (or $+e$). The amount of electric charge on one electron equals $-e$.

A charge 6.2×10^{18} protons is called one coulomb (1 C) of positive charge.

Units for Electric Current

Current is measured in units of amperes, often called amps. The symbol for ampere is A. One amp (1 A) equals the amount of current present when 1 C of positive charge moves between two locations in 1 sec; that is, 1 A = 1 C/sec, or

$$1\,A = \frac{1\,C}{1\,sec}$$

Values of Physical Constants

Symbol	Description	Numerical Value
c	Speed of light in vacuum	299,792,458 m sec^{-1} exactly; 3.0×10^8 m sec^{-1} approximately
—	Speed of sound in air (20°C, 1 atm)	343 m sec^{-1}
h	Planck's constant	6.626×10^{-34} J sec
g	Acceleration of gravity at Earth's surface	9.8 m sec^{-2} approximately
e	Proton charge	1.60×10^{-19} C
$-e$	Electron charge	-1.60×10^{-19} C
M_e	Electron mass	9.11×10^{-31} kg
M_p	Proton (and neutron) mass	1.67×10^{-27} kg
k_B	Boltzmann constant	1.381×10^{-23} J K^{-1}

Sources and Credits for Quotes Used with Permission

Chapter 1

1. Diane Grayson; from "Rethinking the Content of Physics Courses," *Physics Today*, February 2006, 31. © American Institute of Physics.
2. David DiVincenzo; from his talk at the Conference on Laser and Electro-Optics/ Quantum Electronics and Laser Science 2006 Conference, recorded by M.G. Raymer.
3. Nicholas Negroponte; from *Being Digital* (New York: Knopf, 1995).
4. William Gibson; from *NPR Talk of the Nation*, November 30, 1999, Timecode: 27 min 20 sec; © National Public Radio.

Chapter 2

1. Eugene Wigner; from "The Unreasonable Effectiveness of Mathematics in the Natural Sciences," in *Philosophical Reflections and Syntheses (E.P. Wigner: The Collected Works: Part B),* edited by G.G. Emch, Jagdish Mehra, and Arthur S. Wightman (Berlin: Springer-Verlag, 2001).
2. Claude Shannon (a); from "A Mathematical Theory of Communication," *The Bell System Technical Journal* 27 (1948): 379–423, 623–656.
3. Neil Gershenfeld; from *The Physics of Information Technology* (Cambridge, UK: Cambridge University Press, 2000).
4. Claude Shannon (b); ibid.
5. Ray Kurzweil; from *The Law of Accelerating Returns*, http://www.kurzweilai.net (March 7, 2001). © Ray Kurzweil.

Chapter 3

1. Isaac Newton; from *The Principia* (1687).
2. Isaac Asimov; from *The History of Physics* (New York: Walker, 1984), 84.
3. John Donne; from *An Anatomy of the World: The First Anniversary*, by R.S. Bear, translation in Renascence Editions (2003). ©The University of Oregon.

Chapter 4

1. William Thomson (Lord Kelvin); from J. Z. Buchwald, Biography in *Dictionary of Scientific Biography* (New York: Scribner, 1970–1990).

Chapter 5

1. Michael Faraday; Laboratory journal entry #10,040 (March 19, 1849); published in *The Life and Letters of Faraday* (1870) Vol. II, edited by Henry Bence Jones, 253 (Ann Arbor: University of Michigan Library Scholarly Publishing Office, 2005).
2. Michael Faraday; from Philosophical Transactions, 1832, 122: 125–162; V. Experimental Researches in Electricity, by Michael Faraday, Petersburgh, &c.
3. Robert Dexter Conrad, from *Office of Naval Research 50th Anniversary Naval Research Reviews*, Office of Naval Research, One/1996, Vol XLVIII.

Chapter 6

1. George Boole; from *An Investigation of the Laws of Thought* (Cambridge, 1847); re-published by Dover Publications (June 1, 1958); and by Merchant Books (July 19, 2008).

Chapter 7
1. Heinrich Hertz; public domain.

Chapter 8
1. Louis Pasteur; public domain.

Chapter 9
1. Niels Bohr; from his earliest paper on the atomic model published in the Royal Society's Philosophical Transactions; cited in "Niels Henrik David Bohr," by Leon Rosenfeld in *Dictionary of Physics Biography, Vol. 2.* (New York: Scribner, 1970) 240.
2. Isaac Newton; from *Principia Mathematics,* 1687; public domain.
3. Albert Einstein; source unknown.
4. Steven Dutch; from *Science, Pseudoscience, and Irrationalism*, Natural and Applied Sciences, University of Wisconsin–Green Bay. http://www.uwgb.edu/dutchs/index.html.
5. Murray Gell-Mann; from *The Quark and the Jaguar*, by Murray Gell-Mann (New York: W.H. Freeman, 1994).

Chapter 10
1. John Bardeen; from his Nobel Lecture, *Semiconductor Research Leading to the Point Contact Transistor.* © The Nobel Foundation 1956. http://www.nobelprize.org/nobel_prizes.
2. Steve Meloan; from *Toward a Global "Internet of Things,"* in Sun Developer Network (November 11, 2003).
3. John Fowler; quoted by Steve Meloan ibid.
4. Alien Technology; from Alien Technology Whitepaper (2007). © Alien Technology Corporation.

Chapter 11
1. Claude Shannon; from "A Mathematical Theory of Communication," *Bell System Technical Journal*, 27 (1948) 379–423, 623–656.

Chapter 12
1. Albert Einstein (a); from an article by Einstein in *Annalen der Physik*, 1905; English translation of the German original is taken from D. Ter Haar, *The Old Quantum Theory*, (Oxford, U.K.: Pergamon, 1967).
2. Albert Einstein (b); found in a letter of December 12, 1951 to his friend Michele Angelo Besso; quoted by E. Wolf in *Optics News*, 5, no.1 (1979) 39.
3. William Cassarly; from "High-Brightness LEDs," in *Optics and Photonic News*, January 2008, 19. © Optical Society of America.

Chapter 13
1. Ibn al-Haytham; from "Universality and Modernity of Ibn al-Haytham's Thought and Science," by Dr Valérie Gonzalez (2002). http://www.iis.ac.uk. ©The Institute of Ismaili Studies.
2. Tim Berners-Lee; *Hypertext and Our Collective Destiny*, a talk by Tim Berners-Lee, October 12, 1995. © Tim Berners-Lee.

Chapter 14
1. Charles H. Townes; from *How the Laser Happened* (Oxford, U.K.: Oxford University Press, 1999).

2. Theodore Maiman; from Laser Pioneer Interviews, edited by the Lasers and Applications Staff (Torrance, CA: High Tech Publications, 1985).

Chapter 15

1. Charles Kao; from *Technology of Our Times: People and Innovation in Optics and Optoelectronics*, edited by Frederick Su (Bellingham, WA: SPIE Optical Engineering Press, 1990).

Chapter 16

1. Paul Baran; from "Founding Father, an Interview with Paul Baran," *Wired*, March 2001, 145–153.

Glossary

Absolute zero – the lowest possible temperature (0K, −273.15°C, −459.67°F).

Absorption – process in which energy in a light wave is converted to energy stored in a medium.

Acceleration – rate of change of an object's speed or direction of motion.

Accelerometer – device for detecting acceleration.

Add/drop multiplexer – an optical switch that can redirect signals in input channels to different output channels.

Ampere (amp, A) – unit of electric current equal to one coulomb per second, or 6.25×10^{18} electrons per second.

Amplitude modulation (AM) – variation of the amplitude of a carrier wave in proportion to an audio signal.

Amplitude – the difference between a wave's zero displacement and its maximum displacement.

Analog – varying smoothly and continuously in a manner analogous to something else.

Atom – composed of a nucleus with one or more electrons moving around it.

Atomic number – the number of protons in an atom.

Audio range – frequencies corresponding to speech and music: 20 Hz to 20 kHz.

Autonomous – describing a system in which each element controls itself.

Band gap – the forbidden energy region between two electron energy bands in a crystal.

Bandwidth – width of the range of frequencies in a signal or in a channel.

Bar – unit of pressure equal to 1×10^5 N/m^2, or 1×10^5 Pa.

Battery – a chemical device that produces voltage between two terminals.

Baud – unit for data rate, equals 1 bit per second.

Beats – periodic pulsation of a wave's strength resulting from interference of two waves with unequal frequencies.

Binary number system – a system of counting that uses two digits (0, 1) and place values equal to powers of two.

Bipolar transistor – transistor made of opposing diodes.

Bit – a binary digit (0 or 1); also the smallest unit of information.

Boltzmann's constant – a constant relating temperature to kinetic energy in a gas. $k_B = 1.38 \times 10^{-23}$ J/K.

Boolean logic – a mathematical system for performing logic operations.

Brightness – referring to the high power of laser beams, which at the same time have the properties of monochromaticity, directionality, and coherence.

Broadcasting network – a communication network in which all stations share a single physical channel.

Byte – a group of 8 bits.

Cache – a small, very fast random-access memory used to store small groups of data.

Capacitor – an electronic component that stores plus and minus charge on two metal plates separated by a small gap.

Carrier wave – a high-frequency electromagnetic wave with a constant amplitude.

Cavity – *see Light resonator.*

Cell (communications) – a geographical area in which radio signals from a single base station can be received.

Cell (computers) – a region in an electronic memory that can store one bit of information.

Channel – a designated portion of a medium or spectral band of frequencies within which information is broadcast.

Channel bandwidth – the width of the range of frequencies (spectral window) that is allocated to the channel.

Chemical energy – the energy associated with chemical bonds that hold various substances together.

Circuit switching – configuring network switches to send data between two stations by a dedicated, continuous physical channel.

CMOS – complementary metal-oxide semiconductor.

Coherence – the property of having all wave fronts traveling in nearly the same direction and oscillating together in a coordinated manner.

Combinational logic circuit – a logic circuit having no gate outputs sent back to the input of the same gate.

Compact disc – a metalized plastic disk for optical storage of data.

Conduction band – the energy band in a crystal just above the valence band.

Conductor – material with high electrical conductivity.

Convection – the transfer of thermal energy from one location to another by bulk motion of a gas or a liquid.

Coulomb (C) – a unit of electrical charge equal to the charge of 6.2×10^{18} electrons.

CPU – central processing unit, where logic operations take place in a computer.

Crest – a maximum in wave displacement.

Critical angle – the angle of a light ray's direction of travel above which a ray inside a medium will reflect 100% from a boundary.

Cross-talk – interference with one channel by signals from another channel.

Crystal – crystalline solid, a phase of matter in which atoms are held rigidly in a regular pattern.

Current (electric) – rate of flow, measured in amperes, that is, coulombs per second.

Current (liquid) – rate of flow, measured in liters per second.

Cycle – one complete oscillation of a periodic wave, after which the wave returns to its original form.

Damping – the gradual loss of energy of an oscillating object or wave.

Data period – time between data pulses.

Data rate – the number of bits being sent per second.

Decimal number system – a system of counting that uses ten digits (0–9) and place values equal to powers of ten.

Demultiplexer (DMUX) – device that separates light waves on the basis of their having different wavelengths or different arrival times.

Depletion region – a thin, charged, region around a p-n junction.

Diffusion – the random wandering of electrons into and out of different regions in a crystal.

Digit – a single symbol representing a number (e.g., decimal digits are 0, 1, 2, 3, ...9).

Digital – varying discretely or discontinuously.

Digital repeater – an electronic device that receives and amplifies a digital signal, detects it using thresholding, and retransmits a corrected digital signal.

Diode – a one-way current device made with a p-n junction.

Directionality – the property of laser light that allows it to travel in a tight beam over long distances.

Dispersion – the variation of wave speed with frequency.

Displacement – the difference of a wave's value from its equilibrium (zero) value.

Distance – the difference between two positions.

Domain – a microscopic region inside a metal such as iron, within which all atomic current loops are oriented in the same direction.

Dopant or **impurity** – element put into an otherwise pure silicon crystal to alter its properties.

Doping – the introduction of dopants into an otherwise pure semiconductor crystal.

DRAM – dynamic RAM, stable for short periods but needs to be refreshed.

Efficiency – (of a laser) the ratio of laser output power to the input pump power.

Electrical conductivity – indicates the ability of a material to transport (i.e., conduct) electric current.

Electrical conductor – a material through which electric current can easily flow.

Electrical insulator – a material through which electric current cannot easily flow.

Electrical resistance – friction-like phenomenon that decreases the potential energy of an electron as it travels through a material.

Electric charge – the property of particles that determines electrical force between two objects.

Electric circuit – a group of conducting objects through which electrons flow.

Electric current – the flow of positive electric charge (opposite the direction in which electrons flow).

Electric field – agent of a charged object by which it exerts a force on another charged object; represented by a pattern of electric field vectors.

Electric field line – a curved line that indicates the direction of the electric force exerted on a plus-charged object located at a specific position.

Electric potential energy – the potential energy that a charged object has due to its location in an electric field.

Electromagnet – a magnet created by passing electric current through a wire coil.

Electromagnetic (or **EM**) – referring to interacting electrical and magnetic fields.

Electromagnetic spectrum – a continuum of electromagnetic waves arranged by their frequency.

Electromagnetic wave – a coordinated oscillation of electric and magnetic fields that travels from one place to another (e.g., light, radio, microwaves, television waves, and x-rays).

Electromagnetism – phenomena involving electric and magnetic forces.

Electron – an elementary particle with negative charge and mass about 2,000 times less than that of a proton or neutron.

Element – substance made of a single kind of atom.

Emission – process in which energy in an atom is converted to energy in a light wave.

Enabled data latch, or **D-latch** – a read-write memory circuit.

Energy – the capacity of a physical system to do work.

Energy band – a continuous range of possible energies for electrons in a crystal.

Energy gap – a range of energies in which electrons are forbidden.

Energy shell – a group of states or orbits having nearly equal energies.

Exclusion Principle – the quantum rule that only one electron can occupy any given state.

Feedback (laser) – sending an amplifier's output back into its input.

Feedback (logic) – sending a logic gate's output back into its input.

Field – a pattern of vectors, each indicating direction and strength.

Field-effect transistor (FET) – a semiconductor device that conducts electrons when an electric field is applied.

Flash memory – a nonvolatile (semipermanent) form of electronic computer memory.

Force – an influence that can accelerate a material object; that is, impose a change of speed or direction of motion.

Force vector – an arrow that points in the direction of a force, whose length indicates the strength of the force.

Frequency – the number of cycles per second in an oscillation.

Frequency modulation (FM) – variation of the frequency of a carrier wave in proportion to an audio signal.

Frequency multiplexing – distinguishing signal channels by using their distinct frequencies as identifiers.

Frequency spectrum – the strength of signal amplitudes for each frequency.

Friction – the force that occurs between two objects when their surfaces are in contact.

Gain, G – the factor by which the power of a light beam is increased while passing through an amplifying medium.

Gas – phase of matter in which atoms or molecules move freely and fill whatever space is available to them.

Gate (logic) – a device or element that performs a logic operation.

Gate (transistor) – the part of a transistor that controls the current passing through it.

Glass – transparent noncrystalline solid.

Gravitational energy – the potential energy that is stored in an object by pushing it "up" through an opposing gravitational force.

Heat – the process of transferring thermal energy from one object or place to another.

Heat capacity – the amount of thermal energy transferred to an object when its temperature is raised by 1°C.

Heat conduction – the transfer of thermal energy from one object to another by direct contact.

Heat (or thermal) conductivity – the property of a material that indicates its ability to conduct thermal energy.

Hertz – unit of frequency; 1 Hz = 1 cycle per second.

Hole – a place in a crystal (or in an energy band) where an electron could be, but is not.

Hypothesis – a preliminary conjecture or statement about the natural world that can be tested by experiments.

Induced magnet – iron after it has been exposed to magnetic forces from the outside, thereby having its microscopic magnetic domains aligned.

Information – a measure of the decrease in one's uncertainty about the outcome of a certain event; measured in bits.

Information content – the minimum number of binary digits (bits) needed to faithfully represent a set of data.

Input/output (logic) – a bit that can equal 0 or 1, used to represent logic statements or numbers.

Insulator (electronics) – material having very low electrical conductivity.

Insulator (heat) – material having very low thermal conductivity.

Integrated circuit (IC) – a circuit in which all elements are fabricated on a single semiconductor wafer.

Interference – the interaction of two or more waves when they meet.

Internet – the global internetwork.

Internetwork – a network made by connecting two or more distinct networks, which may use different protocols.

Joule (J) – a unit of energy (equal to that delivered in 1 sec by a 1 W power source).

Kinetic energy – the energy associated with movement or motion of objects having mass.

Land – region on surface of a compact disk where no pit is present.

Laser (device) – a coherent light source that operates by stimulated emission.

LASER (process) – Light Amplification by Stimulated Emission of Radiation.

Law – a general rule that describes how some aspect of the natural world behaves, and which appears to be universal in its applicability.

Level of redundancy (network) – the number of distinct pathways that can be used to connect any two nodes in a network.

Light – a traveling wave of oscillating electric and magnetic fields with frequency in the visible range.

Light-emitting diode (LED) – a semiconductor diode that emits light when a voltage creates current passing through it.

Light receiver, or photodetector – a semiconductor crystal that responds to light by increasing its electrical conductivity.

Light resonator (cavity) – an arrangement of two or more mirrors that reflect light back through a gain medium many times, providing feedback.

Liquid – phase of matter in which atoms or molecules are tightly packed but are able to gradually move, so the liquid takes on the shape of its container.

Logic – a process of determining output statements or data from input statements or data.

Logic circuit – a set of connected logic gates used to perform a series of operations.

Logic operation (or gate) – an elementary rule (e.g., NOT, AND, OR) for arriving at a logic outcome.

Longitudinal wave – wave in which the parts of the medium oscillate parallel to the direction the wave is traveling.

Loss – the decrease of a wave's amplitude as it travels through a medium that takes energy from the wave.

Magnetic field – the agent of a magnetic object by which it exerts a force on another magnetic object; represented by a pattern of field vectors.

Magnetic field lines – lines that indicate strengths and directions of a magnetic field.

Magnetic hard drive – a device that stores data on a disk of magnetic material.

Magnetism – the phenomena of magnets and magnetic forces.

Magnetization – the process of making a piece of metal into a magnet by the application of a strong magnetic field to the metal.

Mass – the property of an object that provides its "reluctance" to accelerate when subjected to a force.

Material dispersion – the variation of wave speed with frequency, as a consequence of the intrinsic properties of a material.

Mechanical energy – the energy associated with the overall (not internal) motions or locations of objects.

Medium – a material through which waves can move.

Memory – a computer component that can store records of earlier events.

Memory circuit – a circuit whose present state depends on its past history.

Metal – a crystal having partially empty energy bands and therefore high electrical conductivity.

Microelectrical mechanical systems (MEMS) – systems consisting of tiny movable mirrors and other devices on the scale of approximately 0.1 mm.

Mode – a certain path or pattern that a wave can follow in a wave-guide or in a resonator.

Mode dispersion – the variation of wave speed with frequency as a consequence of the geometrical properties of a wave-guide.

Model – a tentative scheme for explaining and predicting how certain real events occur.

Molecule – two or more atoms held together by chemical bonds.

Monochromaticity – the property of having a single color (frequency).

Multimode fiber – an optical fiber in which light can travel in more than one mode.

Multiplexer (MUX) – device that carries out multiplexing of light waves (combining of several waves of differing wavelength into a single beam).

Multiplexing – the transmission of multiple signals over a single transmission medium, with each being carried by a separate channel.

n-channel FET, or **n-FET** – a field-effect transistor in which an electron-rich channel is created for electron flow.

Net force – the resultant force occurring when two or more forces, possibly in different directions, act on a single object.

Network – an interconnected collection of autonomous computers.

Neutron – elementary particle having the same mass as a proton and zero electric charge.

Node (networks) – a device that transfers data from stations into and out of a network.

Node (waves) – region where interfering waves cancel, so the wave displacement equals zero.

Noise – unwanted randomness in an electrical or optical signal.

Noncrystalline solid – a phase of matter such as glass, in which the atoms are packed tightly but randomly.

Nonvolatile memory – memory that holds its bit values after power is turned off.

n-type semiconductor – a semiconductor doped with electron-adding elements.

Nucleus – protons and neutrons bound tightly together at the center of an atom.

Observation – the outcome of a controlled, repeatable experiment.

Ohm – unit of electrical resistance. One ohm is the resistance of an object that conducts one ampere of current when a voltage of one volt is put across it.

One-time latch – a circuit that can store the value of a bit only once and cannot be reset.

Optical fiber – cylindrical wave-guide for light, usually made of silica glass.

Orbit – region of space in an atom where an electron moves when it has a certain discrete energy.

Packet – a group of about 1,000 bits traveling together through a network.

Packet switching – routing by passing packets from node to node using a channel dedicated to the connection only for the duration of one packet's transmission.

Particle – an elementary object with a definite location in space.

p-channel FET, or **p-FET** – a field-effect transistor in which an electron-deficient channel is created for electrons to flow.

Period – the time for a periodic motion to go from a maximum to a minimum displacement and back to the maximum.

Periodic wave – one that looks the same before and after a certain time period.

Photodiode – a diode used for detecting light.

Photoelectric effect – an electron in a crystal absorbs a photon of frequency f and jumps up across the energy gap if the energy of the photon is greater or equal to the energy of the gap.

Photolithography – a method of lithography used for creating circuits on a silicon crystal.

Photon – discrete "bundle" of light energy.

Photon rate – the average number of photons per second in a light beam.

Pit – tiny indentation in the surface of a compact disk.

Pixel – picture element, many of which make up a picture.

Place value – the value corresponding to each place or position in a number.

Planck's constant – constant in Planck's law, $h = 6.6 \times 10^{-34}$ J $\cdot$ sec/photon.

Planck's law – the relation between light frequency and photon energy: $E_f = h \cdot f$.

p-n junction – region around the contact between a p-type and a n-type semiconductor.

Polarization – the direction (e.g., vertical or horizontal) in which the electric field in an EM wave oscillates as the wave propagates.

Population inversion – the condition of having more atoms in a medium in their higher-energy state than their lower-energy state.

Potential energy – the energy stored in an object, or arrangement of objects, depending on their positions.

Power – the rate at which energy is converted from one form to another or transferred from one place to another.

Powers of 2 – the set of numbers that result from multiplying 2 by itself any number times.

Precision – the fineness with which a measuring process can distinguish between two nearly equal values of a quantity.

Protocol – a set of rules for interpreting a list of binary numbers.

Proton – elementary particle with mass and positive electric charge.

p-type semiconductor – a semiconductor doped with electron-deficient elements.

Pump source – a system that provides a continuous source of power, usually an electrical current that continuously excites atoms in the gain medium to their higher-energy states.

Quantization – the requirement of discreteness of a numerical value representing a property.

Quantum jump – the sudden change of an electron's state that occurs when it gains or loses energy.

Quantum theory – mathematical theory describing electrons and other particles in terms of quantized waves.

Radiation (heat) – the heating or cooling of an object by light waves and infrared waves absorbed or emitted by the object.

RAM – random access memory.

Rate of photon absorption – the number of atoms per second that jump to their higher-energy states as a result of absorbing photons.

Rate of spontaneous (or stimulated) emission – the number of atoms per second that jump from their higher-energy states to a lower-energy state as a result of emitting photons.

Ray pictures – a pictorial representation of a light wave, showing only the wave's direction of travel.

Receiver – a device that converts signals from a communication channel into a useful representation of the original information.

Rectifier – a diode circuit that removes the negative voltage components from a signal.

Reflectivity of a mirror – the fraction of light power (or photons) that reflects from a mirror when a light beam impinges on it.

Refraction – the bending of wave fronts and propagation direction of a wave at a boundary between mediums of different refractive index.

Refractive index – a parameter that indicates the speed of light in a medium.

Relative precision – the fineness of a measurement compared to the range of its most likely value.

Relay - electronic device for relaying or transferring the effect of pressing one switch to another switch.

Repeater – an electronic device that receives and amplifies a signal and retransmits it.

Resistance (electricity) – the degree to which a region resists electric current.

Resistance (matter) – the degree to which a region resists flow of gas or liquid current.

Resistor (electricity) – an object that resists electric current, used to control the flow of current in electrical circuits.

Resistor (matter) – a region of material that resists gas or liquid current.

Resonance – an enclosure with tendency of an oscillating object to oscillate with increasingly large amplitude when driven by a periodic force at the resonance frequency.

Resonance frequency – the natural oscillation frequency of an unforced oscillator.

Resonator – reflecting surfaces (such as mirrors) that reinforce or resonate a wave.

Rule – an apparent regularity of occurrences.

Sampling – measuring a continuously varying voltage or signal at discrete times.

Sampling theorem – theorem stating that a communication channel with a bandwidth B can carry data at a rate up to $2B$.

Semiconductor – material with a small energy gap above its highest filled energy band, creating medium conductivity.

Sequential logic – a logic circuit in which a logic gate's output can feed back to the input of the same gate.

Set-reset latch, or **S-R latch** – a circuit that stores a bit value; can be set and reset.

Side-band pair – a pair of spectral peaks equally separated from a carrier frequency.

Signal bandwidth – the width of the spectral window occupied by the spectral recipe needed to represent a signal.

Silica – silicon dioxide (SiO_2) glass.

Simple harmonic motion (SHM) – periodic motion with a unique frequency, characterized by a smooth, regular exchange of energy between two forms, for example, kinetic and potential.

Simple harmonic wave – a wave whose displacement at each point undergoes simple harmonic motion.

Single-mode fiber – an optical fiber in which light can travel in only one mode.

Solid-state electronics – circuits based on semiconductor crystals.

Sound – the transfer of energy through a material by the vibration of matter.

Sound waves – waves of compression in air or another medium.

Spectral recipe (spectrum) – the distribution (number and strength) of spectral peaks comprising a signal.

Spectrum – pattern of frequencies or colors.

Speed – the rate of change of an object's position or distance.

Speed of light, c – speed at which light waves travel; $c = 3.0 \times 10^8$ m/sec.

Spontaneous emission – the emission of light by an atom when no outside stimulus is present.

SRAM – static RAM; is stable as long is power is provided.

Standing wave – a wave whose crests and troughs do not travel, but simply oscillate between positive and negative values.

State – specification that an electron is moving in a particular orbit and has a particular direction of spin.

Station – a device such as a computer or telephone that is used for sending and receiving data.

Stimulated emission – the emission of light by an atom in an excited energy state when an external light stimulus is present.

Switching network – a network in which each node contains many switches to route the messages coming into it.

Temperature – a measure of the average energy per atom in a gas, liquid, or solid.

Theory – a substantiated system of reasoning, based on a collection of well-tested models and laws, which explains or predicts many particular physical phenomena.

Thermal energy – the internal kinetic and potential energies associated with the random motions of large numbers of atoms making up an object.

Thermal equilibrium – situation when two objects have zero net transfer of thermal energy between them.

Thermodynamics – the study of heat and work.

Three-bit adder – a binary logic circuit that adds two one-bit numbers and a carry-in bit to yield a carry-out and a sum.

Threshold (laser)– the value of pump power at which the gain times the mirror reflectivity is greater than one ($R \times G > 1$), allowing the laser to self-oscillate.

Thresholding – technique for restoring distorted or noisy signals to their original digital levels by comparing them to a reference or threshold value.

Time multiplexing – procedure in which data pulses carrying bits from different senders are interweaved in time before being sent through a channel.

Total internal reflection (TIR) – the phenomenon of 100% reflection when a light wave's travel direction is above a critical angle.

Transistor – a semiconductor electronic "valve" for controlling electron voltage and current.

Transmitter – a device that converts information to a physical form suitable for transmitting on a communication channel.

Transparent medium – a material such as glass that transmits most light.

Transverse wave – wave in which the parts of the medium move perpendicularly to the direction the wave is traveling.

Trough – a minimum in a wave displacement.

Two-bit adder – a binary logic circuit that adds two one-bit numbers to yield a carry-out and a sum.

Unit – a fixed quantity (of length, volume, time, etc.) used as a standard of measurement.

Valence band – the highest-energy band in a crystal that is completely filled with electrons.

Vector – an arrow, the length of which indicates the strength of a field and the direction of which indicates the direction of a field.

Volt (V) – the unit of voltage, corresponding to changing the potential energy of 1 C of charge by 1 J.

Voltage – the change of electric potential energy of a charged object divided by the amount of charge on the object.

Watt (W) – unit of power, equaling one joule per second.

Wave – an organized oscillating pattern that travels from one place to another.

Wave front – a line along the top of the crests (or bottom of the troughs) in a wave, perpendicular to the direction of travel.

Wave-guide – a transparent structure with a high-refractive-index core surrounded by a low-index cladding, which guides light by total internal reflection.

Wavelength – the distance from one wave crest to the next.

Wavelength multiplexing – a scheme in which distinct data streams being carried on light with different wavelengths are combined before being sent through a fiber.

Wave period – the time for a point on a periodic wave to go from a maximum to a minimum displacement and back to the maximum.

Wave speed – traveling speed of a wave.

Wavelength division multiplexing (WDM) – wavelength-division multiplexing, same as wavelength multiplexing.

Work – the amount of energy transferred when moving an object through a certain distance by continuously applying a force.

Index

Note: "fn" indicates footnote; "q" indicates quote by the listed person; **bold** indicates primary definitions